Advances in Global Change Research

Volume 75

This book series is indexed in SCOPUS.

Advances in Global Change Research

Aims and Scopes

This book series is aimed at addressing a range of environmental issues through state-of-the-art and/or interdisciplinary approaches. The books in the series can either be monographs or edited volumes based, for example, on the outcomes of conferences and workshops, or by invitation of experts. The topics that the series can consider publishing include, but are not limited to:

Physical and biological elements of earth system science, in particular

- Climate change
- Biodiversity
- Sea-level rise
- Paleo-climates and paleo-environments

Social aspects of global change

- Environmentally-triggered migrations
- Environmental change and health
- Food security and water availability
- Access to essential resources in a changing world

Economic and policy aspects of global change

- Economic impacts
- Cost-benefit analyses
- Environmental governance
- Energy transition

Methodologies for addressing environmental issues

- Planetary data analysis
- Earth observations from space
- Proxy data analyses
- Numerical modeling
- Statistical analyses

Solutions to global environmental problems

- Sustainability
- Ecosystem services
- Climate services
- Technological/engineering solutions

Books in the series should be at least 200 pages in length, and include a table of contents.

Images supplied in colour will be reproduced in both the print and electronic versions of the book at no cost to the author/editor.

Manuscripts should be provided as a Word document and will be professionally typeset at no cost to the authors/editors.

All contributions to an edited volume should undergo standard peer review to ensure high scientific quality, while monographs should also be reviewed by at least two experts in the field.

Manuscripts that have undergone successful review should then be prepared according to the Publisher's guidelines manuscripts: https://www.springer.com/gp/authors-editors/book-authors-editors/book-manuscript-guidelines

Wame L. Hambira • Ayana Angassa Abdeta
Ditiro B. Moalafhi • Victor K. Muposhi
Ketlhatlogile Mosepele
Editors

Environmental Change and Biodiversity Conservation in sub-Saharan Africa

Volume 1

Springer

Editors
Wame L. Hambira
Botswana University of Agriculture and Natural Resources
Gaborone, Botswana

Ayana Angassa Abdeta
Botswana University of Agriculture and Natural Resources
Gaborone, Botswana

Ditiro B. Moalafhi
Botswana University of Agriculture and Natural Resources
Gaborone, Botswana

Victor K. Muposhi
Botswana University of Agriculture and Natural Resources
Gaborone, Botswana

Ketlhatlogile Mosepele
Botswana University of Agriculture and Natural Resources
Gaborone, Botswana

ISSN 1574-0919 ISSN 2215-1621 (electronic)
Advances in Global Change Research
ISBN 978-3-031-73135-8 ISBN 978-3-031-73136-5 (eBook)
https://doi.org/10.1007/978-3-031-73136-5

This Springer imprint is published by the registered company Springer Nature Switzerland AG
The registered company address is: Gewerbestrasse 11, 6330 Cham, Switzerland

Foreword

Around the world, species are disappearing and natural ecosystems are being degraded at an alarming rate. From the lush of the Congo Basin to the expansive savannas of the rest of Africa, the environmental richness of Africa is unparalleled. Yet, this richness is under threat, facing a myriad of challenges that demand our attention and action. According to the Global Assessment Report on Biodiversity and Ecosystem Services of the Intergovernmental Science-Policy Platform on Biodiversity and Ecosystem Services (IPBES) published in 2019, of an estimated eight million animal and plant species (75% of which are insects), around one million are threatened with extinction—more than ever before in human history. Invasive alien species which are spread by human activities pose major global threats to nature, economies, food security, and human health. Most of these concerns are not yet well addressed with adequate focus and resources.

"Environmental Change and Biodiversity Conservation in Sub-Saharan Africa," Volume I, covers the sub-themes of biodiversity conservation in the Anthropocene and sustainable range and forest management and delves deep into the intricate relationship between the changing environment and the preservation of biodiversity in this region. Edited by experts who have dedicated their lives to understanding and safeguarding Africa's natural heritage, this book serves as a beacon of knowledge and insight in the face of environmental uncertainty.

Within these pages, readers will embark on a journey that explores the complex interplay between human activities, climate change, habitat loss, and the delicate balance of ecosystems. This book touches on marginalized communities' vulnerability, poverty, and hunger; the promotion of gender equality in finding nature-based solutions to climate change; and biodiversity conservation challenges. Through a multidisciplinary approach, the contributors illuminate the interconnectedness of environmental change and biodiversity loss, highlighting both the challenges and the opportunities for conservation efforts in Sub-Saharan Africa. The authors emphasize the great need to understand the positive and negative anthropogenic effects on biodiversity to attain sustainability.

This book calls upon policymakers, researchers, conservation practitioners, and local communities to come together in a shared commitment to protect and sustain the natural heritage of Sub-Saharan Africa for generations to come. As we stand at a critical moment in the history of our planet, "Environmental Change and Biodiversity Conservation in Sub-Saharan Africa," Volume I, serves as a timely reminder of the fragility of our natural world and the profound responsibility we bear as stewards of the Earth. May this book inspire dialogue, inform decisions, and catalyze actions that pave way toward a future where biodiversity thrives and the beauty of Africa's landscapes endures.

Director Scientific Services, Zimbabwe Parks and Wildlife Management Authority
Harare, Zimbabwe
May 2024

Edson Gandiwa

Contents

Contributors

Ayana Angassa Abdeta Botswana University of Agriculture and Natural Resources, Gaborone, Botswana

Lawrence Akanyang University of Agriculture and Natural Resources, Gaborone, Botswana

Oluwagbenga Titus Alao Osun State University, Osogbo, Nigeria

Tsheko Baruti Botswana University of Agriculture and Natural Resources, Gaborone, Botswana

Mpaphi C. Bonyongo Okavango Research Institute, University of Botswana, Maun, Botswana

Permanent Okavango River Basin Water Commission, Gaborone, Botswana

Raban Chanda University of Botswana, Gaborone, Botswana

Admire Chanyandura Chinhoyi University of Technology, Chinhoyi, Zimbabwe

Endale Difabachew Mizan–Tepi University, Mizan-Aman, Ethiopia

Meshack Dludlu University of Eswatini, Kwaluseni, Eswatini

Abdi Khalil Edriss Lilongwe University of Agriculture and Natural Resources (LUANAR), Lilongwe, Malawi

Richard W. S. Fynn Okavango Research Institute, University of Botswana, Maun, Botswana

Hesekia Garekae University of Botswana, Gaborone, Botswana

Reniko Gondo Okavango Research Institute, University of Botswana, Maun, Botswana

Wame L. Hambira Botswana University of Agriculture and Natural Resources, Gaborone, Botswana

Obusitswe Kakambi Department of Forestry and Range Resources, Ministry of Environment and Tourism, Kasane, Botswana

Keotshephile Kashe Okavango Research Institute, University of Botswana, Maun, Botswana

Mengistu Ketema Haramaya University, Dire Dawa, Ethiopia

Ednah Kgosiesele University of Agriculture and Natural Resources, Gaborone, Botswana

Oluwatoyin Dare Kolawole Okavango Research Institute, University of Botswana, Maun, Botswana

Centre for Gender and Africa Studies, University of the Free State, Bloemfontein, South Africa

Linganani Kombani University of Botswana, Gaborone, Botswana

Ishmael Kopong University of Agriculture and Natural Resources, Gaborone, Botswana

Belaineh Legesse Frontieri Consult PLC, Addis Ababa, Ethiopia

Boipuso Legwatagwata University of Agriculture and Natural Resources, Gaborone, Botswana

Joyce Lepetu Botswana University of Agriculture and Natural Resources, Gaborone, Botswana

Summer Mabula University of Agriculture and Natural Resources, Gaborone, Botswana

Kamogelo Makgobota University of Agriculture and Natural Resources, Gaborone, Botswana

Shimane W. Makhabu University of Agriculture and Natural Resources, Gaborone, Botswana

Topoyame Isaac Makoi University of Agriculture and Natural Resources, Gaborone, Botswana

James Maradza Richard Lyons Attorneys, Gaborone, Botswana

Thembinkosi Mathowa University of Agriculture and Natural Resources, Gaborone, Botswana

Joseph E. Mbaiwa Okavango Research Institute, University of Botswana, Maun, Botswana

Ronnie Matlanaga Mmolotsi University of Agriculture and Natural Resources, Gaborone, Botswana

Ditiro B. Moalafhi Botswana University of Agriculture and Natural Resources, Gaborone, Botswana

Goemeone E. J. Mogomotsi Botswana University of Agriculture and Natural Resources, Gaborone, Botswana

University of Botswana, Gaborone, Botswana

Patricia K. Mogomotsi University of Botswana, Gaborone, Botswana

Jema Haji Mohamed School of Agricultural Economics, Haramaya University, Dire Dawa, Ethiopia

Witness Mojeremane University of Agriculture and Natural Resources, Gaborone, Botswana

Belda Mosepele Botswana University of Agriculture and Natural Resources, Gaborone, Botswana

Ketlhatlogile Mosepele Botswana University of Agriculture and Natural Resources, Gaborone, Botswana

Ineelo Mosie Okavango Research Institute, University of Botswana, Maun, Botswana

Iris Mosipuri Botswana University of Agriculture and Natural Resources, Gaborone, Botswana

Naomi Moswete University of Botswana, Gaborone, Botswana

Kudzai S. Mpakairi University of the Western Cape, Cape Town, South Africa

African Leadership University, Kigali, Rwanda

Victor K. Muposhi Botswana University of Agriculture and Natural Resources, Gaborone, Botswana

Chinhoyi University of Technology, Chinhoyi, Zimbabwe

Chiedza Ngonidzashe Mutanga Okavango Research Institute, University of Botswana, Maun, Botswana

Mkululi Ndlovu Zimbabwe Parks and Wildlife Management Authority, Harare, Zimbabwe

Gosiame Neo-Mahupeleng Department of Wildlife and Aquatic Resources, Faculty of Natural Resources, Botswana University of Agriculture and Natural Resources, Gaborone, Botswana

N. Nkonyane University of Eswatini, Kwaluseni, Eswatini

D. Mashego University of Eswatini, Kwaluseni, Eswatini

Diamantino Carlos Pinto Ministry of Agriculture, Maputo, Mozambique

Gofaone Rammotokara Botswana University of Agriculture and Natural Resources, Gaborone, Botswana

Melusi Rampart University of Agriculture and Natural Resources, Gaborone, Botswana

C. Sekonopo University of Eswatini, Kwaluseni, Eswatini

Zibanani Seletlo University of Agriculture and Natural Resources, Gaborone, Botswana

Feyera Senbeta Botswana University of Agriculture and Natural Resources, Gaborone, Botswana

Mpho R. Setlalekgomo University of Agriculture and Natural Resources, Gaborone, Botswana

Keoikantse Sianga Botswana University of Agriculture and Natural Resources, Gaborone, Botswana

Tomás Adriano Sitoe Ministry of Agriculture, Centre for Socioeconomic Studies, Agricultural Research Institute Mozambique, Maputo, Mozambique

Demel Teketay University of Agriculture and Natural Resources, Gaborone, Botswana

Dixon Olutade Torimiro Botswana University of Agriculture and Natural Resources, Gaborone, Botswana

Nelson Motlapele Tselaesele Botswana University of Agriculture and Natural Resources, Gaborone, Botswana

Gaolathe Tsheboeng University of Botswana, Gaborone, Botswana

Koketso Tshireletso University of Agriculture and Natural Resources, Gaborone, Botswana

Chiedza C. Zembe Chinhoyi University of Technology, Chinhoyi, Zimbabwe

Part I
Biodiversity Conservation in the Anthropocene

Chapter 1
Sub-Saharan Africa's Perspectives on Environmental Change and Biodiversity Conservation: An Introduction

Wame L. Hambira, Victor K. Muposhi, Ditiro B. Moalafhi, Ayana Angassa Abdeta, and Ketlhatlogile Mosepele

1.1 Overview

Environmental change significantly alters ecosystems, contributing to biodiversity loss, diminished agricultural productivity, and compromised food security (Habibullah et al. 2022; Muluneh 2021; Sintayehu 2018). Healthy ecosystems and their associated services are crucial for the resilience of societies, economies, and environmental life support functions (Folke et al. 2021; Hernández-Blanco et al. 2022). However, these ecosystems face significant pressure from various direct and indirect drivers of change (Isbell et al. 2023; Jaureguiberry et al. 2022), which undermine their ability to support and contribute to human well-being (Bennett et al. 2023; Krause and Tilker 2022). Similarly, most global change scenario models predict further declines in biodiversity and nature's contributions to people over the coming decades (Baisero et al. 2020; Díaz et al. 2019; Pereira et al. 2010). This is particularly alarming for sub-Saharan Africa, where the loss of biodiversity-dependent ecosystem services disproportionately affects vulnerable and marginalized communities (Archer et al. 2018; Ryan et al. 2016), hindering global initiatives aimed at reducing poverty, hunger, and inequality (Pérez-Peña et al. 2021; Zhao et al. 2024).

Sub-Saharan Africa has diverse aquatic and terrestrial ecosystems that provide multiple services, some utilized through both non-extractive (e.g. ecotourism) and extractive approaches (e.g. fishing, logging). However, both the extractive and non-extractive uses of resources in these ecosystems are exerted to increasing pressure due to evolving consumption patterns, social demands, and other factors (Lindsay

W. L. Hambira (✉) · V. K. Muposhi · D. B. Moalafhi · A. A. Abdeta · K. Mosepele
Botswana University of Agriculture and Natural Resources, Sebele, Gaborone, Botswana
e-mail: whambira@buan.ac.bw; vmuposhi@buan.ac.bw; dmoalafhi@buan.ac.bw; aabdeta@buan.ac.bw; kmosepele@buan.ac.bw

W. L. Hambira et al. (eds.), *Environmental Change and Biodiversity Conservation in sub-Saharan Africa*, Advances in Global Change Research 75,
https://doi.org/10.1007/978-3-031-73136-5_1

et al. 2022). This compromises the resilience of ecosystems, the sustainability of management practices, and their ongoing contributions to human well-being (Hernández-Blanco et al. 2022; IPBES 2022). Understanding the direct and indirect drivers of ecosystem and biodiversity changes is crucial for achieving sustainability, particularly in sub-Saharan Africa where data are often fragmented.

Recognizing the urgency of global action to mitigate escalating biodiversity loss, the international community formulated and established transformative conservation strategies, which were adopted as the Kunming–Montreal Global Biodiversity Framework (Hughes 2023; Schröter et al. 2023; Xu and Wang 2023). The GBF articulates a vision for 2050 wherein biodiversity is valued, conserved, restored, and judiciously utilized, thereby sustaining ecosystem services, promoting planetary health, and delivering benefits essential for all humanity (CBD 2022; Hughes and Grumbine 2023). There is an urgent need for comprehensive, evidence-driven research to address the complexities of biodiversity loss and environmental change in Southern Africa, ensuring alignment with global conservation objectives. This edited volume seeks to consolidate scholarly contributions that offer viable, empirically based strategies for policymakers and data-driven approaches to conservation. The collection highlights the urgency of policy interventions to address and mitigate the drivers of environmental change and biodiversity loss. It underscores the importance of increased collaboration and knowledge sharing among diverse conservation stakeholders, such as academics, practitioners, and policymakers, to foster effective conservation strategies (Rice et al. 2020). Therefore, this compilation serves as a valuable resource for learners, researchers, and policymakers dedicated to the conservation of biodiversity, providing essential insights and tools to guide effective environmental conservation.

1.2 Environmental Change and Biodiversity Conservation in Sub-Saharan Africa: *Towards the 2030 Sustainable Development Agenda*

Sub-Saharan Africa harbours an extraordinary richness in biodiversity, juxtaposed with some of the highest human population densities and profound poverty rates globally (Chapman et al. 2022). This region, critical to global biodiversity conservation, faces unprecedented challenges due to both local and external anthropogenic pressures (Ogwu et al. 2022). The biodiversity in this region, though globally valuable, is increasingly threatened by the insatiable demand for natural resources driven by consumption patterns of affluent nations and the operational demands of multinational corporations resulting in intensive resource extraction activities across the continent (Lindsay et al. 2022). Additionally, the region contends with severe local pressures, notably the exponential growth in human populations, which exacerbates the strain on its natural resources (Baloch et al. 2020; Juju et al. 2020; O'Connell et al. 2019).

It is crucial to acknowledge sub-Saharan Africa's integral role in meeting global biodiversity goals, reflecting on its unique environmental challenges and opportunities (Chapman et al. 2022; Saliu et al. 2023). Recognizing the region's biodiversity not only for its inherent value but also for its essential contributions to human well-being and sustainable development is essential for fostering a sustainable coexistence between humans and nature. However, there is overwhelming evidence demonstrating the concurrent crises of biodiversity loss and climate change (O'Connor et al. 2020; Pfenning-Butterworth et al. 2024). This is exacerbated by unsustainable exploitation of Earth's resources (Pettorelli et al. 2021; Rinaldi 2021). The Living Planet Index (LPI) offers a comprehensive assessment of mammals, birds, fish, reptiles, and amphibians, and reports a 69% decline in monitored wildlife populations since 1970 (Freeman et al. 2022). Approximately half of the threats to these populations are due to habitat loss and disruptions to migratory pathways, consistent with findings from other studies (Freeman et al. 2022; Isbell et al. 2023). This is corroborated by the Intergovernmental Science-Policy Platform on Biodiversity and Ecosystem Services (IPBES) assessment report of 2019 (IPBES 2019; Ruckelshaus et al. 2020). The assessment indicates that habitat conversion, overexploitation of resources, climate change, invasive species, and other factors are threatening nature's ability to sustain human well-being. The IPBES and the Intergovernmental Panel on Climate Change (IPCC) reports, offer foundational perspectives that enhance our understanding of sub-Saharan Africa's unique position in the global biodiversity and sustainability narrative (Borie et al. 2021; Streck 2023).

The GBF addresses these challenges by setting targets to mitigate habitat conversion, resource overexploitation, climate change, and invasive species, aiming to preserve nature's capacity to support human well-being. The GBF aligns with recent assessments indicating that threats such as habitat conversion and climate change jeopardize nature's capacity to support human well-being (Li et al. 2023; Schröter et al. 2023). These international agreements emphasize the critical role of sub-Saharan Africa's biodiversity and ecosystem services, highlighting the necessity for conservation strategies that are both sustainable and equitable (Obura et al. 2023). To effectively address biodiversity loss and meet conservation and sustainability goals, urgent and comprehensive changes in economic, social, and political dimensions are essential (Kedward et al. 2023; Lindsay et al. 2022; Pörtner et al. 2023). Accordingly, there is dire need for an integrated strategy to address this crisis in the Anthropocene for sustainability (Jaureguiberry et al. 2022; Pörtner et al. 2023).

The GBF encompasses four overarching goals that envision a 2050 world in harmony with nature, along with 23 action-oriented targets designed to halt and reverse biodiversity loss by 2030 (CBD 2022; Hughes and Grumbine 2023). Its goals and targets are aligned with several Sustainable Development Goals (SDGs), particularly SDG 15 (Life on Land) and SDG 14 (Life Below Water), focussing on the conservation and sustainable use of terrestrial and aquatic ecosystems, which are core to biodiversity (Andriamahefazafy et al. 2022; Isabelle and Westerlund 2022; Krauss 2022; Parmesan et al. 2022). It also strongly intersects with SDG 13 (Climate Action), leveraging biodiversity and ecosystems as natural solutions to enhance resilience against climate change impacts (Seddon 2022; Smith et al. 2022;

Welden et al. 2021). Beyond these, the GBF synergizes with SDG 1 (No Poverty) and SDG 2 (Zero Hunger) by supporting sustainable livelihoods and food security through ecosystem services (Yang et al. 2020; Yin et al. 2021), see Table 1.1.

More so, SDG 11 (Sustainable Cities and Communities) benefits from urban biodiversity conservation and ecosystem restoration initiatives that enhance living conditions and resilience (Hawken et al. 2021; Kowarik et al. 2020). Through its holistic approach, GBF further supports multiple Sustainable Development Goals as it intersects with SDG 3 (Good Health and Well-Being) by providing natural medicines and promoting cleaner air (Baldwin-Cantello et al. 2023; Das et al. 2021;

Table 1.1 Synergies between selected Sustainable Development Goals (SDGs) and the Kunming–Montreal Global Biodiversity Framework Goals and Targets

Selected SDGs	Description	GBF goals and targets synergies
SDG 1: No Poverty	End poverty in all its forms everywhere.	GBF Goal 1: Address the underlying causes of biodiversity loss by mainstreaming biodiversity across government and society, and GBF Target 1: By 2030, ensure that biodiversity values are integrated into national and local development and poverty reduction strategies
SDG 2: Zero Hunger	End hunger, achieve food security and improved nutrition, and promote sustainable agriculture	GBF Goal 2: Reduce the direct pressures on biodiversity and promote sustainable use. GBF Target 3: By 2030, at least 30% of land and sea areas are conserved through effectively and equitably managed, ecologically representative, and well-connected systems of protected areas and other effective area-based conservation measures and integrated into the wider landscapes and seascapes
SDG 11: Sustainable Cities and Communities	Make cities and human settlements inclusive, safe, resilient, and sustainable	GBF Goal 4: Enhance the benefits to all from biodiversity and ecosystem services. GBF Target 5: By 2030, areas under agriculture, aquaculture, and forestry are managed sustainably, ensuring conservation of biodiversity
SDG 12: Responsible Consumption and Production	Ensure sustainable consumption and production patterns	GBF Goal 2: Reduce the direct pressures on biodiversity and promote sustainable use. GBF Target 6: By 2030, all fish and invertebrate stocks and aquatic plants are managed and harvested sustainably, legally, and applying ecosystem-based approaches, so that overfishing is avoided, recovery plans and measures are in place for all depleted species, fisheries have no significant adverse impacts on threatened species and vulnerable ecosystems, and the impacts of fisheries on stocks, species, and ecosystems are within safe ecological limits

(continued)

Table 1.1 (continued)

Selected SDGs	Description	GBF goals and targets synergies
SDG 13: Climate Action	Take urgent action to combat climate change and its impacts	GBF Goal 1: Address the underlying causes of biodiversity loss by mainstreaming biodiversity across government and society. GBF Target 1: By 2030, ensure that biodiversity values are integrated into national and local development and poverty reduction strategies
SDG 14: Life Below Water	Conserve and sustainably use the oceans, seas, and marine resources for sustainable development	GBF Goal 3: Reduce the direct pressures on biodiversity and promote sustainable use. GBF Target 4: By 2030, pollution, including from excess nutrients, has been brought to levels that are not detrimental to ecosystem function and biodiversity
SDG 15: Life on Land	Protect, restore, and promote sustainable use of terrestrial ecosystems, sustainably manage forests, combat desertification, and halt and reverse land degradation, and halt biodiversity loss	GBF Goal 2: Reduce the direct pressures on biodiversity and promote sustainable use. GBF Target 3: By 2030, at least 30% of land and sea areas are conserved through effectively and equitably managed, ecologically representative, and well-connected systems of protected areas and other effective area-based conservation measures and integrated into the wider landscapes and seascapes

Marselle et al. 2021). Additionally, it contributes to SDG 6 (Clean Water and Sanitation) by maintaining water quality and availability through ecosystem services (Arora and Mishra 2022; Schlattmann et al. 2022). It also advances SDG 12 (Responsible Consumption and Production) by promoting sustainable resource utilization, which aligns with the secondary objective of the Convention on Biological Diversity (Arora and Mishra 2023; Sun et al. 2022). The Global Biodiversity Framework's overarching objectives underscore the pivotal role of resilient and healthy ecosystems in attaining global sustainability (Schröter et al. 2023), given their integral connections to the Sustainable Development Goals (Dube et al. 2021; Hole et al. 2022; Obura et al. 2021; Pires et al. 2021).

1.3 The Contributions

This edited volume is structured into two thematic sections relevant to biodiversity conservation and global environmental change: (i) biodiversity conservation in the Anthropocene and (ii) sustainable range and forest management. Following this introduction, subsequent chapters are organized according to these thematic areas as outlined below.

1.3.1 Part I: Biodiversity Conservation in the Anthropocene

The Anthropocene epoch is characterized by profound human-induced alterations in biodiversity, species abundance, and biotic interactions, which are reshaping ecosystems at unparalleled rates (Malhi 2017). These transformations are primarily driven by escalating demands for resources such as food, water, timber, fibre, fuel, and minerals, leading to widespread habitat destruction, climate change, and an unparalleled decline in global biodiversity (Champagnon et al. 2023).

Mutanga, Gondo, Kolawole, and Mbaiwa underscore the impact of nature-based tourism on biodiversity loss, despite its heavy reliance on natural and wildlife resources. They highlight various stressors originating from the tourism industry, including climate change, land use changes, biological invasions, recreational activities, technological advancements, pollution, and natural resource overexploitation. The chapter emphasizes the need for innovative intervention strategies, such as implementing climate-smart nature-based tourism, monitoring human-induced ecosystem changes at tourism sites, enforcing stringent phytosanitary policies, and adopting visitor management frameworks like carrying capacity and limits of acceptable change. Tomás Adriano Sitoe and Diamantino Carlos Pinto examined community participation in Natural Resource Management Committees in Mapulanguene, Magude District, southern Mozambique, where committee establishment followed resource overuse and poaching. While committee involvement showed benefits, limited community control over local ecotourism projects by the private sector, excluded from the committees, posed a constraint. James Maradza, Raban Chanda, and Naomi Moswete propose Nexus Thinking as an approach to biodiversity conservation, emphasizing mutually beneficial linkages in human-dominated ecosystems, exemplified by interdependencies among conservation, livelihoods, and biodiversity around Mokolodi Nature Reserve in Botswana, advocating for both global perspective and local action.

Keoikantse Sianga, Mpaphi C. Bonyongo, and Richard W. S. Fynn argue that habitat fragmentation and loss, compounded by climate-related factors associated with socio-economic development, have resulted in the decline of crucial seasonal resources necessary for sustaining large-scale migrations of large herbivores. Through analysis of satellite and GPS collar data from various zebra *Equus quagga* populations, along with published information on collared zebra populations in northern Botswana, they demonstrated the potential impact of climate change on zebra movements. Their recommendation emphasizes the importance of maintaining unrestricted movement along migratory routes to safeguard zebra populations from fragmentation caused by climate change, thereby ensuring access to seasonal ranges and promoting genetic diversity across populations. Similarly, Victor K. Muposhi, Mkululi Ndlovu, Kudzai Mpakairi, Keoikantse Sianga, Admire Chanyandura, Chiedza C. Zembe, and Gregory Chiyangwa explored the factors that affect the space-use patterns of key animal species in protected areas using Random Forest and Gradient Boosting Model. They further emphasize the significance of data-driven decisions in conservation and sustainable development within the Anthropocene era. Tsheboeng, Sekonopo, and Kashe assessed the influence of a national botanical garden on plant diversity and population structure by sampling area enclosure and adjacent communal areas. The total density of woody plants was higher in area enclosures compared to communal areas. The findings indicate a healthier structural composition in enclosures, suggesting that expanding such areas could enhance conservation efforts in Botswana.

Belda Mosepele, Ketlhatlogile Mosepele, Ineelo Mosie, and Keoikantse Sianga argue that fish dynamics in flood-pulsed floodplains is essential for the sustainable utilization of fisheries resources. The study indicates that fish species exhibit high mobility in floodplains, with new floods creating foraging opportunities and ecological habitats that contribute to the instability of fish populations in these seasonal environments. For their chapter Wame Hambira, Patricia Mogomotsi, Victor Muposhi, and Goemeone Mogomotsi dive into a topical issue on the contentions surrounding consumptive tourism, particularly the use of regulated hunting as a form of conservation in Botswana. The chapter explores social media users' reactions to the lifting of the hunting ban by Botswana in 2014 which resulted in an uproar calling for the boycotting of the country as a tourist destination. The chapter highlights the views of extremists environmental and conservation activists when it comes to advocating for what they deem to be pro-environmental and pro-sustainable behaviour. The authors argue that the socio-economic and ecological conditions of some African countries require a far more nuanced consideration in terms of priorities for their local growth and development agenda.

1.3.2 Sustainable Range and Forest Management

Forests are crucial for ecological functioning, supporting sustainable livelihoods and economic development (Moore and Schindler 2022). However, they are increasingly threatened by invasive species, natural disasters, land use changes, disease, wildfires, climate change, urbanization, and fragmentation (Lindsay et al. 2022; Reyer et al. 2015). Sustainable forest management is thus imperative to guard the ability of the forest to sustain ecosystem services taking into account the following elements: (i) extent of forest resources; (ii) forest biological diversity; (iii) forest health and vitality; (iv) productive functions of forest resources; (v) protective functions of forest resources; (vi) socio-economic functions of forests; and (vii) legal, policy, and institutional frameworks (Blattert et al. 2022; Kimengsi et al. 2022; Tadesse et al. 2022).

Dixon Olutade Torimiro highlights the significance of integrating ICT in agricultural extension to address challenges in forest management such as inadequate technologies, governance issues, legal and policy challenges, and socio-economic barriers such as user poverty. He proposes an e-extension model aimed at developing a Forest Management Information System that improves data generation, distribution, and utilization, potentially enhancing rural livelihoods through increased income, productivity, poverty alleviation, and living standards. Similarly, Reniko Gondo, Oluwatoyin D. Kolawole, and Chiedza N. Mutanga employed remote sensing and GIS, combined with a random forest classification algorithm, to examine spatiotemporal variations in forest cover in Maun from 1992 to 2022. Their analysis revealed that the Normalized Difference Vegetation Index (NDVI) decreases, and Land Surface Temperature (LST) increases in densely built-up areas, while densely vegetated areas exhibit high NDVI and low LST. To inform decision-making and management strategies for vegetation monitoring under a changing climate in Botswana and sub-Saharan Africa, Linganani Kombani and Ednah Kgosiesele analysed the spatio-temporal changes in vegetation cover. Utilizing the Normalized Vegetation Index (NDVI) from the Advanced Very-High-Resolution Radiometer (AVHRR) satellite, they analysed the spatio-temporal variability of vegetation from 1982 to 2015 using AVHRR NDVI multitemporal time series data. Their findings indicate significant spatial and temporal fluctuations in vegetation cover, yet the interplay of human activities and climate change complicates attributing specific trends to discrete climatic variables.

Feyera Senbeta emphasizes the need for sustainable use and management of Ethiopia's coffee forests through environmental marketing schemes and participatory initiatives. This recommendation follows his findings that these forests are endangered by multiple threats, including commercial agriculture, logging, and climate change. In response to the recurrent threat of forest fires, Keoikantse Sianga, Witness Mojeremane, Ayana Angassa Abdeta, Kamogelo Makgobota, and Summer Mabula advocate for the broader adoption of effective fire management technologies. Their review indicates that fires variably impact plant communities and species, leading to significant losses of forest resources and biodiversity, which in turn disrupt ecosystem services. The chapter by Abdeta et al. further discusses how bush encroachment in arid and semiarid rangelands is influenced by changes in fire regimes, heightened livestock grazing, and climate variability, resulting in the proliferation of indigenous woody plants and introduced alien invasive species. This encroachment modifies the landscape structure, biodiversity, and the quality of ecosystem services derived from these rangelands. Tsheboeng, Nxumalo, Dludlu, and Nkonyane evaluate the effects of bush clearing on species composition, ecological status, and grazing value of understory plants in semi-arid savannah ecosystems with bush-encroached and cleared treatments. They found a higher prevalence of palatable species in cleared areas, suggesting that bush clearing can enhance rangeland conditions in savannah ecosystems.

Joyce Lepetu, Hesekia Garekae, Gofaone Rammotokara, and Topoyame Makoi highlight the pivotal role of non-timber forest products (NTFPs) in supporting the livelihoods of marginalized communities. Their research indicates that 65 plant species are utilized across various socio-demographic groups, underscoring the universal importance of NTFPs as essential livelihood resources that enhance community resilience and reduce vulnerability. Additionally, Endale Difabachew, Abdi-Khalil Edriss, Jema Haji, Belaineh Legesse, and Mengistu Ketema's analysis of forest conservation economics demonstrates that households value increased sustainable harvesting at US$4.96, with governance, economic incentives, and participatory forest management plan (PFMP) playing crucial roles in enhancing forest conservation and service flow.

Witness Mojeremane, Ronnie M. Mmolotsi, Melusi Rampart, Demel Teketay, Obusitswe Kakambi, Topoyame I. Makoi, Ishmael Kopong, Kamogelo Makgobota, Thembinkosi Mathowa, Gosiame Neo-Mahupeleng, and Keoikantse Sianga investigated the effects of various pre-sowing treatments on the germination of *Brachystegia boehmii* and *Piliostigma thonningii* seeds. They discovered that *Brachystegia boehmii* lacks physical dormancy, unlike *Piliostigma thonningii*. Seed germination, a critical stage in the life cycle of flowering plants, significantly influences their population dynamics and survival. In a related study, Shimane W. Makhabu, Koketso Tshireletso, Mpho R. Setlalekgomo, Zibanani Seletlo, Boipuso Legwatagwata, and Lawrence Akanyang examined the effects of four invasive alien species—*Cenchrus biflorus* Roxb., *Verbesina encelioides* (Cav.) Benth. & Hook. f. ex A. Gray, *Prosopis juliflora* (Swartz) DC, and *Salvinia molesta* D. Mitch—on biodiversity and ecosystem services in Botswana, including the Central Kalahari Game Reserve and the Okavango Delta. Their findings indicate significant negative impacts on biodiversity, agriculture, tourism, and fishing.

1.4 Summary

The Anthropocene epoch denotes a period characterized by profound human-induced alterations in biodiversity and ecosystems, driven by escalating demands for resources, leading to widespread habitat destruction, climate change, and biodiversity. Conversely, forests, vital for ecological functioning and sustainable livelihoods, face multifaceted threats including invasive species, natural disasters, and climate change, necessitating sustainable forest management strategies that encompass various elements such as biodiversity, health, productivity, and socio-economic functions. In this context, understanding and addressing the complexities of biodiversity conservation and sustainable forest management are crucial imperatives in navigating the challenges of the Anthropocene era. The themes and subsequent chapters in this volume emphasize the critical need for implementing comprehensive conservation strategies to safeguard biodiversity, ecosystems, and human well-being amidst unparalleled anthropogenic pressures and environmental changes.

References

Andriamahefazafy M, Touron-Gardic G, March A, Hosch G, Palomares MLD, Failler P (2022) Sustainable development goal 14: to what degree have we achieved the 2020 targets for our oceans? Ocean Coast Manag 227:106273. https://doi.org/10.1016/j.ocecoaman.2022.106273

Archer E, Dziba L, Mulongoy KJ, Maoela MA, Walters M, Biggs RO, Cormier Salem M-C, Declerck F, Diaw MC, Dunham AE, Failler P, Gordon C, Harhash KA, Kasisi R, Kizito F, Nyingi W, Oguge N, Osman-Elasha B, Stringer LC et al (2018) The regional assessment report on biodiversity and ecosystem services for Africa: summary for policymakers. IPBES. https://hal.science/hal-02333968

Arora NK, Mishra I (2022) Sustainable development goal 6: global water security. Environ Sustain 5(3):271–275. https://doi.org/10.1007/s42398-022-00246-5

Arora NK, Mishra I (2023) Responsible consumption and production: a roadmap to sustainable development. Environ Sustain 6(1):1–6. https://doi.org/10.1007/s42398-023-00266-9

Baisero D, Visconti P, Pacifici M, Cimatti M, Rondinini C (2020) Projected global loss of mammal habitat due to land-use and climate change. One Earth 2(6):578–585. https://doi.org/10.1016/j.oneear.2020.05.015

Baldwin-Cantello W, Tickner D, Wright M, Clark M, Cornelius S, Ellis K, Francis A, Ghazoul J, Gordon JE, Matthews N, Milner-Gulland EJ, Smith P, Walmsley S, Young L (2023) The triple challenge: synergies, trade-offs and integrated responses for climate, biodiversity, and human wellbeing goals. Clim Pol 23(6):782–799. https://doi.org/10.1080/14693062.2023.2175637

Baloch MA, Danish K, S. U.-D, Ulucak ZŞ (2020) Poverty and vulnerability of environmental degradation in sub-Saharan African countries: what causes what? Struct Chang Econ Dyn 54:143–149. https://doi.org/10.1016/j.strueco.2020.04.007

Bennett J, Marandure T, Hawkins H-J, Mapiye C, Palmer A, Lemke S, Wu L, Moradzadeh M (2023) A conceptual framework for understanding ecosystem trade-offs and synergies, in communal rangeland systems. Ecosyst Serv 61:101533. https://doi.org/10.1016/j.ecoser.2023.101533

Blattert C, Eyvindson K, Hartikainen M, Burgas D, Potterf M, Lukkarinen J, Snäll T, Toraño-Caicoya A, Mönkkönen M (2022) Sectoral policies cause incoherence in forest management and ecosystem service provisioning. Forest Policy Econ 136:102689. https://doi.org/10.1016/j.forpol.2022.102689

Borie M, Mahony M, Obermeister N, Hulme M (2021) Knowing like a global expert organization: comparative insights from the IPCC and IPBES. Glob Environ Chang 68:102261. https://doi.org/10.1016/j.gloenvcha.2021.102261

CBD (2022) Decision adopted by the conference of the parties to the convention on biological diversity 15/4. Kunming-Montreal Global Biodiversity Framework. https://www.cbd.int/doc/decisions/cop-15/cop-15-dec-04-en.pdf

Champagnon J, Elmberg J, Guillemain M, Lavretsky P, Clark RG, Söderquist P (2023) Silent domestication of wildlife in the Anthropocene: the mallard as a case study. Biol Conserv 288:110354. https://doi.org/10.1016/j.biocon.2023.110354

Chapman CA, Abernathy K, Chapman LJ, Downs C, Effiom EO, Gogarten JF, Golooba M, Kalbitzer U, Lawes MJ, Mekonnen A, Omeja P, Razafindratsima O, Sheil D, Tabor GM, Tumwesigye C, Sarkar D (2022) The future of sub-Saharan Africa's biodiversity in the face of climate and societal change. Front Ecol Evol 10. https://doi.org/10.3389/fevo.2022.790552

Das T, Holland P, Ahmed M, Husain L (2021) Sustainable development goal 3: good health and well-being. In: Das T, Nayar PD (eds) South-East Asia eye health: systems, practices, and challenges. Springer, pp 61–78. https://doi.org/10.1007/978-981-16-3787-2_4

del Pérez-Peña MC, Jiménez-García M, Ruiz-Chico J, Peña-Sánchez AR (2021) Analysis of research on the SDGs: the relationship between climate change, poverty and inequality. Appl Sci 11(19):Article 19. https://doi.org/10.3390/app11198947

Díaz SM, Settele J, Brondízio E, Ngo H, Guèze M, Agard J, Arneth A, Balvanera P, Brauman K, Butchart S, Chan KMA, Garibaldi LA, Ichii K, Liu J, Subramanian S, Midgley G, Miloslavich P, Molnár Z, Obura D et al (2019) The global assessment report on biodiversity and ecosystem services: summary for policy makers. Intergovernmental Science-Policy Platform on Biodiversity and Ecosystem Services. https://ri.conicet.gov.ar/handle/11336/116171

Dube K, Chikodzi D, Nhamo G (2021) Making SDGs work to end hunger, sustain energy, resolve climate change, and reverse biodiversity loss. In: Nhamo G, Chikodzi D, Dube K (eds) Sustainable development goals for society Vol. 2: food security, energy, climate action and biodiversity. Springer International Publishing, pp 3–15. https://doi.org/10.1007/978-3-030-70952-5_1

Folke C, Polasky S, Rockström J, Galaz V, Westley F, Lamont M, Scheffer M, Österblom H, Carpenter SR, Chapin FS, Seto KC, Weber EU, Crona BI, Daily GC, Dasgupta P, Gaffney O, Gordon LJ, Hoff H, Levin SA et al (2021) Our future in the Anthropocene biosphere. Ambio 50(4):834–869. https://doi.org/10.1007/s13280-021-01544-8

Freeman R, McRae L, Marconi V, Almond REA, Grooten M (2022) A deep dive into the living planet index: a technical report. (p 21) [a technical report]. WWF. https://www.wwf.nl/globalassets/pdf/lpr/lpr-2022-a-deep-dive-into-the-living-planet-index.pdf

Habibullah MS, Din BH, Tan S-H, Zahid H (2022) Impact of climate change on biodiversity loss: global evidence. Environ Sci Pollut Res 29(1):1073–1086. https://doi.org/10.1007/s11356-021-15702-8

Hawken S, Rahmat H, Sepasgozar SME, Zhang K (2021) The SDGs, ecosystem services and cities: a network analysis of current research innovation for implementing urban sustainability. Sustain For 13(24):Article 24. https://doi.org/10.3390/su132414057

Hernández-Blanco M, Costanza R, Chen H, deGroot D, Jarvis D, Kubiszewski I, Montoya J, Sangha K, Stoeckl N, Turner K, van't Hoff V (2022) Ecosystem health, ecosystem services, and the well-being of humans and the rest of nature. Glob Chang Biol 28(17):5027–5040. https://doi.org/10.1111/gcb.16281

Hole DG, Collins P, Tesfaw A, Barrera L, Mascia MB, Turner WR (2022) Make nature's role visible to achieve the SDGs. Global Sustain 5:e8. https://doi.org/10.1017/sus.2022.5

Hughes AC (2023) The Post-2020 Global Biodiversity Framework: how did we get here, and where do we go next? Intgr Conserv 2(1):1–9. https://doi.org/10.1002/inc3.16

Hughes AC, Grumbine RE (2023) The Kunming-Montreal Global Biodiversity Framework: what it does and does not do, and how to improve it. Front Environ Sci 11. https://doi.org/10.3389/fenvs.2023.1281536

IPBES (2019) Global assessment report on biodiversity and ecosystem services of the Intergovernmental Science-Policy Platform on Biodiversity and Ecosystem Services (Version 1). [object Object]. https://doi.org/10.5281/ZENODO.3831673

IPBES (2022) Summary for policymakers of the thematic assessment of the sustainable use of wild species of the Intergovernmental Science-Policy Platform on Biodiversity and Ecosystem Services (IPBES). In: Fromentin JM, Emery MR, Donaldson J, Danner MC, Hallosserie A, Kieling D, Balachander G, Barron ES, Chaudhary RP, Gasalla M, Halmy M, Hicks C, Park MS, Parlee B, Rice J, Ticktin T, Tittensor D (eds) IPBES secretariat, Bonn, p 44. https://doi.org/10.5281/zenodo.7411847

Isabelle DA, Westerlund M (2022) A review and categorization of artificial intelligence-based opportunities in wildlife, ocean and land conservation. Sustain For 14(4):Article 4. https://doi.org/10.3390/su14041979

Isbell F, Balvanera P, Mori AS, He J-S, Bullock JM, Regmi GR, Seabloom EW, Ferrier S, Sala OE, Guerrero-Ramírez NR, Tavella J, Larkin DJ, Schmid B, Outhwaite CL, Pramual P, Borer ET, Loreau M, Omotoriogun TC, Obura DO et al (2023) Expert perspectives on global biodiversity loss and its drivers and impacts on people. Front Ecol Environ 21(2):94–103. https://doi.org/10.1002/fee.2536

Jaureguiberry P, Titeux N, Wiemers M, Bowler DE, Coscieme L, Golden AS, Guerra CA, Jacob U, Takahashi Y, Settele J, Díaz S, Molnár Z, Purvis A (2022) The direct drivers of recent global anthropogenic biodiversity loss. Sci Adv 8(45):eabm9982. https://doi.org/10.1126/sciadv.abm9982

Juju D, Baffoe G, Dam Lam R, Karanja A, Naidoo M, Ahmed A, Jarzebski MP, Saito O, Fukushi K, Takeuchi K, Gasparatos A (2020) Sustainability challenges in sub-Saharan Africa in the context of the sustainable development goals (SDGs). In: Gasparatos A, Ahmed A, Naidoo M, Karanja A, Fukushi K, Saito O, Takeuchi K (eds) Sustainability challenges in sub-Saharan Africa I: continental perspectives and insights from Western and Central Africa. Springer, pp 3–50. https://doi.org/10.1007/978-981-15-4458-3_1

Kedward K, Ryan-Collins J, Chenet H (2023) Biodiversity loss and climate change interactions: financial stability implications for central banks and financial supervisors. Clim Pol 23(6):763–781. https://doi.org/10.1080/14693062.2022.2107475

Kimengsi JN, Owusu R, Djenontin INS, Pretzsch J, Giessen L, Buchenrieder G, Pouliot M, Acosta AN (2022) What do we (not) know on forest management institutions in sub-Saharan Africa? A regional comparative review. Land Use Policy 114:105931. https://doi.org/10.1016/j.landusepol.2021.105931

Kowarik I, Fischer LK, Kendal D (2020) Biodiversity conservation and sustainable urban development. Sustain For 12(12):Article 12. https://doi.org/10.3390/su12124964

Krause T, Tilker A (2022) How the loss of forest fauna undermines the achievement of the SDGs. Ambio 51(1):103–113. https://doi.org/10.1007/s13280-021-01547-5

Krauss JE (2022) Unpacking SDG 15, its targets and indicators: tracing ideas of conservation. Globalizations 19(8):1179–1194. https://doi.org/10.1080/14747731.2022.2035480

Li Q, Ge Y, Sayer JA (2023) Challenges to implementing the Kunming-Montreal Global Biodiversity Framework. Land 12(12):Article 12. https://doi.org/10.3390/land12122166

Lindsay PA, Anderson SH, Dickman A, Gandiwa P, Harper SH, Morakinyo AB, Nyambe N, O'Brien-Onyeka M, Packer C, Parker AH, Robson AS, Ruhweza A, Sogbohossou A, Steiner KW, Tumenta PN (2022) Shepherding sub-Saharan Africa's terrestrial biodiversity through peak anthropogenic pressure toward a green anthropocene. Annu Rev Environ Resour 47. https://ora.ox.ac.uk/objects/uuid:59c9d0d4-e015-4613-9e11-461d42dcdf29

Malhi Y (2017) The concept of the Anthropocene. Annu Rev Environ Resour 42:77–104. https://doi.org/10.1146/annurev-environ-102016-060854

Marselle MR, Hartig T, Cox DTC, de Bell S, Knapp S, Lindley S, Triguero-Mas M, Böhning-Gaese K, Braubach M, Cook PA, de Vries S, Heintz-Buschart A, Hofmann M, Irvine KN, Kabisch N, Kolek F, Kraemer R, Markevych I, Martens D et al (2021) Pathways linking biodiversity to human health: a conceptual framework. Environ Int 150:106420. https://doi.org/10.1016/j.envint.2021.106420

Moore JW, Schindler DE (2022) Getting ahead of climate change for ecological adaptation and resilience. Science 24;376(6600):1421–1426. https://doi.org/10.1126/science.abo3608. Epub 2022 Jun 23. PMID: 35737793.

Muluneh MG (2021) Impact of climate change on biodiversity and food security: a global perspective—a review article. Agric Food Secur 10(1):36. https://doi.org/10.1186/s40066-021-00318-5

O'Connell MJ, Nasirwa O, Carter M, Farmer KH, Appleton M, Arinaitwe J, Bhanderi P, Chimwaza G, Copsey J, Dodoo J, Duthie A, Gachanja M, Hunter N, Karanja B, Komu HM, Kosgei V, Kuria A, Magero C, Manten M et al (2019) Capacity building for conservation: problems and potential solutions for sub-Saharan Africa. Oryx 53(2):273–283. https://doi.org/10.1017/S0030605317000291

O'Connor B, Bojinski S, Röösli C, Schaepman ME (2020) Monitoring global changes in biodiversity and climate essential as ecological crisis intensifies. Eco Inform 55:101033. https://doi.org/10.1016/j.ecoinf.2019.101033

Obura DO, Katerere Y, Mayet M, Kaelo D, Msweli S, Mather K, Harris J, Louis M, Kramer R, Teferi T, Samoilys M, Lewis L, Bennie A, Kumah F, Isaacs M, Nantongo P (2021) Integrate biodiversity targets from local to global levels. Science 373(6556):746–748. https://doi.org/10.1126/science.abh2234

Obura D, Agrawal A, DeClerck F, Donaldson J, Dziba L, Emery MR, Friedman K, Fromentin J-M, Garibaldi LA, Mulongoy J, Navarrete-Frias C, Reidl PM, Roe D, Timoshyna A (2023) Prioritizing sustainable use in the Kunming-Montreal global biodiversity framework. PLOS Sustain Transform 2(1):e0000041. https://doi.org/10.1371/journal.pstr.0000041

Ogwu MC, Izah SC, Iyiola AO (2022) An overview of the potentials, threats and conservation of biodiversity in Africa. In: Chibueze Izah S (ed) Biodiversity in Africa: potentials, threats and conservation. Springer Nature, pp 3–20. https://doi.org/10.1007/978-981-19-3326-4_1

Parmesan C, Morecroft MD, Trisurat Y, Adrian R, Anshari GZ, Arneth A, Gao Q, Gonzalez P, Harris R, Price J, Stevens N, Talukdarr GH (2022) Achieving the SDGs with biodiversity. Akademie der Naturwissenschaften Schweiz (SCNAT), Forum Biodiversität Schweiz. In: Pörtner H-O, Roberts DC, Tignor M, Poloczanska ES, Mintenbeck K, Alegría A, Craig M, Langsdorf S, Löschke S, Möller V, Okem A, Rama B (eds) . Cambridge University Press. https://www.ipcc.ch/report/ar6/wg2/

Pereira HM, Leadley PW, Proença V, Alkemade R, Scharlemann JPW, Fernandez-Manjarrés JF, Araújo MB, Balvanera P, Biggs R, Cheung WWL, Chini L, Cooper HD, Gilman EL, Guénette S, Hurtt GC, Huntington HP, Mace GM, Oberdorff T, Revenga C et al (2010) Scenarios for global biodiversity in the 21st century. Science 330(6010):1496–1501. https://doi.org/10.1126/science.1196624

Pettorelli N, Graham NAJ, Seddon N, da Cunha M, Bustamante M, Lowton MJ, Sutherland WJ, Koldewey HJ, Prentice HC, Barlow J (2021) Time to integrate global climate change and biodiversity science-policy agendas. J Appl Ecol 58(11):2384–2393. https://doi.org/10.1111/1365-2664.13985

Pfenning-Butterworth A, Buckley LB, Drake JM, Farner JE, Farrell MJ, Gehman A-LM, Mordecai EA, Stephens PR, Gittleman JL, Davies TJ (2024) Interconnecting global threats: climate change, biodiversity loss, and infectious diseases. Lancet Planet Health 8(4):e270–e283. https://doi.org/10.1016/S2542-5196(24)00021-4

Pires APF, Rodriguez Soto C, Scarano FR (2021) Strategies to reach global sustainability should take better account of ecosystem services. Ecosyst Serv 49:101292. https://doi.org/10.1016/j.ecoser.2021.101292

Pörtner H-O, Scholes RJ, Arneth A, Barnes DKA, Burrows MT, Diamond SE, Duarte CM, Kiessling W, Leadley P, Managi S, McElwee P, Midgley G, Ngo HT, Obura D, Pascual U, Sankaran M, Shin YJ, Val AL (2023) Overcoming the coupled climate and biodiversity crises and their societal impacts. Science 380(6642):eabl4881. https://doi.org/10.1126/science.abl4881

Reyer C, Brouwers N, Rammig A, Brook B, Epila J, Grant R (2015) Forest resilience and tipping points at different spatiotemporal scales: approaches and challenges. J Ecol 103: 5–15

Rice WS, Sowman MR, Bavinck M (2020) Using theory of change to improve post-2020 conservation: a proposed framework and recommendations for use. Conserv Sci Pract 2(12):e301. https://doi.org/10.1111/csp2.301

Rinaldi A (2021) Biodiversity 2030: a road paved with good intentions. EMBO Rep 22(6):e53130. https://doi.org/10.15252/embr.202153130

Ruckelshaus MH, Jackson ST, Mooney HA, Jacobs KL, Kassam K-AS, Arroyo MTK, Báldi A, Bartuska AM, Boyd J, Joppa LN, Kovács-Hostyánszki A, Parsons JP, Scholes RJ, Shogren JF, Ouyang Z (2020) The IPBES global assessment: pathways to action. Trends Ecol Evol 35(5):407–414. https://doi.org/10.1016/j.tree.2020.01.009

Ryan CM, Pritchard R, McNicol I, Owen M, Fisher JA, Lehmann C (2016) Ecosystem services from southern African woodlands and their future under global change. Philos Trans R Soc Biol Sci 371(1703):20150312. https://doi.org/10.1098/rstb.2015.0312

Saliu AO, Komolafe OO, Bamidele CO, Raimi MO (2023) The value of biodiversity to sustainable development in Africa. In: Izah SC, Ogwu MC (eds) Sustainable utilization and conservation of Africa's biological resources and environment. Springer Nature, pp 269–294. https://doi.org/10.1007/978-981-19-6974-4_10

Schlattmann A, Neuendorf F, Burkhard K, Probst E, Pujades E, Mauser W, Attinger S, von Haaren C (2022) Ecological sustainability assessment of water distribution for the maintenance of ecosystems, their services and biodiversity. Environ Manag 70(2):329–349. https://doi.org/10.1007/s00267-022-01662-3

Schröter M, Berbés-Blázquez M, Albert C, Hill R, Krause T, Loos J, Mannetti LM, Martín-López B, Neelakantan A, Parrotta JA, Quintas-Soriano C, Abson DJ, Alkemade R, Amelung B, Baptiste B, Barrios E, Djoudi H, Drakou EG, Durance I et al (2023) Science on ecosystems and people to support the Kunming-Montreal Global Biodiversity Framework. Ecosyst People 19(1):2220913. https://doi.org/10.1080/26395916.2023.2220913

Seddon N (2022) Harnessing the potential of nature-based solutions for mitigating and adapting to climate change. Science 376(6600):1410–1416. https://doi.org/10.1126/science.abn9668

Sintayehu DW (2018) Impact of climate change on biodiversity and associated key ecosystem services in Africa: a systematic review. Ecosyst Health Sustain 4(9):225–239. https://doi.org/10.1080/20964129.2018.1530054

Smith P, Arneth A, Barnes DKA, Ichii K, Marquet PA, Popp A, Pörtner H-O, Rogers AD, Scholes RJ, Strassburg B, Wu J, Ngo H (2022) How do we best synergize climate mitigation actions to co-benefit biodiversity? Glob Chang Biol 28(8):2555–2577. https://doi.org/10.1111/gcb.16056

Streck C (2023) Synergies between the Kunming-Montreal Global Biodiversity Framework and the Paris Agreement: the role of policy milestones, monitoring frameworks and safeguards. Clim Pol 23(6):800–811. https://doi.org/10.1080/14693062.2023.2230940

Sun Z, Behrens P, Tukker A, Bruckner M, Scherer L (2022) Global human consumption threatens key biodiversity areas. Environ Sci Technol 56(12):9003–9014. https://doi.org/10.1021/acs.est.2c00506

Tadesse T, Teklay G, Mulatu DW, Rannestad MM, Meresa TM, Woldelibanos D (2022) Forest benefits and willingness to pay for sustainable forest management. Forest Policy Econ 138:102721. https://doi.org/10.1016/j.forpol.2022.102721

Welden EA, Chausson A, Melanidis MS (2021) Leveraging nature-based solutions for transformation: reconnecting people and nature. People Nat 3(5):966–977. https://doi.org/10.1002/pan3.10212

Xu J, Wang J (2023) Analysis of the main elements and implications of the Kunming-Montreal Global Biodiversity Framework. Biodivers Sci 31(4):23020. https://doi.org/10.17520/biods.2023020

Yang S, Zhao W, Liu Y, Cherubini F, Fu B, Pereira P (2020) Prioritizing sustainable development goals and linking them to ecosystem services: a global expert's knowledge evaluation. Geograph Sustain 1(4):321–330. https://doi.org/10.1016/j.geosus.2020.09.004

Yin C, Zhao W, Cherubini F, Pereira P (2021) Integrate ecosystem services into socio-economic development to enhance achievement of sustainable development goals in the post-pandemic era. Geograph Sustain 2(1):68–73. https://doi.org/10.1016/j.geosus.2021.03.002

Zhao X, Zhang K, Zhao R (2024) Sustainable Forest development in the digital era: the impact of internet use on the happiness of forest farmers' families in ecologically fragile ethnic areas of China. Forests 15(3):Article 3. https://doi.org/10.3390/f15030564

Wame L. Hambira (PhD) is an Associate Professor of Environmental Management at the Botswana University of Agriculture and Natural Resources (BUAN). Prof Hambira has served the Intergovernmental Science-Policy Platform for Biodiversity and Ecosystem Services (IPBES) as a scoping expert (the nexus assessment of the interlinkages among biodiversity, water, food and health in the context of climate change as well as the global assessment of the second biodiversity and ecosystem assessment). Prof Hambira has also served as a United Nations Environment Program (UNEP) Review Editor for the Solution Pathways for Transformation of Environmental Systems chapter of the seventh Global Environment Outlook report.

Victor K. Muposhi is a Conservation Scientist and Associate Professor of Wildlife Conservation in the Department of Wildlife and Aquatic Resources at the Botswana University of Agriculture and Natural Resources. His research interests include biodiversity, ecosystems dynamics, spatial ecology, socio-ecological systems, human-wildlife coexistence, and sustainability science.

Ditiro B. Moalafhi is a Hydrologist and water resources management professional with over 20 years of experience. He holds a PhD in Civil and Environmental Engineering (Hydro-climatology) from the University of New South Wales in Sydney Australia, MPhil in Environmental Science, Postgraduate Diploma in Education, and Bachelor of Science all from the University of Botswana. Professor Moalafhi is currently an Associate Professor of water resources management at the Botswana University of Agriculture and Natural Resources, where he is the Head of Wildlife and Aquatic Resources department. He is also the SASSCAL Research Chair in Water Resource Quantity and Availability.

Ayana Angassa Abdeta is a Professor in the Department of Range and Forest Resources—Botswana University of Agriculture and Natural Resources. He received BSc and MSc from the Alemaya University of Agriculture, Ethiopia, PhD from the Norwegian University of Life Sciences, Department of International Environment and Development Studies, Postdoctoral training from both the University of Hohenheim (Food Security Centre) and University of Cologne (Botanical Institute) in Germany, Certificate (2015) in Disaster Risk Management from the Humboldt University of Berlin, Certificate (2020) in Project Management Foundations from the University of Cape Town, and Certificate (2022) in Agroforestry Systems from the University of Freiburg, Germany.

Ketlhatlogile Mosepele (PhD) is the Vice Chancellor of the Botswana University of Agriculture and Natural Resources. He holds the degree of Doctor of Philosophy (PhD) in Fisheries Biology and Management from the University of Bergen, Norway. He also holds the degree of Master of Philosophy (MPhil) and a Postgraduate Diploma in Fisheries Biology and Management from the University of Bergen, Norway. Professor Mosepele obtained his degree of Bachelor of Science (BS) in Agriculture majoring in Fisheries Science from the New Mexico State University, USA. He previously served as a Professor of Fisheries Biology and Management and Deputy Director of the Okavango Research Institute, University of Botswana.

Chapter 2
Nature-Based Tourism-Induced Pressures Contributing to Biodiversity Disturbance and Loss in Sub-Saharan Africa

Chiedza Ngonidzashe Mutanga, Reniko Gondo, Oluwatoyin Dare Kolawole, and Joseph E. Mbaiwa

2.1 Introduction

Biodiversity greatly benefits nature-based tourism (NBT) and is crucial to the sustained growth of the sector (Christ et al. 2003). NBT is one of the tourism sectors with the highest growth rates (Suta et al. 2017). Both mainstream tourism and NBT have grown globally in recent decades (Heffernan 2022). The United Nations World Tourism Organization (UNWTO) confirmed that 2019 was the tenth consecutive year of tourism growth, with increases in international arrivals across all regions (UNWTO 2020). The World Travel and Tourism Council (WTTC) reports that in 2019, sub-Saharan Africa (SSA) was one of the tourism destinations with the fastest tourism growth rates across the globe (WTTC 2019). The diverse landscapes of SSA offer many benefits, including the provision of natural capital on which many people rely, the provision of a wide range of essential ecosystem services, and a major cultural and heritage value (Galvani et al. 2016; Macdonald et al. 2015; Stolton and Dudley 2019). Terrestrial ecosystems are valuable resources for SSA's NBT (Stolton and Dudley 2019).

NBT encompasses a wide range of tourist activities that are frequently commercialized and involve engagement with the natural world distant from the participant's normal surroundings (Weiler and Hall 1992). Wildlife tourism, mass tourism, adventure tourism, low-impact tourism, ecotourism, and sustainable tourism are just a few examples of the numerous types of tourism that fall under the category of NBT (Suta et al. 2017; Olson et al. 2001). These types of tourism rely on undeveloped or wild natural resources, including wildlife species, habitats, scenery, landscapes, and salt and freshwater features (Goodwin 1996).

C. N. Mutanga (✉) · R. Gondo · O. D. Kolawole · J. E. Mbaiwa
Okavango Research Institute, University of Botswana, Maun, Botswana
e-mail: mutangac@ub.ac.bw; rgondo@ub.ac.bw; tkolawole@ub.ac.bw; jmbaiwa@ub.ac.bw

W. L. Hambira et al. (eds.), *Environmental Change and Biodiversity Conservation in sub-Saharan Africa*, Advances in Global Change Research 75,
https://doi.org/10.1007/978-3-031-73136-5_2

NBT positively contributes to the achievement of sustainable development goals (Richardson 1998; Saarinen et al. 2009). It is, therefore, promoted by governments, international lending institutions, and private businesses as an ideal development strategy that combines economic growth with environmental conservation (Duffy 2005). While SSA's natural resources are potentially of immense global value, they are being threatened by increasing multiple anthropogenic pressures, including rapid human population growth and associated demands such as agricultural expansion and increased local and global demand for resources (Lindsey et al. 2022). In another vein, tourism contributes significantly to each of the five main stresses that directly cause biodiversity loss, including habitat change, overexploitation, pollution, invasive alien species, and climate change (Gössling and Hall 2006; Hall 2006). Even if biodiversity is an important component of NBT, the degree to which the latter causes biodiversity loss through land use and land cover change (LULCC), habitat loss and fragmentation, and contribution to climate change may make tourism a harmful industry (Hall 2006). It is, therefore, important to consider both the positive and negative effects of tourism with a view to accomplishing sustainable biodiversity conservation.

Heffernan (2022) argues that NBT can become quite a complex phenomenon that requires careful analysis to understand the many costs and benefits associated with it. For instance, while photographic tourism (a form of NBT) is framed as a nonconsumptive form of tourism, tourism in general is known to be an activity with a heavy carbon footprint and ecological footprint. In contrast to its "eco" framing, ecotourism development has also been shown to increase detrimental ecological effects (Büscher and Fletcher 2017; Lasso and Dahles 2021; Pandya et al. 2022). Samdin et al. (2013) posit that one of the ecological challenges associated with the designation of national parks as public tourism attractions is overuse by tourists. This shows that all forms of tourism are, in some way, consumptive.

Although designating places as protected areas (PAs) and establishing national parks might serve as a method for enhancing benefits and maintaining natural resources, many anthropogenic-related biodiversity losses, such as land clearing, pollution, and climate change, are associated with tourism growth (Gössling et al. 2010). Natural resource management in the case of national parks is thus a highly challenging assignment because of the inherently incompatible values and objectives related to the protection of natural resources and the creation of profitable tourist offerings (Suta et al. 2017). However, while NBT contributes significantly to biodiversity disturbance and loss, there has not been a systematic examination of the different pressures that NBT exerts on the environment. Given the fragmented nature of the literature available on the subject, there is still limited knowledge on the extent of the contribution of NBT to biodiversity disturbance and loss in different PAs in SSA. This, therefore, presents some challenges in proffering pragmatic solutions to NBT management and good governance considering the different pressures from the NBT. The chapter, therefore, draws from the literature to provide a general overview of NBT as a driver of biodiversity disturbance and loss. This is meant to encourage in-depth research on the subject by both academic and industry practitioners. Specifically, this chapter examines NBT-related stressors of biodiversity in SSA. While the second section presents the theoretical framework of the chapter, the third section highlights its methodological approach. Finally, the fourth section presents the results and discussion.

2.1.1 The Drivers-Pressure-State Impact-Response (DPSIR) Framework

This chapter is informed by the Drivers-Pressure-State Impact-Response (DPSIR) framework (Fig. 2.1). The DPSIR framework was developed by the European Environment Agency (EEA) in 1995 (EEA 1995). The framework was built upon the Pressure-State-Response (PSR) framework developed by the Organisation for Economic Co-operation and Development (OECD) in 1993, which also evolved from the Stress-Response (S-R) framework developed by Statistics Canada in 1979 (Gari et al. 2015). However, the PSR framework focused on human-induced pressures and responses and did not effectively factor natural variability into the pressure category. These shortcomings led to the development of the DPSIR framework in 1995, which incorporates natural variability as a pressure on the current state and addresses responses to the impact of changes in state on human well-being (Gari et al. 2015).

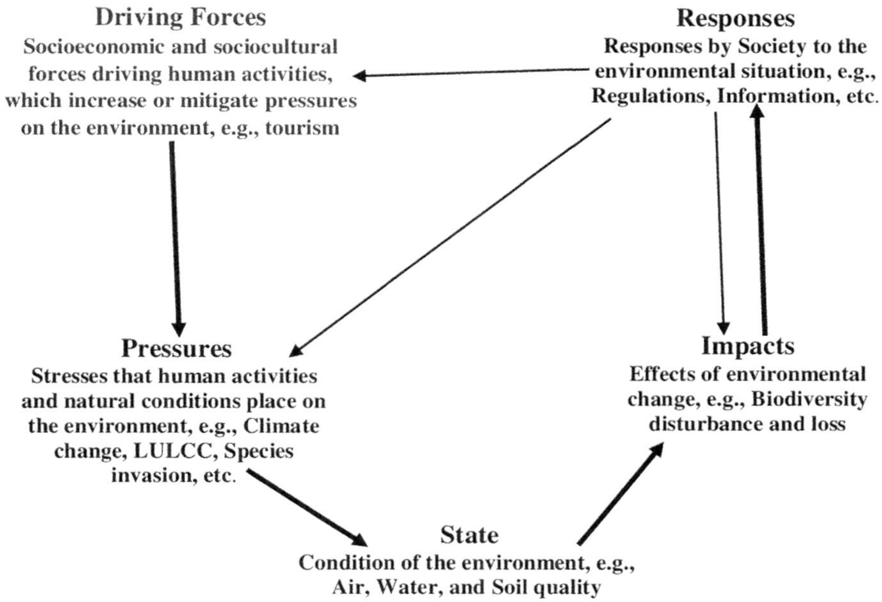

Fig. 2.1 Drivers-Pressure-State Impact-Response (DPSIR) framework. (Adapted from EEA (1999))

The DPSIR framework is used to describe how society and the environment interact (Maxim et al. 2009). According to this framework, sociocultural and economic factors (driving forces) put pressure on the environment and have an impact on the ecosystem's health. This may then prompt a societal response to the ensuing environmental condition (Ness et al. 2010). State in this context refers to the condition of the environment because of the pressure that human activity has placed on it. Impacts connote the effects of environmental degradation on the environment and are highly dependent on the characteristics of the drivers (Maxim et al. 2009). The DPSIR framework is thus a useful tool for elucidating the relationship between the environment and NBT development and tracing the origins and consequences of environmental problems within a specific ecosystem. The framework is frequently used to analyze and evaluate environmental issues with a view to choosing the best course of action. For instance, while Kosamu et al. (2022) used the framework to assess freshwater ecosystems in Lake Malombe, Malawi, Kim et al. (2021) used it in the identification of economic elements affecting global wildfire activities. Additionally, Lin et al. (2007) used the framework to evaluate the impacts of developmental activities on the coastal environment and society in Xiamen, China. The framework was used in this chapter for a better assessment of the relationship between the NBT and its resultant pressures and how these pressures affect the state of the environment, leading to biodiversity disturbance and loss. Overall, the framework helps to capture the general picture of the different pressures exerted by NBT on the environment, which contribute to biodiversity loss and disturbance.

The DPSIR approach can be employed for all ecosystems (Gari et al. 2015), and as such, the examples used in the chapter are from both aquatic and terrestrial systems. The chapter, however, partially adopts the framework by paying attention to the assessment of the pressures exerted by the NBT, how they negatively affect the condition of the environment, and how they impact biodiversity. Society's responses to the resultant environmental situation are, therefore, not covered in this chapter.

2.2 Methods

2.2.1 Study Context

SSA has over 7500 PAs covering over 4.5 million km^2 of land and ocean surface (UNEP-WCMC 2019) (Fig. 2.2). These PAs are part of most of the NBT hotspots in the region and are home to some of the world's iconic species, which include lion (*Panthera leo*), leopard (*Panthera pardus*), rhinoceros (*Diceros bicornis* and *Ceratotherium simum*), elephant (*Loxodonta africana* and *Loxodonta cyclotis*), African buffalo (*Syncerus caffer*), gorilla (*Gorilla* spp.), and chimpanzee (*Pan troglodytes*) (Chapman et al. 2022).

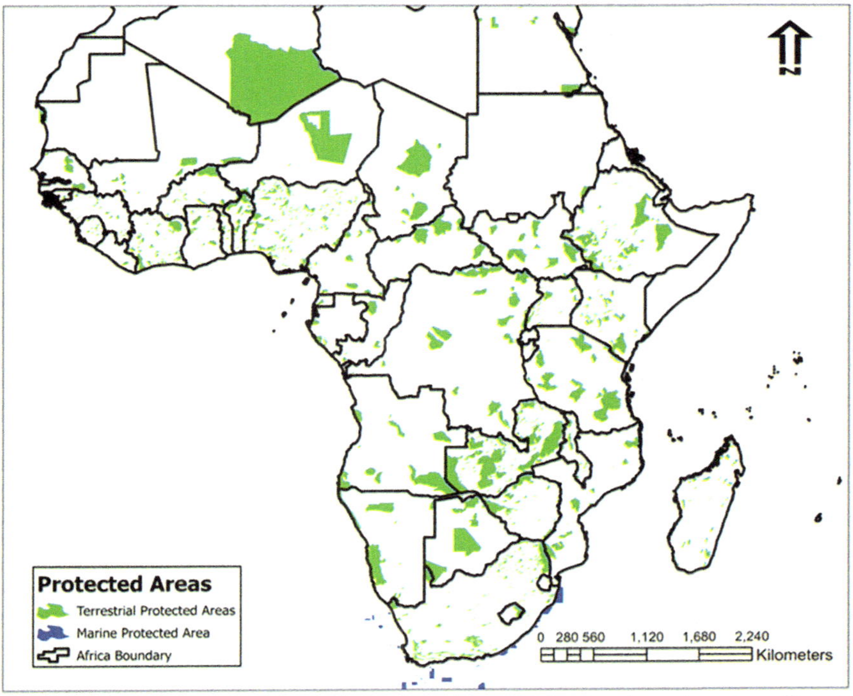

Fig. 2.2 Location of SSA's PAs. Terrestrial PAs that fall under the IUCN's categories I–VI classification for PAs are shown in this map. Cartography: Okavango Research Institute GIS Laboratory

2.2.2 Data Collection and Analysis

A meta-synthesis of scholarly literature on the environmental impacts of tourism was conducted. While the chapter focuses on SSA, scholarly literature sources from other regions were also consulted. This enabled the synthesis of ideas from a larger scholarly community while reflecting more broadly on the target region (Sumanapala and Wolf 2020). While in some instances, the effect of NBT on biodiversity disturbance and loss was perceptible, examples from SSA were very limited. Given this scenario, relevant examples were borrowed from other regions, which, although not specific to SSA, helped to provide an understanding of how the NBT affects biodiversity. The literature search began by producing an extensive list of articles consisting of literature from peer-reviewed journal articles, books, edited book chapters, academic theses, and online sources. Google Scholar, Web of Science, Scopus, and Google were used to source relevant literature. The key search terms and phrases used for the initial search were "anthropogenic factors/drivers of biodiversity loss/disturbance," "threats to biodiversity conservation," "stresses causing biodiversity loss/disturbance," "environmental impacts of nature-based tourism," and

Table 2.1 Distribution of articles per database

Article type	Google	Scopus	Web of Science	Google Scholar	Total
Journal articles	–	64	4	3	71
Book chapters	–	–	–	3	3
Technical reports	9	–	–	4	13
Books	9	–	–	–	9
Academic theses	1	–	–	–	1
Other online sources	3	–	–	–	3
Total	22	64	4	10	100

"environmental impacts of tourism." Where possible, original research and reports were traced and cited, and in cases where information differed between sources, evidence from the academic literature and reports by NGO and nonprofit bodies were prioritized over the information hosted on advocacy websites. For each publication, abstracts that only mentioned pressures or stresses causing biodiversity loss with no specific mention of NBT or tourism were excluded in the screening stage. Furthermore, publications in which the nature of the impact was not adequately explained or did not align with the objective of the chapter were excluded. Where no examples of NBT's role in a specific pressure were found from SSA, a factor approach was used as a search term to enrich the search. A typical example is "Impact of NBT on climate change." Although not specifically addressing SSA, publications with such examples were included in the review. From an initial pool of 386 documents, 100 were finally used in this review (Table 2.1). A descriptive interpretive qualitative approach was used to analyze the data.

2.3 Results and Discussion

Rooted in the DPSIR framework, this section addresses seven major pressures that NBT exerts on the environment, which in turn negatively impact biodiversity. The seven pressures, which include climate change, LULCC, biological invasion, recreational activities, modern technology, pollution, and over-exploitation of natural resources, are therefore analyzed below:

2.3.1 *Climate Change*

Although scientists define climate change differently, it is generally accepted that climate change involves both global warming caused by greenhouse gas (GHG) emissions from human activity and the ensuing widespread changes in weather patterns (IPCC 2013). NBT is an industry with a large carbon footprint. It is noteworthy that the eco-friendliest variant of tourism is consumptive in its own way when consideration is given to this footprint. Heffernan (2019) argues that all tourism is consumptive and that carbon footprint, GHG emissions, water, and other resource use, waste left behind, and infrastructure development must all be factored in when analysing various forms of conservation. While the majority of tourists who travel to SSA to take part in photographic tourism are environmentalists who mean no harm to wildlife or their habitats, they do leave a heavy carbon footprint, increased resource pressures and forever changed natural environments.

The navigation of tourists in any given location constitutes the interaction between tourism and the environment, and human activities that take place because of travel have an impact on the environment (Zhu et al. 2022). It must be noted that the eco-friendliest activities that take place across SSA tend to be GHG and resource intensive. International tourists to the region emit a great deal of aviation greenhouse gases (GHGs) when they travel. On arrival at their destinations, they usually use high-powered diesel four-wheel-drive vehicles to tour the environment. Razzaq et al. (2021) posit that due to the growing trend in industry, transportation has increased, carbon neutrality has become a scarce resource, and climate change has become a harmful factor. Moreover, research has shown that tourists generally use more water and power, eat more, drink more, and generally live in excess than their everyday consumerism when on holidays (Heffernan 2019). NBT is thus at the expense of environmental degradation as the bulk of carbon dioxide, and GHGs are emitted from the tourism sector (Anwar et al. 2021).

As such, NBT contributes to climate change, which in turn affects species and habitats (Hall 2006; United Nations World Tourism Organisation, United Nations Environmental Programme, and World Meteorological Organisation 2008). While neither the exact contribution of the NBT to climate change nor to biodiversity loss is known, climate change projections in SSA indicate a more extreme change in rainforest regions and estimate a 3–4 °C increase in temperature by 2100 (Chapman et al. 2022). For example, Guinea Conakry, a popular tourism destination in West Africa, is clearly warming, with cold extremes decreasing and warm extremes increasing, while total precipitation is decreasing (Aguilar et al. 2009).

2.3.2 Land Use and Land Cover Change

Although nature tourism generally has less impact on the environment and requires less infrastructure than mass tourism (Alderman 1994), the rise in the number of visitors to natural destinations in SSA has made it necessary to expand the number of lodging options and road networks. For instance, the number of lodges in Kenya's Maasai Mara Nature Reserve has increased, going from six in the early 1980s to more than 25 in 2022 (Travel Specialist 2022). NBT development mainly involves human modifications to ecosystems and land cover (Dong et al. 2008). Here, infrastructure development activities necessitated by tourism, such as clearing land and draining and clearing wetlands, have a direct impact on the habitat (Gössling and Hall 2006). Although NBT rarely directly exterminates species, habitat loss and fragmentation caused by tourist-related land use and development contribute to the reduction in species' ranges and extinctions (Gondo et al. 2023; Hall 2006).

Structures in ecologically fragile areas destroy habitats. Access roads are often more destructive than the tourist projects themselves (Ikiara and Okech 2002). The geographic ranges and seasonal distributions of numerous species have shrunk due to the excessive conversion of natural forests into built-up areas (Brooks et al. 2002). For example, the LULC dynamics in the Kilombero Valley Ramsar Site of Tanzania have seen a decline in the Puku antelope population (Bonnington et al. 2010). Similarly, Mundia and Murayama (2009) recorded a decline in wildlife species in the Masai Mara Ecosystem in Kenya.

2.3.3 Biological Invasion

A major impact on biodiversity results from the introduction of alien species into the environment. Invasive species are nonnative plants, animals, or other organisms whose introduction and spread harm native species, ecosystems, the economy, and human health (ICS 2021). Nearly 40% of all animal extinctions since the seventeenth century have been attributed to invasive alien species (Secretariat of the Convention on Biological Diversity 2006). Approximately 480,000 species are estimated to have been intentionally or unintentionally introduced to places outside of their geographic ranges globally (Pimentel et al. 2001). For example, the Salvinia molesta has, in the past, invaded waterways across the world, including Australia, Southeast Asia, the Pacific, and parts of Africa, including Zimbabwe's Lake Kariba and Botswana's Okavango Delta (Kashe et al. 2020; Lee 2001). While its spread is assisted by wind and water movement of vegetative fragments, its dispersal is also invariably through human activities (e.g., boats, aquaria). Vila and Pujadas (2001) posit that a country's visitor population is positively correlated with the prevalence of invasive species.

Increased travel increases the likelihood that new plants, animals, and disease species will be introduced, posing challenges and having a negative influence on environments, including coral reefs, woodlands, and grasslands (Anderson et al. 2015). As an ornamental plant, *Opuntia stricta* cactus, which is native to the Americas, was brought to East Africa decades ago. Since then, it has expanded throughout South Africa and eastern Africa, displacing native species and having a devastating effect on countries' natural resources. The cactus has taken over approximately 60,000 hectares of land in Kruger National Park and approximately 1000 km^2 in Kenya, destroying the habitat for wildlife (Witt 2016). The same scenario applies in West and Southern Africa, where Siam weed (*Eupatorium odoratum*) was introduced from Texas, the United States. Although it has a great ability to fertilize the soil, the weed has since become uncontrollable in the region. It has been reported to be the most problematic invasive species that affects species diversity within protected rainforests in Africa (Struhsaker et al. 2005). Witt (2016) noted that when nonnative plants invade habitats and displace native plant species on which many other species depend, there are knock-on effects for iconic species such as lions and rhinos, which ultimately reduce their tourism value, underscoring the impact of invasive species on biodiversity. This is because the lack of fodder for herbivores reduces the latter's population, which has an adverse effect on predators such as lions. While tourists are most at risk of introducing new invasive species into the region, those who work in the tourism sector also have a greater impact when they grow exotic plants in the gardens of campsites, hotels, or lodges, where some escape cultivation and quickly spread and harm local flora and fauna (Anderson et al. 2015). Invasive species are thus linked to tourist infrastructure, such as the construction of walkways and lodges and the planting of nonnative species in hotel gardens (Anderson et al. 2015).

Anderson et al. (2015) thus conclude that both tourists and the infrastructure associated with tourism serve as carriers of alien invasive species, making tourism one of the major pathways through which invasive species are dispersed. Nature-based tourists participating in activities such as fishing, boating, hiking, camping, and moving between sites can spread invasive species as well (Anderson et al. 2015). These activities also bring nature enthusiasts to ecologically sensitive areas such as parks, biodiversity hotspots and remote wilderness sites. An abundant supply of invasive species is brought in by many tourists, vehicles, and vessels from many geographic locations (Lockwood et al. 2005). In addition, common recreational activities such as hiking, mountain biking, and off-road driving can damage habitats, possibly promoting species invasion (Jauni et al. 2015). When an activity completely or substantially kills the native plants or animal biomass in a locality, it causes disturbance and alters the species' available niches within its ecosystem (Byers 2002). As a result of their better rates of growth and reproduction, invasive species can succeed particularly well in disturbed ecosystems where they can swiftly colonize disturbed areas (Britton-Simmons and Abbott 2008).

Invasive species can cling to clothing, animal fur, camping gear, firewood, vehicles, and watercraft and be transported to new locations (ICS 2021). Aquatic invasive species have the potential to severely impact watersheds, affecting food webs, diminishing water quality, and outcompeting native plants and animals for food and habitat. Recreational fishing and pleasure boats are major vectors of the spread of aquatic invasive species (AIS). AIS can be attached to boat hulls, propellers, trailers, and equipment and can also be transported in live wells or ballast tanks that contain standing water. Animals such as invasive mussels have tiny larvae that can live in water for up to 30 days and, as adults, attach to virtually any hard surface (Anderson et al. 2015). Invasive fish compete with and prey upon native fish species, carry parasites and diseases, and can greatly alter food webs and ecosystems (ICS 2021).

2.3.4 Recreational Activities

Visitors' use of natural areas can have an adverse effect on the environment by disturbing wildlife, introducing and dispersing exotic species, polluting water, eroding soil, and harming the natural environment's flora and fauna (Newsome et al. 2012). NBT activities, which include walking and hiking, camping, getting close to and personal with wildlife, and motorized activities, have a variety of negative effects on the environment and natural resources (Buckley 2004). For instance, Kenya's natural ecosystems, such as Maasai Mara Game Reserve, have suffered greatly because of the recreational activities associated with the sharp increase in visitor numbers, with animal populations rapidly declining because of this (Travel Specialist 2022). Walking and trekking damage the tracks, the soil, and the plants. Numerous consequences, such as multiple treading, track widening, root exposure, and soil erosion/track deepening, are widespread even though many of these impacts can be mitigated by wise trail design and material selection (Newsome 2013). Another frequent influence on vegetation and soil is trampling, which might engender a decrease in vegetation height, soil compaction, soil loss, decreased soil moisture, loss of organic matter, and loss of ground cover vegetation. Trampling can also result in the loss of native plant species, alteration in the composition of vegetation, and the introduction of weeds and pathogens (Sumanapala and Wolf 2020). Additionally, tourist camping has negative effects on biotic ecosystem structure, habitat loss, and biodiversity because of wood harvesting, among other things (Freedman et al. 1996).

Activities involving powerboats have an impact on aquatic birds both directly and indirectly as well as in more subtle ways (Mosisch and Arthington 2004). In addition, reefs are often damaged by the activities of careless tourists as divers. Furthermore, dive/snorkel boat operators may throw their anchors into corals, and local entrepreneurs often break off pieces of corals to sell as souvenirs (Christ et al. 2003).

Conventional passenger vehicles, four-wheel drives, and other off-road vehicles used in safaris to promote terrestrial tourism have an impact on wild animals and their habitat in PAs (Hall 1994). Certain tourist activities have brought negative effects, such as the crushing of vegetation by vehicles and harassment of predators. Wolf et al. (2013) showed that disturbance from passing tourist traffic, vehicle-related mortality, habitat modification, and disturbed biotic relationships, such as the rise of strong competitors, may negatively influence bird communities residing in ecosystems next to roadways. As a result of the effects of trampling, altered environmental conditions, and competition with species that benefit from disturbance, vegetation communities along roadways suffer significantly from edge effects (Wolf and Croft 2014). In Kenya, uncontrolled shell and coral collection coupled with physical damage to coral reefs contributes to the decay of marine parks on the coast (Sindiyo and Pertet 1984).

Animals and their habitats may react negatively to an increase in visitor interaction with wildlife, including different degrees of disturbance and harassment (Hammitt and Cole 1998). If disturbance from tourism continues or occurs frequently, changes in a biological organization may result in adjustments to population size, age–sex composition, species community composition, species richness, and species diversity (Liley and Sutherland 2007). Additionally, the negative impacts of recreational activities might cause stress for a wide variety of species. Depending on its duration, frequency, and magnitude, a stressor can have negative side effects such as immune deficiencies, developmental delays, weight loss, or decreased reproductive success. Stressors are unpredictable or uncontrollable stimuli that evoke a physiological or behavioral reaction (Szott 2020). Recreational

Table 2.2 Effects of recreational activities on wild animals

Effect	Causes	Natural areas of investigation in SSA	Sources
Changes in animal behaviors, e.g., aggression, foraging, resting, parenting behavior, habituation	Human disturbance in form of high human presence, hunting, provision of food by tourists, and observation by tourists, tourist vehicles, increased tourist presence in the form of accommodation	Dzanga-Sangha Protected Areas and Dzanga-Ndoki, Central African Republic; Amboseli National Park, Kenya; South African protected areas, i.e., Letsatsing Game Reserve, Pilanesberg National Park, Madikwe Game Reserve, Mabula Game Reserve; Virunga volcanoes in East Africa, along the northern border of Rwanda, the DRC, and Uganda	Hunninck et al. (2018), Millspaugh et al. (2007), Szott (2020), Blom et al. (2004)
Transmission of diseases from humans to animals	Close contact, contamination of food	Bwindi Impenetrable National Park and Kibale National Park, Uganda; Mahale Mountains National Park, Tanzania; Virunga volcanoes	Weber et al. (2020), Kaur et al. (2008)
Weakening of social bond	Tourist infrastructure and traffic	Soysambu Conservancy and Lake Nakuru National Park, Kenya; Kruger National Park, South Africa	Belton et al. (2018), Muller et al. (2019)
Spatial movement and changes in habitat use	Tourist presence and vehicles	The Amboselie ecosystem, Kenya, comprising Mount Kilimanjaro, Chyulu Hills, Tsavo West National Park and the Kenya/Tanzania borderland landscape; Chobe National Park, Botswana; Pilanesberg National Park and Madikwe Game Reserve, South Africa	Ngene et al. (2017), Stokke and du Toit (2002), Woolley et al. (2009), Szott (2020)

activities can lead to animals emigrating from their natural habitat, thereby altering population structures and enabling exotic species to establish themselves within the habitat (Muntifering et al. 2019). Table 2.2 summarizes some of the ways in which recreational activities cause animal disturbance.

2.3.5 *Modern Technology*

Technological advancement implies increased mobility due to radical global and local transport developments and intensified use of expensive and high-tech outdoor recreation equipment (Elmahdy et al. 2017). As tourism uses energy-intensive technologies for its operations, findings have shown that tourism plays a significant role in environmental degradation (Zhu et al. 2022). Moreover, while technological advancement in the transport industry is improving the accessibility of tourist destinations in SSA, it has also proved to be a challenge to sustainable tourism (Scott and Gössling 2015). Low-cost carriers are also boosting tourism in remote or less advantaged locations (Costa and Almeida 2015). The construction of roads and bridges has a direct positive impact on tourism flows; it is perceived as a main reason behind the emergence of peripheral tourism destinations (Aguila and Ragot 2014) where the NBT is typically located. Thus, the two main effects of technology on the ecosystem are pollution and the depletion of natural resources. A resource becomes easily depleted when it is used up more quickly than when it is replenished. Aquifer depletion, deforestation, soil erosion, and excessive resource usage are a few examples of resource depletion (Edinburgh Sensors 2021).

2.3.6 Pollution

Tourism is the main factor contributing to environmental pollution (Gulistan et al. 2020) through littering, water, and air pollution. Pollution, as a problem associated with NBT, can have negative consequences for biodiversity conservation (Gössling 2002).

Littering occurs when waste or abandoned items are improperly disposed of in the wrong places. It is made worse in remote regions where it can be logistically challenging to collect rubbish (such as on mountains and in water bodies), which results in the decimation of innumerable animals (Christ et al. 2003). Littering can lead to injury or death of animals if they become tangled in debris, if they accidentally step on littered materials such as broken glass and pins or if they consume some of the litter's ingredients.

Air pollution results when dangerous or excessive levels of gases such as carbon dioxide, carbon monoxide, sulfur dioxide, nitric oxide, and methane are released into the Earth's atmosphere. Two main effects of air pollution are its negative effects on animal health and global warming. As such, the subject of tourist-related pollution is now widely discussed in academic tourism circles (Zhang and Lu 2022). Some academics contend that the current economic expansion and development caused by the tourism industry results in pollution and environmental deterioration (Zeng et al. 2021; Xiong et al. 2022). This is because emissions of air pollution caused by the tourism industry are higher than those of other service industries. This might be because the tourism sector uses much energy (Xiong et al. 2022), and practically all its sectors (such as catering, transportation, commerce, and lodging) have seen a growth in their need for heavy energy, which has led to significant air pollution emissions. Tourism and the use of transport are closely related and responsible for most of the gas emissions into the atmosphere (Ştefănică 2017). While air transport is the primary contributor to gas emissions into the atmosphere, emissions from all modes of transport are strongly connected with acid rains, global warming, and serious phenomena of local pollution (Ştefănică 2017). Ahmad and Ma (2022) found that the combustion of fuels needed for lodging, transportation, and land use is responsible for over 70% of the carbon emissions produced by the tourism industry. Higham et al. (2016) also found that the demand for travel, dining, lodging, and other tourism-related services dramatically increases emissions of air pollutants worldwide.

On the other hand, water pollution is the contamination of water sources such as lakes, rivers, oceans, and groundwater and is typically caused by human activities, which commonly result in eutrophication and the destruction of ecosystems. While most cruise lines have begun to implement comprehensive waste management programs, waste disposal from cruise ships has been problematic, as they have limited capacity to carry all their waste until they reach their home port in addition to the limited incentive (and capacity) that destination ports have to accommodate periodic discharges (Christ et al. 2003). Additionally, the development of lodging, recreational amenities, and other amenities frequently results in a rise in sewage contamination. Wastewater does harm plants and animals by polluting the lakes and oceans around popular tourist destinations. For instance, coral reefs are mined for building materials and are also harmed by sewage runoff, which encourages the growth of algae that cover the filter-feeding corals, making it difficult for them to live (Holden 2000).

Darkoh and Mbaiwa (2014) noted that the Okavango Delta in Botswana was gradually under the threat of tourism where the capacity limits for tourism growth were being exceeded in certain parts of the area. The resultant ecological effects included habitat alteration due to the construction of tracks and roads in PAs and the construction of airstrips, as well as disturbance of wildlife by boats and vehicular movements. Other ecological effects were noise pollution caused by motorized boats, small planes, and visitors; sewerage discharge into the water; and littering and rubbish buildup. Darkoh and Mbaiwa (2014) argued that NBT could worsen the threats to biodiversity loss and disturbance in the delta if it was not properly managed and controlled.

2.3.7 Overexploitation of Natural Resources

Overexploitation of resources causes great harm to the environment. The very presence of tourists, especially in excessive numbers, leads to resource strain, which eventually leads to environmental degradation. The tourist accommodation sector overuses natural resources such as water, land, wood, and food, which damages the local environment and compromises the long-term sustainability of the tourism industry (Ikiara and Okech 2002). Deforestation from tourism (through construction, use of wood for fires, etc.) can have direct implications for habitat conservation and watershed management and can also increase the likelihood of landslides (Christ et al. 2003).

2.4 Conclusion

Guided by the DPSIR framework, this chapter analyses the seven major pressures that the NBT exerts on the environment, leading to biodiversity disturbance and loss. The pressures include climate change, LULCC, biological invasion, recreational activities, modern technology, pollution, and overexploitation of natural resources. NBT has a tremendous impact on the environment, especially because of its interaction with wild species and populations, which are highly sensitive to human disturbance. Some of the negative impacts of NBT, as outlined in this chapter, include trail (soil) erosion; air, noise, and water pollution; littering; and a decline in plant and animal varieties and species. The intensity of tourist development and use, ecosystem resilience, long-term versus short-term tourism planning, and the degree of modification of the tourism sites all affect how much of an impact these pressures have. Biodiversity is affected by choices made by tourists regarding their activities, such as hiking, boating, and sports. Trends in the tourism marketplace also determine the effects on biodiversity. For instance, some attractive nature-tourism destinations in biodiversity-rich areas experience a pronounced growth phase in accommodation development with serious negative impacts on biodiversity.

To reduce tourism-related anthropogenic threats to biodiversity, the chapter recommends the promotion of innovative and practically oriented intervention strategies, which include implementing climate-smart NBT, paying close attention to human-induced ecosystem changes at tourism sites, stringent phytosanitary policy implementation and interventions in tourism development guided by visitor management frameworks such as the carrying capacity and limits of acceptable change, among others. Although the detrimental effects of tourism on environmental conservation are widely documented, more empirical research on tourism-related anthropogenic threats in continental Africa is needed to develop more effective context-specific intervention strategies.

References

Aguila GM, Ragot R (2014) Ecotourism industry in Ilijan Batangas City, Philippines: assessing its effects as a basis of proposed tourism development plan. Q J Bus Stud 1(1):24–35

Aguilar E, Aziz Barry A, Brunet M, Ekang L, Fernandes A, Massoukina M et al (2009) Changes in temperature and precipitation extremes in western Central Africa, Guinea Conakry, and Zimbabwe, 1955–2006. J Geophys Res Atmos 114(D2):1

Ahmad N, Ma X (2022) How does tourism development affect environmental pollution? Tour Econ 28(6):1453–1479

Alderman CL (1994) The economics and the role of privately owned lands used for nature tourism, education, and conservation. In: Munasinghe M, McNeely J (eds) Protected area economics and policy: linking conservation and sustainable development. World Bank and IUCN, pp 273–317

Anderson LG, Rocliffe S, Haddaway NR, Dunn AM (2015) The role of tourism and recreation in the spread of nonnative species: a systematic review and meta-analysis. PLoS One 10(10):e0140833

Anwar A, Siddique M, Dogan E, Sharif A (2021) The moderating role of renewable and nonrenewable energy in environment-income nexus for ASEAN countries: evidence from method of moments quantile regression. Renew Energy 164(9):56–967

Belton LE, Cameron EZ, Dalerum F (2018) Social networks of spotted hyaenas in areas of contrasting human activity and infrastructure. Anim Behav 135:13–23

Blom A, Cipolletta C, Brunsting AM, Prins HH (2004) Behavioral responses of gorillas to habituation in the Dzanga-Ndoki National Park, Central African Republic. Int J Primatol 25(1):179–196

Bonnington C, Steer MD, Lamontagne-Godwin J, Owen N, Grainger M (2010) Evidence for local declines in Tanzania's puku antelope (*Kobus vardoni* Livingstone, 1857) population between 1999 and 2003. Afr J Ecol 48(4):1139–1142

Britton-Simmons KH, Abbott KC (2008) Short-and long-term effects of disturbance and propagule pressure on a biological invasion. J Ecol 96:68–77

Brooks TM, Mitttermeier RA, Fonseca CG, Rylands AB, Konstant RW, Flick P, Pilgrim J, Oldfield J, Martin G, Hilton-Taylor C (2002) Habitat loss and extinction in the hotspots of biodiversity. Conserv Biol 16:909–923

Buckley R (2004) Impacts of ecotourism on birds. In: Buckley R (ed) Environmental impacts of ecotourism. CAB International, Cambridge, pp 187–289

Büscher B, Fletcher R (2017) Destructive creation: capital accumulation and the structural violence of tourism. J Sustain Tour 25(5):651–667

Byers JE (2002) Impact of non-indigenous species on natives enhanced by anthropogenic alteration of selection regimes. Oikos 97(3):449–458

Chapman CA, Abernathy K, Chapman LJ, Downs C, Effiom EO, Gogarten JF et al (2022) The future of sub-Saharan Africa's biodiversity in the face of climate and societal change. Front Ecol Evol 744:790552

Christ C, Hilel O, Matus S, Sweeting J (2003) Tourism and biodiversity: mapping tourism's global footprint. Conservation International, Washington, DC

Costa V, Almeida C (2015) Low-cost carriers, local economy and tourism development at four Portuguese airports. A model of cost-benefit analysis. J Spat Organ Dynam 3(4):245–261

Darkoh MB, Mbaiwa JE (2014) Okavango delta-a Kalahari oasis under environmental threats. J Biodivers Endang Spec 2(4):1–6

Dong R, Yu L, Liu G (2008) Impact of tourism development on land-cover change in a matriarchal community in the Lugu Lake area. Int J Sustain Dev World Ecol 15(1):28–35

Duffy R (2005) Global environmental governance and the challenge of shadow states: the impact of illicit sapphire mining in Madagascar. Dev Chang 36(5):825–843

Edinburgh Sensors (2021) The impact of technology on the environment and how environmental technology could save our planet. Available: https://edinburghsensors.com/news-and-events/environmental-technology-the-impact-of-technology-on-the-environment-and-environmental-technology/. Accessed 30 Nov 2022

Elmahdy YM, Haukeland JV, Fredman P (2017) Tourism megatrends: a literature review focused on nature-based tourism. MINA fagrapport 42. Norwegian University of Life Sciences Faculty of Environmental Sciences and Natural Resource Management

European Environmental Agency (EEA) (1995) Europe's environment: the Dobris assessment. European Environment Agency, Copenhagen

European Environmental Agency (EEA) (1999) Environmental indicators: typology and overview. European Environmental

Freedman B, Zelazny V, Beaudette D, Fleming FG, Gerrow J, Woodley S (1996) Biodiversity implications of changes in the quantity of dead organic matter in managed forests. Environ Rev 4(3):238–265

Galvani AP, Bauch CT, Anand M, Singer BH, Levin SA (2016) Human–environment interactions in population and ecosystem health. Proc Natl Acad Sci 113(51):14502–14506

Gari SR, Newton A, Icely JD (2015) A review of the application and evolution of the DPSIR framework with an emphasis on coastal social-ecological systems. Ocean Coast Manag 103:63–77

Gondo R, Kolawole OD, Mfundisi KB (2023) Land use and land cover changes along the Boteti-Thamalakane River system in Ngamiland District, Botswana. Sci Afr 20:e01595

Goodwin H (1996) In pursuit of ecotourism. Biodivers Conserv 5(3):277–291
Gössling S (2002) Global environmental consequences of tourism. Glob Environ Chang 12:283–302
Gössling S, Hall CM (eds) (2006) Tourism and global environmental change. Routledge, London
Gössling S, Hall CM, Peeters P, Scott D (2010) The future of tourism: can tourism growth and climate policy be reconciled. A climate change mitigation perspective. Tour Recreat Res 35(2):119–130
Gulistan A, Tariq YB, Bashir MF (2020) Dynamic relationship among economic growth, energy, trade openness, tourism, and environmental degradation: fresh global evidence. Environ Sci Pollut Res Int 27(12):13477–13487
Hall SJ (1994) Physical disturbance and marine benthic communities: life in unconsolidated sediments. Oceanogr Mar Biol Annu Rev 32:179–239
Hall CM (2006) Tourism, biodiversity and global environmental change. In: Gössling S, Hall CM (eds) Tourism and global environmental change: ecological, economic, social and political interrelationships. Routledge, London, pp 211–226
Hammitt W, Cole D (1998) Wildland recreation: ecology and management. Wiley
Heffernan A (2019) El discurso oficial de las autoridades marroquíes y sus aliados en la ecologización de sus políticas públicas: modernización del autoritarismo, inserción internacionaly lucha contra el cambio climático. Relac Int 42:157–173
Heffernan A (2022) Accounting for climate change in community-based natural resource management: reflections on wildlife conservation in Namibia. J South Afr Stud 48:1–14
Higham J, Cohen SA, Cavaliere CT, Reis A, Finkler W (2016) Climate change, tourist air travel and radical emissions reduction. J Clean Prod 111:336–347
Holden A (2000) Environment and tourism. Routledge, London
Hunninck L, Ringstad IH, Jackson CR, May R, Fossøy F, Uiseb K et al (2018) Being stressed outside the park-conservation of African elephants (*Loxodonta africana*) in Namibia. Conserv Physiol 6(1):cox080
ICS (2021) Best Management Practices, Tourism and Invasive Species: A Pocket Guide for Tourism Operators, Guides and Staff. Invasive Species Council of BC, 2021 Edition. Available: https://bcinvasives.ca/wp-content/uploads/2021/12/Tourism-BMP-Booklet-WEB.pdf, Accessed 30 Nov 2022
Ikiara M, Okech C (2002) Discussion Paper No. 19 of 2002 on impact of tourism on environment in Kenya: status and policy
Intergovernmental Panel on Climate Change (IPCC) (2013) The physical science basis. Contribution of working group I to the fifth assessment report of the intergovernmental panel on climate change, 1535, 2013
Jauni M, Gripenberg S, Ramula S (2015) Non-native plant species benefit from disturbance: a meta-analysis. Oikos 124(2):122–129
Kashe K, Heath R, Heath A, Teketay D, Thupe BO (2020) Potential impact of alien invasive plant species on ecosystem services in Botswana: a review on *Prosopis juliflora* and *Salvinia molesta*. In: Sustainability in developing countries: case studies from Botswana's journey towards 2030 agenda, pp 11–31
Kaur T, Singh J, Tong S, Humphrey C, Clevenger D, Tan W et al (2008) Descriptive epidemiology of fatal respiratory outbreaks and detection of a human-related metapneumovirus in wild chimpanzees (*Pan troglodytes*) at Mahale Mountains National Park, Western Tanzania. Am J Primatol 70(8):755–765
Kim YS, Rodrigues M, Robinne FN (2021) Economic drivers of global fire activity: a critical review using the DPSIR framework. Forest Policy Econ 131:102563
Kosamu IBM, Makwinja R, Kaonga CC, Mengistou S, Kaunda E, Alamirew T, Njaya F (2022) Application of DPSIR and tobit models in assessing freshwater ecosystems: the case of Lake Malombe, Malawi. Water 14(4):619
Lasso AH, Dahles H (2021) A community perspective on local ecotourism development: lessons from Komodo National Park. Tour Geogr 23:1–21

Lee WG (2001) Introduced plants, negative effects of. In: Encyclopedia of biodiversity, pp 501–515
Liley D, Sutherland WJ (2007) Predicting the population consequences of human disturbance for ringed plovers *Charadrius hiaticula*: a game theory approach. Ibis 149(1):82–94
Lin T, Xue XZ, Lu CY (2007) Analysis of coastal wetland changes using the "DPSIR" model: a case study in Xiamen, China. Coast Manag 35(2–3):289–303
Lindsey PA, Anderson SH, Dickman A, Gandiwa P, Harper S, Morakinyo AB et al (2022) Shepherding sub-Saharan Africa's wildlife through peak anthropogenic pressure toward a green Anthropocene. Annu Rev Environ Resour 47(1):91–121
Lockwood JL, Cassey P, Blackburn T (2005) The role of propagule pressure in explaining species invasions. Trends Ecol Evol 20(5):223–228
Macdonald EA, Burnham D, Hinks AE, Dickman AJ, Malhi Y, Macdonald DW (2015) Conservation inequality and the charismatic cat: *Felis felicis*. Global Ecol Conserv 3:851–866
Maxim L, Spangenberg JH, O'Connor M (2009) An analysis of risks for biodiversity under the DPSIR framework. Ecol Econ 69(1):12–23
Millspaugh JJ, Burke T, Van Dyk GUS, Slotow ROB, Washburn BE, Woods RJ (2007) Stress response of working African elephants to transportation and safari adventures. J Wildl Manag 71(4):1257–1260
Mosisch T, Arthington A (2004) The impacts of recreational power boating on freshwater ecosystems. In: Buckley R (ed) Environmental impacts of ecotourism. CABI Publishing, NewYork, pp 125–154
Muller Z, Cuthill IC, Harris S (2019) Giraffe (*Giraffa camelopardalis*) social networks in areas of contrasting human activity and lion density. Ethology 125(10):702–715
Mundia CN, Murayama Y (2009) Analysis of land use/cover changes and animal population dynamics in a wildlife sanctuary in East Africa. Remote Sens 1(4):952–970
Muntifering JR, Linklater WL, Naidoo R, Uri-Khob S, Preez PD, Beytell P et al (2019) Sustainable close encounters: integrating tourist and animal behaviour to improve rhinoceros viewing protocols. Anim Conserv 22(2):189–197
Ness B, Anderberg S, Olsson L (2010) Structuring problems in sustainability science: the multi-level DPSIR framework. Geoforum 41(3):479–488
Newsome D (2013) An 'ecotourist's recent experience in Sri Lanka. J Ecotour 12(3):210–220
Newsome D, Moore S, Dowling R (2012) Natural area tourism: ecology, impacts, and management. Channel View Books, UK
Ngene S, Okello MM, Mukeka J, Muya J, Njumbi S, Isiche J (2017) Home range sizes and space use of African elephants (*Loxodonta africana*) in the Southern Kenya and Northern Tanzania borderland landscape. Int J Biodivers Conserv 9(1):9–26
Olson DM, Dinerstein E, Wikramanayake ED, Burgess ND, Powell GV, Underwood EC et al (2001) Terrestrial ecoregions of the world: a new map of life on earth. Bioscience 51(11):933–938
Pandya R, Dev HS, Rai ND, Fletcher R (2022) Rendering land touristifiable: (eco) tourism and land use change. Tour Geogr 25:1–17
Pimentel D, McNair S, Janecka J, Wightman J, Simmonds C, O'connell C et al (2001) Economic and environmental threats of alien plant, animal, and microbe invasions. Agric Ecosyst Environ 84(1):1–20
Razzaq A, Sharif A, Najmi A, Tseng ML, Lim MK (2021) Dynamic and causality interrelationships from municipal solid waste recycling to economic growth, carbon emissions and energy efficiency using a novel bootstrapping autoregressive distributed lag. Resour Conserv Recycl 166:105372
Richardson JA (1998) Wildlife utilisation and biodiversity conservation in Namibia: conflicting or complementary objectives? Biodivers Conserv 7:549–559
Saarinen J, Becker FO, Manwa H, Wilson D (eds) (2009) Sustainable tourism in southern Africa: local communities and natural resources in transition. Channel View Publications
Samdin Z, Aziz YA, Radam A, Yacob MR (2013) Sustainability of ecotourism resources at Taman Negara National Park: contingent valuation method. Int J Bus Soc 14(2):235–244
Scott D, Gössling S (2015) What could the next 40 years hold for global tourism? Tour Recreat Res 40(3):269–285

Secretariat of the Convention on Biological Diversity (2006) Global biodiversity outlook 2. Montreal

Sindiyo DM, Pertet FN (1984) Tourism and its impact on wildlife conservation in Kenya. Indus Environ 7(1):14–19

Ştefănică M (2017) Environmental impact of transportation in the tourism industry-dimensions and actions. In: Studies and scientific researches. Economics edition. (25)

Stokke S, Du Toit JT (2002) Sexual segregation in habitat use by elephants in Chobe National Park, Botswana. Afr J Ecol 40(4):360–371

Stolton S, Dudley N (2019) The New Lion economy: unlocking the value of lions and their landscapes. Equilibrium Research, Bristol

Struhsaker TT, Struhsaker PJ, Siex KS (2005) Conserving Africa's rainforests: problems in protected areas and possible solutions. Biol Conserv 123(1):45–54

Sumanapala D, Wolf ID (2020) Think globally, act locally: current understanding and future directions for nature-based tourism research in Sri Lanka. J Hosp Tour Manag 45:295–308

Suta N, Hrnjic A, Banda A (2017) Natural resources management in tourism: dimensions and impact of tourist offer in the Southeastern Europe National Parks. In: Financial environment and business development. Springer, Cham, pp 537–562

Szott I (2020) The impact of wildlife tourism on African elephants, Loxodonta africana, in South Africa (Doctoral dissertation, Liverpool John Moores University)

Travel Specialist (2022) The impact of wildlife tourism in Africa. https://www.thetravelspecialists.net.au/impact-of-wildlife-tourism-in-africa/

UNEP-WCMC (2019) The World Database on Protected Areas (WDPA) (UNEP-WCMC & IUCN). Available: www.protecdplanet.net. Accessed 18 Oct 2022

United Nations World Tourism Organisation, United Nations Environmental Programme, & World Meteorological Organisation (2008) Climate change and tourism: Responding to global challenges. Madrid

UNWTO (2020) 'International Tourism Growth Continues to Outpace the Global Economy', 20 January 2020, available at https://www.unwto.org/international-tourism-growth-continues-to outpace-the-economy, retrieved 7 April 2022

Vila M, Pujadas J (2001) Socioeconomic parameters influencing plant invasions in Europe and North Africa. In: McNeely JA (ed) The great reshuffling: human dimensions of invasive alien species. IUCN Biodiversity Policy Coordination Division, Gland, pp 75–79

Weber A, Kalema-Zikusoka G, Stevens NJ (2020) Lack of rule-adherence during mountain gorilla tourism encounters in Bwindi Impenetrable National Park, Uganda, places gorillas at risk from human disease. Front Public Health 8:1

Weiler B, Hall CM (1992) Special interest tourism: in search of an alternative. In: Special interest tourism: in search of an alternative, pp 199–204

Witt A (2016) Tourism and biodiversity loss: the threat of invasive species. Available: https://travindy.com/2016/05/tourism-and-biodiversity-loss-the-threat-of-invasive-species/#. Accessed 02 Dec 2022

Wolf ID, Croft DB (2014) The impacts of tourism hotspots on vegetation communities show a higher potential for self-propagation along roads than hiking trails. J Environ Manag 143(1):173–185

Wolf ID, Hagenloh G, Croft DB (2013) Vegetation moderates impacts of tourism usage on bird communities along roads and hiking trails. J Environ Manag 129(15):224–234

Woolley LA, Millspaugh JJ, Woods RJ, Van Rensburg SJ, Page BR, Slotow R (2009) Intraspecific strategic responses of African elephants to temporal variation in forage quality. J Wildl Manag 73(6):827–835

WTTC (2019) The economic impact of global wildlife tourism: travel and tourism as an economic tool for the protection of wildlife. World Travel Tourism Council, London

Xiong C, Khan A, Bibi S, Hayat H, Jiang S (2022) Tourism subindustry level environmental impacts in the US. Curr Issue Tour 26(6):1–19

Zeng J, Wen Y, Bi C, Feiock R (2021) Effect of tourism development on urban air pollution in China: the moderating role of tourism infrastructure. J Clean Prod 280:124397

Zhang J, Lu Y (2022) Exploring the effects of tourism development on air pollution: evidence from the panel smooth transition regression model. Int J Environ Res Public Health 19(14):8442

Zhu S, Luo Y, Aziz N, Jamal A, Zhang Q (2022) Environmental impact of the tourism industry in China: analyses based on multiple environmental factors using novel quantile autoregressive distributed lag model. Econ Res-Ekonomska Istraživanja 35:1–27

Chiedza Ngonidzashe Mutanga is a senior research fellow in sustainable tourism at the Okavango Research Institute, University of Botswana. She holds a DPhil in Tourism and Wildlife Management from the Chinhoyi University of Technology, Zimbabwe. Her research interests include sustainable tourism development, with a focus on wildlife tourism, nature-based tourism, community-based tourism and community livelihoods, ecotourism, protected area tourism, community-based natural resource management, and wildlife resource conservation.

Reniko Gondo holds a PhD degree from the University of Botswana. He earned his MSc and BSc degrees from the Bindura University of Science Education and Zimbabwe Open University (Zimbabwe), respectively. Currently, he serves as a senior research fellow at the Okavango Research Institute (ORI), University of Botswana (UB). Gondo worked as a geography teacher for ten years in Zimbabwe. Additionally, he served as a part-time lecturer for 3 years at the Bindura University of Science Education. Gondo joined the Okavango Research Institute, University of Botswana, in February 2016. He authored over 20 publications focusing on the interface between Indigenous and Western knowledge in the governance of natural resources. Gondo is a reviewer for *Water Resources Management*, *Scientific African*, *Sustainable Water Resources Management*, and *Geography and Sustainability*.

Oluwatoyin Dare Kolawole is a professor of rural development at the Okavango Research Institute (ORI), University of Botswana. He is a research fellow at the Centre for Gender and Africa Studies, University of the Free State, South Africa. He is also an adjunct faculty at the College of Health and Social Sciences in the Eastern University, St. Davids, Pennsylvania. He works at the interface of science, policy, and agriculture with major interests in politics of knowledge, sustainable agriculture, natural resources conservation, and community development. He was the University of Canterbury Visiting Canterbury Fellow in 2014 and a CODESRIA laureate in 1994 and 2006.

Joseph E. Mbaiwa is a professor of tourism studies at the Okavango Research Institute (ORI), University of Botswana. He holds a PhD in Recreation, Park & Tourism Sciences from the Texas A&M University (USA). Joe Mbaiwa has over 20 years' experience in tourism and natural resource management research and consultancy experience in Botswana. He has widely published on issues of tourism development, rural livelihoods, and conservation.

Chapter 3
Adopting a Nexus Thinking Framework for Biodiversity Conservation and Sustainability in the Anthropocene

James Maradza, Raban Chanda, and Naomi Moswete

3.1 Introduction

The strategy of integrating human resources into natural resource conservation is not new in Africa. A recount of colonial history revealed that legendary communities such as those of the Maasai-Mara, Ngorongoro and Amboseli areas in East Africa practised sustainable utilization of natural resources (Murphree 2009). Similarly, in precolonial Botswana, communities satisfactorily managed their own resources according to their traditional *dikgosi* customs (Arntzen et al. 2007; Phuthego and Chanda 2004). However, today, critical observation and literature reveal that most Sub-Saharan African communities are battling with loss of biodiversity and environmental degradation, impacting the well-being of the poor more, especially those whose livelihoods depend on natural resources (Schlossberg et al. 2019). Anthropogenic activities such as urbanization, agriculture and fuel wood extraction in marginalized rural communities have been the most influential drivers of loss of biodiversity and environmental degradation, resulting in dwindling natural resources both in absolute numbers and diversity (Pimm et al. 2014). This is worrisome and requires urgent interventions based on systems or nexus thinking frameworks to protect biodiversity and biodiversity-dependent livelihoods and ecosystem services.

J. Maradza (✉)
Richard Lyons, Gaborone, Botswana

R. Chanda
Harvest University, Lusaka, Zambia
e-mail: rchanda@harvestuniversity.edu.zm

N. Moswete
Department of Environmental Science, University of Botswana, Gaborone, Botswana
e-mail: moatshen@ub.ac.bw

W. L. Hambira et al. (eds.), *Environmental Change and Biodiversity Conservation in sub-Saharan Africa*, Advances in Global Change Research 75,
https://doi.org/10.1007/978-3-031-73136-5_3

Over the years, collective resource management and rural development theory have been on the drive to address the challenges of poverty, inequality and environmental degradation due to anthropogenic activities (Suich et al. 2015). Although community-based natural resource management (CBNRM) is hailed for the success of collective management theory (Fabricius et al. 2004), as it devolves authority for ecosystem management to local communities (Fabricius and Collins 2007), its applicability is limited to the management and utilization of common pool resources (Mbaiwa 2011). Conversely, the model does not apply to resources owned and managed privately in areas proximate to local communities. It is the quest to pre-empt or mitigate potential or actual conflict between the common property and private property resource management regimes that necessitates the need to search for models that recognize and forge mutually beneficial linkages for the sustainable coexistence of the two property regimes. As such, this chapter presents and proposes nexus thinking—akin to system thinking—as a collaborative resource governance strategy. As reported in Maradza (2019) and Maradza et al. (2022), NT as a concept was first developed as a superior approach to the understanding of interlinkages among water, energy and food (WEF). According to Schlör et al. (2021), the nexus is defined as a concept of systems thinking. Although there is no universal definition of "system thinking", at best, it encompasses thinking of a whole system other than its parts, dealing with linkages and observing patterns rather than content (Alcamo 2017). That is, systems thinking helps to describe the linkages in the NT framework approach. Therefore, nexus thinking is akin to systems thinking. The nexus concept has been promoted by the WEF and the Global Water Partnership (GWP) as a complementary governance framework to integrated water resource management (IWRM) and how its implication shifted the WEF nexus in favour of a more sustainable, equitable and sustainable utilization of resources. The approach was illustrated empirically through a WEF nexus case in Switzerland (Kellner 2023) and was invoked in the conception of the UN Sustainable Development Goals (SDGs) (Benson et al. 2015). Thus, NT has applications at higher socioecological and geographical scales. However, it is here being applied at a local, case study scale based on the principle of thinking globally while acting locally, on the assumption that if the framework was widely adopted across global space alongside kindred frameworks, such as CBNRM and IWRM, the cumulative beneficial socioeconomic and ecosystem effects in the Anthropocene (as defined by Steffen et al. 2011) would be significant. NT not only provides a new insight into categorizing nexus research but also allows assessing its quality from a comprehensive nexus (systems thinking) perspective that ushers in valuable and systematic understanding research within the sphere of sustainability. The study explores NT as a collaborative model that promotes linkages compatible with socioecological situations such as that of the MNR on the outskirts of the City of Gaborone in Botswana juxtaposed to a rural community of even name. The chapter asserts that sustainable solutions to complex environmental challenges are found in the greater understanding and consideration of linkages and interdependencies among different sectors that interface in the socioeconomic and ecological system. Therefore, the chapter explores adopting the NT framework as a means to strengthen biodiversity conservation, ecotourism and livelihoods for communities proximate to private protected areas, with MNR as the case study.

3.2 Sustainability: History, Resources and Development

Wiersum (1995) asserts that the concept of sustainability was originally coined in forestry, implying the following: never harvest more than what the forest yields in new growth. This is an apt use of the concept in the context of fauna and flora, which together constitute renewable natural capital. In this context, sustainability is interchangeable with conservation or the wise use of renewable natural resources (Fabricius et al. 2004). However, the concept assumed a broader meaning in 1987 when the Brundtland Commission equated it to the newly coined term of sustainable development (Brundtland 1985). Defined as "development that meets the needs of the present generation without compromising the ability of future generations to meet their own needs", sustainable development was hailed as a development paradigm that could resolve the rising tension between aspirations of mankind towards a better life vis-à-vis the limitations imposed by nature and the dangers of environmental degradation (Kuhlman and Farrington 2010). The fundamental principles of sustainability have been anchored on striking a balance between socioeconomic benefits and conservation of natural resources to safeguard the welfare of future generations (Castellani and Sala 2010; Fabricius and Collins 2007; Fabricius et al. 2013; Snyman 2013).

Many environmentalists and conservationists recognize that the future prospects of protected areas are limited without the involvement and support of local communities (Castellani and Sala 2010). Wildlife-related benefits have an impact on people's attitudes towards living with nature. Unfortunately, poverty has become a foremost societal challenge (Snyman 2013): becoming the cause and consequence of environmental degradation. Although poverty and environmental degradation seem to be two distinct issues, a strong relationship exists between them. The Millennium Ecosystem Assessment (MA) study demonstrated the precarious state of ecosystems and biodiversity at the global level (cf. MA 2005). Biodiversity conservation challenges are usually associated with household income (Castellani and Sala 2010).

In most instances, people with higher assets and high income are more likely to participate in a biodiversity conservation initiative than their lower income counterparts whose livelihoods may depend on the unavoidable exploitation of biodiversity (Snyman 2013). The understanding derived from the connection between biodiversity conservation and biodiversity benefits (income) implies that the priority of conservation initiatives should focus on poverty reduction and local livelihood improvement. Local communities should be integrated into biodiversity mainstreaming to generate economic benefits and, in turn, effectively conserve ecosystems and natural habitats and preserve viable population species. It is the positive attitude towards environmental conservation that upholds the goals of environmental sustainability, as manifested by the local community's appreciation of nature. Keeping in perspective interlinkages existing between protected areas and the local people and in view of the importance of the linkages to livelihoods, ecotourism and conservation, the question is how benefits can be channelled to the local communities in a privately owned protected ecotourism conservation area.

The chapter adopts the NT framework as an alternative solution to the CBNRM framework, which is more applicable to common property natural resource management (Maradza 2019; Maradza et al. 2022). The NT model has the potential to offer equitable distribution of benefits of conservation from the support and involvement of local communities without upsetting the functioning of privately owned and managed resources. The model is also ideal for enhancing a private nature reserve's corporate social responsibility towards the local community, thereby increasing synergies and promoting resource security (Maradza et al. 2022). When mutually beneficial linkages among private protected areas and local livelihoods are clearly elaborated and appreciated, protection of the natural environment and its biodiversity is assured. To the authors' knowledge, this study is arguably the first to apply the NT model to the governance of privately owned nature reserves juxtaposed to common property resources and associated settlements (cf. Maradza 2019).

The aim of the chapter is to explore the applicability of NT to the linkages of conservation, local livelihoods and ecotourism development in Botswana and using Mokolodi nature reserve (MNR) as a case study. Specifically, the main objectives of the chapter are:

- To establish the characteristics of conservation, ecotourism and local livelihoods in and around the MNR before exploring the relationships among them (cf. Table 3.1).
- To determine the extent of interdependency among conservation, ecotourism and local livelihoods using nexus thinking for the MNR (cf. Table 3.2).

Based on the aim and research objectives, the chapter seeks to answer the following questions: (a) What characterizes conservation at MNR? (b) What characterizes ecotourism in MNR? (c) What characterizes the livelihoods of communities around MNRs (d) To what extent are the characteristics of conservation, ecotourism, and local livelihoods interdependent?

3.3 The Study Area

3.3.1 Mokolodi Nature Reserve

The advent of protected areas in Botswana commenced during the colonial era in 1885 (Mbaiwa 2005). After independence, the constitution passed through several modifications (Adams et al. 2003). The first changes were instituted through The State Land Act of 1966 and The Tribal Land Act of 1968 (Adams et al. 2003). These

Table 3.1 Conceptualized Components Interdependence related to ecotourism, conservation and livelihoods

Component	Characteristics	Interdependent characteristics	
		Ecotourism	*Local livelihoods*
Conservation	Anti-poaching, fencing, debushing	Tourist income	Formal employment
	Monitoring erosion and bush encroachment, fire management	Tourist activities, e.g. bush braai, tracking of wildlife	Job creation and formal employment
	Game capturing and off take	Tourist facilities	Informal employment/ cultural activities
	Rhino population and movement monitoring	Tourist guiding	Formal employment
	Water point distribution, waterhole provision for wild animals	Wildlife viewing	Agropastoralism and farming/livestock
Ecotourism	*Characteristics*	*Conservation*	*Local livelihoods*
	Tourist activities (e.g. bush braai)	Vegetation monitoring	Formal employment
	Tourist types (e.g. local, international, educational)	Biodiversity protection activities	Informal employment (e.g. curios and or/farm produce)
	Tour guiding	Environmental education	Provision of services (e.g. natural herbs)
	Tour facilities	Capacity building	Agropastoralism farming
	Tour income	Anti-poaching	Formal employment
Local Livelihoods	*Characteristics*	*Conservation*	*Ecotourism*
	Informal employment	Rhino population and movement monitoring	Rhino tracking
	Formal employment	Erosion and bush encroachment, debushing for fire management and trails especially in wet season	Tour income
	Agropastoral farming	Water point distribution, game restocking	Tourist activities
	Provision of services, e.g. natural herbs	Vegetation monitoring	Number of tourists per unit time

Data source: Maradza (2019)

statutes promulgated three categories of land tenure—tribal land, state land and freehold land (Adams et al. 2003). Between 1966 and 1972, over 15,000 square kilometres of state land were alienated and sold as freehold land both to Europeans and Botswana (White 1999). The MNR land was then owned by Richard Transfeldt and Mmaphatse and sold to Ian and Gwithian Kirby in 1972 for cattle ranching (MNR 2003). In 1994, the land was donated to the Mokolodi Wildlife Foundation (MWF) to promote ecotourism and conservation of biodiversity (MWF 2016).

Table 3.2 Variables and corresponding proxies or indicators for each objective, the associated questions and their appropriate measurement scales

Objective	Research question	Variable	Variable indicator/proxy	Measurement scale
1. To establish the characteristics of conservation, ecotourism and local livelihoods in and around the MNR.	1.1 What characterizes conservation at MNR?	1.1.1 Conservation activity	Animal patrols per day Animal census frequency Vegetation inventory frequency	Nominal/ interval/ Ordinal
	1.2 What characterizes eco-tourism in MNR?	1.2.1 Ecotourism activity	Tour guiding Visitor behaviour control/ monitoring Visitor numbers per day/week	Nominal/ interval
	1.3 What characterizes livelihoods of communities around MNR?	1.3.1 Livelihoods activity	Engaging in arable farming Livestock rearing (cattle, goats, sheep and poultry) Formal employment Informal employment	Nominal
2. To determine the extent of interdependency among conservation, ecotourism and local livelihoods using nexus thinking for the MNR.	2.1 To what extent are the characteristics of conservation, ecotourism and local livelihoods interdependent?	2.1.1 linkages (NT framework)	Tour guiding vs local employment vs. conservation Tourist behaviour monitoring vs. local employment vs. conservation Vegetation monitoring vs. local employment vs. ecotourism Animal tracking vs. local employment vs. conservation	Nominal/ ordinal/ interval

Data sources: Maradza (2019)

In Botswana, these protected areas (conservancies) are owned by private organizations known as Trusts. Likewise, a voluntary Board of Trustees meets on a quarterly basis to oversee the running of various projects in MNR. The foundation oversees the management and functioning of MNR management and the Education Centre. MNR's revenue is accrued from tourism related-activities, donations and grant support from the United Nations Development Programme (UNDP) and Global Environmental Facility Small Grant Programme. The funds are utilized to advance environmental education, manage and conserve wildlife and natural resources, create employment and develop and maintain the reserve's infrastructure. The conservation team carries out day-to-day management activities such as patrols, providing supplementary feeding, and protecting wildlife from poaching. MNR's annual financial statements are not only available to the general public through the public gazette but are also audited by independent auditors.

3.3.2 Mokolodi Nature Reserve Community

In many African countries, the establishment of protected areas saw the removal of local people from their ancestral land, with little to no compensation, displacing many communities and restricting resource use and loss of land for farming (Mbaiwa 2005). Unlike MNR, it was established under private ownership. The Mokolodi community, traditionally, did not have exclusive access and use of MNR land or have any claim against it. That is why the community embraced the ecotourism transition. Although the community was formally recognized in 2006, it existed long before the establishment of MNR. Available records suggest that as early as 1933, people were already inhabitants in the area (MWF 2010). The current residents of this community are second- and third-generation original inhabitants. As such, they have witnessed various changes, notably the transformation of a once cattle ranch employing only seven people to a thriving nature reserve with a staff complement of over 100 people, the majority of which are from the local community (MWF 2016). In view of this livelihood transition, there is sustainable interaction between the reserve and the local community.

The chief (Kgosi) holds overall traditional power in the community and is the custodian of the community's values, norms and beliefs, and he presides over traditional gatherings popularly known as the *kgotlas* (Fabricius et al. 2013). A *kgotla* (meeting and meeting place) is a traditional gathering where societal and environmental issues are conveyed, discussed, shared and resolved by community leaders. Likewise, Kgosi Boitshoko Rasethogwane of Mokolodi Community oversees all development activities in the village, from informing the community to the delegation of key people who oversee the running of such activities. The chief works in tandem with the Village Development Committee. The VDCs were established by a Presidential directive of 1968 with the goal of implementing development programs in villages (Fabricius et al. 2013). In the village, VDCs are responsible for advancing all development matters as well as coordinating institutions' activities. The overall administration of the community is under the local chief and the VDC.

3.3.3 Mokolodi Nature Reserve Stakeholders

The stakeholders of MNR consisted of staff of MNR; Mokolodi Community; South East District Council; Non-Governmental Organizations—Kalahari Conservation Society, Birdlife Botswana, UNDP; Mokolodi Chief and VDC Chair; relevant government departments: Department Wildlife and National Parks, Department of Environmental Affairs and Department of Tourism.

3.4 Nexus Thinking: Linkages Between Livelihoods, Conservation and Ecotourism

3.4.1 Ecotourism

Over the last decade, ecotourism has made remarkable progress with various initiatives, such as sustainable development, innovative eco lodges, community involvement and education. The term ecotourism is credited to Hector Ceballos-Lascuráin, who coined the term "ecotourism" in 1983 and defined it as environmentally responsible, enlightening travel and visitation to undisturbed natural areas with the goal of enjoying and appreciating nature as well as any other accompanying cultural features both past and present that enhance conservation, low visitor impact and provide for beneficially active socioeconomic involvement of local people (Ceballos-Lascuráin 1996). This placed ecotourism into four principles of responsible tourism (Harrison 2008): minimizing environmental impacts, respecting host cultures, maximizing benefits to local communities and enhancing tourists' satisfaction.

This theory postulates ecotourism as promoting not only environmental protection but also local poverty reduction, employment creation and economic development. Therefore, ecotourism is opposed to mass tourism (bad tourism) and prides itself as a form of alternative tourism (good tourism). This view is further elaborated by Page and Dowling (2002), who expounded that ecotourism is a niche form of tourism that is nature-based, ecologically sustainable, environmentally educative and locally beneficial. In Botswana, rural communities are largely characterized by high levels of unemployment and poverty, and they rely heavily on natural resources. Ecotourism assists these rural communities by meeting their socioeconomic and environmental needs through biodiversity conservation initiatives (Campbell et al. 2013).

3.4.2 *Nature and Concept of Livelihood–Conservation–Ecotourism Linkage*

It is a common phenomenon that local people in any rural society rely on products and services from their immediate natural environment to meet basic survival needs (Phutego and Chanda 2004). Livelihood consideration is central to any local (rural) socioeconomic development project, either through employment, education, or corporate social responsibility. However, community welfare benefits are not about plundering natural resources but rather creating socioeconomic and ecological footprints that help to improve local livelihoods. A suitable approach is necessary to ensure that the local community derives such benefits from natural resources (whether government or privately owned) in a win-win situation (Karki 2013). It is the premise of the study to try and address the gap created by the absence of an approach akin to CBNRM that could be used to diffuse or pre-empt the actual or potential conflicts or antagonisms between local communities and nearby privately or government managed natural resources. As the interaction and connection kicks in, valuable insights are shared and implemented, addressing poverty, loss of biodiversity and environmental degradation (Bazilian et al. 2011).

The driving force behind the sequence of different activities is the bond or link that connects the respective components (Groenfeldt 2010). According to Salafsky and Wollenberg (2000), the link closes the loop, making the system self-perpetuating. The linked activities manage any perceived threats by providing more attractive livelihood options, thereby inhibiting local communities from carrying out damaging livelihood activities. When the value of biodiversity to the local community is enhanced, the local people naturally take measures necessary to mitigate both internal and external threats to such beneficial biological diversity (Worboys et al. 2015). The inherent notion is that the local stakeholders living around MNRs, with no control over biodiversity, are given opportunities to benefit directly from biodiversity conservation. Such opportunities foster development and instil a positive culture towards care for the natural environment. Key species such as rhinoceros play a crucial role in attracting tourists and guaranteeing the flow of revenue, which is then transferred to conservation or education. The benefits act as incentives for livelihood improvement, deterring any possible damage to the natural environment (Phutego and Chanda 2004). Therefore, the concept of conservation is considered successful because the local people are able to appreciate, share and enjoy the socioeconomic and environmental benefits attached to and derived from it (Worboys et al. 2015).

3.4.3 Applying Nexus Thinking to Mokolodi Nature Reserve

The conservation–ecotourism–rural livelihood nexus reflects three important components tied together by socioeconomic and ecological activities to address the challenges of poverty, environmental degradation and depletion of natural resources. These socioeconomic and ecological activities promote interaction, collaboration and coordination between the local village and MNR. The sustainability and stability of these components (conservation–ecotourism–rural livelihoods) are assured at best by treating these components fairly and equally. Stable quantities of biodiversity proportional to and within the carrying capacity of the park avert environmental degradation. In the same vein, ensuring accessibility of biodiversity to ecotourists strengthened ecotourism development and boosted local employment opportunities, guaranteeing the improvement of livelihoods. Generally, a sustainable ecotourism venture guarantees stable revenue, which is ploughed back into the nature reserve as wages for employees, funding anti-poaching activities and other projects that would enhance employment and transform the livelihoods of the local people (Bach et al. 2012). As people realize that their sources of livelihoods are directly linked to biodiversity and ecotourism, they positively respond to any conservation initiatives. As such, MNR is entrusted to fully develop and maintain ecotourism activities such as animal tracking, game viewing, and bush braai, which attract tourists and boost revenue inflow. However, any disruption to any of the components has a negative effect on one or all the components. For instance, any reduction in the number and type of tourists visiting a place limits job opportunities and results in a loss of ecotourism income. Poor income entailed that the funding for anti-poaching and monitoring activities is reduced and ultimately the community loses income through lost employment, subjecting the people to poverty (Croes and Vanegas 2008). The NT model appreciates the value of interactions and interdependence between human and biological components of socioecological systems.

3.5 Methodology

3.5.1 Target Population

The study targeted the staff of Mokolodi Nature Reserve and the people of the neighbouring local community. The current population of Mokolodi Community is approximately 1624 (CSO 2011). To manage area navigation and safety, the researcher and two research assistants were accompanied by MNR tour guides. A Global Positioning System, i.e. GPS map, Garmin: GPS®76CSx, was used to determine the UTM coordinates of the MNR and the surrounding community. The chief gave permission and guidance on the residences of community members.

3.5.2 Research Framework

The triangulation methodology was adopted as the research framework for generating and analysing data as per the views of Creswell and Clark (2007). The triangulation methodology draws inferences from both qualitative and quantitative information. The approach helps to gain insights into the issues of benefits sharing, conservation and ecotourism. The data were analysed through the process of open coding and verbal synthesis (Creswell and Clark 2007). The use of open coding helped to classify all the collected data and compared it with other parts of the data set. The use of various methods helped to account for biases and limitations associated with the use of one single method of data collection (Mertens 2004). It also served as a verification process to increase the validity, credibility and reliability of the study results.

3.5.3 Data Collection

The study used both primary and secondary data collection methods. The survey research approach was used to collect the primary data. The data were collected using a semi-structured questionnaire consisting of a mixture of open- and closed-ended questions to obtain both quantitative and qualitative responses. Secondary data (past and present literature) were used as points of reference for analysis. The information was sourced directly from MNRs during field work. The survey approach was appropriate because it enabled the researcher to effectively measure the characteristics of identified variables and determine the extent of interdependency between such identified variables.

3.5.4 Sampling

The sampled size (n) was determined using the following Yamane (1967) formula (cited in Neuman 2007) with a 90% confidence level and a margin of error of 10%, where n is the sample size, N is the population size and e is the margin of error:

$$n = \frac{N}{1 + N(e)^2}$$

According to CSO (2011), the Mokolodi Community had a population of 624 people, and the average household size per household was 3.76. The population was divided by 3.76 to determine the number of household units, 624/3.76 = 165.96. Then, the Taro Yamane formula was used to calculate the sample size from the total number of households as reflected below.

$$n = \frac{N}{1+N(e)^2} = n = \frac{165.96}{1+165.96(0.1)^2} = 62.4\,\text{Households}$$

Due to logistical and financial challenges, a sample size of 62 households was targeted for the research survey, representing approximately 37% of the total number of households in the community chosen for the study. The study used systematic random sampling to select the respondents.

With respect to MNR management staff, purposive sampling was used to select respondents. This is a non-random sampling technique whereby the researcher solicits information through an interview guide or research questionnaire from people who are willing to give information by virtue of their knowledge or experience in their field of study for the convenience of the researcher who applies his own judgement to select the informants (Moswete 2009). At the time of the research, the number management staff at MNR was 10. By virtue of their experience, a research questionnaire was administered to all of them to solicit and elicit knowledge and insights on conservation and ecotourism.

3.5.5 *Data Analysis*

3.5.5.1 Conservation-Related Activities

The research was developed around the linkages among conservation, ecotourism and local livelihoods in and around the MNR. To establish the characteristics of conservation, ecotourism and local livelihoods, three variables were identified and measured for analysis (Table 3.1). The variable indicators or proxies with respect to conservation variable activities are animal patrols per day, animal census frequency and vegetation inventory frequency. Data were operationalized by six items consisting of a 5-point Likert-type scale from 1 to 5, where 1 = strongly disagree; 2 = disagree; 3 = neutral; 4 = agree and 5 = strongly agree. A nominal scale, where a yes or no answer was used, as well as open-ended questions where respondents provided their reasoning to substantiate their answers.

3.5.5.2 Ecotourism-Related Activities

The proxies for ecotourism-related activities (Table 3.1) are tour guiding, visitor behaviour control/monitoring and visitor numbers per day/week. In this particular area, the level of concern was to adhere to the principles of ecotourism, namely, monitoring visitor behaviour, maintaining visitor numbers and providing tour guidance. A 5-point Likert-type scale was used to measure the variable's activity from 1 to 5, where 1 = strongly disagree, 3 = neutral, and 5 = strongly agree. The nominal scale of measurement was used where a yes or no answer was given as well as open ended questions where respondents provided their reasoning to substantiate their answers.

3.5.5.3 Livelihood-Related Activities

With regard to livelihood-related activity (Table 3.1), the proxies included engaging in arable farming, livestock rearing (cattle, goats, sheep and poultry), formal employment, and informal employment. The constructs provided general information and consisted of scale/nominal and ordinal measurement questions. This section consisted of both open-ended and closed-ended questions designed to provide information about local people's means of livelihood.

3.5.5.4 Proxies of Conservation, Ecotourism and Livelihoods

The raw data with respect to the three variables (Table 3.1) were converted into meaningful data through the application of rational and critical thinking. The data were arranged and analysed numerically on characteristics of conservation (conservation-related activities), ecotourism (ecotourism-related activities) and rural livelihoods (livelihood-related activities) (Table 3.1). The generated data were sorted and categorized as nominal, ordinal and scale. Cross tabulation was used to create indices through the use of a 5-point Likert scale with responses ranging from strongly agree to strongly disagree. Editing was implemented to facilitate the detection of errors and omissions to achieve accurate, consistent and uniform data. Coding was used to put data into manageable categories consistent with the research problem. To establish the characteristics of conservation, ecotourism and local livelihoods, the Statistical Package for Social Sciences (SPSS) Software Version 22 was used to make the necessary classifications. Classification of the collected data ensured that responses were arranged according to class or common characteristics, and tabulation helped with the arrangement and display of summary data into logical format (Tables 3.3 and 3.4).

Table 3.3 Mean scores of linkages variables

Variable*	Mean	*N*	Common view
Income link (income linkage to MNR)	2.31	62	Slight link
Relationship (among community, MNR environment and ecotourism)	4.47	62	Strongly agree
Linkage (between MNR and Mokolodi villagers)	2.15	62	Agree

Notes: Scales: Income link: 1 (not at all), 2 (slightly), 3 (moderately), 4 (strongly; relationship: 1 (strongly disagree), 2 (disagree), 3 (neutral), 4 (agree), 5 (strongly agree); linkage: 1 (strongly agree), 2 (agree), 3 (neutral), 4 (disagree), 5 (strongly disagree)

Table 3.4 Correlation coefficient (r) matrix revealing nexus patterns

	Variable correlations (r) and p			
Variables	Income link	P	Relationship	P
Income link	1			
Relationship	0.542	0.000*	1	
Linkage	0.131	0.002*	0.107	0.01**

$*p < 0.01$; $**p < 0.05$; $n = 62$

3.5.5.5 Extent of the Interdependencies

To determine the extent of interdependency among conservation, ecotourism and local livelihoods using NT for MNR, a mean score analysis was used to determine the extent of the connections. The test examined the linkages among income link, relationship and linkage variables (Table 3.2). The whole process involved compiling data in Excel and SPSS. Having the data in this format provided concise descriptions, explanations on means score analysis and correlation linkage analysis of the research findings, building an analytical framework of the research problem. The gathered data were summarized into descriptive statistics of means and correlation linkages, which were used to present the socioeconomic and ecological characteristics of the conservation, ecotourism and rural livelihoods nexus.

3.6 Major Findings

3.6.1 Nexus Thinking Framework Between Conservation, Ecotourism and Livelihood

The research used both mean score analysis and a correlation matrix to determine the extent of the relationship between the variables of conservation, ecotourism and livelihoods. Quantitative data were derived from the research questionnaire on the variables income link, relationship and linkage measured using ordinal scales. The results of the descriptive and correlation analysis are presented in Tables 3.3 and 3.4 and interpreted in the subsections that follow.

3.6.2 Linkages Between Conservation and Community Livelihoods

The income variable refers to the financial gain obtained by the local community from their day-to-day livelihood activities connected to MNR. The linkage variable refers to perceived connection or interconnectedness between MNR and the Mokolodi community; such linkages are esteemed to improve livelihoods of the community and MNR's conservation of biodiversity and ecotourism. A total of 62 respondents were requested to indicate the extent to which their incomes were linked to the MNR (Table 3.3). The response was measured through a 4-point Likert scale, namely, *not at all*, *slightly*, *moderately*, and *strongly*. The mean score analysis outcome was 2.3, representing "slightly" on the Likert scale. Similarly, the linkage variable was measured through a 5-point Likert scale, namely, *strongly agree*, *agree*, *neutral*, *disagree*, and *strongly disagree*. The 62 respondents were requested to confirm whether a linkage framework approach (connection between the local community and MNR) is necessary for the sustainable conservation of biodiversity and improvement of their livelihoods. The average score outcome of the linkage notion was 2.15 (Table 3.3), which shows that the local community agreed that the linkage framework would improve their livelihoods and MNR's conservation of biodiversity, albeit only slightly.

A further analysis of the correlation between perceived income and the linkage variables using the correlation analysis (Table 3.4) shows that there is a positive correlation. The correlation coefficient (r) of 0.131 ($p = 0.002$) suggests that there is a statistically significant relationship between the variables of income and linkage between MNR and local community. This would suggest that although the linkage between the Mokolodi community's income and MNR was considered slight, it was nevertheless perceived to be beneficial both to the local community and MNR's conservation and ecotourism activities. The perceived benefits by the local community can be attributed to the value of the MNR's Education Center. The Education Centre prides itself as an acclaimed resource centre for the local community in terms of skills development and knowledge empowerment through workshops, seminars, and environmental conservation awareness campaigns (MWF 2016). Such a knowledge base has not only been instrumental in conserving biodiversity but also essential for the development of ecotourism as a sustainable livelihood initiative from employment opportunities to ecotourism entrepreneurship.

3.6.3 Linkage Between Income and Community Livelihood Activities

There is a positive correlation between perceived income and relationship variables, as shown in Table 3.4. The p value of 0.00 on the correlation matrix shows that there is a statistically significant relationship of $r = 0.542$. This suggests that there could be a self-reinforcing association between MNR's financial benefits to the community (though currently perceived as slight) (Table 3.3) and MNR's primary activities of environmental conservation and ecotourism. The linkage connections are further buttressed by the views echoed by Mokolodi Community households, who concurred that mutually beneficial linkages existed among MNR conservation and ecotourism and community livelihood activities.

The collected data for the linkage variable were measured through a 5-point Likert scale. The respondents indicated that they agreed that the linkage framework between MNR and the local community promotes sustainable conservation of biodiversity and improves the livelihoods of the local people. As suggested by the mean score analysis results, "agree" was positioned as the mean on the Likert scale. The relationship variables and corresponding analysis (Table 3.2) were designed to indicate the perceived degree of relation among the local community, MNR natural environment and ecotourism. The relationship variable was measured through a five-point Likert scale. The respondents were requested to indicate their level of agreement on the relationship existing among the components. Based on the data gathered, the mean score is approximately 4.5, representing "strongly agree" when rounded upwards on the Likert scale.

To assess the linkage between the two variables, a correlation analysis was performed. The correlation matrix shows that there is a positive correlation between these variables ($r = 0.107$; $p = 0.01$). The p value on the correlation matrix shows that there is a statistically significant relationship between these variables, suggesting that the linkages among MNR's natural environment, Mokolodi Village cultural activities and environmental conservation and eco-tourism at MNR were self-reinforcing (Table 3.4).

3.6.4 Interdependency Between Conservation, Ecotourism and Local Livelihoods

The study sought to determine the extent of the interdependency among conservation, ecotourism and local livelihoods. The variables (income link, relationship and linkage) statistically established significant positive correlations in this study (Table 3.4). There was a strong positive interdependency between perceived income links and linkages (Table 3.4). This finding is logical in view of the fact that the more employment opportunities available at MNRs, the more available jobs are for local residents, benefiting locals economically. That is, when economic benefits from conservation and ecotourism accrue to residents, they would perceive ecotourism in particular and MNR in general as their source of sustainable livelihood. In this study, approximately 80% of the workforce at MNR is from the local community (MWF 2016–17). Employment is viewed as the most tangible and direct benefit the local community has had from MNR; hence, the income gained has played an important role in supporting the livelihoods of the local people.

Although more benefits are realized by people directly attached to MNR, nonetheless, income earned by people working at MNR is spent within the community, indirectly benefiting not only their extended families but also the whole community. This economic transition experienced by the local people has resulted in a cultural shift towards conservation and entrepreneurship initiatives. The young people mobilized themselves and formed traditional dance troupes for entertaining tourists and patrons of the community guest house serving traditional cuisine for tourists. However, potential negative outcomes have been noted, especially by elderly adults, who used to obtain firewood and traditional medicine from the reserve, and the area is now fenced off. It was, however, suggested that the reserve could periodically provide such resources to the local people to buttress interrelations.

There was a strong statistically significant correlation between perceived income link and relationship. As local residents benefit from MNR by showcasing their cultural activities, a positive mutual relationship between the community and MNR would be developed. This is in addition to the direct employment opportunities just alluded to. This finding makes sense in that the more and improved the mutual relationship among the components, the more the benefits accrued to both MNR and local residents. Ultimately, the benefit of the linkage notion would drive local residents to value the conservation of flora and fauna for their economic benefit. Similarly, studies have shown that when local residents benefit from ecotourism in their area, they naturally develop positive attitudes towards the care of nature and natural environments (Snyman 2013).

3.7 Conclusion and Recommendations

This study enhances our understanding of the concept of NT, akin to system thinking and its applicability to the case study, where different components are interdependent. Environmental education promotes good relations and neighbourliness between MNR and the local community. Notably, this integration (linkage) keeps the components thriving. However, the success of ecotourism and conservation at MNR would not have been possible without the effort and support of the Mokolodi Community. Similar sentiments have been echoed by Lai and Nepal (2006), who affirm that ecotourism development and success rely heavily on the support and willingness of local communities to participate in park-related activities. MNR also avails platforms where the local community showcases their cultural activities—traditional dance, cuisine and artefacts. The establishment of MNR builds bridges between nature and the local community, thereby promoting coexistence. This shows that tourism benefits accrued by the local community motivate them to develop a positive culture towards MNR, and the net result has been unwavering support by the local community towards MNR's conservation–ecotourism initiatives. The linkages observed have the potential to unlock conservation, ecotourism and livelihood enterprises and transform Botswana's livelihoods and value for nature.

The study adds to the literature on NT and its application to PPAs and GPAs, which are currently very limited. The study's findings have afforded environmental planners and other stakeholders' insight into the potential of NT in addressing loss of biodiversity, poverty and environmental degradation in MNR-like settings. The results of the study have further enhanced the understanding of the linkage notion and how these connections could be developed to promote human–wildlife interaction. In conclusion, the NT approach could be consciously promoted to forge mutually beneficial relationships between private nature conservation and livelihoods of proximate communities in Botswana, in the Southern African Development Community (SADC) Region or beyond. This would in turn contribute to sustainable livelihood, biodiversity and ecosystem management in the Anthropocene.

The study findings show that NT is a unique, innovative and interactive framework that promotes linkages of different components. Engagement is key in the mainstream biodiversity economy; the study recommends that stakeholders and MNR engage in interactive partnership agreements that would enhance local village investment and entrepreneurship. The enhanced benefits will deter any adverse environmental behaviour, such as poaching, thereby promoting sustainable biodiversity conservation.

References

Adams M, Kalabamu F, White R (2003) Land tenure policy and practice in Botswana-Governance lessons for southern Africa. J fur Entwicklungspolitik 19(1):55–74

Alcamo J (2017) Systems thinking for advancing a nexus approach to water, soil and waste. J Environ Econ Manag 67(2):189–208

Arntzen J, Buzwani B, Setlhogile T, Kgathi DL, Motsolapheko MK (2007) Community-based resource management, rural livelihoods and environmental sustainability. Centre for Applied Research, Gaborone

Bach H, Bird J, Clausen TJ, Jensen KM, Lange RB, Taylor R, Wolf A (2012) Transboundary river basin management: addressing water, energy and food security. Scientific report. Mekong River Commission, Lao PDR, Lao

Bazilian M, Rogner H, Howells M, Hermann S, Arent D, Gielen D, Tol RS (2011) Considering the energy, water and food nexus: towards an integrated modelling approach. Energy Policy 39(12):7896–7906

Benson D, Gain A, Rouillard J (2015). Water governance in a comparative perspective: from IWRM to a 'nexus' approach? Water Altern 8(1):753–773

Brundtland GH (1985) World commission on environment and development. Environ Policy Law 14(1):26–30

Campbell SJ, Kartawijaya T, Yulianto I, Prasetia R, Clifton J (2013) Comanagement approaches and incentives improve management effectiveness in the Karimunjawa National Park, Indonesia. Mar Policy 41:72–79

Castellani V, Sala S (2010) Sustainable performance index for tourism policy development. Tour Manag 31(6):871–880

Ceballos-Lascuráin H (1996) Tourism, ecotourism, and protected areas: the state of nature-based tourism around the world and guidelines for its development. IUCN

Creswell JW, Clark VP (2007) Designing and conducting mixed methods research. Sage Publishers, London

Croes R, Vanegas M Sr (2008) Cointegration and causality between tourism and poverty reduction. J Travel Res 47(1):94–103

CSO (2011) 2011 Botswana population and housing census. Statistics Botswana, Gaborone

Fabricius C, Collins S (2007) Community-based natural resource management: governing the commons. Water Policy 9(S2):83–97

Fabricius C, Koch E, Magome H, Turner S (2004) Rights, resources and rural development. In: Community-based natural resource management in Southern Africa. Earthscan, London

Fabricius C, Koch E, Turner S, Magome H (2013) Rights resources and rural development: community based natural resource management in Southern Africa. Routledge, London
Groenfeldt D (2010) The next nexus: environmental ethics, water management and climate change. Water Altern 3(3):575
Harrison D (2008) Pro-poor tourism: A critique. Third World Q 29(5):851–868
Karki ST (2013) Do protected areas and conservation incentives contribute to sustainable livelihoods? A case study of Bardia National Park, Nepal. J Environ Manag 128:988–999
Kellner E (2023) Identifying leverage points for shifting Water-Energy-Food nexus cases towards sustainability through the Networks of Action Situations approach combined with systems thinking. Sustain Sci 18(1):135–152
Kuhlman T, Farrington J (2010) What is sustainability? Sustain For 2(11):3436–3448
Lai PH, In addition, Nepal SK (2006) Local perspectives of ecotourism development in Tawushan Nature Reserve, Taiwan. Tour Manag 27(6):1117–1129
MA (Millennium Ecosystem Assessment) (2005) Ecosystems and human well-being: synthesis. Island Press, Washington, DC
Maradza J (2019) Investigating the applicability of nexus thinking to private protected areas: the case of Mokolodi Nature Reserve, Southeast of Botswana. Gaborone: University of Botswana. (Msc) Thesis, University of Botswana, Gaborone 2017/18. http://ithuteng.ub.bw/handle/10311/2066
Maradza J, Chanda R, Moswete NN (2022) Perspectives on the applicability of Nexus thinking to private protected areas: a case study of Mokolodi Nature Reserve, Botswana. In: Saarinen J (ed) Geographies of tourism and global change: Southern African perspectives on sustainable tourism management: tourism and changing localities. Springer, Cham, pp 153–168
Mbaiwa JE (2005) Wildlife resource utilization at Moremi Game Reserve and Khwai community area in the Okavango Delta, Botswana. J Environ Manag 77(2):144–156
Mbaiwa J (2011) The effects of tourism development on the sustainable utilization of natural resources in the Okavango Delta, Botswana. Curr Issue Tour 14(3):251–273
Mertens DM (2004) Research methods in education and psychology: integrating diversity with quantitative and qualitative approaches, 2nd edn. Sage Publications, Thousand Oaks
Mokolodi Nature Reserve (2003) A brief history of Mokolodi Nature Reserve part 1. Friends Mokolodi Newslett 23:4–5
Mokolodi Wildlife Foundation (2016) Mokolodi 2016–17 annual review: adapting in maturity. Mokolodi Nature Reserve, Gaborone
Moswete NN (2009) Stakeholder perspectives on the potential for community-based ecotourism development and support for the Kgalagadi Transfrontier Park in Botswana. University of Florida
Murphree MW (2009) The strategic pillars of communal natural resource management: benefit, empowerment and conservation. Biodivers Conserv 18:2551–2562
MWF (Mokolodi Wildlife Foundation) (2010). Mokolodi: Wild Dogs Visist Mokolodi, Local Fires Test Mokolodi Differences. Mokolodi Nature Reserve, Gaborone
Neuman LW (2007) Basics of social research: qualitative and quantitative approaches, 2nd edn. Pearson Education, Inc, Boston
Page SJ, Dowling RK (2002) Ecotourism. Prentice Hall, New York
Phuthego T, Chanda R (2004) Traditional ecological knowledge and community-based natural resource management: lessons from a Botswana wildlife management area. Appl Geogr 24(1):57–76
Pimm SL, Jenkins CN, Abell R, Brooks TM, Gittleman JL, Joppa LN, Raven PH, Roberts CM, Sexton JO (2014) The biodiversity of species and their rates of extinction, distribution, and protection. Science 344(6187):1246752
Salafsky N, Wollenberg E (2000) Linking livelihoods and conservation: a conceptual framework and scale for assessing the integration of human needs and biodiversity. World Dev 28(8):1421–1438

Schlör H, Märker C, Venghaus S (2021) Developing a nexus systems thinking test–a qualitative multi and mixed methods analysis. Renew Sustain Energy Rev 138:110543

Schlossberg S, Chase MJ, Sutcliffe R (2019) Evidence of a growing elephant poaching problem in Botswana. Curr Biol 29(13):2222–2228

Snyman S (2013) Household spending patterns and flow ecotourism income into communities around Liwonde National Park, Malawi. Dev South Afr 30(4–5):640–658

Steffen W, Persson A, Deutsch L, Zalasiewicz J, Williams M, Richardson K, Crumley C, Crutzen P, Folke C, Gordon L, Molina M, Ramanathan V, Rockström J, Scheffer M, Schellnhuber HJ, Svedin U (2011) The Anthropocene: from global change to Planetary Stewardship. AMBIO J Hum Environ 40(7):739–761

Suich H, Howe C, Mace G (2015) Ecosystem services and poverty alleviation: a review of the empirical links. Ecosyst Serv 12:137–147

White R (1999) Livestock and land tenure in Botswana. In: Conference on new directions in land tenure

Wiersum KF (1995) 200 years of sustainability in forestry: lessons from history. Environ Manag 19:321–329

Worboys GL, Lockwood M, Kothari A, Feary S, Pulsford I (eds) (2015) Protected area governance and management. Anu Press

Yamane Y (1967) Mathematical formulae for sample size determination. Scientific Research Publishing. www.scirp.org

James Maradza holds an MSc in Environmental Science from the University of Botswana. He has researched and published on perspectives of nexus thinking towards private protected areas. His area of research interest is premised on environmental degradation issues, sustainable livelihoods, and tourism development: exploring sound and sustainable innovative and inventive opportunities that collaborates with socio-economic and environmental challenges in ways that interlink people with nature for sustainable development in Southern Africa.

Raban Chanda is a Human Geographer with long teaching and research experience in Botswana and Zambia. His area of expertise includes environmental assessments, sustainable livelihoods, socio-economic dimensions of environmental change and natural resource management, and climate change adaptation. He is a full Professor at the University of Botswana.

Naomi Moswete is a Senior Lecturer in the Department of Environmental Science, University of Botswana. Her research interests include human geography, tourism as a strategy for rural development, community-based tourism, transboundary conservation areas, sustainable ecotourism, wildlife, parks–people relationships, heritage resource management, and cultural tourism. She is a member of the *CABI Tourism Cases* editorial board and a sectional editor in *Frontiers in Conservation Science Journal.*

Chapter 4
Mechanisms Underlying Ungulate Migration in African Savannas: Insights from Plains Zebra (*Equus quagga*) Migrations in Botswana and the Implications for Conservation Under Climate Change

Keoikantse Sianga, Mpaphi C. Bonyongo, and Richard W. S. Fynn

4.1 Introduction

Large herbivore migrations between discrete seasonal habitats are a global phenomenon (Berger 2004; Harris et al. 2009). Scientific research evidence to date suggests that these seasonal movements are driven by temporal fluctuations in forage quantity and quality and surface water availability (McNaughton 1985; Fryxell and Sinclair 1988; Williamson et al. 1988; Murray 1995; Hobbs et al. 2008; Bennitt et al. 2022) and by the need to obtain high-quality forage for growth and reproduction (Parker et al. 2009). However, in many parts of the world, migrations have either been severely disrupted or lost entirely because of anthropogenic disturbances (Harris et al. 2009), which have resulted in fragmentation of broad functional landscapes (Fynn and Bonyongo 2011).

In Africa, the establishment of fences across several ecosystems has modified and limited connectivity (fragmentation) across habitats, resulting in severe impacts on wildlife species. For example, habitat fragmentation through fences has disrupted the adaptive movements of the blue wildebeest (*Connochaetes taurinus*) and

K. Sianga (✉)
Botswana University of Agriculture and Natural Resources, Gaborone, Botswana
e-mail: ksianga@buan.ac.bw

M. C. Bonyongo
The Permanent Okavango River Basin Water Commission, Gaborone, Botswana
e-mail: casper.bonyongo@okacom.org

R. W. S. Fynn
University of Botswana - Okavango Research Institute, Maun, Botswana
e-mail: rfynn@ub.ac.bw

W. L. Hambira et al. (eds.), *Environmental Change and Biodiversity Conservation in sub-Saharan Africa*, Advances in Global Change Research 75,
https://doi.org/10.1007/978-3-031-73136-5_4

plains zebra (*Equus quagga*) in drought years between the Kgalagadi ecosystem and the Okavango Delta in Botswana (Mbaiwa and Mbaiwa 2006) and zebra and blue wildebeest migrations in the Athi-Kapiti plains—Nairobi National Park ecosystem in Kenya (Bhola et al. 2012; Ogutu et al. 2013). Mbaiwa and Mbaiwa (2006) demonstrated that hundreds of migratory wildebeest and zebra populations and other large herbivores were trapped in these fences during their establishment. Similarly, fences are known to have disrupted the migrations of zebra and blue wildebeest migrations in Kruger National Park in South Africa (Whyte and Joubert 1988), Etosha National Park in Namibia (Berry and Jenssen 1997), the Loita Plains—Mara ecosystem in Kenya (Serneels and Lambin 2001; Serneels et al. 2001), and the Tarangire National Park—Simanjiro Plains ecosystem in Tanzania (Morrison et al. 2016). However, fences have not affected the migrations of zebra populations in our study area.

Southern Africa has lost almost all of its previously diverse large herbivore migrations. This includes the vast springbok (*Antidorcas marsupialis*) (Kgalagadi ecosystem, Botswana), black wildebeest (*Connochaetes gnou*), and blesbok (*Damaliscus pygargus*) migrations in South Africa (Harris et al. 2009), as well as the extensive nomadic movements of red hartebeest *Alcelaphus buselaphus*, zebra, and blue wildebeest in the Kgalagadi ecosystem (Williamson and Williamson 1984; Fynn and Bonyongo 2011). The Kruger National Park dry season and wet season ranges have been fragmented because the protected area only occurs in the wet season range, whereas the dry season range against the Western escarpment is now under settlement and agriculture (Fynn and Bonyongo 2011). However, Botswana, which still retains extensive connectivity between seasonal habitats and even between countries, is likely the last remaining country in southern Africa where some large migrations still occur (Fynn and Bonyongo 2011). This chapter provides a synthesis of the seasonal movement patterns of plains zebra in northern Botswana and explores the possible impacts of climate change on these movements.

4.2 Methodology

4.2.1 Landscapes in Northern Botswana

The landscapes in northern Botswana are characterized by a semiarid climate with mean annual rainfall ranging between 500 and 600 mm in the Okavango, Makgadikgadi and Chobe Enclave regions, most of which falls between November and April (Botswana Meteorological Services 2022). The northern conservation area of Botswana is part of the Kavango Zambezi Transfrontier conservation areas, which borders parts of Botswana, Namibia, Zambia, and Zimbabwe, providing a vast mosaic of unfragmented wetlands, woodlands, and saline grasslands (Fynn and Bonyongo 2011; Naidoo et al. 2016). One of the key features of northern Botswana is that its landscapes still contain most of their historical wet and dry season ranges.

Generally, apart from the Makgadikgadi region, the landscapes have remained unmodified by artificial water, such that herbivores move away from woodland and grassland systems during the dry season. For example, the Savuti-Mababe-Linyanti ecosystem in northern Botswana is still pristine and functional, in contrast to most protected areas in Africa (Sianga and Fynn 2017; Sianga et al. 2017; Vittoz et al. 2020). The reason for functional connectivity in the region is a good network of protected areas (Fig. 4.1).

Vegetation in northern Botswana includes wetland habitats fringed by riverine habitats in the Okavango Delta, Linyanti Swamps, Chobe Riverfront, and Boteti

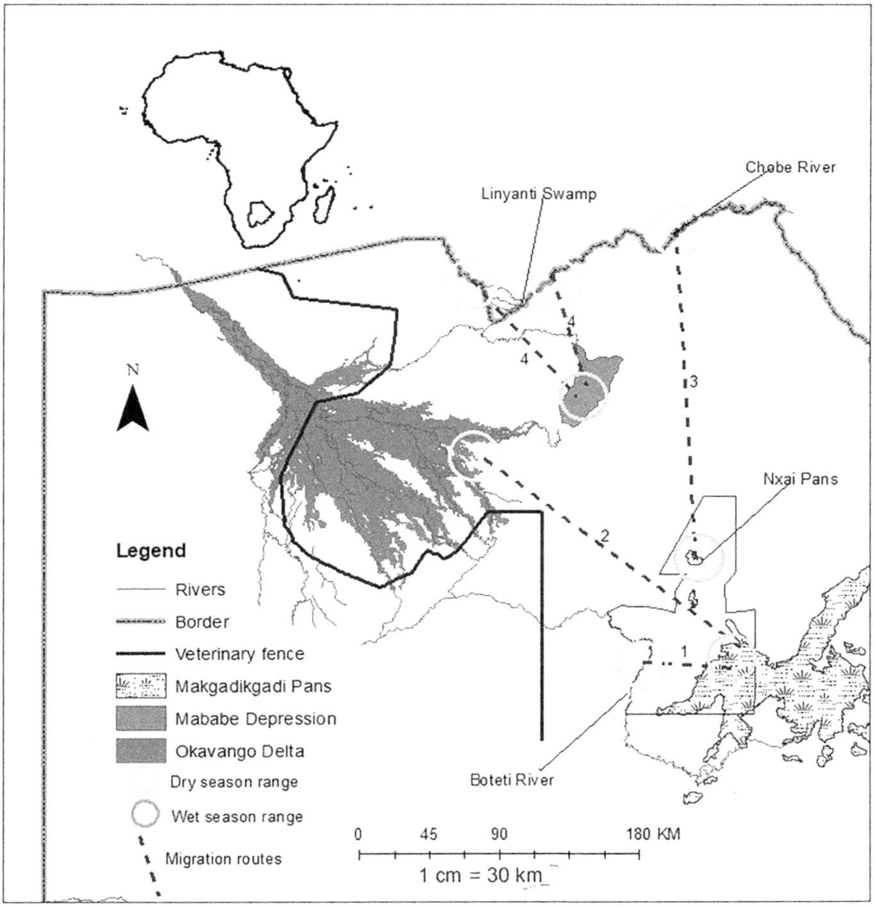

Fig. 4.1 Zebra migration routes in northern Botswana; (1) Boteti River–Makgadikgadi Pans migration; (2) Okavango Delta–Makgadikgadi Pans migration, (3) Chobe River floodplains–Nxai Pans migration, and (4) Linyanti Swamps–Mababe Depression migration. (Data sources: Brooks 2005; Bradley 2012; Bartlam-Brooks et al. 2011; Naidoo et al. 2016; Joos-Vandewalle 2000; Sianga 2014; Cartography: Keoikantse Sianga)

River regions; dryland woodland habitats away from permanent water, and vast open saline grasslands in the Mababe Depression, Makgadikgadi and Nxai Pans, far from permanent water (Weare and Yalala 1971; Teter 2007; Aarrestad et al. 2014; Revermann et al. 2016; Sianga and Fynn 2017; Pellacani et al. 2020; Vittoz et al. 2020; Burrough 2022).

The wetland habitats occur as deep-flooded grasslands dominated by *Oryza longistaminata* and *Vossia cuspidata*, intermediate flooded grasslands dominated by *Panicum repens* and shallow grasslands dominated by *Cynodon dactylon* (Fynn et al. 2015). Riparian woodland is closed woodland dominated by *Lonchocarpus capassa*, *Diospyros mespiliformis*, *Ficus sycomorus*, *Kigelia africana*, *Combretum hereroense*, and *Croton megalobotrys*, and open riparian dominated by *Combretum imberbe*. Dryland woodlands dominated by a mosaic of mopane and sandveld woodlands occur on alluvial soils and river washed Kalahari sands, respectively (Sianga and Fynn 2017). Baikiaea forests (teak woodlands) occur in deep aeolian Kalahari sands in the northern edge of the Mababe Depression near the Ghoha Hills and extend northeast into Zimbabwe (Sianga and Fynn 2017). In the eastern section, the Mababe Depression is dominated by heavy clay soils (Teter 2007) and occupied by open grassland and sparse savanna (Sianga and Fynn 2017). South East of the Mababe Depression, typical Kalahari sandveld mixed with patches of mopane becomes the dominant vegetation, extending south to Nxai Pan and Makgadikgadi Pans (Sianga and Fynn 2017; Burrough 2022).

4.2.2 Data Collection and Analyses

Scientific research on a total of 38 zebra mares collared with GPS-enabled collars for a period of 2 years was synthesized. For example, 21 zebra mares were collared in the southeastern Okavango Delta (Bartlam-Brooks et al. 2011), eight in the Chobe River floodplains (Naidoo et al. 2016), and nine in the Makgadikgadi regions (Brooks 2005). Six zebra mares from six different herds in the Savuti-Mababe-Linyanti ecosystem, northern Botswana, were fitted and tracked with GPS-enabled collars (four satellite and two data logger collars; Africa Wildlife Tracking, Pretoria, South Africa) in November 2011 at the onset of the wet season for a full year. Mares were collared rather than stallions to minimize risks of collar damage during intrasexual fighting (Brooks and Harris 2008). Zebra were sedated with 7 mg M99 (etorphine hydrochloride), 80 mg azaperone, and 2500 IU Hyalase and reversed with 21 mg naltrexone after collar fitting. Mares were observed until they rejoined their herds.

Collars were programmed to take six fixes per day every 4 h for 2 years. A total of 7797 fixes were used for analyses. We used location data from each collared zebra to calculate seasonal 95% kernel utilization distributions (UDs) using the adehabitatHR package (Calenge 2007) in R version 3.3.0 (RCore-Team 2013). The reference bandwidth was used as a smoothing parameter to estimate the UDs (Calenge 2007). Published studies on the seasonal movement patterns of zebra populations (Bartlam-Brooks et al. 2011; Bradley 2012; Naidoo et al. 2016) were reviewed and synthesized in combination with a new study on the movement patterns of zebra populations in northern Botswana.

4.3 Results and Discussion

Analyses of collar data from the collared herds and published studies (Bartlam-Brooks et al. 2011; Bradley 2012; Naidoo et al. 2016) in northern Botswana yielded key insights into seasonal migrations of zebra populations along unfragmented ecological gradients. Zebra populations in northern Botswana displayed geographically distinct seasonal ranges. For example, studies reveal four distinct zebra migrations in northern Botswana (Fig. 4.1). These include the following: (i) from the Boteti River in the dry season to Makgadikgadi Pans saline grasslands in the wet season (~140 km round-trip, Brooks 2005; Bradley 2012), (ii) from the Okavango Delta floodplains and grasslands in the dry season to Makgadikgadi Pans saline grasslands in the wet season (~588 km round-trip, Bartlam-Brooks et al. 2011), (iii) from the Chobe River floodplains in the dry season to the Nxai Pan saline grasslands in the wet season (~500 km round-trip, Naidoo et al. 2016), and (iv) from the Linyanti Swamp region in the dry season to Mababe Depression saline grasslands in the wet season, 120 km round-trip (Joos-Vandewalle 2000; Fynn et al. 2014; Sianga 2014), as shown in Figs. 4.1 and 4.2.

All the recorded migratory zebra populations in this region migrate to mineral-rich saline grasslands during the wet season (Joos-Vandewalle 2000; Brooks 2005; Bartlam-Brooks et al. 2011; Bradley 2012; Sianga 2014; Naidoo et al. 2016), which likely meet their elevated resource requirements for growth and reproduction, as has also been documented elsewhere in Africa (Murray 1995; Hopcraft et al. 2010). The core wet season range for the Linyanti Swamp-Mababe Depression migration in the southern part of the Mababe Depression is characterized by Vachellia-dominated grasslands on silty clay soils dominated by high-quality annual grasses (e.g. 18.3 mg/kg phosphorus, 1683 mg/kg potassium, 203.3 mg/kg sodium, and 1765 mg/kg calcium) on the edge of the sump of the Mababe Depression, with taller grasses on clay soils occurring further into the sump relative to wetland habitats (e.g. 17.7 mg/kg phosphorus, 2715 mg/kg potassium, and 134 mg/kg sodium) (Sianga and Fynn 2017). These two adjacent habitats (Vachellia-dominated grasslands and tall open grasslands) have the highest soil sodium content of all habitats in the Savuti-Mababe-Linyanti ecosystem (Sianga and Fynn 2017).

In addition, tall open grasslands have the highest soil calcium content, while Vachellia-dominated grasslands have some of the highest soil phosphorus contents (Sianga and Fynn 2017). This high soil fertility translates into higher protein and phosphorus contents in grasses than in other habitats in the Savuti-Mababe-Linyanti ecosystem (Fynn et al. 2014). The Vachellia-dominated grasslands around the southern edge of the sump of the Mababe Depression are dominated by short, soft-leaved grasses such as *Urochloa trichopus* and *Chloris virgata*. These grasses are

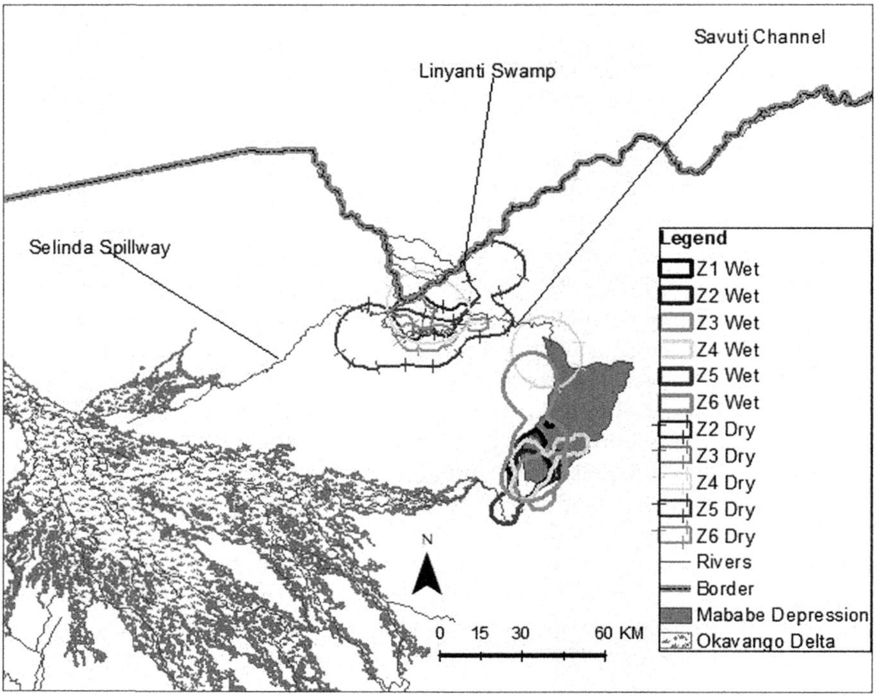

Fig. 4.2 Wet and dry season ranges for six zebra herds in the Savuti-Mababe-Linyanti ecosystem, northern Botswana, in 2011. Zebra selects for the Mababe Depression during the wet season and the Linyanti Swamps region during the dry season. Z1–6 denotes zebra 1 to 6. (Data sources: Sianga 2014; Cartography: Keoikantse Sianga)

highly digestible owing to their low amount of indigestible stem, which combined with their high mineral concentrations in leaf tissue likely allows higher levels of energy and nutrient intake (Wilmshurst et al. 2000). The *Cenchrus ciliaris* of the tall open grasslands may allow greater intake of calcium and sodium (Fynn et al. 2014).

Similarly, the Makgadikgadi and Nxai pans are palaeolake systems situated on soils of lacustrine origin offering relatively short, nutritive grasses and hence key core wet season habitats for the Boteti River-Makgadikgadi Pans migration (Brooks 2005; Bradley 2012), Okavango Delta-Makgadikgadi Pans migration (Bartlam-Brooks et al. 2011), and Chobe River-Nxai Pan migration (Naidoo et al. 2016). Similar to the Serengeti ecosystem, where zebra and wildebeest migrate to saline grasslands during the wet season (Sinclair 1979; Murray 1995), zebra populations in northern Botswana migrate into saline palaeolake systems during the wet season. This is an adaptive strategy to access higher-quality forage for the period of calving and lactation (Joos-Vandewalle 2000; Brooks 2005; Bartlam-Brooks et al. 2011, 2013; Sianga 2014; Bennitt et al. 2022). In combination, these palaeolake systems, such as the southern Mababe Depression, Makgadikgadi, and Nxai Pans, likely provide some of the highest quality forage in northern Botswana. Thus, foraging adaptively in these palaeolake systems during the wet season would likely enable optimal nutrient intake at this critical time of growth and reproduction, when demands for minerals, protein, and energy are greatly elevated (Taylor 1985; Murray 1995; Hopcraft et al. 2010). Selection for protein- and mineral-rich grasslands is a common strategy for herbivores during the wet season (Kreulen 1975; Murray 1995; Fynn and Bonyongo 2011). In the Central Kalahari Game Reserve in Botswana, blue wildebeest and several other herbivores favor saline pan grasslands during the wet season (Selebatso et al. 2018).

Selection for the high mineral and protein content of saline grasslands of lacustrine origin (palaeolake systems) as the wet season range appears to be the mechanism driving wet season home range selection in all four zebra migrations documented in northern Botswana. This is supported by research done in Etosha (Namibia) wildebeest and zebra (Gasaway et al. 1996; Fynn and Bonyongo 2011). In a region dominated by infertile Kalahari sands, these endorheic palaeolake systems have clay and nutrients concentrated by river deposits and evaporative processes well above the levels found in the matrix of Kalahari sands (Sianga and Fynn 2017; Selebatso et al. 2018). Similarly, zebra, wildebeest, and Thomson's gazelles in the Serengeti ecosystem, Tanzania, migrate to the saline grasslands of the Serengeti plains during the wet season, where nutrients have been concentrated by volcanic ash deposits combined with low-rainfall conditions (Sinclair 1979; Murray 1995).

The migration of zebra populations into their dry season ranges may have been linked to permanent water availability and access to sufficient adequate-quality forage in wetland habitats and adjacent woodlands when water had dried up inland and forage quality had declined in most other habitats (Vesey-FitzGerald 1960; Prins and Beekman 1989; Fynn et al. 2014). Wetlands are critical dry season habitats for wild and domestic herbivores across Africa because they provide a reliable reserve of green forage in bulk when forage in other habitats is dry and often depleted (Fynn et al. 2015). Thus, synthesis of these zebra migrations demonstrates predictable and consistent patterns of seasonal habitat selection, with all four migrations selecting high-quality forage in saline regions during the wet season and retreating to extensive wetland systems (or river systems) and adjacent woodlands for the dry season where they can find dependable water and adequate-quality forage. However, despite these documented zebra migratory routes, research has also demonstrated that some zebra populations in northern Botswana exhibit sedentary movement patterns. For example, Bartlam-brooks et al. (2011) demonstrated that some of the collared zebra herds remained in the Okavango Delta for the entire year, utilizing the poor-quality forage in the wetland habitats (Cronberg et al. 1996). Similarly, in the Central Kalahari Game Reserve, some wildebeest are sedentary, while others migrate, which appears to be affected by artificial water provision (Selebatso et al. 2017). Similarly, in the Serengeti, some wildebeest and zebra do not migrate (Serneels et al. 2001).

Similarly, research has demonstrated that the mechanisms underlying migration are twofold. Growing fetuses in the final months before parturition place large demands for energy, protein, and minerals on the mother, which, if not met, results in small and weak calves that never catch up on larger calves and are vulnerable to disease and predation. In addition, after birth, calves need a good milk supply from the mothers and high-quality forage to grow. Saline pan grasslands provide these minerals. Sodium is especially critical because it enables herbivores to use fresh high-quality forage rich in potassium while still maintaining the right Na:K ratios in their diet (Kaspari 2020). Other minerals, such as calcium and phosphorus, are essential for lactation and for cellular processes. This means that mothers must find wet season ranges where forage is highly digestible and rich in energy, protein, and minerals. Such a resource has been termed a fecundity resource (Fynn and Provenza 2023). Once the rains have ceased, grasses dry out, lose quality, and are depleted, resulting in the quantity and quality of forage becoming extremely limiting, leading to declining body condition and, in drought years, to starvation. In regions with wetlands, forage remains plentiful and green throughout the dry season. Thus, wetlands provide critical reserves of reliable adequate quality forage to prevent loss of body condition or starvation and population collapse during drought years (a green reserve resource and buffer resource, Fynn and Provenza 2023). In addition, regrowth of grazed or burned grasses in wetlands provides a bridge of high-quality forage over the late dry season, which supports fetal development and maintenance of body condition in the final months before parturition. This is known as a bridging resource (Fynn and Provenza 2023). Thus, wetlands are a critical mechanism driving migration during the dry season. High-rainfall areas perform the same function as wetlands, e.g. the northern Serengeti/Mara woodlands. Consequently, African wetlands are important dry season ranges for wild and domestic herbivores across Africa (Fynn et al. 2015). Thus, migrations are driven by the need to obtain high-quality fecundity resources during the wet season and reserve and buffer resources during the dry season.

4.4 The Importance of Sustained Connectivity

The migratory behavior of zebra populations in northern Botswana is particularly important in enhancing adaptive access to forage resources seasonally. However, a combination of anthropogenic and climate change effects is likely to threaten the existence of these migrations. For example, the Okavango Delta-Makgadikgadi Pans migration had previously been disrupted by a Setata veterinary fence, which has subsequently been removed, and the migration has resumed (Bartlam-Brooks et al. 2011). Habitat fragmentation through the erection of veterinary control fences in some parts of northern Botswana in the early 1950–1980 disrupted emergency long-range movements of wildebeest populations from the Schwelle in the Kalahari in southern Botswana to the Boteti River and Lake Ngami during severe droughts (Williamson and Williamson 1984; Williamson et al. 1988; Mbaiwa and Mbaiwa 2006; Ferguson and Hanks 2010). In the Boteti Region, a fence erected along the Boteti River in 2005 strangled approximately 300 migrating zebra, which were cut off from their wet season grazing areas in the Makgadikgadi Pans (Gibson 2010). Similarly, the artificial water points (AWPs) established in Chobe National Park (Fullman and Child 2013) and in CT11, a Wildlife Management Area bordering Makgadikgadi National Park to attract wildlife, and between 2014 and 2016 to attract wildlife species during the dry season will probably result in the disruption of migratory zebra populations in the region. Research in the region demonstrates that some zebra herds in this migratory population are becoming attracted to the AWPs relative to the natural water points during the dry season (Bennitt et al. 2022). As a result, reliance on these AWPs by zebra will eventually lead to an adaptive sedentary movement pattern by these populations (Selebatso et al. 2017; Perkins 2018, 2020). However, it must be noted that the use of AWPs for water provision to wildlife in dryland areas could be useful for wildlife management purposes (Owen-Smith 1996), provided that their numbers and distribution are carefully planned (Mwakiwa et al. 2013). To ensure connectivity of the zebra seasonal ranges, the establishment of fences and AWPs in the region should be limited. Sustained connectivity between wet and dry season ranges of migratory populations is therefore important for the existence of zebra populations in northern Botswana.

The current predictions are that Angola and northern Botswana will become proportionally hotter and drier than the world average change (Engelbrecht et al. 2015). This means that large herbivores will have to increasingly rely on floodplains of the Okavango Delta, which can provide reserve and bridging resources to adapt to drought and forage shortages (Fynn and Provenza 2023). Therefore, if wildlife is fenced off from access to the Delta, the effects of climate change will have much more dramatic effects on their populations. Additionally, more variable and less predictable rainfall under climate change will affect the ability of large herbivores to migrate between seasonal ranges because they rely on forage and water from rainfall to move through the migration corridors (Smuts 1978; Bartlam-Brooks et al. 2013). Climate change is predicted to bring variability and unreliable rainfall patterns that will induce more variable rainfall, which may change the timing of leaving a seasonal range. It is anticipated that with the impacts of climate change, there will be a need to increase the need for migration to the dry season range. Climate change may also delay the timing of migration to the wet season range if rains start later. If climate change reduces inflows of water from Angola, the size of the Okavango Delta may shrink, reducing the availability of floodplains.

Therefore, we emphasize the critical role that the floodplains of the Okavango Delta, Linyanti Swamps, and Chobe River play in maintaining viable wildlife populations in northern Botswana, especially under climate change. There are, however, threats to the Okavango Delta, Linyanti Swamps, and Chobe River floodplains, which include increasing water abstraction in the catchment region and Namibia in the dry season for crop production and development (Fynn et al. 2015). There are also plans to develop dams for hydroelectric power generation in the Okavango Catchment that could have disastrous effects on the water supply to the Okavango Delta. Under climate change, precipitation, which has been noted as a facilitator for these migrations (McNaughton 1990; Bartlam-Brooks et al. 2013), will become variable and geographically unpredictable (Robinson et al. 2009). To enhance adaptation to climate change, sustained connectivity across ecosystems in northern Botswana is particularly important in enhancing flexibility in forage access. Although Bartlam-Brooks et al. (2013) demonstrate that zebra can quickly adjust their migrations and nomadic behavior to changing environmental conditions, which is expected under climate change, fragmentation will still be a threat. This highlights the importance of ensuring connectivity across landscapes in northern Botswana. Therefore, zebra need to have access to large areas and to high-quality saline ranges for the wet season and wetlands for the dry season, especially for drought years, which are expected to become more frequent and severe under climate change.

4.5 Conclusion

This chapter synthesizes the migratory patterns of zebra populations in northern Botswana. Each migration encompasses a dry season range with permanent water and floodplains and adjacent woodlands and a saline grassland-based wet season range of lacustrine origin. Factors that concentrate nutrients in landscapes at large scales, such as ancient lake systems in the Kalahari Basin or volcanic deposits and landscape catenas elsewhere, appear to be key factors in determining the wet season ranges of migratory herbivores in Africa. Thus, our study highlights key factors determining the functionality of seasonal home ranges, providing an understanding of the mechanisms driving migration in African herbivore populations. In addition, our results emphasize the importance of extensive wetland systems as dry season ranges for large herbivores because of their ability to provide reliable, adequate-quality forage and water for wildlife over the dry season and in droughts. This functional role will become increasingly important under climate change but is threatened by socioeconomic development in their catchment areas. Additionally, keeping migratory corridors open to maintain habitat connectivity between the functional seasonal home ranges of various herbivores is essential.

Acknowledgements We are grateful to the German Ministry of Education and Research (BMBF), which provided funding for the migration research in the Linyanti Swamp-Mababe Depression as part of the Future Okavango project. We appreciate Dr Rob Jackson's darting the zebra during collar installation and removal. The Department of Wildlife and National Parks and the Ministry of Environment, Wildlife and Tourism have our gratitude for allowing us to undertake this research in northern Botswana.

References

Aarrestad PA, Hytteborn H, Masunga G et al (2014) Vegetation: between soils and herbivores. In: Elephants and Savanna woodland ecosystems: a study from Chobe National Park, Botswana, pp 61–88

Bartlam-Brooks HLA, Bonyongo MC, Harris S (2011) Will reconnecting ecosystems allow long-distance mammal migrations to resume? A case study of a zebra *Equus burchelli* migration in Botswana. Oryx 45:210–216

Bartlam-Brooks HL, Beck PS, Bohrer G et al (2013) In search of greener pastures: using satellite images to predict the effects of environmental change on zebra migration. J Geophys Res Biogeosci 118:1427–1437

Bennitt E, Bradley J, Bartlam-Brooks HL et al (2022) Effects of artificial water provision on migratory blue wildebeest and zebra in the Makgadikgadi Pans ecosystem, Botswana. Biol Conserv 268:109502
Berger J (2004) The last mile: how to sustain long-distance migration in mammals. Conserv Biol 18:320–331
Berry HH, Jenssen J (1997) Aspects of wildebeest *Connochaetes taurinus* ecology in the Etosha National Park-a synthesis for future management. Modoqua 1997:137–148
Bhola N, Ogutu JO, Piepho HP et al (2012) Comparative changes in density and demography of large herbivores in the Masai Mara Reserve and its surrounding human-dominated pastoral ranches in Kenya. Biodivers Conserv 21:1509–1530
Bradley JT (2012) The effect of environmental variability on the foraging behaviour of plains zebra (*Equus quagga*) in the Makgadikgadi, Botswana. University of Bristol, Bristol
Brooks CJ (2005) The foraging behaviour of Burchell's zebra (*Equus burchelli* antiquorum). University of Bristol
Brooks CJ, Harris S (2008) Directed movement and orientation across a large natural landscape by zebras, Equus burchelli antiquorum. Anim Behav 76(2):277–285
Burrough SL (2022) The Makgadikgadi basin. In: Landscapes and landforms of Botswana. Springer, Cham, pp 77–90
Calenge C (2007) Exploring habitat selection by wildlife with adehabitat. J Stat Softw 22:1–19
Cronberg G, Gieske A, Martins E et al (1996) Major ion chemistry, plankton and bacterial assemblages of the Jao/Boro River, Okavango Delta, Botswana: the swamps and flood plains. Arch Hydrobiol Suppl bd Algol stud 107:335–407
Engelbrecht F, Adegoke J, Bopape MJ et al (2015) Projections of rapidly rising surface temperatures over Africa under low mitigation. Environ Res Lett 10:085004
Ferguson K, Hanks J (2010) Fencing impacts: a review of the environmental, social and economic impacts of game and veterinary fencing in Africa with particular reference to the Great Limpopo and Kavango-Zambezi Transfrontier Conservation Areas. Mammal Research Institute, Pretoria, p 329
Fryxell JM, Sinclair ARE (1988) Causes and consequences of migration by large herbivores. Trends Ecol Evol 3:237–241
Fullman TJ, Child B (2013) Water distribution at local and landscape scales affects tree utilization by elephants in Chobe National Park, Botswana. Afr J Ecol 51:235–243
Fynn RWS, Bonyongo MC (2011) Functional conservation areas and the future of Africa's wildlife. Afr J Ecol 49:175–188
Fynn RWS, Provenza F (2023) Functional adaptive resources for large herbivores in African savannas: an ecological-gradient based framework. Front Conserv Sci 4:31
Fynn RWS, Chase M, Röder A (2014) Functional habitat heterogeneity and large herbivore seasonal habitat selection in northern Botswana. S Afr J Wildl Res 44:1–15
Fynn RWS, Murray-Hudson M, Dhliwayo M et al (2015) African wetlands and their seasonal use by wild and domestic herbivores. Wetl Ecol Manag 23:559–581
Gasaway WC, Gasaway KT, Berry HH (1996) Persistent low densities of plains ungulates in Etosha National Park, Namibia: testing the food-regulating hypothesis. Can J Zool 74:1556–1572
Gibson D (2010) Environmental assessments of Botswana's fences. In: *Fencing impacts:* a review of the environmental, social and economic impacts of game and veterinary fencing in Africa with particular reference to the Great Limpopo and Kavango-Zambezi Transfrontier Conservation Areas, pp 121–127
Harris G, Thirgood S, Hopcraft JGC et al (2009) Global decline in aggregated migrations of large terrestrial mammals. Endanger Species Res 7:55–76
Hobbs NT, Galvin KA, Stokes CJ et al (2008) Fragmentation of rangelands: implications for humans, animals, and landscapes. Glob Environ Change 18:776–785
Hopcraft JGC, Olff H, Sinclair ARE (2010) Herbivores, resources and risks: alternating regulation along primary environmental gradients in savannas. Trends Ecol Evol 25:119–128

Joos-Vandewalle ME (2000) Movement of migratory zebra and wildebeest in northern Botswana. University of the Witwatersrand, South Africa
Kaspari M (2020) The seventh macronutrient: how sodium shortfall ramifies through populations, food webs and ecosystems. Ecol Lett 23:1153–1168
Kreulen D (1975) Wildebeest habitat selection on the Serengeti plains, Tanzania, in relation to calcium and lactation: a preliminary report. Afr J Ecol 13:297–304
Mbaiwa JE, Mbaiwa OI (2006) The effects of veterinary fences on wildlife populations in Okavango Delta, Botswana. Int J Wilderness 3:17–23
McNaughton SJ (1985) Ecology of a grazing ecosystem: the Serengeti. Ecol Monogr 55:259–294
McNaughton SJ (1990) Mineral nutrition and seasonal movements of African migratory ungulates. Nature 345:613–615
Morrison TA, Link WA, Newmark WD et al (2016) Tarangire revisited: consequences of declining connectivity in a tropical ungulate population. Biol Conserv 197:53–60
Murray MG (1995) Specific nutrient requirements and migration of wildebeest. In: Sinclair ARE, Arcese P (eds) Serengeti II: dynamics, management and conservation of an ecosystem. Chicago University Press, Chicago, pp 231–256
Mwakiwa E, de Boer WF, Hearne JW et al (2013) Optimization of wildlife management in a large game reserve through waterpoints manipulation: a bioeconomic analysis. J Environ Manag 114:352–361
Naidoo R, Chase MJ, Beytell P et al (2016) A newly discovered wildlife migration in Namibia and Botswana is the longest in Africa. Oryx 50:138–146
Ogutu JO, Owen-Smith N, Piepho HP et al (2013) Changing wildlife populations in Nairobi National Park and adjoining Athi-Kaputiei Plains: collapse of the migratory wildebeest. Open Conserv Biol 7:11–26
Owen-Smith N (1996) Ecological guidelines for waterpoints in extensive protected areas. S Afr J Wildl Res 26:107–112
Parker KL, Barboza PS, Gillingham MP (2009) Nutrition integrates environmental responses of ungulates. Funct Ecol 23:57–69
Pellacani F, Romanens R, Mainga A et al (2020) Plant community diversity in the Chobe Enclave, Botswana: insights for functional habitat heterogeneity for herbivores. Koedoe 62:1–17
Perkins JS (2018) Southern Kalahari piospheres: looking beyond the sacrifice zone. Land Degrad Dev 29:2778–2784
Perkins JS (2020) Changing the scale and nature of Artificial Water Points (AWP) use and adapting to climate change in the Kalahari of Southern Africa. Sustainability in developing countries: case studies from Botswana's journey towards 2030 Agenda, p 51–89
Prins HHT, Beekman JH (1989) A balanced diet as a goal for grazing: the food of the Manyara buffalo. Afr J Ecol 27:241–259
RCore-Team (2013) R: a language and environment for statistical computing. R Foundation for Statistical Computing, Vienna, Austria. Viewed 11 April 2015
Revermann R, Gomes AL, Gonçalves F et al (2016) Vegetation database of the Okavango Basin. Phytocoenologia 46:103–104
Robinson RA, Crick HQ, Learmonth JA et al (2009) Travelling through a warming world: climate change and migratory species. Endanger Species Res 7:87–99
Selebatso M, Bennitt E, Maude G et al (2017) Water provision alters wildebeest adaptive habitat selection and resilience in the Central Kalahari. Afr J Ecol 56:225–234
Selebatso M, Bennitt E, Maude G et al (2018) Water provision alters wildebeest adaptive habitat selection and resilience in the Central Kalahari. Afr J Ecol 56:225–234
Serneels S, Lambin EF (2001) Impact of land-use changes on the wildebeest migration in the northern part of the Serengeti–Mara ecosystem. J Biogeogr 28:391–407
Serneels S, Said MY, Lambin EF (2001) Land cover changes around a major east African wildlife reserve: the Mara Ecosystem (Kenya). Int J Remote Sens 22:3397–3420
Sianga K (2014) Habitat use by buffalo and zebra in relation to spatial and temporal variability of resources in the Savuti-Mababe-Linyanti ecosystem of northern Botswana. Mphil Thesis. University of Botswana, Botswana

Sianga K, Fynn R (2017) The vegetation and wildlife habitats of the Savuti-Mababe-Linyanti ecosystem, northern Botswana. Koedoe 59:1–16

Sianga K, Van Telgen M, Vrooman J et al (2017) Spatial refuges buffer landscapes against homogenization and degradation by large herbivore populations and facilitate vegetation heterogeneity. Koedoe 59:1–13

Sinclair ARE (1979) The eruption of the ruminants. In: Sinclair ARE, Norton-Griffiths (eds) Serengeti: dynamics of an ecosystem. Chicago University Press, Chicago, pp 82–103

Smuts GL (1978) Interrelations between predators, prey, and their environment. J Biosci 28:316–320

Taylor RD (1985) The Response of Buffalo *Syncerus caffer* to the Kariba LakeShore Grasslands (*Panicum repens*) in Matusadona National Park. PhD Thesis, University of Zimbabwe, Zimbabwe

Teter K (2007) Paleoenvironmental reconstruction of Paleolake Mababe, northwestern Botswana from sediment chemistry and biological productivity data. MSc Thesis. Oklahoma State University, Stillwater. United States of America.

Vesey-FitzGerald DF (1960) Grazing succession among East African game animals. J Mammal 41:161–172

Vittoz P, Pellacani F, Romanens R et al (2020) Plant community diversity in the Chobe Enclave, Botswana: insights for functional habitat heterogeneity for herbivores. Koedoe 62:1–17

Weare PR, Yalala A (1971) Provisional vegetation map of Botswana. Bots Notes Rec 3:131–147

Whyte CSJ, Joubert IJ (1988) Blue wildebeest population trends in the Kruger National Park and the effects of fencing. S Afr J Wildl Res 18:78–87

Williamson D, Williamson J (1984) Botswana's fences and the depletion of Kalahari wildlife. Oryx 18:218–222

Williamson D, Williamson J, Ngwamotsoko KT (1988) Wildebeest migration in the Kalahari. Afr J Ecol 26:269–280

Wilmshurst JF, Fryxell JM, Bergman CM (2000) The allometry of patch selection in ruminants. Proc R Soc Lond B Biol Sci 267:345–349

Keoikantse Sianga is a senior lecturer at the Botswana University of Agriculture and Natural Resource. He holds a BSc in Conservation Biology and Wildlife Management and an MPhil and a PhD in Natural Resource Management. He worked as Project Assistant at the University of Botswana, Project Manager (Communities Living Among Wildlife Sustainably, CLAWS). His research interests include functional and vegetation heterogeneity and plant-wildlife interactions, fire ecology and management, human-wildlife interactions, and spatial ecology. Dr Sianga has published eight articles in peer reviewed journals and seven book chapters.

Mpaphi C. Bonyongo is an experienced ecologist with vast experience in research and research management. Over time, Dr Bonyongo amassed vast experience in mobilizing resources and managed several large-scale multi-disciplinary research for development projects. He is a competent researcher and research methods trainer who mentored and transformed young researchers into competent researchers. Dr Bonyongo is an active researcher currently holding an adjunct senior research scholar position at the University of Botswana's Okavango Research Institute and serving as senior scientific officer at the Permanent Okavango River Basin Water Commission.

Richard W. S. Fynn (Nat. Dip. Nat. Con, BSc, MSc, PhD). He worked as a post doc (2003-2005) at UKZN, postdoc (2005-2009) at the Colorado State University, and senior research scholar (2009-2019) and associate professor (2019–present) at the Okavango Research Institute (ORI). His research interests include plant-herbivore interactions, grazing ecosystem ecology, functional habitat heterogeneity and rangeland management, and conservation science. Key interests are in developing optimal grazing management strategies for cattle ranches and in developing more ethically acceptable and sustainable conservation strategies where local communities have a central role. He has published 57 journal articles and 5 book chapters.

Chapter 5
Herbivore Space-Use and Habitat Selection in Human-Mediated Ecosystems: *Implications on Restoration Initiatives*

Victor K. Muposhi, Mkululi Ndlovu, Kudzai S. Mpakairi, Keoikantse Sianga, Admire Chanyandura, Chiedza C. Zembe, and Gregory Chiyangwa

5.1 Introduction: Conservation in the Anthropocene

The Anthropocene era, marked by human dominance over the planet's ecosystems, has brought about a new era of conservation efforts (Darimont et al. 2015; Maxwell et al. 2020). Considering escalating pressures on Earth's resources, conservationists have transitioned from acting with mere purpose to striving for tangible impact (Butchart et al. 2010; Di Marco et al. 2018). The roots of conservationism trace back to the late nineteenth century, when naturalists or early conservationists

V. K. Muposhi (✉)
Department of Wildlife and Aquatic Resources, Botswana University of Agriculture and Natural Resources, Gaborone, Botswana

School of Wildlife and Environmental Sciences, Chinhoyi University of Technology, Chinhoyi, Zimbabwe
e-mail: vmuposhi@buan.ac.bw

M. Ndlovu · G. Chiyangwa
Scientific Services, Zimbabwe Parks, and Wildlife Management Authority, Harare, Zimbabwe

K. S. Mpakairi
Institute of Water Studies, Department of Earth Sciences, University of the Western Cape, Cape Town, South Africa

School of Wildlife Conservation, African Leadership University, Kigali, Rwanda

K. Sianga
Department of Wildlife and Aquatic Resources, Botswana University of Agriculture and Natural Resources, Gaborone, Botswana
e-mail: ksianga@buan.ac.bw

A. Chanyandura · C. C. Zembe
School of Wildlife and Environmental Sciences, Chinhoyi University of Technology, Chinhoyi, Zimbabwe

W. L. Hambira et al. (eds.), *Environmental Change and Biodiversity Conservation in sub-Saharan Africa*, Advances in Global Change Research 75,
https://doi.org/10.1007/978-3-031-73136-5_5

focused on preserving wild spaces and wildlife, often advocating for the establishment of national parks and protected areas (Huntley et al. 2019). However, as industrialization took hold, conservation efforts shifted toward resource management and sustainability (Marvier et al. 2012; Mondal and Palit 2021). Conservation initiatives today, in the Anthropocene, must be holistic and consider how humans are impacting the earth globally (Kueffer and Kaiser-Bunbury 2014; Storch et al. 2022). Conservationists face a multitude of challenges, including climate change, habitat destruction, pollution, and invasive species, among others (Loehle and Li 1996; Hellmann et al. 2008; Caro et al. 2012; Walsh et al. 2016; Dueñas et al. 2021). To address these issues, conservationists must move beyond simply acting with purpose to acting for impact (McIntosh et al. 2017; Bull et al. 2021; Pressey et al. 2021).

Acting with purpose means having a general goal or mission in mind when undertaking conservation efforts (Madliger et al. 2017). An organization dedicated to conservation, for instance, might work to preserve an endangered species or rebuild a damaged ecosystem. Although this strategy can be helpful, it frequently fails to result in significant change. Acting with purpose can lead to unfocused efforts and limited influence if there are no precise aims and observable outcomes (McIntosh et al. 2017). In contrast, acting for impact involves setting clear goals and measuring progress towards achieving those goals (Pressey et al. 2021). It may be argued that a conservation organization might establish a goal for expanding the population of an endangered species by a certain percentage rather than just working to protect it (Sanderson 2006). Conservationists can monitor the effectiveness of their efforts and alter their strategy by establishing precise targets (Cimon-Morin et al. 2014; Drira et al. 2019).

Conservationists in the Anthropocene are making a significant impact by harnessing technology to collect and analyse data (Berger-Tal and Lahoz-Monfort 2018; Hahn et al. 2022). Remote sensing, satellite imagery, and other appropriate technologies have been adopted widely to monitor changes in land use, climate, and biodiversity over time (Hahn et al. 2022). By analysing this data, conservationists can uncover trends and patterns that inform their conservation plans. To achieve sustainable goals and make a genuine impact, it is essential to embrace and willingly take risks and explore new techniques (Bennett et al. 2018; Turnbull et al. 2021). Conservation efforts must adapt to changing circumstances and new challenges (Fowler et al. 2020). The Anthropocene presents unique challenges for conservationists (Landres et al. 2020; Storch et al. 2022) and acting with purpose is no longer sufficient—we must act for impact (Bull et al. 2021; Pressey et al. 2021). It is necessary to set clear goals, use technology to collect and analyse data, collaborate with local communities, and be willing to take chances and explore new ideas. Conservationists may assist conserve the planet's valuable ecosystems and secure a sustainable future for future generations by doing so.

5.2 Herbivore Space Use in Human-Mediated Ecosystems

Herbivores play an essential role in maintaining ecosystem balance and diversity. However, their space use and habitat selection have been greatly affected by human activities such as logging and hunting (Brodie et al. 2015; Martínez-Ramos et al. 2016; Ximenes Pinho et al. 2017). It is vital to understand how herbivores adapt to these human-mediated habitats to establish effective conservation measures. Herbivore space utilization in human-mediated habitats faces numerous challenges (Storch et al. 2022). Some of the challenges include habitat loss and fragmentation (Bartlett et al. 2016; Püttker et al. 2020), interspecific competition (Sietses et al. 2009; Mpakairi et al. 2017; Brazeau and Schamp 2019), predation (Davies et al. 2021), and anthropogenic disturbances (Rich et al. 2016; Storch et al. 2022). These challenges can significantly affect herbivore population dynamics, behaviour, and spatial distribution. An example of such effects is the occurrence of ecological traps, where herbivores relocate to suboptimal habitats due to the absence of favourable alternatives, leading to reduced fitness and reproductive rates (Battin 2004; Hale and Swearer 2016; le Roex et al. 2020). Similarly, human actions may create novel environments that are unfavourable for some herbivore species while promoting the survival of others (Lindenmayer et al. 2008).

Herbivores adopt several adaptation techniques to deal with these challenges, including changes in behaviour, physiology, and morphology. Some herbivores may change their foraging patterns to take advantage of new feed resources or to avoid herbivory hotspots (de Boer et al. 2015). They may also adjust their daily activity patterns to minimize human disturbance (Benoit et al. 2020). Moreover, herbivores may select different habitats based on their ecological requirements and the availability of resources (Codron et al. 2016). Habitat selection is a complex process that involves multiple factors, such as forage quality and quantity and palatability, predator and competitor densities, vegetation structure, and proximity to water sources (Wilmshurst et al. 2000; Avgar et al. 2020; Fu et al. 2022). Herbivores may need to modify their habitat selection criteria in human-mediated ecosystems to account for anthropogenic variables such as roads, fences, mining areas, and settlements, among others.

5.3 Approaches to Studying Herbivore Space Use and Habitat Selection

Herbivores are an important component of many ecosystems, and their space use and habitat selection play an important role in regulating ecosystem function. Researchers analyse these factors using several techniques or methodologies, including direct field observations and telemetry (Bastille-Rousseau et al. 2018), modelling (Kimanzi et al. 2014; Milner et al. 2021), spoor and faecal samples

(Bohmann et al. 2013), stable isotope analysis (Bortoluzzi et al. 2021), and camera traps (Agha et al. 2018; Green et al. 2020). Each method has its strengths and limitations, and choosing the appropriate method depends on the research question and the species being studied. Ultimately, a combination of methods will provide the most comprehensive understanding of herbivore space use and habitat selection. Direct observations involve visual monitoring of herbivores to determine their movement patterns and habitat use. This method is labour-intensive but provides detailed information about individual animals' behaviour (Soares-Bolzan et al. 2021).

Telemetry involves fitting radio collars or other tracking devices to herbivores to collect data on the movements and behaviours of wildlife species (Güler and Übeyli 2002). It involves fitting radio collars or other tracking devices to herbivores. These devices enable researchers to remotely track animal activities, providing information on home range size, movement patterns, and habitat preference (Patel et al. 2015; Northrup et al. 2022). This technology has enabled researchers to track the movements of individual animals in real-time over large spatial scales, providing insights into the environmental factors that influence their behaviour (Cooke 2008). Telemetry allows researchers to get vital insights into the ecological processes that drive herbivore behaviour and population dynamics. Such insights play a pivotal role in enhancing conservation efforts and informing management strategies aimed at preserving these invaluable species.

The species distribution model (SDM) is another method for studying herbivore space usage and habitat selection. Researchers apply SDMs to predict the distribution of species across geographic areas based on environmental variables such as vegetation type, topography, and water availability (Phillips et al. 2006a, b). These models aim to capture the ecological niche of a species by identifying the environmental conditions that are associated with its presence or absence. Similarly, SDMs can be used to determine distribution, identify crucial habitats, and evaluate how changing habitats would affect herbivores (Wijeyakulasuriya et al. 2020; Milner et al. 2021). However, modelling relies on assumptions and simplifications, and the accuracy of the results depends on the quality of the data used in constructing the models (Elith et al. 2011; Scharf and Buderman 2020).

Camera traps are increasingly being used to study herbivore space use and habitat selection, especially for elusive or nocturnal species (Green et al. 2020). Camera traps are motion-activated cameras that capture images of wildlife passing through a monitored area. By analysing these images, researchers can determine the identity of the species, their movement patterns, feeding behaviour, and habitat preferences (Agha et al. 2018; Tarugara et al. 2019, 2021). Camera traps are non-invasive and can collect data over long periods, making them ideal for monitoring populations or assessing the impact of habitat changes (Rich et al. 2016). A summary of these and other approaches is outlined in Table 5.1.

Table 5.1 Summary of approaches used in studying herbivore space use and habitat selection

Approach	Description	References
Field observations	Involve direct monitoring of herbivores in their natural habitats Researchers may use techniques such as visual surveys, camera traps, and radio telemetry to track and observe herbivore movements, feeding patterns, and habitat preferences This approach provides valuable information on real-time behaviour and habitat selection	Bukovinszky et al. (2005), Bonnet et al. (2015), Garnick et al. (2018)
GPS Tracking:	Global Positioning System (GPS) tracking involves fitting herbivores with GPS collars or tags that record their precise locations at regular intervals Data can be used to analyse movement patterns, home range sizes, and habitat use over extended periods GPS tracking provides detailed information on individual and population-level space use and habitat selection	Samuel and Kenow (1992), Aarts et al. (2008), Martin et al. (2009), Bastille-Rousseau et al. (2018)
Resource Selection Functions (RSFs)	Statistical models used to analyse the relationship between animal presence or resource use and habitat characteristics Approach involves collecting data on herbivore locations and environmental variables such as vegetation type, topography, and water availability Researchers identify factors that influence habitat selection by comparing the observed resource use to the availability of different habitat attributes	Boyce and McDonald (1999), Boyce et al. (2002), Northrup et al. (2013)
Remote Sensing	Techniques utilize satellite imagery and aerial photography to analyse landscape-level habitat characteristics Approach provides information on vegetation types, land cover, and habitat heterogeneity over large areas Researchers can examine how herbivores respond to different habitat features across the landscape by combining remote sensing data with herbivore presence or movement data	Homer et al. (1993), Osborne et al. (2001), Kerr and Ostrovsky (2003), Mugo et al. (2010), Rechsteiner et al. (2017)

(continued)

Table 5.1 (continued)

Approach	Description	References
Stable Isotope Analysis	Non-invasive method used to study herbivore diet composition and habitat use indirectly Researchers can determine the types of plants or habitats the animals have been utilizing by analysing stable isotopes (e.g. carbon, nitrogen) in herbivore tissues such as hair, bone, or faeces Approach provides insights into long-term habitat preferences and dietary patterns	Phillips (2001), Bearhop et al. (2004), Inger and Bearhop (2008), Katzenberg and Waters-Rist (2018)
Experimental Studies	Involve manipulating habitat conditions or resources to understand how herbivores respond to changes in their environment Researchers may create artificial food plots or water sources in specific locations to assess herbivore preferences and use Approach helps to elucidate the influence of specific habitat factors on space use and selection	Meadows and Campbell (1972), Butman (1987), Aikins and Kikuchi (2001)
Modelling Approaches	Mathematical and computational models, such as agent-based models and habitat suitability models Used to simulate and predict herbivore space use and habitat selection Models integrate data on habitat characteristics, herbivore behaviour, and landscape structure to understand how animals interact with their environment Modelling approaches can provide valuable insights into future habitat preferences and help inform conservation and management strategies	Elith et al. (2011), Phillips et al. (2006a, b), Teng et al. (2022)

Understanding how herbivores use space and select habitats in human-mediated ecosystems is crucial for developing effective conservation and management strategies (Jung 2017). Moreover, considering the specific needs and preferences of different herbivore species can help prioritize conservation efforts and avoid unintended consequences. Overall, herbivore space-use and habitat selection in human-mediated ecosystems are complex and dynamic processes that depend on multiple factors, both natural and anthropogenic (Börger et al. 2008). By studying these processes, we can gain valuable insights into the ecological mechanisms underlying herbivore adaptation to human activities and identify opportunities to promote their conservation and coexistence with humans.

In natural ecosystems, animal space use is a function of the interplay of movement and the environment as mediated by individual characteristics, individual state, and the external environment (Börger et al. 2008; Viana et al. 2018). However, in human-mediated ecosystems, habitat selection and animal movement patterns are often modified by human activities (Ramos et al. 2016). Successful restoration programs initiatives in modified ecosystems, e.g. recreational parks (Gutzwiller et al. 2017; Klanjšček et al. 2018), require empirical studies on mechanisms that drive habitat use and selection by wildlife species (Thurfjell et al. 2014). Incorporating anthropogenic disturbances and biotic factors (e.g. competition and vegetation) into species distribution modelling improves the predictive power of the models (Lewis et al. 2017a, b). This information is considered essential for mapping the spatial distribution of priority areas for conservation and the enactment of sound conservation policies integral to sustainable protected area management (Muposhi et al. 2016a, b).

5.4 Roan Antelope Case Study

In this chapter, a case study in which satellite telemetry data was applied to explore roan antelope (*Hippotragus equinus*) home-range sizes, and habitat selection in Umfurudzi Recreational Park, Zimbabwe (Box 5.1, Fig. 5.1). The case study uses modelling techniques to provide a comprehensive understanding of roan antelope space use and habitat selection as a framework for the species' strategic restoration. It is based on two broad objectives: (i) to establish the home ranges of roan antelope; (ii) to model the suitability of habitat patches as influenced by various environmental variables and consider the implications for restoration initiatives in Umfurudzi Recreational Park.

Hypothetically, roan antelope home ranges and space use are simultaneously influenced by three broad factors: biotic, abiotic, and anthropogenic disturbances. Here, abiotic factors include surface water availability, rainfall, temperature, and topography as these impose physiological constraints that determine movement patterns and habitat selection. Whereas biotic factors include vegetation (type and productivity) and the presence of competitors, which limit the suitability of habitat patches (Chirima et al. 2013). As a result, parameter estimates were anticipated to reflect preference for habitat patches with optimum high-quality feed and avoidance of patches with high populations of competitors, given that the roan antelope is a specialist grazer that is sensitive to competition (Havemann et al. 2016; Knoop and Owen-Smith 2006).

Box 5.1: Overview of Umfurudzi Recreational Park. (Source: Muposhi et al. 2014, 2016a, b)

Umfurudzi Recreational Park in Zimbabwe, which has been restored through a private public partnership arrangement.

Location: latitude: 16.9257° S, Longitude: 31.8563° E, Mashonaland Central province, North-eastern, Zimbabwe, Africa.

Area: 760 km^2

Altitude: 740–1020 m

Climate: mean annual rainfall (650 mm), 90% of the precipitation falls between November and April. Temperatures range from 8 °C in winter to 41 °C in summer.

Habitat: Miombo woodland, rocky hills, riverine vegetation, and some vlei patches.

Late 1980s: Unprecedented decline in population sizes of most wildlife species with some occurring in isolated small herds prior to the onset of the restoration program in 2011.

Restoration efforts: Reintroduction of several wildlife species (e.g. African elephant (*Loxodonta africana*), Cape buffalo (*Syncerus caffer*), eland (*Taurotragus oryx*), wildebeest (*Connochaetes taurinus*), sable (*Hippotragus niger*), tsessebe (*Damaliscus lunatus lunatus*), waterbuck (*Kobus ellipsiprymnus*), and zebra (*Equus quagga*).

Roan antelope: little information was available on the existence of roan antelope in this park as there were no prior sightings. In 2016, a small remnant herd of six roan antelope (one adult female, one adult male, three female sub-adults, and one sub-adult male) was discovered in the northern section of the park.

5.4.1 Roan Antelope Relocation Data

To provide spatial information on this gregarious antelope species, an adult female roan antelope from the sole and remnant roan antelope herd in Umfurudzi Recreational Park was fitted with a GPS collar device, SAT2084 on 5 October 2016. This procedure was conducted by certified veterinarians and approved by the Zimbabwe Parks and Wildlife Management Authority (ZPWMA). The device was programmed to transmit satellite readings after every 5 hours, and these were downloaded for monitoring purposes. Some GPS fixes were missed after time-out due to poor orientation or bad sky view. Eventually, roan antelope location data ($n = 3925$) were collected between October 2016 to March 2019.

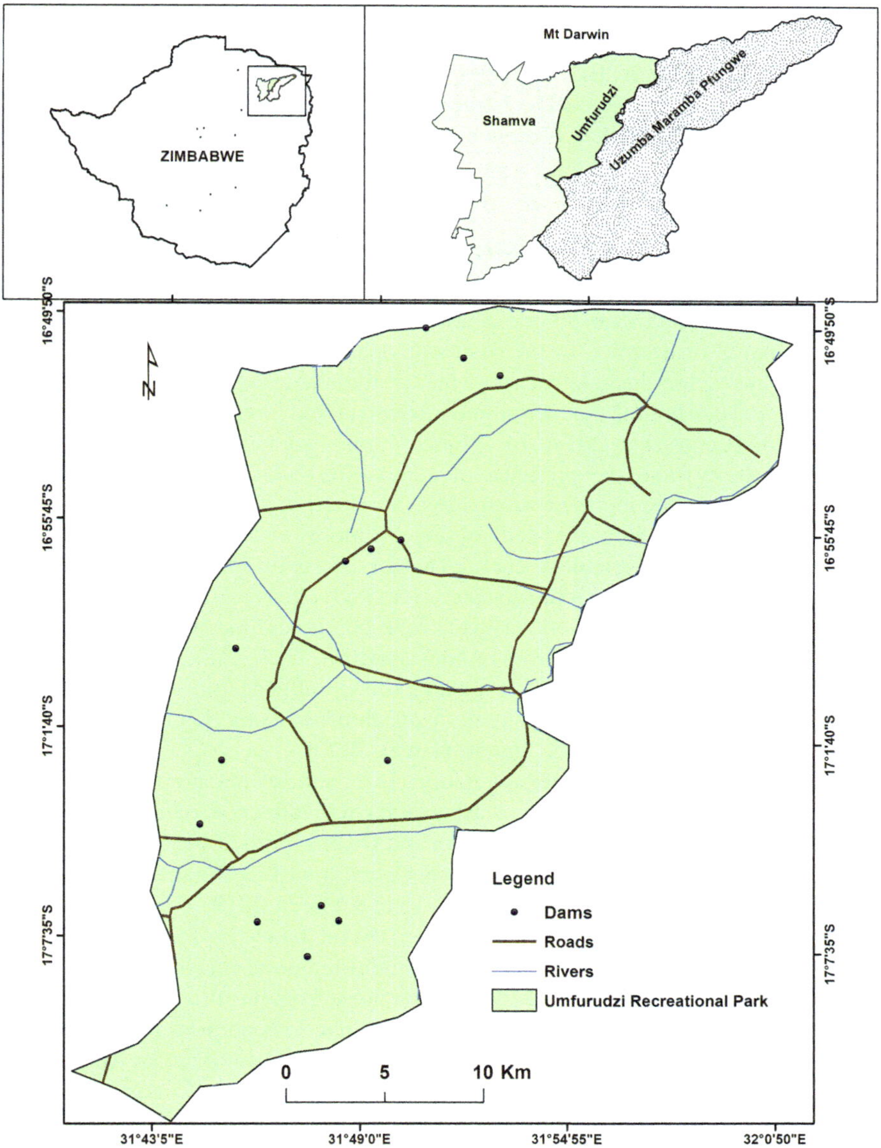

Fig. 5.1 Map of Umfurudzi Recreational Park and the surrounding areas in northeast Zimbabwe

5.4.2 Home Range Estimation

The home range of the roan antelope was estimated from the GPS telemetry data. The data collected between October 2016 to March 2019 had a fix rate of 95%. Such a fix rate ensured the data were suitable for use in estimating the home-range of roan antelope (Girard et al. 2002; Mitchell et al. 2019). Specifically, the Kernel Density Estimate (KDE) at 95% to compute the space utilized by the roan antelope between October 2016 to March 2019 was used. The seasonal changes in the space utilized for the wet and dry seasons. The home ranges were estimated using the adehabitatHR package in R software (Calenge and Fortmann-Roe 2015).

5.4.3 Overview of Environmental Variables

We used three broad factors, biotic, abiotic, and anthropogenic disturbances to predict the spatial distribution of the roan antelope in Umfurudzi Recreational Park. Here, abiotic factors include surface water availability, and slope as these impose physiological constraints that determine movement and foraging patterns, and habitat selection (Lewis et al. 2017a, b). Whereas biotic factors include vegetation (type and productivity) and the presence of competitors which limit suitable habitat patches and consequently their occupation (Mpakairi et al. 2017). Here, anthropogenic activities include roads, and mining (Muposhi et al. 2016a, b). A total of eleven environmental variables were used namely Normalized Difference Vegetation Index (NDVI), landcover, slope, distance from kudu, distance from zebra, distance from sable, distance from rivers, distance from boreholes, distance from dams, distance from roads, and distance from mining activities. These variables were included in the analysis based on their observed and known influence on the distribution of most ungulate species in heterogenous savannah ecosystems (Kimanzi et al. 2014; Chamaillé-Jammes et al. 2016; Tagwireyi et al. 2020).

NDVI was used as a proxy for vegetation cover, and an indicator for habitat quality (cover, space, and feed), a critical aspect known to influence roan antelope foraging ecology and habitat selection (Knoop and Owen-Smith 2006; Havemann et al. 2022). NDVI was derived from Landsat 8 Operational Line Imager (OLI) surface reflectance collected between October 2016 and March 2019. The surface reflectance images were composite, and we derived the median value for all these images in Google Earth Engine (GEE). The use of multi-temporal images and using their median value were used to reduce the effect of noise from the difference in illumination conditions, phenological changes and atmospheric conditions (Mpakairi et al. 2022). Landcover was derived from the Sentinel-2 landcover data for Africa available freely from https://2016africalandcover20m.esrin.esa.int/. Landcover was included in this study since ungulate movement is known to be driven by landcover dynamics (Kimanzi et al. 2014, 2015).

The slope was derived from the digital elevation model (DEM) provided by Shuttle Radar Topographic Mission (SRTM). The slope was included as an environmental variable because vegetation patterns and water availability are closely aligned to slope characteristics (Albaba 2014). South-facing slopes have been observed to be more xeric (hot and drier) which creates a microclimate that promotes high plant diversity (Balfour and Bond 1993). However, most species have been observed to avoid areas with steep slopes owing to the high energy costs of utilizing these areas (Kimanzi et al. 2014, 2015). The slope was calculated in ArcMap 10.1 using the spatial analyst package (Daniels 2006).

Distance from rivers, dams, and boreholes was included in the analysis as roan antelope is water dependent and prefers foraging in habitat patches in proximity to water sources (Kimanzi et al. 2014; Smit et al. 2007). These metrics can be used to measure vegetation greenness and water availability which are key elements in the distribution of ungulates in savannah ecosystems (Chapungu and Nhamo 2016; Szabó et al. 2020). In addition, roan antelope movement and space use has been observed to follow vegetation and water (Wilson and Hirst 1977; Havemann et al. 2016, 2022). Rivers tend to create microhabitats which are often preferred by most species during the dry season even when they have dried up, as such, rivers were incorporated as an environmental predictor. This preference is mostly due to the availability of palatable forage during drier months of the year compared to other habitat patches (Muposhi et al. 2014). The distance from rivers, dams, and borehole data was digitized from very high resolution on the Google Earth platform (www.googleearth.com), and their distances were calculated using the Euclidean distance function in ArcMap 10.1 (ArcGIS 2021).

Interspecific competition is a key driver in the distribution of most ungulates (Mpakairi et al. 2017). Data on the occurrence of possible roan antelope competitors (i.e. kudu, sable, and zebra) were obtained from the systematic monitoring of annual data on species occurrence at Umfurudzi Recreational Park for the period under study. The Euclidean distance from kudu, sable, and zebra were included as a proxy of the interspecific competition that exists between roan antelope and other species. This was premised on the understanding that species will compete for forage and other resources if they occupy the same space as compared to when they are further apart (Harrington et al. 1999; Havemann et al. 2022). The distance from kudu, sable, and zebra were calculated in ArcGIS using the Euclidean distance of each species from its sighting. The Euclidean distance from roads was included to show the likely effect of linear infrastructure and developments that are commonly found along roads in protected areas (Kimanzi et al. 2015). Most species, including ungulates, have been observed to have a high flight initiation distance in areas close to roads as compared to areas further away from the road. The Euclidean distance from roads was calculated in the ArcGIS environment using the road data digitized from high-resolution imagery available on the Google Earth platform.

Mining was included as an environmental factor because the associated activities can fragment natural habitats and isolate populations (Sonter et al. 2018). This can make it difficult for roan antelope to transverse landscapes in search of feed and water (Sonter et al. 2018). Mining activities in Umfurudzi Recreational Park are widespread. Euclidean distance from mines was included in the analysis to understand the effect of mining activities on roan distribution. The data for all mining activities were collected in the field using a handheld GPS with an error of ±4 m. Given that the environmental variables had varying spatial resolution, the variable layers were projected and resampled to the same spatial resolution before their use in the model (Fig. 5.2).

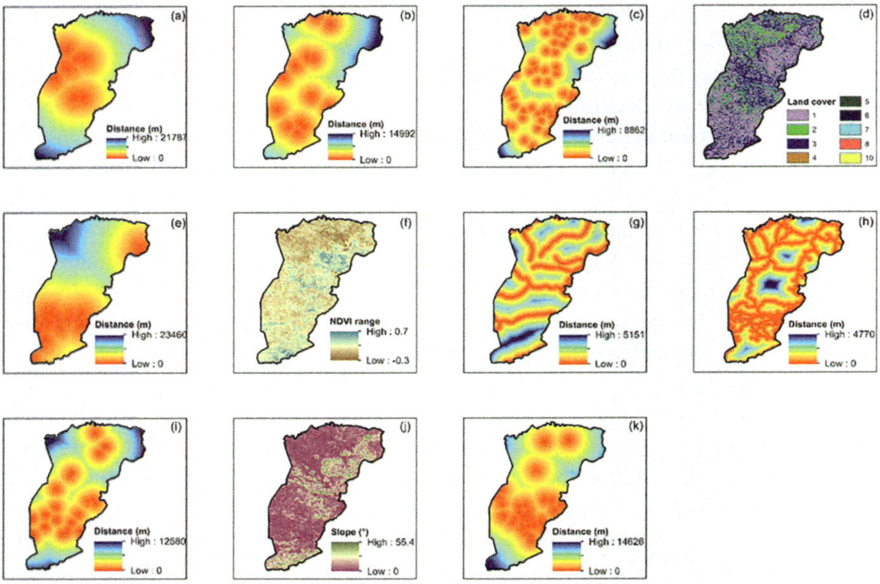

Fig. 5.2 The environmental variables that were included in our modelling process and include (**a**) distance from boreholes, (**b**) distance from dams, (**c**) distance from Kudu, (**d**) Land cover: *1, 2, 3, 4, 5, 6, 7, 8, 10 denotes tree cover, shrub cover, grassland, cleared patches, riparian vegetation, sparse vegetation, bare land, built-up, and water*, (**e**) distance from mining activities, (**f**) Normalized Difference Vegetation Index (NDVI), (**g**) distance from the road, (**h**) distance from rivers, (**i**) distance from sable, (**j**) slope, and (**k**) distance from zebra

5.4.4 Multicollinearity of Environmental Variables

The eleven environmental variables were assessed for multicollinearity to mitigate model overfitting using the variance inflation factor (VIF). Considering multicollinearity assessment, when confronted with a VIF surpassing the critical threshold of 10 (Mpakairi et al. 2017; Naimi and Araújo 2016), it is imperative to scrutinize the variables with high VIF values and explore potential remedies such as variable transformation or exclusion. By addressing the issue of excessive VIF, the reliability and validity of the findings are enhanced while ensuring the integrity of the statistical inferences (Chattefuee and Hadi 2006). In this case study, all the environmental variables had a VIF of less than 5 and were retained for the model (Table 5.2).

5.4.5 Modelling Framework

To model the potential distribution of roan antelope in Umfurudzi Recreational Park, an ensemble model based on two classification and regression algorithms that were the Gradient Boosting Model (GBM) and Random Forest (RF) algorithm was

Table 5.2 The variance inflation factor (VIF) of the environmental variables used for modelling

Variable	VIF score
NDVI	1.84
Slope	1.38
Landcover	1.14
Distance from mining activities (m)	1.53
Distance from rivers (m)	2.65
Distance from boreholes (m)	2.53
Distance from dams (m)	3.59
Distance from roads (m)	1.11
Distance from sable (m)	4.84
Distance from zebra (m)	4.78
Distance from kudu (m)	4.70

used. These models were selected in this study based on their robustness in modelling species distribution since they can handle non-linear data while avoiding overfitting (Dube et al. 2015). The ensemble model built using the GBM and RF was based on all roan antelope predictions that had an area under the receiver under the operating curve (ROC) above 0.7. Models that have a ROC above 0.5 are more than random (Gwitira et al. 2015). The ensemble model was used since it selects parts of the candidate models (i.e. RF and GBM) that are highly accurate to build a highly predictive model (Thuiller et al. 2009). The GBM, RF, and ensemble models were calibrated using 80% ($n = 3140$) of the roan antelope location data and the remaining 20% ($n = 785$) was used for model validation using ROC and the True Skill Statistic (TSS) (Chattefuee and Hadi 2006). The RF, GBM, and ensemble models were run in R using the biomod2 package (Thuiller et al. 2009).

5.4.6 Roan Antelope Home Ranges

Based on the relocations data for the study period, roan antelope had an overall home range of 119.6 km^2 (Fig. 5.3). It was observed that roan antelope traversed a bigger area during the dry season (118.8 km^2) compared to the wet season (85.9 km^2) during the study period (Fig. 5.4). The habitat patches utilized by the roan antelope were mainly inside the park, however, some habitat patches outside of the park, within a 2 km buffer from the park boundary were also traversed. The average space utilized outside the park during the collaring period was 4.3 km^2. However, this varied with season, as it was mainly high in the dry season (8 km^2) when compared to the wet season (6 km^2).

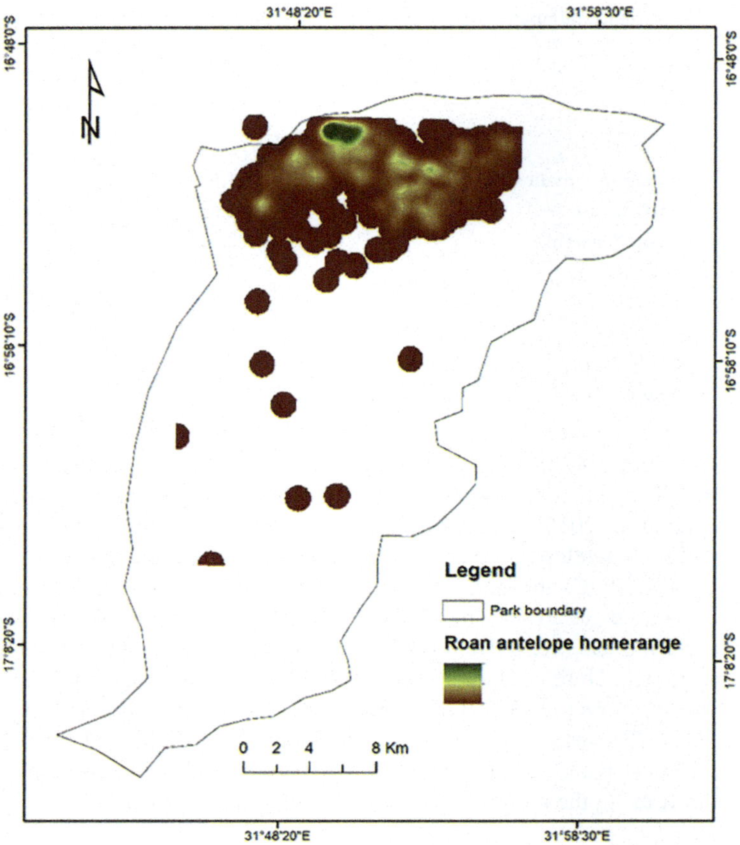

Fig. 5.3 The overall home range of roan antelope for the period 2016–2019 in Umfurudzi Recreational Park, Zimbabwe

5.4.7 Modelling Performance and Variable Importance

The Random Forest and Gradient Boosting Models that were used to build the ensemble model were able to predict the roan antelope distribution with high accuracy (TSS > 0.78 and AUC > 0.94). The final ensemble model used in this study was able to predict 98% (AUC = 0.99 and TSS = 0.89) of the variability in the potential distribution of roan antelope in Umfurudzi Recreational Park.

The variability in the distribution of the roan antelope was mainly driven by distance from mining activities, and their proximity to Kudu, boreholes, and Zebra. Proximity to sable antelope had the least effect on roan antelope distribution along with some of the biophysical variables such as NDVI, slope, and landcover (Fig. 5.5).

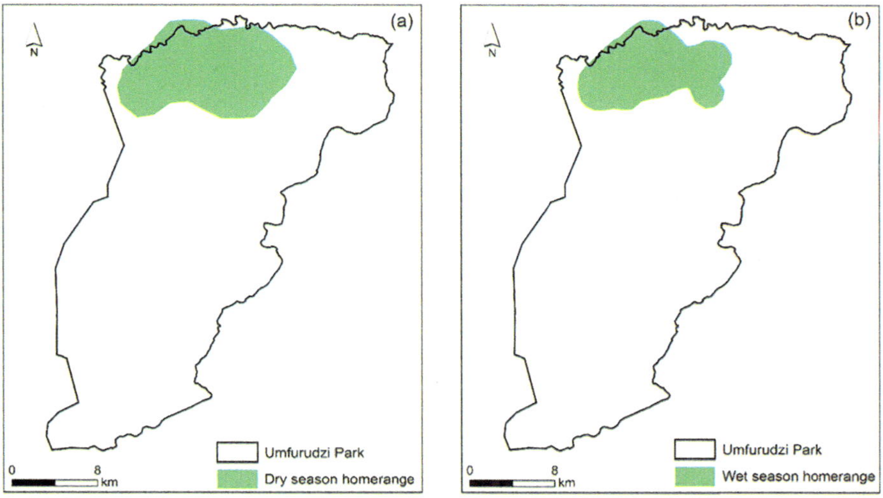

Fig. 5.4 The home range of roan antelope for (**a**) the dry and (**b**) wet season period from 2016–2019 in Umfurudzi Recreational Park, Zimbabwe

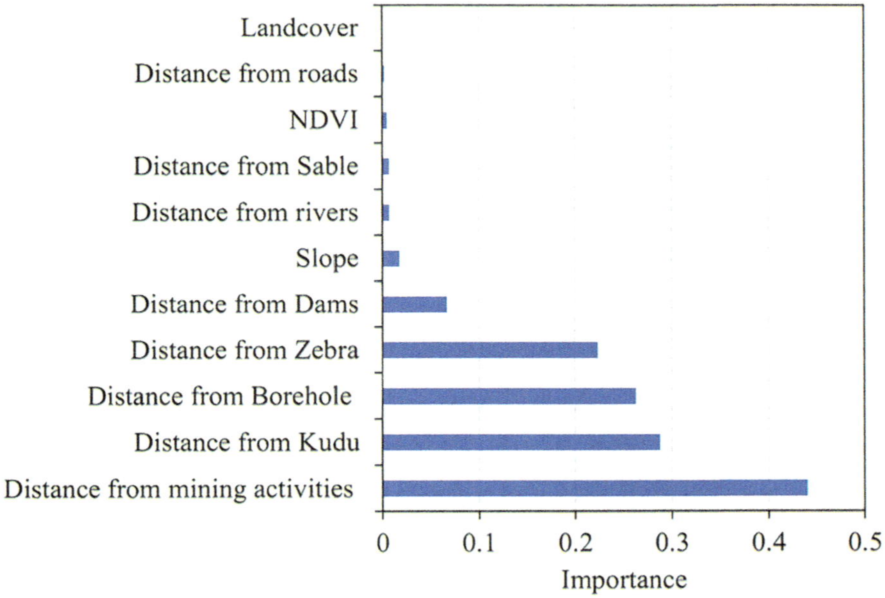

Fig. 5.5 The variable importance of the ensemble model predicting the distribution of roan antelope in Umfurudzi Recreational Park

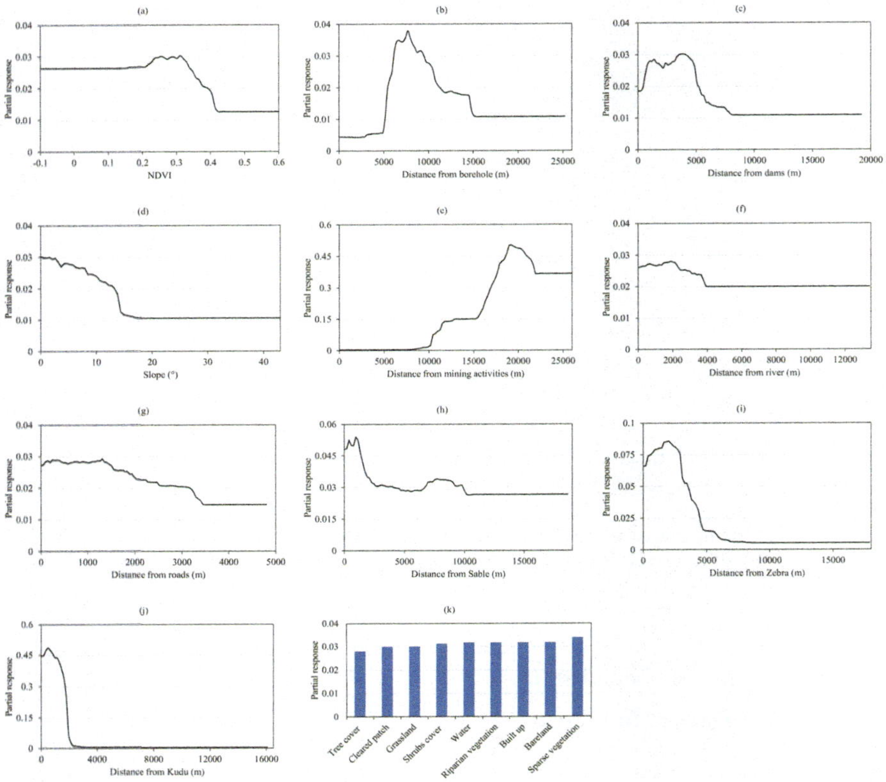

Fig. 5.6 The partial response of roan antelope spatial distribution to (**a**) Normalised Difference Vegetation Index (NDVI), (**b**) Normalised Difference Water Index (NDWI), (**c**) slope, (**d**) distance from kudu, (**e**) distance from zebra, (**f**) distance from sable, (**g**) distance from roads, (**h**) distance from rivers, and (**i**) landcover

Roan antelope preferred areas that had sparse vegetation, flat (slope <10°), close to kudu (0–2 km), Zebra (0–8 km), roads (0–3.3 km), and rivers (>4 km). However, roan antelope selectively utilized areas that were 6.5–9 km away from sable antelope. Increasing NDVI, Normalised Difference Water Index (NDWI), and distance from roads did not influence roan distribution (Fig. 5.6).

5.4.8 Probabilistic Roan Antelope Distribution

The probabilistic distribution of roan antelope varied across Umfurudzi Recreational Park. A high probability of roan antelope presence was noted in the northern parts of the park bordering Shamva rural communities and partly Chesa small-scale farming communities. However, the central parts of the Umfurudzi Recreational Park

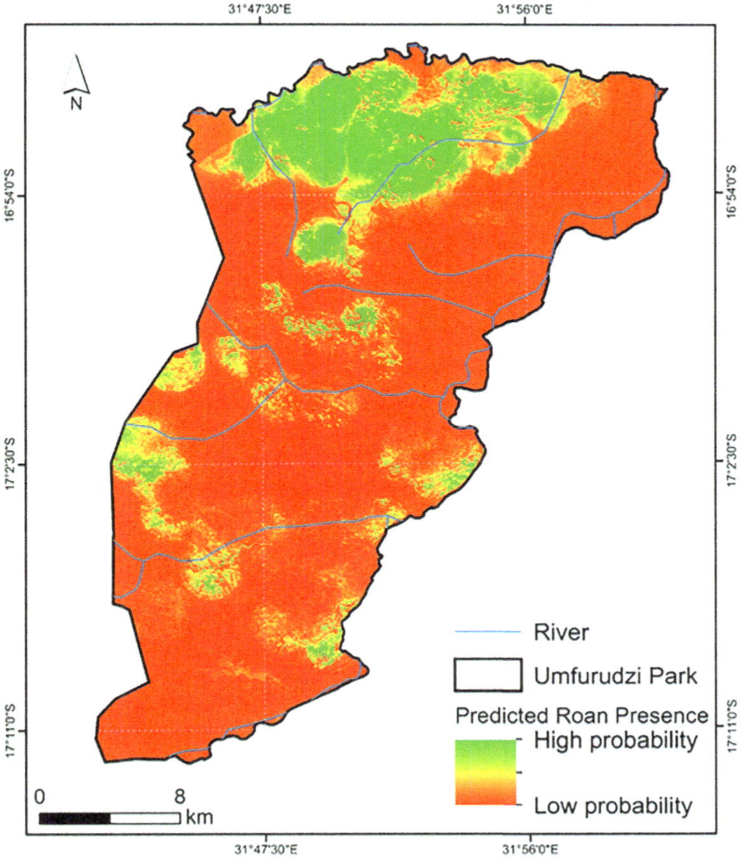

Fig. 5.7 Probabilistic distribution of the roan antelope in Umfurudzi Recreational Park, Zimbabwe based on Random Forest and Gradient Boosting Model

had the least probability of roan presence as modelled using the eleven environmental variables (Fig. 5.7).

5.5 Discussion

The analysis of telemetry data revealed distinct seasonal variations in the home ranges of roan antelope within Umfurudzi Recreational Park, which aligns with findings from other landscapes (Perrin and Taolo, 1998; Kimanzi and Wanyingi 2014). In this case study, several environmental factors that could influence the space-use patterns and habitat selection of the roan antelope were examined. The results indicated that approximately 25% of Umfurudzi Recreational Park was predicted to provide suitable habitat for these antelopes. However, the impact of predation, including the presence of hyenas, leopards, and lions, on shaping the habitat selection or preference of roan antelope remains unresolved as it was not specifically investigated in this study.

5.5.1 Roan Antelope Home Range Sizes

The roan antelope home range sizes concurred with those observed in other savannah ecosystems where the individuals traverse more areas during the dry season compared to the wet season (Perrin and Taolo 1998). The wet and dry seasons can have a significant impact on the home range sizes of herbivores (Ofstad et al. 2016; Dakwa et al. 2020). During the wet season, when water and vegetation are abundant, herbivores may have smaller home ranges as they can find sufficient resources within a smaller area (Dailey et al. 2007; Ofstad et al. 2019). In contrast, during the dry season, when resources are scarce and concentrated in certain areas, herbivores may need to cover a larger area to find enough feed and water to survive (Valeix 2011; Van Beest et al. 2011; Viana et al. 2018). Similarly, most herbivores are known to traverse longer distances during the dry season to find water and new grazing areas. These movements are heavily influenced by the availability of water sources (Kernohan et al. 2001; Naidoo et al. 2012).

Home range sizes can also vary greatly depending on the type of habitat that the individual species occupy (Ofstad et al. 2019). For example, herbivores in open grasslands may have larger home ranges than those in forested areas, as they need to cover more ground to find feed and avoid predators (Ofstad et al. 2016; Van Moorter et al. 2016). Roan antelope are known to prefer occupying open wooded and open grassland in most savannah ecosystems (Wilson and Hirst 1977; Knoop and Owen-Smith 2006; Kimanzi et al. 2014; Havemann et al. 2022). The northern sections of Umfurudzi Recreational Park have extensive open woodland and grassland patches which seem to be preferred the most by the roan antelope.

The home range sizes of herbivores can also be influenced by other factors, such as predation risk (McLoughlin and Owen-Smith 2003) and human disturbance (Van Beest et al. 2011; Kimanzi and Wanyingi 2014; Benitez et al. 2022). In areas where there is a high risk of predation, herbivores may have larger home ranges to avoid areas where predators are more likely to be found (McLoughlin and Owen-Smith 2003; Huggler et al. 2022). The main source of human disturbances in Umfurudzi Recreational Park is mining and to some extent subsistence hunting. We argue that in areas where there is a high level of anthropogenic disturbance such as human presence and mining activities, herbivore species (i.e. roan antelope) may have smaller home ranges as they are forced to stay closer to their preferred grazing areas to avoid human activity (Kimanzi et al. 2015; Havemann et al. 2016).

5.5.2 Roan Antelope Habitat Selection

The spatial distribution of herbivores is largely determined by the availability and quality of their preferred habitat (Muposhi et al. 2016a, b; Bastille-Rousseau et al. 2018; Avgar et al. 2020). In this chapter, Random Forest and Gradient Boosting Models were used to explore the various factors that could influence habitat selection by a specialist grazer, roan antelope. The probabilistic distribution model for the roan antelope was mainly driven by distance from mining activities, their proximity to kudu, boreholes (artificial water points), and zebra. However, proximity to sable antelope as well as characteristics associated with the biophysical environment such as NDVI (i.e. habitat quality), slope, and land cover (i.e. habitat type) had less of an impact on habitat suitability.

Although feed availability and quality are the most critical factors that influence habitat selection by herbivores, there was a weak association with habitat quality (i.e. NDVI) in this study. These findings are contrary to other studies where high-quality forage, which provides the necessary nutrients for growth, reproduction, and survival influenced habitat selection (Kimanzi et al. 2014). In some cases, NDVI has been found to provide a useful means to assess forage quality where crude protein is a limiting resource (Ryan et al. 2012). However, if NDVI measurements are too coarse to detect microhabitats producing high-quality forage, or if the diet is heavily influenced by forage items that are weakly connected with landscape greenness, it may not be a reliable proxy for diet quality (Creech et al. 2016; Gautam et al. 2019). As a result, while NDVI can provide a sense of production, including the greenness of habitat patches, it has limits in predicting dry grassland habitat quality (Weber et al. 2018).

Roan antelope chose and preferred habitat patches away from mining activities, which were the primary source of disturbance in Umfurudzi Recreational Park. These findings support the hypothesis that roan antelope are sensitive to disturbances and tend to avoid such habitat areas to optimize their fitness (Storch et al. 2022). This supports the idea that anthropogenic disturbances (i.e. mining activities) severely limit the suitability of habitat patches for wild herbivores in natural ecosystems (Butchart et al. 2010; Sonter et al. 2018). Noise from mining operations is perceived as a threat, and this would influence how the roan antelope choose habitat patches in the park. However, anthropogenic disruptions can cause ecological traps and impair the fitness of wild herbivores (Battin 2004; Hale and Swearer 2016). Although the effects of poaching or unlawful harvesting were not investigated in this study, this factor may alter roan antelope spatial distribution in the park, as has been documented in South Africa (le Roex et al. 2020).

Humans can be considered as prevalent apex predators (Lewis et al. 2017a, b) and as such present a significant source of risk for wild herbivores in natural ecosystems (Gaynor et al. 2022). It is argued that anthropogenic disturbances exacerbate the roan antelope's landscape of fear, thus shaping the way habitat patches are selected and used in the park (Laundré et al. 2010; Davies et al. 2021; Moleón and Sánchez-Zapata 2023). To minimize the risk of mortality or displacement, herbivores may avoid areas with high levels of disturbance and choose habitats with low levels of disturbance regardless of their suitability (Ciuti et al. 2012). Anthropogenic disturbances must be considered in purpose-driven conservation planning and restoration attempts in human-mediated ecosystems because they alter the suitability of habitat patches for wildlife species (Smith et al. 2019; Fremout et al. 2020).

Although predation was not considered an environmental variable, Umfurudzi Recreational Park does feature resident predators such as spotted hyenas (*Crocuta crocuta*), which offer a risk factor and may influence roan antelope habitat choices as documented in other studies (e.g. McLoughlin and Owen-Smith 2003; Crosmary et al. 2012). Predators can lower herbivore survival and reproductive success; hence, it is hypothesized that roan antelope may have chosen habitats that provide cover, concealment, and escape routes to lessen the risk of predation (Crosmary et al. 2012). As previously noted, the roan antelope may have chosen open habitat patches with optimum visibility to detect prospective predators and escape swiftly (Sirot and Pays 2011).

Herbivores may avoid areas with high competition or select habitats with a low density of conspecifics to reduce the competition for resources (Mpakairi et al. 2017). In this study, roan antelope chose and used habitat areas that were also used by potential competitors like kudu and zebra. This was inconsistent with the view that competition for resources, such as feed and water tend to influence the habitat selection and suitability for most herbivores (Brazeau and Schamp 2019). However, it is hypothesized that spatially nested resource partitioning could contribute to coexistence of the roan antelope with other herbivores through niche separation (Chirima et al. 2013; Owen-Smith et al. 2015).

Herbivore habitat selection and suitability have been demonstrated to be highly influenced by habitat heterogeneity, or diversity in habitat traits (Muposhi et al. 2016a, b). Herbivores, according to research, prefer environments with a high degree of heterogeneity, which provides a diverse range of resources and microhabitats (Mason and Fortin 2017). It is postulated that changing habitat patches can provide herbivores with a variety of feed sources, such as diverse plant species or plant sections, to help them achieve their nutritional needs (Muposhi et al. 2016a, b). A mosaic of microhabitats can provide diverse circumstances for herbivores to suit their physiological requirements (Massé and Côté 2012). Some herbivores, for example, may require cooler, shady locations to regulate their body temperature, while others may require open, high-visibility areas to detect predators (Ferry et al. 2020).

Table 5.3 Summary of factors influencing the ranging patterns and habitat selection by roan antelope in human-mediated ecosystems

Factor	Notes	References
Habitat	Roan antelopes prefer open grasslands and savannas with scattered trees and shrubsRoan antelope seldom select dense forests and patches with little or no water as they reduce their fitness	Avgar et al. (2020), Battin (2004), Dakwa et al. (2020), Kimanzi et al. (2014), Ofstad et al. (2016), Püttker et al. (2020)
Water availability	Roan antelopes require access to water sources such as rivers, streams, and waterholes. The distribution of roan antelope is often influenced by the availability of these water sources	Redfern et al. (2003, 2005), Havemann et al. (2022), Knoop and Owen-Smith (2006), Valls-Fox et al. (2018)
Competition	Roan antelopes may be outcompeted by other herbivores such as zebras, wildebeest, and giraffes for feed and water resources	Harrington et al. (1999), Havemann et al. (2022)
Predators	The distribution of roan antelope is influenced by the presence of predators such as lions, leopards, and hyenas. They tend to avoid areas where predators are abundant as this shapes their landscape of fear	Battin (2004), Laundré et al. (2010), McLoughlin and Owen-Smith (2003)
Anthropogenic disturbances	Human activities such as hunting, poaching, habitat destruction, fire, and mining activities can impact the spatial distribution of roan antelope in natural ecosystems. Roan antelope are also sensitive to disturbances such as noise and vehicle traffic	Butchart et al. (2010), Forssman and Root-Bernstein (2018), Kimanzi et al. (2015), Kimanzi and Wanyingi (2014), Kuijper et al. (2015), Sonter et al. (2018)
Grazing pressure	The distribution of roan antelope may be influenced by the intensity of grazing pressure from other herbivores and livestock	Harrington et al. (1999), Knoop and Owen-Smith (2006), Wilson and Hirst (1977)

5.6 Ramifications for Conservation and Management

5.6.1 Space-Use Patterns by Herbivores in Human-Mediated Ecosystems

Overall, the availability of feed, water, and other resources had a significant impact on how herbivores use their space (i.e. their home ranges), which can change significantly across seasons (Redfern et al. 2003, 2005; Havemann et al. 2022). Understanding the factors that influence home range sizes is crucial for effective conservation efforts and the successful reinforcement of small populations through reintroductions (Perrin and Taolo 1998). This assists in identifying critical areas necessary for the survival and persistence of the species, guiding management decisions focused on habitat protection and preservation (Gutzwiller et al. 2017; Mondal and Palit 2021). Overall, habitat heterogeneity plays a big part in the distribution and abundance of herbivores, and knowing how it affects these species can help conservation and management efforts to keep their habitats suitable for animals (Lindenmayer et al. 2008; Thurfjell et al. 2014; Martínez-Ramos et al. 2016).

This chapter clearly shows that habitat selection is a complex process impacted by a variety of factors. Some of the essential elements that govern the spatial distribution of herbivores include habitat availability and quality, anthropogenic disturbances (such as mining activities of various types and illegal harvesting), competition, and predation risk (Table 5.3).

5.6.2 *Incorporating Purpose in Restoration Planning and Implementation*

To achieve meaningful conservation outcomes, the purpose must be incorporated into wildlife restoration strategy and implementation (McIntosh et al. 2017; Schwartz et al. 2018). Purpose-driven approaches provide a strategic framework that aligns restoration efforts with broader conservation goals, societal values, and ecosystem health (White and Tuttle 2013; Cimon-Morin et al. 2014). This chapter argues that by defining clear goals and considering ecological, social, and economic factors, targeted restoration ensures that action is guided by a comprehensive understanding of ecosystem dynamics and stakeholder needs.

There is a need to emphasize the importance of engagement with local communities, stakeholders, and indigenous knowledge, promoting inclusivity and long-term sustainability. Paying attention to the important considerations for conservation project planning should enable practitioners to work within any framework and adapt tools to their individual management situation (Campellone et al. 2018). To maximize conservation impact in the face of anthropogenic challenges and foster resilient ecosystems for future generations, purpose must be incorporated into the systematic planning and execution of wildlife restoration projects (McIntosh et al. 2017; Smith et al. 2019).

5.7 Conclusion

Home range sizes and habitat selection are crucial factors that need to be considered when planning reintroduction programs for herbivores. These initiatives seek to restore animal populations in regions where they have been extirpated or where populations have sharply declined. It's crucial to comprehend the extent of herbivores' home ranges since it can help determine how many animals can be successfully reintroduced into a given area. The choice of habitat is particularly important for reintroduction projects since it guarantees that the wildlife species are released to suitable habitats with enough resources. Suitable habitats can provide a range of feed sources, cover, and shelter that can support the animals' physiological needs.

The risk of predation, disease, and human–wildlife conflict linked with reintroduction projects can also be decreased by choosing appropriate habitats and home range sizes. Additionally, it can support the establishment of herbivore populations in the wild. Because of this, the success of herbivore reintroduction initiatives depends on careful consideration of home range sizes and habitat selection, which must be based on in-depth research and monitoring of the target species. Making more data-driven judgments about conservation and sustainable development in the Anthropocene will require an understanding of the factors that affect the space-use patterns of important wildlife species in protected areas.

Integrating purpose into animal restoration planning and implementation within a recreational park with mining activity presents a unique challenge. It is necessary to strike a balance between the ecological goals of wildlife restoration and the economic demands of mining operations, as well as the recreational needs of park visitors. Purpose-driven approaches in this context involve carefully considering the impacts of mining activities on wildlife habitats and developing mitigation strategies.

Fostering collaboration among stakeholders, such as mining firms, park management, local communities, and conservation groups, is essential for integrating different viewpoints. This would guarantee the creation of a thorough restoration plan that considers ecological and socio-economical concerns. We may work toward a harmonious coexistence between mining operations, wildlife conservation, and recreational park use by adding purpose.

Acknowledgements We are grateful for the chance to work on this project and the logistical help received from the Umfurudzi Recreational Park management committee, Pioneer Africa, and Umfurudzi Park workers, in particular the former Park Manager Jan Stander.

References

Aarts G, MacKenzie M, McConnell B, Fedak M, Matthiopoulos J (2008) Estimating space-use and habitat preference from wildlife telemetry data. Ecography 31:140–160

Agha M, Batter T, Bolas EC, Collins AC, Gomes da Rocha D, Monteza-Moreno CM, Preckler-Quisquater S, Sollmann R (2018) A review of wildlife camera trapping trends across Africa. Afr J Ecol 56:694–701

Albaba I (2014) The effects of slope orientations on vegetation characteristics of Wadi Alquf forest reserve (WAFR) West Bank-Palestine. Int J Agric Soil Sci 2:118–125

ArcGIS E (2021) 10.1 Redlands, Califonia: ESRI

Avgar T, Betini GS, Fryxell JM (2020) Habitat selection patterns are density dependent under the ideal free distribution. J Anim Ecol 89:2777–2787

Balfour DA, Bond WJ (1993) Factors limiting climber distribution and abundance in a Southern African Forest. J Ecol 81:93–100

Bartlett LJ, Newbold T, Purves DW, Tittensor DP, Harfoot MB (2016) Synergistic impacts of habitat loss and fragmentation on model ecosystems. Proc R Soc Lond B Biol Sci 283:20161027

Bastille-Rousseau G, Murray DL, Schaefer JA et al (2018) Spatial scales of habitat selection decisions: implications for telemetry-based movement modelling. Ecography 41:437–443

Battin J (2004) When good animals love bad habitats: ecological traps and the conservation of animal populations. Conserv Biol 18:1482–1491

Bearhop S, Adams CE, Waldron S et al (2004) Determining trophic niche width: a novel approach using stable isotope analysis. J Anim Ecol 73:1007–1012

Benitez L, Kilian JW, Wittemyer G et al (2022) Precipitation, vegetation productivity, and human impacts control home range size of elephants in dryland systems in northern Namibia. Ecol Evol 12:e9288

Bennett NJ, Whitty TS, Finkbeiner E et al (2018) Environmental stewardship: a conceptual review and analytical framework. Environ Manag 61:597–614

Benoit L, Hewison AJM, Coulon A et al (2020) Accelerating across the landscape: the energetic costs of natal dispersal in a large herbivore. J Anim Ecol 89:173–185

Berger-Tal O, Lahoz-Monfort JJ (2018) Conservation technology: the next generation. Conserv Lett 11:e12458

Bohmann K, Schnell IB, Gilbert MTP (2013) When bugs reveal biodiversity. Mol Ecol 22:909–911

Bonnet OJ, Meuret M, Tischler MR et al (2015) Continuous bite monitoring: a method to assess the foraging dynamics of herbivores in natural grazing conditions. Anim Prod Sci 55:339–349

Börger L, Dalziel BD, Fryxell JM (2008) Are there general mechanisms of animal home range behaviour? A review and prospects for future research. Ecol Lett 11:637–650

Bortoluzzi J, Darby J, Jackson AL (2021) Technological advances in stable isotope analysis provide even more insights into animal ecology. J Anim Ecol 90:1394–1397

Boyce MS, McDonald LL (1999) Relating populations to habitats using resource selection functions. Trends Ecol Evol 14:268–272

Boyce MS, Vernier PR, Nielsen SE, Schmiegelow FK (2002) Evaluating resource selection functions. Ecol Model 157:281–300

Brazeau HA, Schamp BS (2019) Examining the link between competition and negative co-occurrence patterns. Oikos 128:1358–1366

Brodie JF, Giordano AJ, Zipkin EF, Bernard H, Mohd-Azlan J, Ambu L (2015) Correlation and persistence of hunting and logging impacts on tropical rainforest mammals. Conserv Biol 29:110–121

Bukovinszky T, Potting RP, Clough Y, Van Lenteren JC, Vet LE (2005) The role of pre-and post-alighting detection mechanisms in the responses to patch size by specialist herbivores. Oikos 109:435–446

Bull JW, Strange N, Smith RJ et al (2021) Reconciling multiple counterfactuals when evaluating biodiversity conservation impact in social-ecological systems. Conserv Biol 35:510–521

Butchart SH, Walpole M, Collen B, Van Strien A, Scharlemann JP, Almond RE, Baillie JE, Bomhard B, Brown C, Bruno J, Carpenter KE (2010) Global biodiversity: indicators of recent declines. Science 328:1164–1168

Butman CA (1987) Larval settlement of soft-sediment invertebrates: the spatial scales of pattern explained by active habitat selection and the emerging role of hydrodynamical processes. Oceanogr Mar Biol 25:113–165

Calenge C, Fortmann-Roe S (2015) adehabitatHR: Home Range Estimation. Https://Cran.r--Project.Org/Web/Packages/AdehabitatHR/Index.Html

Campellone RM, Chouinard KM, Fisichelli NA et al (2018) The iCASS platform: nine principles for landscape conservation design. Landsc Urban Plan 176:64–74

Caro T, Darwin J, Forrester T et al (2012) Conservation in the Anthropocene. Conserv Biol 26:109–113

Chamaillé-Jammes S, Charbonnel A, Dray S et al (2016) Spatial distribution of a large herbivore community at waterholes: an assessment of its stability over years in Hwange National Park, Zimbabwe. PLoS One 11:e0153639

Chapungu L, Nhamo L (2016) An assessment of the impact of climate change on plant species richness through an analysis of the Normalised Difference Water Index (NDWI) in Mutirikwi Sub-catchment, Zimbabwe. S Afr J Geomat 5:244–268

Chattefuee S, Hadi AS (2006) Regression analysis by example. Wiley

Chirima GJ, Owen-Smith N, Erasmus BFN et al (2013) Distributional niche of relatively rare sable antelope in a South African savanna: habitat versus biotic relationships. Ecography 36:68–79

Cimon-Morin J, Darveau M, Poulin M (2014) Towards systematic conservation planning adapted to the local flow of ecosystem services. Glob Ecol Conserv 2:11–23

Ciuti S, Northrup JM, Muhly TB et al (2012) Effects of humans on behaviour of wildlife exceed those of natural predators in a landscape of fear. PLoS One 7:e50611

Codron D, Codron J, Sponheimer M et al (2016) Within-population isotopic niche variability in savanna mammals: disparity between carnivores and herbivores. Front Ecol Evol 4:15

Cooke SJ (2008) Biotelemetry and biologging in endangered species research and animal conservation: relevance to regional, national, and IUCN Red List threat assessments. Endanger Species Res 4:165–185

Creech TG, Epps CW, Monello RJ et al (2016) Predicting diet quality and genetic diversity of a desert-adapted ungulate with NDVI. J Arid Environ 127:160–170

Crosmary WG, Makumbe P, Côté SD et al (2012) Vulnerability to predation and water constraints limit behavioural adjustments of ungulates in response to hunting risk. Anim Behav 83:1367–1376

Dailey M, Gitelman AI, Ramsey FL et al (2007) Habitat selection models to account for seasonal persistence in radio telemetry data. Environ Ecol Stat 14:55–68

Dakwa KB, Cuthill IC, Harris S (2020) Seasonal variation in the selection and use of habitats by large herbivores at Mole National Park, Ghana. West Afr J Appl Ecol 28:132–139

Daniels WD (2006) On-site verification of slope: spatial analyst curvature function. In: ASRI international user conference

Darimont CT, Fox CH, Bryan HM et al (2015) The unique ecology of human predators. Science 349:858–860

Davies AB, Cromsigt JPGM, Tambling CJ et al (2021) Environmental controls on African herbivore responses to landscapes of fear. Oikos 130:171–186

de Boer WF, Van Oort JWA, Grover M et al (2015) Elephant-mediated habitat modifications and changes in herbivore species assemblages in Sabi Sand, South Africa. Eur J Wildl Res 61:491–503

Di Marco M, Venter O, Possingham HP et al (2018) Changes in human footprint drive changes in species extinction risk. Nat Commun 9:4621

Drira S, Ben Rais Lasram F, Ben Rejeb Jenhani A et al (2019) Species-area uncertainties impact the setting of habitat conservation targets and propagate across conservation solutions. Biol Conserv 235:279–289

Dube T, Mutanga O, Abdel-Rahman EM et al (2015) Predicting Eucalyptus spp. stand volume in Zululand, South Africa: an analysis using a stochastic gradient boosting regression ensemble with multi-source data sets. Int J Remote Sens 36:3751–3772

Dueñas MA, Hemming DJ, Roberts A et al (2021) The threat of invasive species to IUCN-listed critically endangered species: a systematic review. Glob Ecol Conserv 26:e01476

Elith J, Phillips SJ, Hastie T et al (2011) A statistical explanation of MaxEnt for ecologists. Divers Distrib 17:43–57

Ferry N, Cordonnier M, Hulot FD et al (2020) Heterogeneity of water physico-chemical characteristics in artificially pumped waterholes: do African herbivores drink at the same locations and does it lead to interference competition? J Arid Environ 173:104014

Forssman N, Root-Bernstein M (2018) Landscapes of anticipation of the other: ethno-ethology in a deer hunting landscape. J Ethnobiol 38:71–87

Fowler BT, Rieder E, Jackson S et al (2020) Harnessing a multifaceted stewardship framework: a bare necessity for parks and protected areas. Parks Stewardship Forum 36(3):494

Fremout T, Thomas E, Gaisberger H et al (2020) Mapping tree species vulnerability to multiple threats as a guide to restoration and conservation of tropical dry forests. Glob Chang Biol 26:3552–3568

Fu M, Pan H, Song X et al (2022) Back-and-forth shifts in habitat selection by giant pandas over the past two decades in the Daxiangling Mountains, southwestern China. J Nat Conserv 66:126129

Garnick S, Barboza PS, Walker JW (2018) Assessment of animal-based methods used for estimating and monitoring rangeland herbivore diet composition. Rangel Ecol Manag 71:449–457
Gautam H, Arulmalar E, Kulkarni MR et al (2019) NDVI is not reliable as a surrogate of forage abundance for a large herbivore in tropical forest habitat. Biotropica 51:443–456
Gaynor KM, McInturff A, Brashares JS (2022) Contrasting patterns of risk from human and non-human predators shape the temporal activity of prey. J Anim Ecol 91:46–60
Girard I, Ouellet JP, Courtois R et al (2002) Effects of sampling effort based on GPS telemetry on home-range size estimations. J Wildl Manag 66:1290–1300
Green SE, Rees JP, Stephens PA et al (2020) Innovations in camera trapping technology and approaches: the integration of citizen science and artificial intelligence. Animals 10:132
Güler NF, Übeyli ED (2002) Theory and applications of biotelemetry. J Med Syst 26:159–178
Gutzwiller KJ, D'Antonio AL, Monz CA (2017) Wildland recreation disturbance: broad-scale spatial analysis and management. Front Ecol Environ 15:517–524
Gwitira I, Murwira A, Zengeya FM et al (2015) Modelled habitat suitability of a malaria causing vector (*Anopheles arabiensis*) relates well with human malaria incidences in Zimbabwe. Appl Geogr 60:130–138
Hahn NR, Bombaci SP, Wittemyer G (2022) Identifying conservation technology needs, barriers, and opportunities. Sci Rep 12:4802
Hale R, Swearer SE (2016) Ecological traps: current evidence and future directions. Proc R Soc Lond B Biol Sci 283:20152647
Harrington R, Owen-Smith N, Viljoen PC et al (1999) Establishing the causes of the roan antelope decline in the Kruger National Park, South Africa. Biol Conserv 90:69–78
Havemann CP, Retief TA, Tosh CA et al (2016) Roan antelope *Hippotragus equinus* in Africa: a review of abundance, threats, and ecology. Mamm Rev 46:144–158
Havemann CP, Retief TA, Collins K et al (2022) Home range and habitat use of roan antelope *Hippotragus equinus* in Northern Botswana. J Arid Environ 196:104648
Hellmann JJ, Byers JE, Bierwagen BG et al (2008) Five potential consequences of climate change for invasive species. Conserv Biol 22:534–543
Homer CG, Edwards TC, Ramsey RD et al (1993) Use of remote sensing methods in modelling sage grouse winter habitat. J Wildl Manag 1:78–84
Huggler KS, Holbrook JD, Hayes MM et al (2022) Risky business: how an herbivore navigates spatiotemporal aspects of risk from competitors and predators. Ecol Appl 32:e2648
Huntley BJ, Bej P, Pinto PV et al (2019) Biodiversity conservation: history, protected areas and hotspots. In: Biodiversity of Angola: science and conservation: a modern synthesis, pp 495–512
Inger R, Bearhop S (2008) Applications of stable isotope analyses to avian ecology. Ibis 150:447–461
Jung TS (2017) Extralimital movements of reintroduced bison (*Bison bison*): implications for potential range expansion and human-wildlife conflict. Eur J Wildl Res 63:35
Katzenberg MA, Waters-Rist AL (2018) Stable isotope analysis: a tool for studying past diet, demography, and life history. In: Biological anthropology of the human skeleton, pp 467–504
Kernohan BJ, Gitzen RA, Millspaugh JJ (2001) Analysis of animal space use and movements. In: Radio tracking and animal populations. Academic Press
Kerr JT, Ostrovsky M (2003) From space to species: ecological applications for remote sensing. Trends Ecol Evol 18:299–305
Kimanzi JK, Wanyingi JN (2014) The declining endangered Roan antelope population in Kenya: what is the way forward? In: Conference papers in science 2014, vol 2014, p 1
Kimanzi JK, Sanderson RA, Rushton SP (2014) Habitat suitability modelling and implications for management of roan antelopes in Kenya. Afr J Ecol 52:111–121
Kimanzi JK, Sanderson RA, Rushton SP et al (2015) Spatial distribution of snares in Ruma National Park, Kenya, with implications for management of the roan antelope *Hippotragus equinus* langheldi and other wildlife. Oryx 49:295–302
Klanjšček J, Geček S, Marn N et al (2018) Predicting perceived level of disturbance of visitors due to crowding in protected areas. PLoS One 13:e0197932

Knoop MC, Owen-Smith N (2006) Foraging ecology of roan antelope: key resources during critical periods. Afr J Ecol 44:228–236

Kueffer C, Kaiser-Bunbury CN (2014) Reconciling conflicting perspectives for biodiversity conservation in the Anthropocene. Front Ecol Environ 12:131–137

Kuijper DPJ, Bubnicki JW, Churski M et al (2015) Context dependence of risk effects: wolves and tree logs create patches of fear in an old-growth forest. Behav Ecol 26:1558–1568

Landres P, Hahn BA, Biber E et al (2020) Protected area stewardship in the Anthropocene: integrating science, law, and ethics to evaluate proposals for ecological restoration in wilderness. Restor Ecol 28:315–327

Laundré JW, Hernández L, Ripple WJ (2010) The landscape of fear: ecological implications of being afraid. Open J Ecol 3:1

le Roex N, Dreyer C, Ferreira SM (2020) Poaching creates ecological traps within an iconic protected area. Anim Conserv 23:250–259

Lewis JS, Farnsworth ML, Burdett CL et al (2017a) Biotic and abiotic factors predicting the global distribution and population density of an invasive large mammal. Sci Rep 7(7):1–12

Lewis PM, Burns GL, Jones D (2017b) Response and responsibility: humans as apex predators and ethical actors in a changing societal environment. Food Webs 12:49–55

Lindenmayer DB, Fischer J, Felton A et al (2008) Novel ecosystems resulting from landscape transformation create dilemmas for modern conservation practice. Conserv Lett 1:129–135

Loehle C, Li BL (1996) Habitat destruction and the extinction debt revisited. Ecol Appl 6:784–789

Madliger CL, Franklin CE, Hultine KR et al (2017) Conservation physiology and the quest for a 'good' Anthropocene. Conserv Physiol 5:cox003

Martin J, Tolon V, Van Moorter B et al (2009) On the use of telemetry in habitat selection studies. In: Telemetry: research, technology, and applications. Nova Science Publishers, Hauppauge, pp 37–55

Martínez-Ramos M, Ortiz-Rodríguez IA, Piñero D et al (2016) Anthropogenic disturbances jeopardize biodiversity conservation within tropical rainforest reserves. Proc Natl Acad Sci USA 113:5323–5328

Marvier M, Lalasz R, Kareiva P (2012) Conservation in the Anthropocene beyond solitude and fragility. In: The Breakthrough Institute - conservation in the Anthropocene, vol 2, pp 4–9

Mason THE, Fortin D (2017) Functional responses in animal movement explain spatial heterogeneity in animal–habitat relationships. J Anim Ecol 86:960–971

Massé A, Côté SD (2012) Linking habitat heterogeneity to space use by large herbivores at multiple scales: from habitat mosaics to forest canopy openings. For Ecol Manag 285:67–76

Maxwell SL, Cazalis V, Dudley N et al (2020) Area-based conservation in the twenty-first century. Nature 586:217–227

McIntosh EJ, Pressey RL, Lloyd S et al (2017) The impact of systematic conservation planning. Annu Rev Environ Resour 42:677–697

McLoughlin CA, Owen-Smith N (2003) Viability of a diminishing roan antelope population: predation is the threat. Anim Conserv 6:231–236

Meadows PS, Campbell JI (1972) Habitat selection by aquatic invertebrates. Adv Mar Biol 10:271–382

Milner JE, Blackwell PG, Niu M (2021) Modelling and inference for the movement of interacting animals. Methods Ecol Evol 12:54–69

Mitchell DLJ, White PCL, Arnold KE (2019) The trade-off between fix rate and tracking duration on estimates of home range size and habitat selection for small vertebrates. PLoS One 14:e0219357

Moleón M, Sánchez-Zapata JA (2023) Extending the dynamic landscape of fear in a human-dominated world. Trends Ecol Evol 38:215–216

Mondal S, Palit D (2021) Challenges in natural resource management for ecological sustainability. In: Natural resources conservation and advances for sustainability, pp 29–59

Mpakairi KS, Ndaimani H, Tagwireyi P et al (2017) Missing in action: species competition is a neglected predictor variable in species distribution modelling. PLoS One 12:e0181088

Mpakairi KS, Dube T, Farai D et al (2022) Spatio-temporal variation of vegetation heterogeneity in groundwater-dependent ecosystems within arid environments. Ecol Inform 69:101667

Mugo R, Saitoh SI, Nihira A et al (2010) Habitat characteristics of skipjack tuna (Katsuwonus pelamis) in the western North Pacific: a remote sensing perspective. Fish Oceanogr 19:382–396

Muposhi VK, Chanyandura A, Gandiwa E et al (2014) Post-release monitoring of diet profile and diet quality of reintroduced African buffalo (*Syncerus caffer*) in Umfurudzi Park, Zimbabwe. Trop Conserv Sci 7:440–456

Muposhi VK, Chademana TC, Gandiwa E et al (2016a) Edge effects: impact of anthropogenic activities on vegetation structure and diversity in western Umfurudzi Park, Zimbabwe. Afr J Ecol 54:450–459

Muposhi VK, Gandiwa E, Chemura A et al (2016b) Habitat heterogeneity variably influences habitat selection by wild herbivores in a semi-arid tropical savanna ecosystem. PLoS One 11:e0163084

Naidoo R, du Preez P, Stuart-Hill G et al (2012) Factors affecting intraspecific variation in home range size of a large African herbivore. Landsc Ecol 27:1523–1534

Naimi B, Araújo MB (2016) Sdm: a reproducible and extensible R platform for species distribution modelling. Ecography 39:368–375

Northrup JM, Hooten MB, Anderson CR et al (2013) Practical guidance on characterizing availability in resource selection functions under a use–availability design. Ecology 94:1456–1463

Northrup JM, Vander Wal E, Bonar M et al (2022) Conceptual and methodological advances in habitat-selection modeling: guidelines for ecology and evolution. Ecol Appl 32:e02470

Ofstad EG, Herfindal I, Solberg EJ et al (2016) Home ranges, habitat and body mass: simple correlates of home range size in ungulates. Proc R Soc Lond B Biol Sci 283:20161234

Ofstad EG, Herfindal I, Solberg EJ et al (2019) Use, selection, and home range properties: complex patterns of individual habitat utilization. Ecosphere 10:e02695

Osborne PE, Alonso JC, Bryant RG (2001) Modelling landscape-scale habitat use using GIS and remote sensing: a case study with great bustards. J Appl Ecol 38:458–471

Owen-Smith N, Martin J, Yoganand K (2015) Spatially nested niche partitioning between syntopic grazers at foraging arena scale within overlapping home ranges. Ecosphere 6:1–17

Patel SH, Morreale SJ, Panagopoulou A et al (2015) Changepoint analysis: a new approach for revealing animal movements and behaviors from satellite telemetry data. Ecosphere 6:1–13

Perrin MR, Taolo C (1998) Home range, activity pattern and social structure of an introduced herd of roan antelope in KwaZulu-Natal, South Africa. Afr J Wildl Res 28:27–32

Phillips DL (2001) Mixing models in analyses of diet using multiple stable isotopes: a critique. Oecologia 127:166–170

Phillips SB, Aneja VP, Kang D et al (2006a) Maximum entropy modeling of species geographic distributions. Int J Glob Environ Issues 6:231–259

Phillips SJ, Anderson RP, Schapire RE (2006b) Maximum entropy modeling of species geographic distributions. Ecol Model 190:231–259

Pressey RL, Visconti P, McKinnon MC et al (2021) The mismeasure of conservation. Trends Ecol Evol 36:808–821

Püttker T, Crouzeilles R, Almeida-Gomes M et al (2020) Indirect effects of habitat loss via habitat fragmentation: a cross-taxa analysis of forest-dependent species. Biol Conserv 241:108368

Ramos A, Petit O, Longour P et al (2016) Space use and movement patterns in a semi-free-ranging herd of European Bison (*Bison bonasus*). PLoS One 11:e0147404

Rechsteiner C, Zellweger F, Gerber A et al (2017) Remotely sensed forest habitat structures improve regional species conservation. Remote Sens Ecol Conserv 3:247–258

Redfern JV, Grant R, Biggs H et al (2003) Surface-water constraints on herbivore foraging in the Kruger National Park, South Africa. Ecology 84:2092–2107

Redfern JV, Grant CC, Gaylard A et al (2005) Surface water availability and the management of herbivore distributions in an African savanna ecosystem. J Arid Environ 63:406–424

Rich LN, Miller DAW, Robinson HS et al (2016) Using camera trapping and hierarchical occupancy modelling to evaluate the spatial ecology of an African mammal community. J Appl Ecol 53:1225–1235

Ryan SJ, Cross PC, Winnie J et al (2012) The utility of normalized difference vegetation index for predicting African buffalo forage quality. J Wildl Manag 76:1499–1508

Samuel MD, Kenow KP (1992) Evaluating habitat selection with radio-telemetry triangulation error. J Wildl Manag 1:725–734

Sanderson EW (2006) How many animals do we want to save? The many ways of setting population target levels for conservation. Bioscience 56:911–922

Scharf HR, Buderman FE (2020) Animal movement models for multiple individuals. Wiley Interdiscip Rev Comput Stat 12:e1506

Schwartz MW, Cook CN, Pressey RL et al (2018) Decision support frameworks and tools for conservation. Conserv Lett 11:e12385

Sietses DJ, Faupin G, de Boer WF et al (2009) Resource partitioning between large herbivores in Hustai National Park, Mongolia. Mamm Biol 74:381–393

Sirot E, Pays O (2011) On the dynamics of predation risk perception for a vigilant forager. J Theor Biol 276:1–7

Smit IPJ, Grant CC, Devereux BJ (2007) Do artificial waterholes influence the way herbivores use the landscape? Herbivore distribution patterns around rivers and artificial surface water sources in a large African savanna park. Biol Conserv 136:85–99

Smith RJ, Bennun L, Brooks TM et al (2019) Synergies between the key biodiversity area and systematic conservation planning approaches. Conserv Lett 12:e12625

Soares Bolzan AM, Szymczak LS, Nadin L et al (2021) What, how, and how much do herbivores eat? The Continuous Bite Monitoring method for assessing forage intake of grazing animals. Ecol Evol 11:9217–9226

Sonter LJ, Ali SH, Watson JEM (2018) Mining and biodiversity: key issues and research in conservation science. Proc R Soc Lond B Biol Sci 285:20181926

Storch D, Šímová I, Smyčka J et al (2022) Biodiversity dynamics in the Anthropocene: how human activities change equilibria of species richness. Ecography 2022:1

Szabó L, Deák B, Bíró T et al (2020) NDVI as a proxy for estimating sedimentation and vegetation spread in artificial lakes-monitoring of spatial and temporal changes by using satellite images overarching three decades. Remote Sens 12:1468

Tagwireyi P, Wenga T, Ndaimani H et al (2020) Environmental correlates of cheetah (Acinonyx jubatus) space-use in a Savanna landscape. Afr J Wildl Res 50:157–166

Tarugara A, Clegg BW, Gandiwa E et al (2019) Measuring body dimensions of leopards (*Panthera pardus*) from camera trap photographs. PeerJ 7:e7630

Tarugara A, Clegg BW, Gandiwa E et al (2021) The effect of competing carnivores on the feeding behaviour of leopards (*Panthera pardus*) in an African savanna. Ecol Evol 11:7743–7753

Teng Y, Zhang Z, Zhang S et al (2022) Habitat suitability analysis and ecological corridor designs for red deer (Cervus elaphus) in the southern Greater Khingan Mountains. Shengtai Xuebao 42:5990–6000

Thuiller W, Lafourcade B, Engler R et al (2009) BIOMOD - a platform for ensemble forecasting of species distributions. Ecography 32:369–373

Thurfjell H, Ciuti S, Boyce MS (2014) Applications of step-selection functions in ecology and conservation. Mov Ecol 2:1–12

Turnbull JW, Clark GF, Johnston EL (2021) Conceptualising sustainability through environmental stewardship and virtuous cycles—a new empirically-grounded model. Sustain Sci 16:1475–1487

Valeix M (2011) Temporal dynamics of dry-season water-hole use by large African herbivores in two years of contrasting rainfall in Hwange National Park, Zimbabwe. J Trop Ecol 27:163–170

Valls-Fox H, Chamaillé-Jammes S, de Garine-Wichatitsky M et al (2018) Water and cattle shape habitat selection by wild herbivores at the edge of a protected area. Anim Conserv 21:365–375

Van Beest FM, Rivrud IM, Loe LE et al (2011) What determines variation in home range size across spatiotemporal scales in a large browsing herbivore? J Anim Ecol 80:771–785

Van Moorter B, Rolandsen CM, Basille M et al (2016) Movement is the glue connecting home ranges and habitat selection. J Anim Ecol 85:21-31

Viana DS, Granados JE, Fandos P et al (2018) Linking seasonal home range size with habitat selection and movement in a mountain ungulate. Mov Ecol 6:1–11

Walsh JR, Carpenter SR, Van Der Zanden MJ (2016) Invasive species triggers a massive loss of ecosystem services through a trophic cascade. Proc Natl Acad Sci USA 113:4081–4085

Weber D, Schaepman-Strub G, Ecker K (2018) Predicting habitat quality of protected dry grasslands using Landsat NDVI phenology. Ecol Indic 91:447–460

White PS, Tuttle JP (2013) Ecological sustainability as the fourth landmark in the development of conservation ethics. Conserv Biol 27:952–957

Wijeyakulasuriya DA, Eisenhauer EW, Shaby BA et al (2020) Machine learning for modeling animal movement. PLoS One 15:e0235750

Wilmshurst JF, Fryxell JM, Bergman CM (2000) The allometry of patch selection in ruminants. Proc R Soc Lond B Biol Sci 267:345–349

Wilson DE, Hirst SM (1977) Ecology and factors limiting Roan and Sable antelope populations in South Africa. Wildl Monogr 54:3–111

Ximenes Pinho B, Dáttilo W, Leal IR (2017) Structural breakdown of specialized plant-herbivore interaction networks in tropical forest edges. Glob Ecol Conserv 12:1–8

Victor K. Muposhi is a Conservation Scientist and an Associate Professor of Wildlife Conservation in the Department of Wildlife and Aquatic Resources at the Botswana University of Agriculture and Natural Resources. His research interests include biodiversity, ecosystems dynamics, spatial ecology, socio-ecological systems, human-wildlife coexistence, and sustainability science.

Mkululi Ndlovu is a Senior Ecologist with Zimbabwe Parks and Wildlife Management Authority. He has research interests in wildlife ecology and management, vegetation and habitat management, sustainable use, invasive species management, and human-wildlife conflicts.

Kudzai S. Mpakairi is a Researcher and Geospatial Scientist affiliated with the University of the Western Cape and African Leadership University. He has research interests in agroecosystems, machine learning, spatial ecology, climate change, and conservation.

Keoikantse Sianga is a Senior Lecturer at the Botswana University of Agriculture and Natural Resource. He holds a BSc in Conservation Biology and Wildlife Management and as MPhil and a PhD in Natural Resource Management. Previously, he has worked as Project Assistant at the University of Botswana as well as Project Manager – Communities Living Among Wildlife Sustainably (CLAWS). His research interests include functional and vegetation heterogeneity and plant-wildlife interactions, fire ecology and management, human-wildlife interactions, and spatial ecology. Dr Sianga has published four articles in peer reviewed journals and seven book chapters.

Admire Chanyandura is a Researcher and Lecturer in the School of Wildlife and Environmental Conservation at the Chinhoyi University of Technology, Zimbabwe. His research interests include biodiversity conservation, wildlife management, ecology, and climate change.

Chiedza C. Zembe is Conservation Scientist and freelance environmental consultant. Her research interests include environmental and social impact assessment, ecosystems dynamics, and wildlife ecology and management.

Gregory Chiyangwa iis a Senior Ecologist with Zimbabwe Parks and Wildlife Management Authority. He has research interests in wildlife ecology and management, ecosystems dynamics, socio-ecological systems, invasive species management and community based natural resources management.

Chapter 6
The Influence of Area Enclosure on the Conservation of Plant Species in an Urban Ecosystem

G. Tsheboeng, C. Sekonopo, and K. Kashe

6.1 Introduction

Woody plant populations are widely threatened by anthropogenic activities resulting in their degradation due to unsustainable over-exploitation (Hedge and Enters 2000). The degradation of the woodland population will negatively affect ecosystem functioning (Chapin et al. 2000; Pisani et al. 2021) and threaten the provision of ecosystem services which humans depend on (Brooks et al. 2006). Ecosystems services are benefits derived by human beings from ecosystems including plant ones (Oguh et al. 2021). To sustain benefits from forest ecosystems there is a need for the conservation of plant species which are responsible for producing the ecosystem services which human beings depend on. Approaches to conserving forest ecosystems include the establishment of protected areas. Protected areas play a critical role of in situ conservation of species on which other organisms including human beings depend on for the provision of ecosystem services (Pisani et al. 2021). There are several approaches to establishing in situ conservation areas. An example of those approaches is the use of area exclosure (Mengistu et al. 2005; Bahiru 2008; Gebrehiwot and Veen 2014; Teketay et al. 2018). An area enclosure is the in situ conservation of plants in which human and animal impact is excluded (Aerts et al. 2009; Teketay et al. 2010). It is regarded as the most affordable and convenient natural restoration approach to degraded lands (Teketay et al. 2018).

G. Tsheboeng (✉) · C. Sekonopo
University of Botswana, Department of Biological Sciences, Gaborone, Botswana
e-mail: tsheboengg@ub.ac.bw

K. Kashe
University of Botswana, Okavango Research Institute, Maun, Botswana
e-mail: kkashe@ub.ac.bw

W. L. Hambira et al. (eds.), *Environmental Change and Biodiversity Conservation in sub-Saharan Africa*, Advances in Global Change Research 75,
https://doi.org/10.1007/978-3-031-73136-5_6

In Africa, the use of area exclosures is mostly prevalent in Ethiopia (Abebe et al. 2006; Birhane et al. 2017; Mengistu 2011). Previous studies assessing the impact of this approach on the conservation of plant species have found that it is an adequate strategy that was able to enhance their diversity, density, and population structure (Mengistu et al. 2005; Bahiru 2008; Grebrehiwot and Veen 2014; Teketay et al. 2018). For instance in Ethiopia Mengistu et al. (2005) found that area exclosures successfully promoted the restoration of degraded plant populations. It was found that species diversity and density were higher in area exclosures when compared to open/communal grazing lands (Mengistu et al. 2005). Furthermore, the author also found that woody plants in area exclosure had a healthy population structure characterised by many individuals in the lower diameter size classes and seedlings when compared to communal areas. In another study, still in Ethiopia in Kallu woreda (Bahiru 2008), it was also found that area exclosures are successful in restoring degraded lands. There was higher species richness, density, diversity, frequency, and healthy population structure in area exclosures than in communal areas (Bahiru 2008).

Previous studies in Botswana show that woody plant populations are threatened by degradation from over-exploitation (Fox et al. 2017; Sebego et al. 2019; Akinyemi and Kgomo 2019; Pelane et al. 2023). The threats of vegetation over-exploitation are from over-dependence from the local communities (Pelane et al. 2023) due to limited options for livelihood sustenance. Pelane et al. (2023) found that there was high utilisation of woody species for firewood in Foley and Makomoto communities which threatens long-term sustainability of woody plant populations. Elsewhere, in the Chobe National Park ecosystem, it has been found that a threat to woody vegetation is mainly from the destruction/over-exploitation by elephants (Skarpe et al. 2004). It was found that increased selective browsing pressure from elephants in Chobe National Park led to a change in the species composition of woody plants with *Vachellia/Senegalia* species replaced by *Capparis tomentosa* and *Combretum mossambicense* (Skarpe et al. 2004). Furthermore, in the Linyanti area, it was found that elephant damage contributed to 16% of mortality in *Senegalia nigrescens* (Teren and Owen-Smith 2010). In Moremi Game reserve, it was found that woody plant species such as *Diospyroys mespiliformis* and *Philenoptera violacea* were characterised by unhealthy population structures which could have resulted from the impacts of elephants (Tsheboeng and Murray-Hudson 2013). The findings from these studies show that game reserves and national parks still expose the vegetation to destruction and over-exploitation by herbivores, especially elephants. Based on the findings of studies elsewhere (Mengistu et al. 2005; Bahiru 2008; Grebrehiwot and Veen 2014; Teketay et al. 2018), area exclosure approach could be the most affordable and efficient strategy to use for restoration of degraded vegetation in Botswana.

Despite the foregoing, in Botswana the adoption of area exclosure in the conservation and restoration of degraded vegetation approach is low. Currently, there is only one 9 ha national botanical garden established in Gaborone (Isaiah and Komi 2015). As elsewhere (Mengistu et al. 2005; Bahiru 2008; Grebrehiwot and Veen 2014; Teketay et al. 2018), the primary purpose of this botanical garden is to conserve threatened plant species and promote their restoration by excluding anthropogenic influence (Isaiah and Komi 2015). However, since the establishment of the national botanical garden as an area exclosure, there are no studies that assessed its influence on the restoration of woody plant populations.

Previous studies in Botswana on the influence of area exclosure on plant conservation have found that it enhances the diversity, richness, regeneration, and population structure of plant species (Flyman 1999; Käller 2003; Teketay et al. 2016; Neelo et al. 2015; Teketay et al. 2018 and Kashe et al. 2023). However, these studies varied in terms of spatial coverage with Neelo et al. (2015), Teketay et al. (2018), and Kashe et al. (2023) carried out in Maun, which is located in the northwest of Botswana. Therefore, their findings are limited in providing information on the influence of the national botanical garden on the restoration of woody plant populations. This is primarily due to variations in the biotic and abiotic environment influencing population dynamics between the northwestern and southern parts of Botswana. These factors are known to influence plant species composition, regeneration, and population structure in the northern and southern parts of Botswana. For instance, the study by Neelo et al. (2015) was conducted in the Island Safari Lodge which was considered an area exclosure which is found adjacent to the Thamalakane River. Therefore, the study was mainly focused on riparian vegetation in contrast to those found in the national botanical garden which is a savannah ecosystem. The diversity, regeneration, and population structure of riparian plant species in Thamalakane River in the distal region of the Okavango Delta is mainly influenced by flooding (Ringrose and Matheson 2001) while in the national botanical garden, the climate is semi-arid (Johnsson 2004). These climatic differences will result in variations in ecosystem functioning in these two areas leading to different species composition, regeneration, and population structure. In semi-arid environments, moisture is a limiting factor to the recruitment of woody species due to seedlings dying back (Mlambo and Nyathi 2004).

The study by Teketay et al. (2018) focused mainly on *Colophospermum mopane* woodland on the Okavango Research Institute campus which represented an area exclosure. The impact of *C. mopane* as a highly allelopathic species (Timberlake 1999) may negate the positive influence of area exclosure on the restoration of plant populations by inhibiting recruitment in other species due to increased mortality of their seedlings (Lukhele and van Ryssen 2003). In addition to this, with the study site closer to Thamalakane River associated site conditions such as water depth will also influence plant species composition, regeneration, and population structure (Lite et al. 2005). Due to this, the results of the study (Teketay et al. 2018) cannot be generalised to our current study area due to geographical limitations and environmental heterogeneity.

The recent study by Kashe et al. (2023) was conducted in Maun international airport compound which was fenced ten years ago. The study area was also dominated by *C. mopane* with other species found in very low densities (Kashe et al. 2023). For instance, the density of *C. mopane* was 1971 individuals/ha inside the airport and 6201 individuals/ha outside. Except for *Vachellia erioloba* (150 individuals/ha inside and 130 individuals/ha outside), other species had comparatively very low densities (Kashe et al. 2023). For instance, the density for *Vachellia tortilis* (25 inside and 44 outside), *Phyllanthus reticulatus* (seven inside and 0 outside), and *Croton megalobotrys* (one inside and 0 outside) were very low when compared to that of *C. mopane*. Therefore, the ecosystem functioning and plant population dynamics of the study area are mainly influenced by *C. mopane*. It is on this basis that the findings from Kashe et al. (2023) cannot be generalised to the national botanical garden due to differences in prevailing environmental conditions.

Studies in the southern part of Botswana on area exclosures were Flyman (1999), Käller, (2003), and Teketay et al. (2016) all conducted in the Mokolodi nature reserve. All these studies were conducted in an area that had a history of livestock utilisation (Schroder 2001) and at the time of data collection it was entirely used for wildlife conservation (Flyman 1999; Käller 2003; Teketay et al. 2016). Therefore, the response of the vegetation both the historical and current use may limit the generalization of the findings to an area whose main disturbance to vegetation is anthropogenic as is found within the settlements/homesteads. Furthermore, the studies in the Mokolodi nature reserve on the seedlings' dynamics and not population structure (Flyman 1999; Käller 2003) except for the study by Teketay et al. (2016).

Other than the geographical limitations, all the studies in Botswana were conducted in quasi-area exclosures (Flyman 1999; Käller 2003; Neelo et al. 2015; Teketay et al. 2018 and Kashe et al. 2023). The study sites for all these studies were not primarily demarcated as area exclosures. For instance, the study by Teketay et al. (2018) was conducted in the Okavango Research Institute compound while that of Neelo et al. (2015) was conducted in the Island safari compound. A recent study by Kashe et al. (2023) was conducted at Maun international airport. All the aforementioned study sites are quasi-area exclosures as they have their primary uses which are not to primarily exclude the anthropogenic and animal influence on plant species. As a result, there still exists a knowledge gap on the influence of the national botanical garden on species diversity, regeneration, and population structure. To address this knowledge gap, our study assessed and compared plant population structure between national botanical gardens (area exclosure) and communal areas.

6.2 Materials and Methods

6.2.1 Study Area

The study was conducted in the National Botanical Garden (24° 40′ S, 25° 56′ E (24° 66′ S, 25° 97′ E) in Gaborone, Village (Isiah and Komi 2015) and adjacent communal area in Tlokweng and west of the Botanical Garden (Fig. 6.1). The botanical garden was established for the sole purpose of conserving plants. The garden was established in 1968 in a 9 ha field (Isaiah and Komi 2015). The National Botanical Garden and communal area represented exclosure and open area, respectively. The outside of the National Botanical Garden represented a disturbed

Fig. 6.1 Map of the study areas. (Map adapted from: Google Earth map)

communal area. This is because a lot of human activities such as agriculture, settlement, land clearing, and population growth take place owing to its proximity to the capital city Gaborone. The inside of the Botanical Garden represents an area with very low disturbances because it is not exposed to human activities. The national botanical garden is characterised by a hot rainy season from October to March, and a long dry season from April to September. The rainfall is erratic and reaches a maximum of 485 mm. The country in general is susceptible to drought. Summer days are often hot and can reach a temperature of 44 °C (Isiah and Komi 2015).

6.2.2 Data Collection

Data collection was done between February 2022 and March 2022. In the National Botanical Garden, a GPS was used to locate and mark 53 potential plots which were 20 m away from each other. Thirty plots were randomly selected in each site (Botanical garden and communal area). A 'RANDBETWEEN' function in Microsoft Excel was used to give a list of 30 integers which were the number of plots for field data collection (Tsheboeng et al. 2017). In each area, a 10 m × 10 m plot was measured and marked using range poles. Plant species were identified and their percentage cover was estimated in each plot using Braun-Blanquet (Ellenberg and Mueller-Dombois 1974). A Vernier caliper was used to measure the diameter of each woody plant at about 1.3 m above ground height. For big trees, the circumference was measured and used to calculate the diameter. For multi-stemmed plants, the circumference of each stem was measured, and an average was calculated.

6.2.3 *Data Analyses*

6.2.3.1 Population Structure

Total species were counted in each site to determine species richness. In addition to this, evenness, diversity, density, frequency, dominance, and importance value index were calculated in each site.

Species diversity was determined through the Shannon–Wiener diversity index (Shannon 1948) as follows:

$$= -\sum_{i=1}^{S} Pi \ln Pi$$

where H' = Shannon–Wiener diversity index, S = species richness, P_i = proportion of S made up of the ith species.

Evenness was determined using the Shannon–Wiener equitability index. This was calculated as:

$$H' = -\sum Pi \ln Pi$$

where E = Shannon–Wiener evenness, H' = Shannon–Wiener diversity index and S = species richness.

The similarity in plant species composition between the two sites was assessed through Jaccard's similarity index (Krebs 1989). Mean density and Relative density were calculated for each site following Kent and Coker (1992). In addition, the mean frequency for each species was calculated at each site (Neelo et al. 2015). From the mean frequency, we calculated the Relative frequency as follows (Neelo et al. 2015):

$$\text{Relative Frequency} = \left(\text{Frequency of species } A \div \text{Frequency of all species}\right) \times 100$$

Basal area (BA)/dominance was calculated using the following formula: BA (m^2) = $\pi \times$ [DBH (cm)]^2/40000, where $\pi = 3.14$. Relative basal area/dominance was calculated as (Neelo et al. 2015):

$$\text{Relative basal area} = \left(\text{Basal area of species } A \div \text{Total basal area}\right) \times 100$$

Relative dominance, relative frequency, and relative density were summed to give the Importance Value Index (Kent and Coker 1992). The following diameter classes (X-axis) were plotted against density (Y-axis) in a bar chart to assess the pattern of plant populations structure:

1 = 0–5, 2 = 5–10, 3 = 10–15, 4 = 15–20, 5 = 20–25, 6 = 25–30, 7 = 30–35, 8 = >35 cm (Teketay et al. 2018). Mean canopy cover was calculated for each species in the two sites.

6.2.4 *Statistical Data Analysis*

6.2.4.1 Population Structure

We compared the means of frequency, density, evenness, diversity, and the basal area between botanical gardens and communal areas using a student *t-test* to determine whether they differed significantly or not. The significant level was set at $p = 0.05$.

In addition to the patterns of population structure, we used linear regression to infer the recruitment of each species in both study areas (Sop et al. 2011). To do this, a correlation coefficient was determined from the relationship between the midpoint (mi) of DBH and the density of individuals (N_i). The density of individuals was log-transformed by $\ln(N_i + 1)$ since some classes had zero individuals. After transformation, regression analysis was conducted between ln(mi) and $\ln(N_i + 1)$. The correlation coefficients were interpreted following Shackleton (1993):

(i) A negative slope showed ongoing recruitment
(ii) A positive slope showed a lack of recruitment

6.2.4.2 Plant Communities

Plant communities for both areas were classified through hierarchical cluster analysis (Dufrêne and Legendre 1997). Comparison in plant species composition for the different communities was done through multi-response permutation procedures (McCune and Grace 2002) and Bray–Curtis ordination (Beals 1984). Indicator species analysis was used to determine indicator values for each species in a given plant community (Dufrêne and Legendre 1997). The Monte Carlo test was used to test for the significance of the indicator value determined for each species. The indicator value is the sum of the relative frequency and relative abundance of each species in its respective community (Dufrêne and Legendre 1997). Student's t-test was used to compare mean plant canopy cover for selected species between communal and exclosure areas.

6.3 Results

6.3.1 Diversity

There was higher species richness (21) in area exclosure than in communal area (12). Likewise, there was higher genera richness (17) in area exclosure than in communal area (11) (Tables 6.1 and 6.2). In terms of family richness, in the communal area, Fabaceae had the highest species richness with three species followed by Bignoniaceae, Tiliaceae, and Combretaceae with two species each. Burseraceae and Anacardiaceae each had one species (Table 6.1). Fabaceae still had the highest species richness (11) inside the area exclosure followed by Combretaceae and Tiliaceae with two species each. Other families had one species each (Table 6.2). In the open area, the *Grewia* genus had the highest species richness with two species while other genera had one species each (Table 6.1). Inside the area exclosure, genera *Combretum* and *Grewia* had the highest species richness with two species each. Other genera only had one species (Table 6.2).

Table 6.1 A list of species and their associated population parameters in a communal area

Species	Family	Density	RD	Frequency	RF	Dominance	RDO	IVI
Vachelia tortilis	Fabaceae	376	69.7	86.6	41.3	49122.1	72.4	183.4
Dichrostachys cinerea	Combretaceae	30	5.5	20	9.53	1686.2	2.48	17.5
Sclerocarya birrea	Combretaceae	26	4.9	13.3	6.34	13734.3	20.2	31.4
Senegalia mellifera	Rhamnaceae	20	3.7	20	9.53	424.243	0.62	13.8
Combretum hereroense	Tiliaceae	20	3.7	13.3	6.34	313.119	0.46	10.5
Grewia flavescens	Fabaceae	16	3.1	13.3	6.34	42.1627	0.06	9.5
Rhigozum brevispinosum	Fabaceae	10	1.9	10	4.76	456.369	0.67	7.33
Commiphora angolensis	Fabaceae	10	1.9	10	4.76	335.241	0.49	7.15
Grewia monticola	Burseraceae	10	1.9	6.7	3.19	29.2178	0.04	5.13
Ziziphus mucronata	Bignoniaceae	6	1.2	6.7	3.19	434.306	0.64	5.03
Terminalia sericea	Tiliaceae	6	1.2	6.7	3.19	86.952	0.12	4.51
Azanza garckeana	Anacardiaceae	6	1.2	3.3	1.6	1182.06	1.74	4.51
Total		536						

Table 6.2 A list of species and their associated population parameters in an area exclosure

Species	Family	Density	RD	Frequency	RF	Dominance	RDO	IVI
Vachellia tortilis	Fabaceae	80	13.6	46.6	14.6	12099.05	32.3	60.5
Senegalia erubescens	Fabaceae	73	12.4	40	12.5	5670.385	15.1	40
Rhigozum brevispinosum	Bignoniaceae	57	9.6	20	6.26	1264.685	3.38	19.2
Senegalia mellifera	Fabaceae	53	9.04	33.3	10.4	2092.089	5.59	25
Combretum hereroense	Combretaceae	53	9.04	23.3	7.29	658.3006	1.76	18
Dichrostachys cinerea	Fabaceae	50	8.5	23.3	7.29	293.5426	0.78	16.5
Tarchonanthus camphoratus	Asteraceae	43	7.3	10	3.13	505.5608	1.35	11.7
Ziziphus mucronate	Rhamnaceae	27	4.5	13.3	4.16	1092.737	2.92	11.5
Sclerocarya birrea	Anacardiaceae	23	4	13.3	4.16	2522.967	6.74	14.9
Colophospermum mopane	Fabaceae	23	4	10	3.13	3887.325	10.3	17.5
Grewia flavescens	Tiliaceae	23	4	13.3	4.16	812.3962	2.17	10.3
Grewia monticola	Tiliaceae	20	3.4	20	6.26	54.45061	0.14	9.8
Senegalia senegal	Fabaceae	13	2.3	10	3.13	563.3923	1.5	6.93
Boscia albitrunca	Capparaceae	10	1.7	10	3.13	5324.816	14.2	19
Commiphora pyrachanthoides	Burseaceae	10	1.7	10	3.13	203.7661	0.54	5.37
Azanza garckeana	Malvaceae	6	1.1	6.6	2.06	35.86652	0.09	3.25
Combretum imberbe	Combretaceae	6	1.1	3.33	1.04	3.822271	0.01	2.15
Brachystegia tamarindoides	Fabaceae	6	1.1	3.33	1.04	119.4132	0.31	2.45
Peltophorum africanum	Fabaceae	3	0.6	3.33	1.04	107.233	0.28	1.92
Spirostachys Africana	Euphorbiaceae	3	0.6	3.33	1.04	8.960084	0.02	1.66
[a]*Eucalyptus sp*	Myrtaceae	3	0.6	3.33	1.04	60.31858	0.16	1.8
Total		585						

[a]Exotic species

6.3.2 Comparison of Plant Species Composition Between Area Exclosure and Communal Area

Ten species were common on both sites ($S_j = 45\%$) while 11 were exclusive to area exclosure. The open area did not have any species exclusive to it. Some of the species exclusive to the closed areas were *Peltophorum africanum*, *C. mopane*, and *Boscia albitrunca*, while *Terminalia sericea* was exclusively found in open sites. In terms of genera, 14 were common in both sites ($S_j = 50\%$), seven were found in an area exclosure only while one was found in the open site. Some of the genera found exclusively in closed areas were *Peltophorum*, *Brachystegia*, and *Spirostachys* while *Terminalia* was found in open sites only. In terms of comparison of families, eight were common between the sites ($S_j = 67\%$) while four were exclusive to the closed area and there were none exclusive to the open site. Families found only in the closed areas were Capparaceae, Asreraceae, Euphorbiaceae, and Myrtoideae.

6.3.2.1 Diversity and Evenness

Mean species diversity (0.78 ± 0.08) was significantly (p = 0.001136) higher in closed area than in open one (0.43 ± 0.07). Mean evenness was not significantly different (p = 0.32034) between closed (0.80 ± 0.03) and open sites (0.90 ± 0.03). Mean density was higher in area exclosure (585 individuals ha^{-1}) than in the communal (536 individuals ha^{-1}) areas (Tables 6.1 and 6.2). In the open areas, the species with high density (individuals ha^{-1}) were *V. tortilis* (376), *Dichrostachys cinerea* (30), *Sclerocarya birrea* (26), *Senegalia mellifera* (20), and *Combretum hereroense* (20) (Table 6.1).

The species with high density (individuals ha^{-1}) area exclosure were *V. tortilis* (80), *Senegalia erubescens* (73), *Rhigozum brevispinosum* (57), *S. mellifera* (53), and *C. hereroense* (53) (Table 6.2). In open areas, the most frequent species were *V. tortilis* (86.6%), *S. mellifera* (20.0%), *D. cinerea* (20.0%), *Grewia flavescens* (13.3%), and *C. hereroense* (13.3%) (Table 6.1). The most frequent species in area exclosure were *V. tortilis* (46.6%), *S. erubescens* (40%), *S. mellifera* (33.3 *C. hereroense* (23.3%), and *Grewia monticola* (20.0%) (Table 6.2).

In communal areas, the dominant species were *V. tortilis* (49122.1), *D. cinerea* (1686.2), *S. birrea* (13734.3), *Azanza garckeana* (1182.06), and *R. brevispinosum* (456.369) (Table 6.1). The dominant (cm^2 ha^{-1}) species in area exclosure were *V. tortilis* (12099), *S. erubescens* (5670.385), *B. albitrunca* (5324.816), *C. mopane* (3887.325), and *S. mellifera* (2092.089) (Table 6.2).

6.3.2.2 Importance Value Index

In communal area, some of the species with the highest IVIs were *V. tortilis* (183.4%), *D. cinerea* (17.5%), *S. mellifera* (13.8%), *C. hereroense* (10.5%), and *G. flavescens* (9.5%) (Table 6.1). Some of the species with the highest IVIs in areas area exclosure were *V. tortilis* (60.5%), *S. erubescens* (40%), *S. mellifera* (25%), *B. albitrunca* (19%), and *C. hereroense* (18%) (Table 6.2).

6.3.3 Population Structure and Regeneration Status

Woody plants in the exclosure area showed a reverse-J-shaped population structure (Fig. 6.2a) while in the open area, they had a bell-shaped structure with higher density in the lower and middle size classes and low density in the higher size classes (Fig. 6.2b).

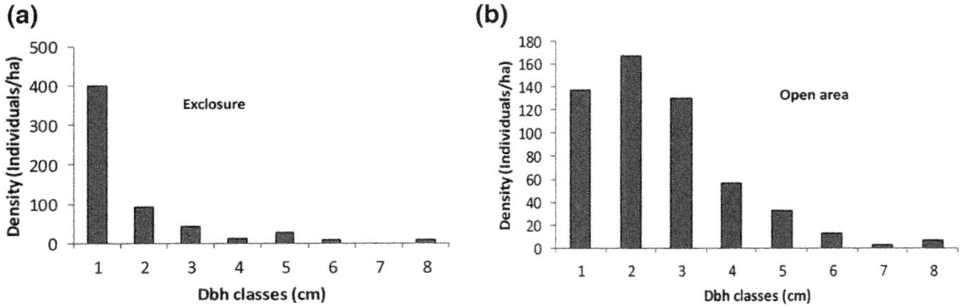

Fig. 6.2 Comparison of overall plant population structure between area exclosure and open area. The size classes are 1 = 0–5, 2 = 5–10, 3 = 10–15, 4 = 15–20, 5 = 20–25, 6 = 25–30, 7 = 30–35, 8 = >35 cm

Vachellia tortilis had a U-shaped population structure in area exclosure (Fig. 6.3a) while in an open area, it was characterised by a bell-shaped structure (Fig. 6.3b). *Dichrostachys cinerea* had two classes in both area exclosure (Fig. 6.3c) and open area (Fig. 6.3d).

Senegalia mellifera was characterised by the first three diameter classes in area exclosure (Fig. 6.4a) while in the open area, it had the first two diameter classes (Fig. 6.4b). *Sclerocarya birrea* was characterised by an irregular structure with missing size classes in both sites (Fig. 6.4c, d).

6.3.3.1 Recruitment

All the species show an ongoing recruitment in area exclosure. In open areas, only *S. birrea* shows a lack of recruitment. The recruitment is higher in area exclosure than in open areas as shown by higher negative correlation coefficients in the former than in the latter (Table 6.3). Comparing the overall recruitment between the two sites shows that area exclosure has higher ongoing recruitment than the open area (Table 6.3).

6.3.3.2 Plant Communities

There were three communities identified for both sites. These were *Vachellia tortilis–Grewia flavescens*, *Senegalia erubescens–Rhigozum brevispinosum*, and *Dichrostachys cinerea–Combretum hereroense* (Table 6.4). Of these, *Vachellia tortilis–Grewia flavescens* (dominated by communal area species) was found in the communal area, while *Dichrostachys cinerea–Combretum hereroense* and *Senegalia erubescens–Rhigozum brevispinosum* were found only in area exclosure (Table 6.4).

Vachellia tortilis–Grewia flavescens and *Senegalia erubescens–Rhigozum brevispinosum* were significantly highly ($T = -23.2$, $p < 0.001$) separated in terms of species composition (Table 6.5 and Fig. 6.5).

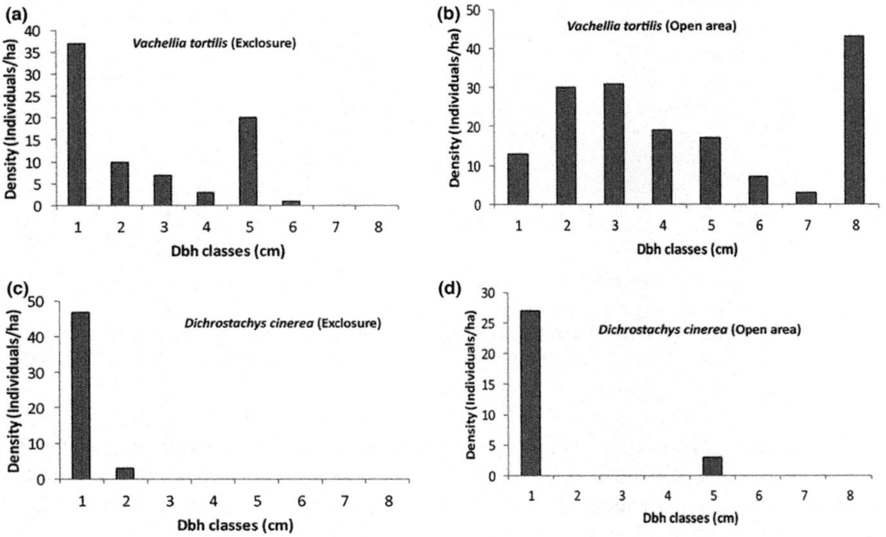

Fig. 6.3 Comparison of population structure of *Vachellia tortilis* and *Dichrostachys cinerea* between area exclosure and open area. The size classes are 1 = 0–5, 2 = 5–10, 3 = 10–15, 4 = 15–20, 5 = 20–25, 6 = 25–30, 7 = 30–35, 8 = >35 cm

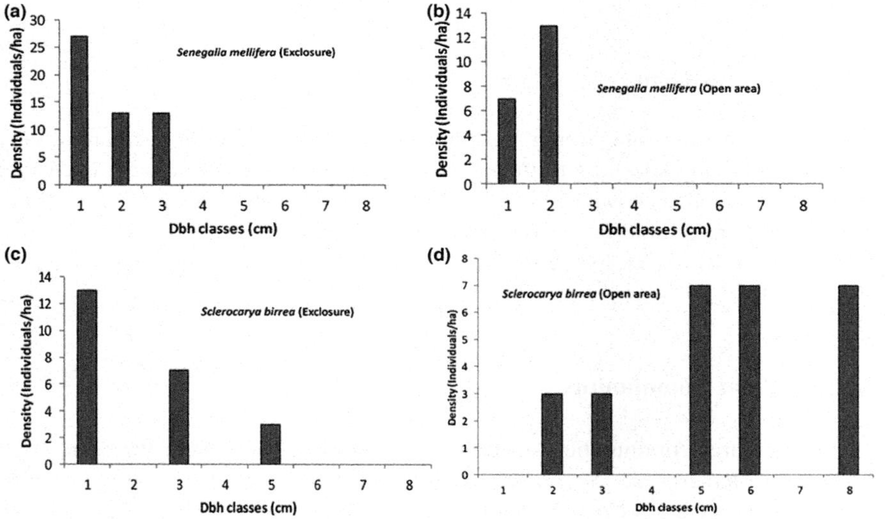

Fig. 6.4 Comparison of population structure of *Senegalia mellifera* and *Sclerocarya birrea* between area exclosure and open area. The size classes are 1 = 0–5, 2 = 5–10, 3 = 10–15, 4 = 15–20, 5 = 20–25, 6 = 25–30, 7 = 30–35, 8 = >35 cm

Table 6.3 Regression analysis slope coefficients and R^2 for each species

Species	Exclosure	R^2	Open area	R^2
Vachellia tortilis	−1.20	0.64	−0.19	0.040
Dichrostachys cinerea	−1.38	0.82	−0.98	0.57
Senegalia mellifera	−1.48	0.74	−0.95	0.65
Sclerocarya birrea	−0.82	0.49	0.45	0.20
Overall	−1.85	0.82	−1.35	0.65

Table 6.4 Plant communities in both area exclosure and communal area

Species	Family	Indicator value	*P*
Vachellia tortilis–Grewia flavescens (Communal)			
Vachellia tortilis	Fabaceae	86.1	0.0002
Grewia flavescens	Tiliaceae	13.2	0.4255
Sclerocarya birrea	Anacardiaceae	9.0	0.6519
Azanza garckeana	Malvaceae	8.1	0.3877
Grewia monticola	Tiliaceae	7.4	0.8268
Commiphora pyrachanthoides	Burseraceae	3.2	1.0000
Peltophorum africanum	Fabaceae	2.7	1.0000
Senegalia erubescens–Rhigozum brevispinosum (Exclosure)			
Senegalia erubescens	Fabaceae	86.1	0.0002
Rhigozum brevispinosum	Bignoniaceae	26.6	0.0368
Senegalia mellifera	Fabaceae	23.5	0.1612
Combretum imberbe	Combretaceae	10.0	0.1600
Tarchonanthus camphoratus	Asteraceae	9.1	0.1776
Vachellia senegal	Fabaceae	5.9	0.6251
Dichrostachys cinerea–Combretum hereroense (Exclosure)			
Dichrostachys cinerea	Fabaceae	50.3	0.0006
Combretum hereroense	Combretaceae	48.3	0.0006
Boscia albitrunca	Brassicaceae	23.1	0.0132
Ziziphus mucronata	Rhamnaceae	13.3	0.2597
Colophospermum mopane	Fabaceae	12.7	0.1438
Spirostachys Africana	Euphorbiaceae	7.7	0.3931

Table 6.5 Pairwise comparisons of plant communities from area exclosure and communal area

Plant communities	*T*	*P*
Vachellia tortilis–Grewia flavescens and *Senegalia erubescens–Rhigozum brevispinosum*	−23.2	0.000000001
Senegalia erubescens–Rhigozum brevispinosum and *Dichrostachys cinerea–Combretum hereroense*	−9.2	0.00000177
Dichrostachys cinerea–Combretum hereroense and *Vachellia tortilis–Grewia flavescens*	21.1	0.000000001

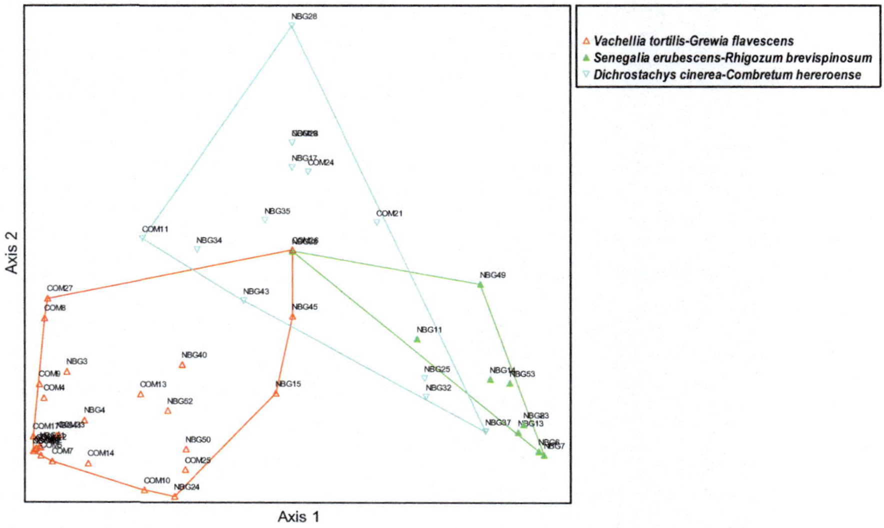

Fig. 6.5 Bray–Curtis ordination of plant communities in communal and exclosure areas

Comparatively, there was a high overlap in species composition between *Senegalia erubescens–Rhigozum brevispinosum* and *Dichrostachys cinerea–Combretum hereroense* (Fig. 6.5); hence, there was less separation between them ($T = -9.2$, $p < 0.001$). Likewise, there was an overlap in species composition between *Dichrostachys cinerea–Combretum hereroense* and *Vachellia tortilis–Grewia flavescens* (Fig. 6.5) even though the difference was still relatively higher ($T = -21.1$, $p < 0.001$) when compared to that between *Senegalia erubescens–Rhigozum brevispinosum* and *Dichrostachys cinerea–Combretum hereroense* (Table 6.4 and Fig. 6.5).

6.3.4 Comparison of Plant Canopy Cover Between Communal and Exclosure Areas

Vachellia tortilis had a significant mean canopy cover in the communal area than in area exclosure (Fig. 6.6). Mean canopy cover for *Senegalia mellifera* was significantly ($p = 0.0246$) higher in area enclosure (mean = 9) than in communal area (mean = 2). Other species did not differ significantly ($p > 0.05$). *Senegalia erubescens*, *Boscia albitrunca*, and *Vachellia senegal* were not statistically compared since they were not represented on either side (Fig. 6.6).

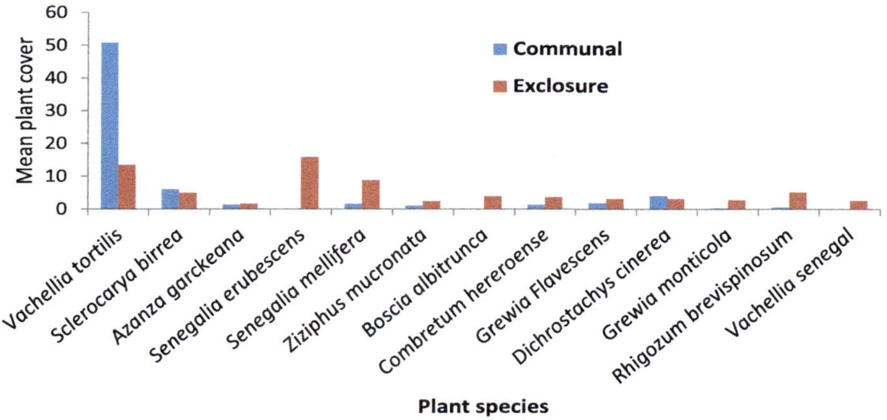

Fig. 6.6 Comparison of plant canopy cover between communal and exclosure areas

6.4 Discussion

The national botanical garden had higher species richness, families, and diversity than the communal area. It was also found that there was a significant difference between plant communities found in communal areas and those in area exclosures. *Senegalia erubescens–Rhigozum brevispinosum* and *Dichrostachys cinerea–Combretum hereroense* plant communities were found in area exclosure. Conversely, *Vachellia tortilis–Grewia flavescens* was found in communal areas. The disparity in plant species diversity and community composition between the two sites could be due to continuous anthropogenic disturbances including cutting down trees, land clearing, and heavy browsing mainly by domesticated animals in communal areas. This implies that more species were cut from the communal area and some of them were either completely exhausted or their abundance severely reduced such as *Senegalia erubescens*, *Rhigozum brevispinosum*, and *Vachellia senegal*. Therefore, the area exclosure could be providing refuge for these species protecting them from over-exploitation. In our study though not quantified, we observed dead stumps of trees indicating that in the communal area, the community is involved in tree cutting which could be for harvesting or land clearing. In other studies, tree cutting mainly for firewood and building materials harvesting has also been found to lead to low diversity of woodland species (Shibru and Balcha 2004; Kuma and Shibru 2015; Neelo et al. 2013; Denbel 2021). Therefore, higher species diversity in botanical gardens suggests that more species are protected from anthropogenic disturbances unlike in the communal areas (Asmare and Gure 2019) and browsing from large herbivores (Kashe et al. 2021). In communal areas, populations of woody plants are influenced mainly by anthropogenic disturbances (Neelo et al. 2013). These results also concur with the findings from other studies (e.g. Neelo et al. 2015; Kuma and Shibru 2015; Teketay et al. 2018; Asmare and Gure 2019; Denbel 2021) in which area exclosures promoted high species richness and a healthy population structure.

There was no significant difference in evenness between exclosure and communal areas. However, even though the evenness values were not significantly different, the communal area had higher evenness. These results contradict findings from other studies elsewhere in which area exclosures exhibited higher evenness compared to communal areas (e.g. Mapaure and Ndeioma et al. 2011; Neelo et al. 2015; Teketay et al. 2018). Mapaure and Ndeioma (2011) and Teketay et al. (2018) found that area exclosure had a higher evenness than communal areas due to low utilisation pressure. The same observation was also made by Neelo et al. (2015) in which area exclosure with no anthropogenic and browsing effects had higher evenness when compared to communal areas. Kashe et al. (2021) also posit that low browsing pressure in area exclosure leads to higher evenness when compared to communal areas.

The disparity in the evenness between our study and others elsewhere could be due to few species in the conservation area contributing more abundance hence resulting in low evenness. For instance, in our study species such as *V. tortilis*, *S. erubescens*, and *R. brevispinosum* had disproportionately higher relative density when compared to *A. garckeana*, *Commiphora pyrachanthoides*, and *P. africanum*. Furthermore, *V. tortilis* and *S. mellifera* are also the only species that had significantly higher canopy cover in communal and exclosure areas respectively. This can also explain the lack of statistical significance in evenness as it indicates that most of the species have equal abundance between the two areas. It is expected that over time, these species will also increase in abundance as a result of persistent conservation leading to an increase in evenness. Conversely, in the communal areas except for *V. tortilis* which had a disproportionately higher relative density and significantly higher canopy cover, other species had relative density and canopy of almost the same magnitude ranging indicating almost equal abundance. This implies that the majority of the woodland species in communal areas are utilised in the same magnitude. Alternatively, the higher abundance of *V. tortilis* could be an indicator of disturbance in communal areas (Ellery and Ellery 1997). Likewise, in conservation areas the higher abundance of *V. tortilis* could also be an indicator of historical disturbance (Helm and Witkowski 2012) with this species showing that the woodland vegetation is still in the pioneer successional stage on its recovery path (Kuma and Shibru 2015).

Generally, except for *V. tortilis* all the species had a higher Importance Value Index (IVI) in area exclosure than in the communal one. The difference in the IVI values between communal and exclosure sites could be caused by a variation among woody plant species composition (Senbeta and Teketay 2003). *Vachellia tortilis* had higher IVIs in both study sites and therefore could suggest that it has a higher competitive edge than other species in both study sites. Other ecologically important species (IVI ≥20) in area exclosure were *S. erubescens* and *S. mellifera* while in the communal area, it was *S. birrea*. IVI analysis helps in determining species that have to be prioritised for conservation. Species with low IVI (≤20) need higher conservation interventions while those with high IVI (≥20) need sustainable management (Birhanu et al. 2021). In our case, 75% of species in communal areas should be prioritised for conservation. These are *R. brevispinosum*, *Ziziphus mucronata*,

G. flavescens, *G. monticola*, *T. sericea*, *A. garckeana*, *S. mellifera*, *C. hereroense*, and *D. cinerea*. In enclosed areas, 82% of species require to be prioritised for conservation. These include *S. birrea*, *A. garckeana*, *S. erubescens*, and *C. mopane*.

The overall population structure in the area exclosure showed an inverted J-shape. This shows a healthy and stable population structure. A healthy population structure implies that in the future should any disturbance occur the individual in the small size classes can be recruited into larger size classes to help stabilise the community of that particular species (Helm and Witkowski 2012). Therefore, our results show that overall area exclosure contributes positively towards protecting woody plants from disturbances and exploitation. This is consistent with studies done elsewhere in which it was found that area exclosure promotes healthy regeneration of woody plant species (Neelo et al. 2013; Teketay et al. 2018; Asmare and Gure 2019; Denbel 2021). Even though the majority of species exhibit this pattern there are a few species that have an unhealthy population structure inside the conservation area. This could be because of past disturbances preceding the fencing of the area exclosure.

On the contrary, the overall population structure in communal areas showed more individuals in the second diameter class of 5–10 cm. The first and third-diameter classes also have a high density of individuals when compared to the rest of the classes. This pattern shows that the woody plants in the communal areas also have the potential of recovering from anthropogenic disturbance. In communal areas, human disturbances particularly browsing and over-harvesting are usually the major reasons for the declining density of individuals in the larger size classes. When we assessed each species' population structure in an open area, we found that some species in the communal areas exhibited a J-Shape pattern with more individuals in the larger size classes than in small size classes indicating unhealthy population structures. The species with this pattern were *S. birrea* with missing size classes at 0–5 cm, 15–20 cm, and 30–35 cm, and *S. mellifera* with missing size classes at 10–15 cm, 15–20 cm, 20–25 cm, 30–35 cm, and >35 cm. This might be a result of heavy browsing pressure leading to the selective removal of seedlings and juveniles (Asmare and Gure 2019). The missing size classes and J-shaped pattern could also be indicator of the selective removal of individuals for human use. *A. garckeana* was characterised by only two size classes (15–20 cm and 30–35 cm). The causes of this are unknown but it could be an indication of over-dependence on this species. Even in the conservation area *A. garckeana* is characterised by only one size class (0–5 cm). The reasons for this should be assessed in future studies. In the meantime, the conservation of this species should be prioritised.

In area exclosure, *V. tortilis* was characterised by a U-shaped population structure with more individuals in the lower and upper size classes. This indicates that there are some years in which there is no recruitment of this species (Gurmessa et al. 2012). Lack of recruitment in some years could be due to low moisture which is not enough to support the survival of the seedlings and their eventual recruitment in the next growth stages (Burke 2006). *Dichrostachys cinerea* was characterised by only two size classes 0–5 cm and 5–10 cm while *S. mellifera* had three size classes of 0–5 cm, 5–10 cm, and 10–15 cm. *S. birrea* had missing size classes in 5–10 cm, 15–20 cm, and all the classes above 20–25 cm. This is an indication of recruitment bottlenecks in these species. Possible reasons for this include persistent dry conditions leading to the mortality of seedlings before they are recruited (Burke 2006), seedling predation (Marone et al. 2000), dormancy of seeds to avoid germinating during dry conditions (Fenner 2012), and competition from herbaceous species which inhibit the survival of seedlings and their recruitment (Hipondoka and Versfeld 2006). Future studies should investigate the causes of the current population structure in an area of exclosure to help with coming up with intervention strategies that can promote recruitment and a healthy population structure.

6.5 Conclusion

We conclude that area exclosure is a viable strategy for the conservation of woody vegetation. This is because our results show that the area exclosure was characterised by higher species diversity, richness, and density when compared to the communal area. Furthermore, plant species in the area exclosure showed an overall healthy population structure when compared to the communal area. Despite this, individual species show unhealthy population structure with missing size classes suggesting recruitment bottlenecks. Therefore, this implies that plants in area exclosure also need to be monitored and strategies devised to help improve their population structure. We would like to recommend the following in order to promote the conservation of woodland species in this area:

1. Species with low IVI values should be prioritised for conservation in communal areas. These include *R. brevispinosum*, *Z. mucronata*, *G. flavescens*, *G. monticola*, *T. sericea*, *A. garckeana*, *S. mellifera*, *C. hereroense*, and *D. cinerea*.
2. In area exclosure in situ conservation of woody plant species should continue. The management of this area should monitor the recruitment of species with low IVI values such as *S. birrea*, *A. garckeana*, *S. erubescens*, and *C. mopane*.
3. Creation of community awareness in Tlokweng on the need to conserve woody plant species for sustainable utilization.
4. Beyond the current study area, there is a need to establish more area exclosures in Botswana to help in the conservation of woody plant species.

References

Abebe MH, Oba G, Angassa A, Weladji RB (2006) The role of area enclosures and fallow age in the restoration of plant diversity in northern Ethiopia. Afr J Ecol 44(4):507–514

Aerts R, Nyssen J, Haile M (2009) On the difference between "exclosures" and "enclosures" in ecology and the environment. J Arid Environ 73(8):762

Akinyemi FO, Kgomo MO (2019) Vegetation dynamics in African drylands: an assessment based on the Vegetation Degradation Index in an agro-pastoral region of Botswana. Reg Environ Chang 19(7):2027–2039

Asmare MT, Gure A (2019) Effect of exclosure on woody species diversity and population structure in comparison with adjacent open grazing land: the case of Jabi Tehnan district north western Ethiopia. Ecosyst Health Sustain 5(1):98–109

Bahiru KM (2008) Enclosure the enclosure as a viable option for the rehabilitation of degraded lands and biodiversity conservation: the case of kallu Kallu woreda, Southern welloWello. Addis ababa Ababa university school of graduate studies, Ethiopia, pp 1–99

Beals EW (1984) Bray-Curtis ordination: an effective strategy for analysis of multivariate ecological data. In: Advances in ecological research, vol 14. Academic Press, pp 1–55

Birhane E, Mengistu T, Seyoum Y, Hagazi N, Putzel L, Rannestad MM, Kassa H (2017) Exclosures as forest and landscape restoration tools: lessons from Tigray Region, Ethiopia. Int For Rev 19(4):37–50

Birhanu T, Mohammed AS, Mekonnen AB (2021) Floristic composition, structure and regeneration status of woody plants in church forests of Dangila, Northwestern Ethiopia. Cogent Food Agric 7(1):1911438

Brooks TM, Mittermeier RA, Da Fonseca GA, Gerlach J, Hoffmann M, Lamoreux JF, Mittermeier CG, Pilgrim JD, Rodrigues AS (2006) Global biodiversity conservation priorities. Science 313(5783):58–61

Burke A (2006) Savanna trees in Namibia—factors controlling their distribution at the arid end of the spectrum. Flora-Morphol Distrib Funct Ecol Plants 201(3):189–201

Chapin Iii FS, Zavaleta ES, Eviner VT, Naylor RL, Vitousek PM, Reynolds HL, Hooper DU, Lavorel S, Sala OE, Hobbie SE, Mack MC (2000) Consequences of changing biodiversity. Nature 405(6783):234–242

Denbel T (2021) Woody species diversity and structure of protected woodlands adjacent to free grazing land woodland at Dugda woreda, Oromia, Ethiopia. Int J Forest Res 2021:1–10

Dufrêne M, Legendre P (1997) Species assemblages and indicator species: the need for a flexible asymmetrical approach. Ecol Monogr 67(3):345–366

Ellenberg D, Mueller-Dombois D (1974) Aims and methods of vegetation ecology. Wiley, New York, p 547

Ellery K, Ellery W (1997) Plants of the Okavango Delta: a field guide. Tsaro

Fenner MW (2012) Seed ecology. Springer Science & Business Media

Flyman MV (1999) The fate of woody plant seedlings in the presence and absence of large herbivores in a semi-arid savanna. Master of Philosophy Thesis, University of Botswana, Gaborone, Botswana

Fox JT, Vandewalle ME, Alexander KA (2017) Land cover change in Northern Botswana: the influence of climate, fire, and elephants on Semi-Arid Savanna Woodlands. Land 6(4):73

Gebrehiwot T, Veen A (2014) The effect of enclosures in rehabilitating degraded vegetation: a case of Enderta district, Northern Ethiopia. Forest Res Open Access 3(4):1000128

Gurmessa F, Soromessa T, Kelbessa E (2012) Structure and regeneration status of Komto Afromontane moist forest, East Wollega Zone, west Ethiopia. J For Res 23:205–216

Hegde R, Enters T (2000) Forest products and household economy: a case study from Mudumalai Wildlife Sanctuary, southern India. Environ Conserv 27(3):250–259

Helm CV, Witkowski ETF (2012) Characterising wide spatial variation in the population size structure of a keystone African savanna tree. For Ecol Manag 263:175–188

Hipondoka MHT, Versfeld WD (2006) Root the root system of Terminalia sericea shrubs across a rainfall gradient in a semi-arid environment of Etosha National Park, Namibia. Ecol Indic 6(3):516–524

Isaiah AM, Komi T (2015) Grasses of the Botswana National Botanical Garden. South African National Biodiversity Institute (SANBI)

Jonsson P (2004) Vegetation as an urban climate control in the subtropical city of Gaborone, Botswana. Int J Climatol 24(10):1307–1322

Käller A (2003) Growth pattern and reproduction of woody vegetation in a semi-arid savanna in southern Botswana. Committee of Tropical Ecology, Minor Field Study, p 86

Kashe K, Teketay D, Heath A, Mmusi M, Mathope T, Galelebalwe MK, Tsholofelo C, Mpofu C (2021) Diversity, population structure and regeneration status of woody species in different habitats in Maun Educational Park, northern Botswana. J For Res 26(4):294–302

Kashe K, Teketay D, Mmusi M, Galelebalwe MK (2023) Restoration of diversity and regeneration of woody species through area exclosure: the case of Maun International Airport in northern Botswana. Bothalia-Afr Biodivers Conserv 53(1):1–12

Kent M, Coker P (1992) Vegetation description and analysis, a practical approach. Belhaven Press, London

Krebs C (1989) Ecological methodology. Har-per Collins Publishers

Kuma M, Shibru S (2015) Floristic composition, vegetation structure, and regeneration status of woody plant species of Oda Forest of Humbo Carbon Project, Wolaita, Ethiopia. J Bot 2015:1

Lite SJ, Bagstad KJ, Stromberg JC (2005) Riparian plant species richness along lateral and longitudinal gradients of water stress and flood disturbance, San Pedro River, Arizona, USA. J Arid Environ 63(4):785–813

Lukhele MS, Van Ryssen JBJ (2003) The chemical composition and potential nutritive value of the foliage of four subtropical tree species in southern Africa for ruminants. S Afr J Anim Sci 33(2):132–141

Mapaure I, Ndeinoma A (2011) Impacts of local-level utilization pressure on the structure of mopane woodlands in Omusati region, Northern Namibia. African J Plant Sci 5(5):305–313

Marone L, De Casenave JL, Cueto VR (2000) Granivory in southern South American deserts: conceptual issues and current evidence. Bioscience 50(2):123–132

McCune B, Grace JB (2002) Analysis of ecological communities. MjM Software, Gleneden Beach

Mengistu A (2011) The role of area closures for soil and woody vegetation rehabilitation in Kewot District, North Shewa (Doctoral dissertation, MSc Thesis, Addis Ababa University, Addis Ababa, Ethiopia)

Mengistu T, Teketay D, Hulten H, Yemshaw Y (2005) The role of enclosures in the recovery of woody vegetation in degraded dryland hillsides of central and northern Ethiopia. J Arid Environ 60(2):259–281

Mlambo D, Nyathi P (2004) Seedling recruitment of Colophospermum mopane on the Highveld of Zimbabwe: a research note. South Afr For J 2004(202):45–53

Neelo J, Teketay D, Masamba W, Kashe K (2013) Diversity, population structure, and regeneration status of woody species in dry woodlands adjacent to Molapo farms in northern Botswana. Open J Forest 3(04):138

Neelo J, Teketay D, Kashe K, Masamba W (2015) Stand structure, diversity, and regeneration status of woody species in open and exclosed dry woodland sites around Molapo farming areas of the Okavango Delta, Northeastern Botswana. Open J Forest 5(04):313

Oguh CE, Obiwulu ENO, Umezinwa OJ, Ameh SE, Ugwu CV, Sheshi IM (2021) Ecosystem and ecological services; need for biodiversity conservation—a critical review. Asian J Biol 11(4):1–14

Pelane GL, Mulalu MI, Mmopelwa G, Stone MT (2023) Rural community dependency on woodland resources: the case of Foley and Makomoto settlements in Central-East Botswana. S Afr Geogr J 105:1–20

Pisani D, Pazienza P, Perrino EV, Caporale D, De Lucia C (2021) The economic valuation of ecosystem services of biodiversity components in protected areas: a review for a framework of analysis for the gargano national park. Sustain For 13(21):11726

Ringrose S, Matheson W (2001) Spatial characteristics of riparian woodlands in the distal Okavango Delta. Botswana Notes Rec 33(1):101–114

Schroder B (2001) Mokolodi nature reserve vegetation assessment 2001 report. Mokolodi Nature Reserve Conservation Department, Mokolodi

Sebego RJ, Atlhopheng JR, Chanda R, Mulale K, Mphinyane W (2019) Land use intensification and implications on land degradation in the Boteti area: Botswana. Afr Geograph Rev 38(1):32–47

Senbeta F, Teketay D (2003) Diversity, community types, and population structure of woody plants in Kimphee forest, a unique nature reserve in Southern Ethiopia. Ethiop J Biol Sci 2(2):169–187

Shackleton CM (1993) Demography and dynamics of the dominant woody species in a communal and protected area of the eastern Transvaal Lowveld. S Afr J Bot 59(6):569–574
Shannon CE (1948) A mathematical theory of communication. Bell Syst Tech J 27(3):379–423
Shibru S, Balcha G (2004) Composition, structure and regeneration status of woody species in Dindin Natural Forest, Southeast Ethiopia: an implication for conservation. Ethiop J Biol Sci 3(1):15–35
Skarpe C, Aarrestad PA, Andreassen HP, Dhillion SS, Dimakatso T, du Toit JT, Duncan JH, Hytteborn H, Makhabu S, Mari M, Marokane W (2004) The return of the giants: ecological effects of an increasing elephant population. AMBIO J Hum Environ 33(6):276–282
Sop TK, Oldeland J, Schmiedel U, Ouedraogo I, Thiombiano A (2011) Population structure of three woody species in four ethnic domains of the sub-sahel of Burkina Faso. Land Degrad Dev 22(6):519–529
Teketay D, Lemenih M, Bekele T, Yemshaw Y, Feleke S, Tadesse W, Moges Y, Hunde T, Nigussie D (2010) Forest resources and challenges of sustainable forest management and conservation in Ethiopia. In: Degraded forests in eastern Africa. Routledge, pp 31–75
Teketay D, Geeves G, Kopong I, Mojeremane W, Sethebe B, Smith S (2016) Diversity, stand structure and regeneration status of woody species, and spatial cover of herbaceous species in Mokolodi Nature Reserve, Southeastern Botswana. Int J Biodivers Conserv 8(8):180–195
Teketay D, Kashe K, Madome J, Kabelo M, Neelo J, Mmusi M, Masamba W (2018) Enhancement of diversity, stand structure and regeneration of woody species through area exclosure: the case of a mopane woodland in northern Botswana. Ecol Process 7(1):1–15
Teren G, Owen-Smith N (2010) Elephants and riparian woodland changes in the Linyanti region, northern Botswana. Pachyderm 47:18–25
Timberlake JR (1999) *Colophospermum mopane*: an overview of current knowledge. In: African plants: biodiversity, taxonomy, and uses. Proceedings of the 1997 AETFAT congress, Harare, Zimbabwe and Royal Botanic Gardens, Kew, London
Tsheboeng G, Murray-Hudson M (2013) Spatial variation of the population size structure of selected riparian tree species in the Okavango Delta, Botswana. Biodivers Ecol 5:341–350
Tsheboeng G, Murray-Hudson M, Kashe K (2017) Regeneration status of riparian tree species in two sites that differ in land-useland use in the Okavango Delta, Botswana. J For Res 28(5):1073–1082

G. Tsheboeng is a Senior Lecturer/Researcher in Plant Ecology at the University of Botswana in the Department of Biological Sciences. His research interests include phytosociology, plant population structure, and soil seedbanks dynamics.

C. Sekonopo research interests include plant ecology and conservation.

K. Kashe is a Senior Research Fellow in Plant Ecology at the University of Botswana based at Okavango Research Institute in Maun. His research interests include (i) dry land vegetation inventory and impacts of invasive plant species and (ii) monitoring the impacts of land-use change on plant populations and communities.

Chapter 7
Stability of a Fish Community in a Seasonal Floodplain of the Okavango Delta, Botswana

B. Mosepele, K. Mosepele, I. Mosie, and K. Sianga

7.1 Introduction

Some organisms undergo seasonal migration as an intrinsic part of their life history strategy to strike a balance between reproduction and survival (Begon et al. 1986; Begon and Mortimer 1986; Ricklefs 1990). These movements occur at temporal and spatial scales at varying magnitudes (Begon et al. 1986). One of the basic tenets in population dynamics is that the degree of spatial and temporal variations within a community structure depends on either environmental fluctuations or the inherent stability of a population (Ricklefs 1990). Therefore, understanding the spatial and temporal scales of fish movements (especially in pulsing systems) is an important building block in determining fish population dynamics and community processes (Begon and Mortimer 1986; Lucas and Baras 2000).

Migratory life histories are very common in fishes (Hillborn and Walters 1992). Some fish undergo seasonal migrations between wet and dry season dry habitats to maximize their success in subsequent generations (Welcomme 1979; Lowe-McConnell 1987; Hillborn and Walters 1992). The timing of these dispersal migrations is generally species specific (Welcomme 1979, 1985). Ephemeral or unpredictable patchy environments (Begon and Mortimer 1986), characteristic of

B. Mosepele
Department of Wildlife and Aquatic Resources, Faculty of Natural Resources,
Botswana University of Agriculture and Natural Resources (BUAN), Gaborone, Botswana
e-mail: bmosepele@buan.ac.bw

K. Mosepele · K. Sianga (✉)
Botswana University of Agriculture and Natural Resources (BUAN), Gaborone, Botswana
e-mail: kmosepele@buan.ac.bw; ksianga@buan.ac.bw

I. Mosie
Okavango Research Institute, University of Botswana (UB), Maun, Botswana
e-mail: imosie@ub.ac.bw

W. L. Hambira et al. (eds.), *Environmental Change and Biodiversity Conservation in sub-Saharan Africa*, Advances in Global Change Research 75,
https://doi.org/10.1007/978-3-031-73136-5_7

seasonal floodplains, are key factors affecting fish community structure (Lowe-McConnell 1987). These seasonal, environmental fluctuations have engendered a broad spectrum of life history responses by different floodplain species (Lowe-McConnell 1987). According to Lowe-McConnell (1987), these ecological attributes are based on phyletic patterns, levels of diversity, levels of environmental complexity, temporal patterns, behavioural attributes, population dynamics, factors controlling population size and trophic specializations. While earlier studies have been conducted on the ecology of the Okavango Delta fish (Merron and Bruton 1988; Merron 1991, 1993a, b), very little attention has been given to these ecological attributes.

Protecting biodiversity is an essential yet challenging problem accentuated by climate change and increasing anthropogenic pressures on ecosystems (Jetschke 2002). Biodiversity, ecosystem functioning, and stability are key questions in floodplain fish community dynamics (Danet et al. 2021). Generally, biodiversity has a damping effect on ecosystem sensitivity (Oliveira et al. 2022), where in the narrow sense, biodiversity refers to a "species relative abundance in a community" (Jetschke 2002). Understanding these attributes in fisheries ecology facilitates the integration of climate change into fisheries management (Danet et al. 2021). At the heart of these questions is understanding the mechanisms that structure ecological communities (Norris 2003). The key focus of this study was to highlight the biodiversity dynamics in a seasonal floodplain fish species community. It is also relevant to assess the stability of a fish community in two microhabitats in a seasonal part of the Okavango Delta.

7.2 Materials and Methods

7.2.1 Study Area

The Nxaraga floodplains (Fig. 7.1) are seasonally flooded by floodwaters that originate from the Angolan highlands (Mladenov et al. 2008). According to Krah et al. (2004), the seasonal floodplains in this part of the delta are usually dry between December and May. This suggests that the local rainfall season between November and February (Bonyongo and Mubyana 2004) is not sufficient for flooding floodplains. The main flooding months are between May and December (Bonyongo and Mubyana 2004), when the maximum flooded area is reached annually between August and September (Krah et al. 2004). However, there is interannual variability in the timing of the maximum flooded area, where Høberg et al. (2002) observed that the maximum flooded area occurs in June. The vegetation in the area is composed of aquatic, primary floodplain, secondary grassland, island grassland, and riverine woodland communities (Krah et al. 2006). According to Ashton et al. (2003a), this aquatic vegetation creates microhabitats for some aquatic macroinvertebrates in flooded floodplains. Shallow lagoons in the study area are characterized by anoxic conditions where water lilies reduce water aeration and act as "stabilizers"

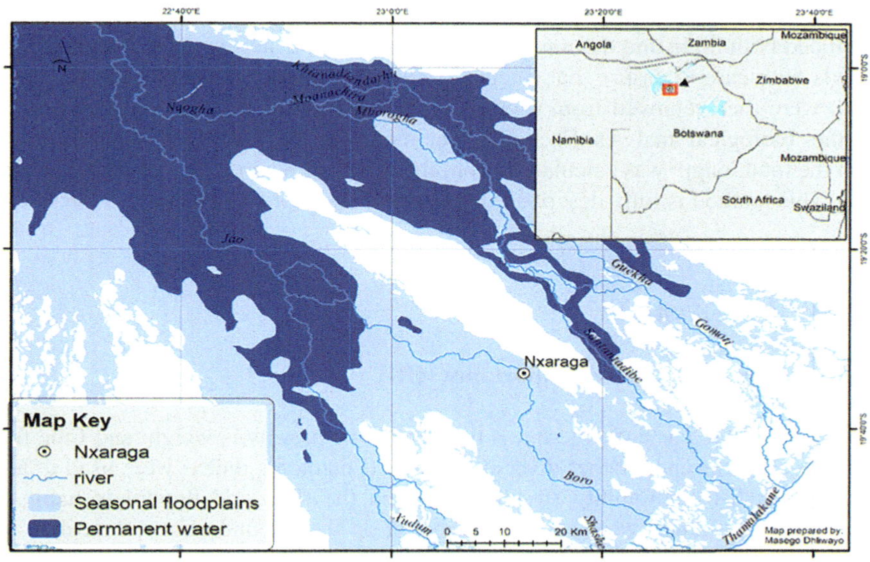

Fig. 7.1 Map of the study area

on the water surface (Ashton et al. 2003b). Nxaraga, a distal seasonal floodplain of the Okavango Delta, is characterized by a mosaic of grassy floodplain habitats, seasonal pools, lagoons, and a relatively more permanent riverine system.

According to McCarthy and Ellery (1995), the seasonal floodplain is characterized by poorly defined channel systems (such as the Boro River) consisting of sinuous depressions. The channel bed is heavily vegetated, with negligible bedload. The Boro River is a seasonally flowing mainstream channel (Merron 1993a). According to Biggs (1976), *Miscanthus junceus*, *Vossia cuspidata*, *Echinochloa stagnina* and sedges such as *Cyperus articulatus* are some of the dominant marginal vegetation species in the area. Plant species such as *Ottelia ulvifolia, Ceratophyllum demersum* and *Lagarosiphon ilicifolius* are among the dominant submerged aquatic vegetation in the river channel. According to Cronberg et al. (1996), the mean depth in the seasonal floodplain ranges from 0.05 to 0.15 m. The total area of the floodplain has been estimated to be 15 ha by Lindholm (2002) at peak floods (end of June).

7.2.2 Data Collection

Standardized experimental fishing using research nets was conducted in the river channel and floodplain habitats. Study sites were selected randomly where both habitats (i.e. rivers and floodplains) were represented in the surveys. The nets were set overnight for approximately 13 hours and then lifted the following morning at the same sites throughout the study period. The day setting also involved setting the nets in the morning and lifting them in the early afternoon. Therefore, the sampling included both night and day samples. Subsequently, the nets were placed in plastic bowls with care to ensure that catches from either net did not mix with the other. Fish were then removed from each individual net and kept in separate bowls for further biological analysis. Each fish specimen was identified to the species level, and the total length was calculated (from the tip of the snout to the tip of the caudal fin). This method is fully described in Mosepele and Kolding (2003).

7.2.3 Data Analysis

7.2.3.1 Index of Relative Importance (IRI)

IRI (IRI = $f(n\% + w\%)$), where n is the number of prey, w is weight and f the frequency of occurrence. Other diversity indices (Shannon's index) were used to test changes in fish community structure between the river and floodplain habitats. Species in this study were categorized based on their abundance as described by Rader (1994), where A = abundant (>100 individuals collected); C = common (20–100 individuals collected); and R = rare (<20 individuals collected). Subsequently, species whose abundance was categorized as rare (R) were left out of any further analysis.

7.2.3.2 Univariate Analysis

Assessment of the stability of the fish species assemblage was performed using a coefficient of variation (CV) similar to that of Mosepele et al. (2017) for the Okavango Delta. CV allows for scaling population variation by the mean, which permits comparison of populations with different mean abundances, thus making it less ambiguous than other metrics (Grossman et al. 1990). CV percentage values were classified into equal quartiles as highly stable (CV $\leq$ 25%), moderately stable (25% $<$ CV $\leq$ 50%), moderately fluctuating (50% $<$ CV $\leq$ 75) and fluctuating (>75%) based on Oberdorff and Porcher (1992). Other metrics used to assess assemblage/community dynamics include %IRI, which, according to Hart et al. (2002), is a good measure of abundance (relative importance). The relative biomass, diversity (Shannon's index), catch per unit effort (CPUE), and net CPUE were calculated via PASGEAR (Kolding and Skaalevik 2010).

7.2.3.3 Multivariate Analysis

Principal Component Analysis (PCA)
The environmental effects on fish community stability were assessed using PCA implemented in the Windows program, which performs a nonparametric

multivariate analysis of ecological data (PcORD). PCA is an unconstrained multivariate tool used to assess the relationships between species assemblages and their environment (Leps and Smilauer 2003). The data were log-transformed (log $(x + 1)$) to minimize the range and skew of the distributions (Cantu and Winemiller 1997). According to Jolliffe (2002), this standardization is important for reducing "the problem of scale dependence from PCA". The Euclidean distance was used as the dissimilarity measure, while p was estimated by a Monte Carlo test using 999 permutations. This test (Monte Carlo) is conducted to evaluate the statistical significance of the relationship between biological and environmental variables (Ter Braak and Smilauer 2002; Leps and Smilauer 2003).

7.3 Results

7.3.1 River and Floodplain Assemblages

A total of 7369 fish specimens comprising 47 species belonging to 11 fish families were collected during the study period (Table 7.1). Cichlidae and Cyprinidae were the two dominant families in both the river and floodplain assemblages. These two families contributed 12 and six species, respectively, to the river assemblage. Furthermore, the two families also contributed eight and 11 species, respectively, to the floodplain assemblage. Cyprinidae was the dominant family in the floodplain assemblage, while Cichlidae was the dominant family in the river assemblage. *Schilbe intermedius* was the most important species in both the river and floodplain assemblages (Fig. 7.2) based on the %IRI (approximately 28 and 19%, respectively). *Sargochromis greenwoodii* was the least important species (approximately 0,001%IRI) in the river assemblage, while *Serranochromis macrocephalus* was the least important species (approximately 0,0002%IRI) in the floodplain assemblage, as summarized in Table 7.1. Generally, cichlids were absent in the floodplain habitat, while cyprinids were generally more abundant in the floodplain habitat.

7.3.2 Dynamics at the Habitat Level

There was seasonal variability in fish biomass, relative abundance (%IRI), catch rate (CPUE), diversity, and stability (CV) in both habitats within the same period (Fig. 7.3). The fish river biomass peaked in August, while a corresponding peak in the floodplain fish biomass was observed in September. The floodplain fish assemblage CPUE (nos./set) decreased from May to June and then remained stable until October, when it increased marginally to November. Conversely, the river fish assemblage CPUE (nos./set) increased marginally from May until November. Generally, there was a decrease in the diversity (H') and mean relative abundance (%IRI) of the floodplain fish species assemblage from May until November, while

Table 7.1 Total species composition from the study area (river and floodplain habitats), where %IRI is used to indicate the most important species in the fish species assemblage

Family	Species	Abundance		%IRI	
		River	Floodplain	River	Floodplain
Alestidae	*Brycinus lateralis*	0	3	0	0,00126
Anabantidae	*Ctenopoma multispine*	1	6	0,00103	0,01541
Cichlidae	*Tilapia sparrmanii*	168	327	19,94585	18,0058
	Coptodon rendalli	22	16	0,88066	0,4334
	Oreochromis andersonii	16	15	0,7847	0,1546
	Serranochromis angusticeps	7	7	0,1121	0,07002
	Sargochromis codringtonii	12	0	0,07201	0
	Oreochromis macrochir	3	0	0,02112	0
	Sargochromis carlottae	2	0	0,01085	0
	Tilapia ruweti	1	0	0,00806	0
	Serranochromis thumbergi	2	0	0,00537	0
	Serranochromis macrocephalus	1	0	0,00299	0
	Serranochromis robustus	1	0	0,00244	0
	Pharyngochromis acuticeps	0	5	0	0,00515
	Serranochromis altus	0	2	0	0,00182
	Sargochromis greenwoodii	1	0	0,00101	0
	Pseudocrenilabrus philander	0	1	0	0,00021
	Serranochromis macrocephalus	0	1	0	0,00019
Clariidae	*Clarias ngamensis*	20	21	3,59607	2,17341
	Clarias gariepinus	28	11	9,91511	0,42978
	Clarias liocephalus	1	0	0,00105	0
Cyprinidae	*Barbus paludinosus*	396	818	15,28701	9,60451
	Barbus bifrenatus	246	607	2,54022	4,48295
	Barbus poechii	17	43	0,17975	0,18554

	Barbus thamalakanensis	12	332	0,06139	4,85912
	Barbus afrovernayi	0	38	0	0,11866
	Barbus fasciolatus	0	13	0	0,02356
	Barbus multilineatus	3	96	0,00606	0,57836
	Barbus haasianus	3	182	0,00606	1,29967
	Barbus barnardi	0	7	0	0,00395
	Barbus radiatus	0	1	0	0,0002
	Coptostomabarbus wittei	0	18	0	0,03044
Distichodontidae	*Hemigrammocharax multifasciatus*	0	13	0	0,01954
Hepsetidae	*Hepsetus cuvieri*	41	79	5,67561	7,35897
Mochokidae	*Synodontis thamalakanensis*	11	8	0,16969	0,02596
	Synodontis leopardinus	13	11	0,11854	0,0402
	Synodontis macrostigma	1	4	0,00164	0,00867
Mormyridae	*Marcusenius altisambesi*	189	136	11,98241	2,78338
	Petrocephalus okavangensis	78	6	0,78891	0,00472
	Cyphomyrus discorhynchus	11	2	0,03568	0,00104
	Squeakers (Synodontis sp.)	4	2	0,0254	0,00313
	Pollimyrus castelnaui	4	27	0,01288	0,02113
	Synodontis macrostoma	1	0	0,00137	0
Procatopodidae	*Aplocheilichthys johnstoni*	0	1451	0	17,50369
	Aplocheilichthys hutereaui	0	868	0	10,50976
	Aplocheilichthys katangae	0	80	0	0,39088
Schilbeidae	*Schilbe intermedius*	207	589	**27,74697**	**18,85093**
Total		1523	5846	100	100

The values in (Table 7.1) indicate the abundance classification according to Rader (1994)

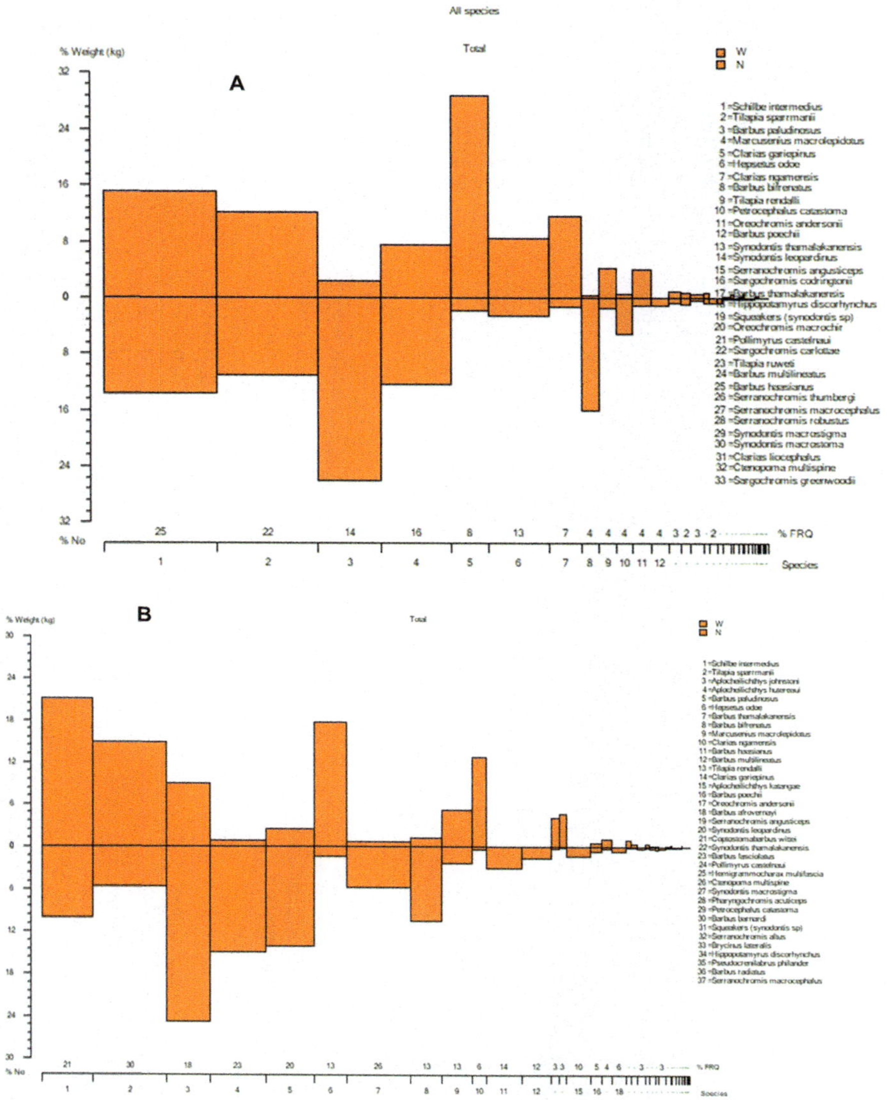

Fig. 7.2 IRI illustrating the species composition of the (**a**) river and (**b**) floodplain species assemblages in the study area

the opposite trend was observed for the river assemblage. Fluctuations in assemblage stability in both habitats followed a similar trend (Fig. 7.3), where the floodplain assemblage was significantly more unstable at the 90% confidence level (Table 7.2). Variability in the other metrics (i.e. diversity, %IRI, biomass, and CPUE) was not significant ($p > 0.05$), as summarized in Table 7.2.

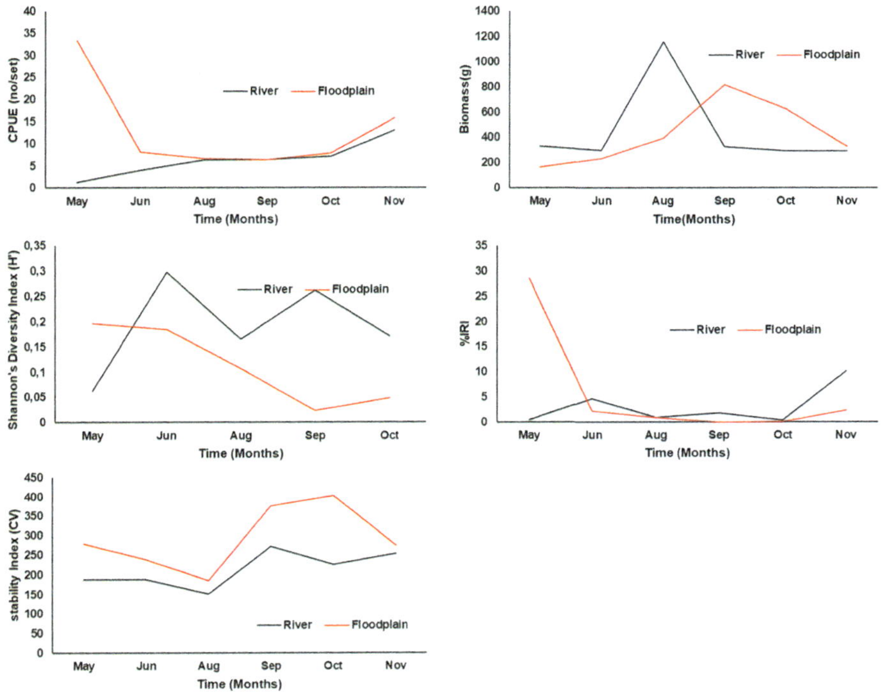

Fig. 7.3 Habitat partitioning of fish species assemblages based on five community metrics (biomass, CPUE, diversity, %IRI and stability) in the study area

7.3.3 Dynamics at the Species Level

Floodplain habitat Dynamic changes in fish relative importance (based on %IRI) within the fish community were observed in the floodplain habitat (Fig. 7.4). Generally, *Tilapia sparrmanii* dominated the fish community during the last 3 months of the year. Its dominance was more pronounced in October (IRI = 81.4%). The two most important species in November were larger fish species, while the third most important species was a smaller species. Conversely, the top three fish species in May were smaller fish, with *A. johnstoni* being the most important species (IRI = 39.3%). These fluctuations in fish community structure are observed for species such as *A. hutereaui,* which moved from being the second most important species in May to being the fourth most important species in June. *S. intermedius,* from fourth in May to first in June and third in August, disappeared from the top five in September and October and then reappeared as the second most important species in November. The dominance of *S. intermedius* in the fish community was most pronounced in September, with a %IRI value of 76.4. *Hepsetus cuvieri* is another species that underwent dynamic changes in relative importance in September, October, and November.

Table 7.2 Results of single-factor ANOVA summarizing the differences in several community indices between the river and floodplain fish species assemblages in the study area

Index	FP_{mean} (± SE)	R_{mean} (± SE)	p
%IRI	5.740 ± 127.108	3.054 ± 14.564	0.593
Biomass (g)	428.850 ± 61 377.481	449.310 ± 120 972.396	0.909
CPUE (g/set)	13.160 ± 112.128	6.360 ± 15.183	0.171
Stability (CV)	293.167 ± 6704.167	212.667 ± 2103.867	0.062*
Diversity (H')	0.130 ± 0.008	0.220 ± 0.010	0.164

p values with an asterisk indicate a significance level of 90%
FP floodplain, *R* river

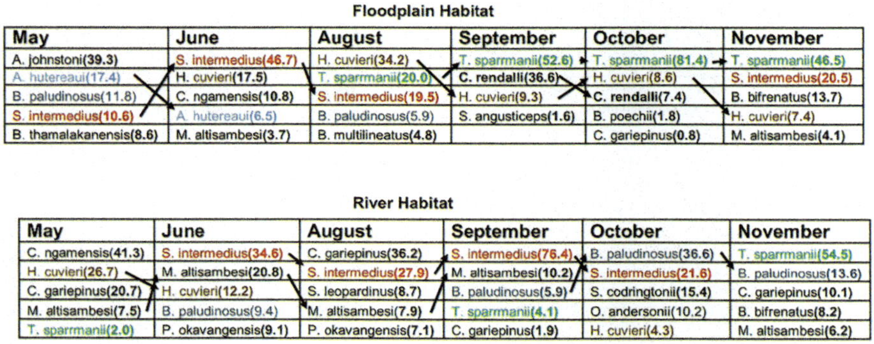

Fig. 7.4 Dynamic seasonal changes in the fish community structure illustrating %IRI changes in the top five species in the fish species assemblage by habitat. The values in parentheses are %IRI

River habitat Figure 7.4 illustrates the dynamic changes in the five most important species in this habitat between May and November. Generally, this habitat is dominated by larger fish species in May, June, and August. Catfishes, which are the top predators, dominated the fish community in May (*Clarias ngamensis*) and August (*C. gariepinus*). *S. intermedius*, which is also a top predator, dominated the fish community in June and September. *S. intermedius* underwent dynamic changes from June (the top species in the community) to August (the third most important species) and from September (the most important species) to October (the second most important species).

Although these differences were not significant according to the ANOVA ($p = 0.43$), populations of the same species within the river habitat were relatively more unstable than were populations of the same species within the floodplain habitat (Fig. 7.5a). *M. altisambesi* and *H. cuvieri* are two examples of this observation, as summarized in Table 3. While *M. altisambesi* populations in the river habitat were "moderately fluctuating" (CV = 67), they were highly fluctuating (CV = 100) in the floodplain. Similarly, *H. cuvieri* populations were "moderately stable" (CV = 38) in the river habitat and "moderately fluctuating" (CV = 71) in the floodplains. Furthermore, within the river habitat, species whose abundance was categorized as rare (R) generally had "highly fluctuating" populations (CV > 75). The only

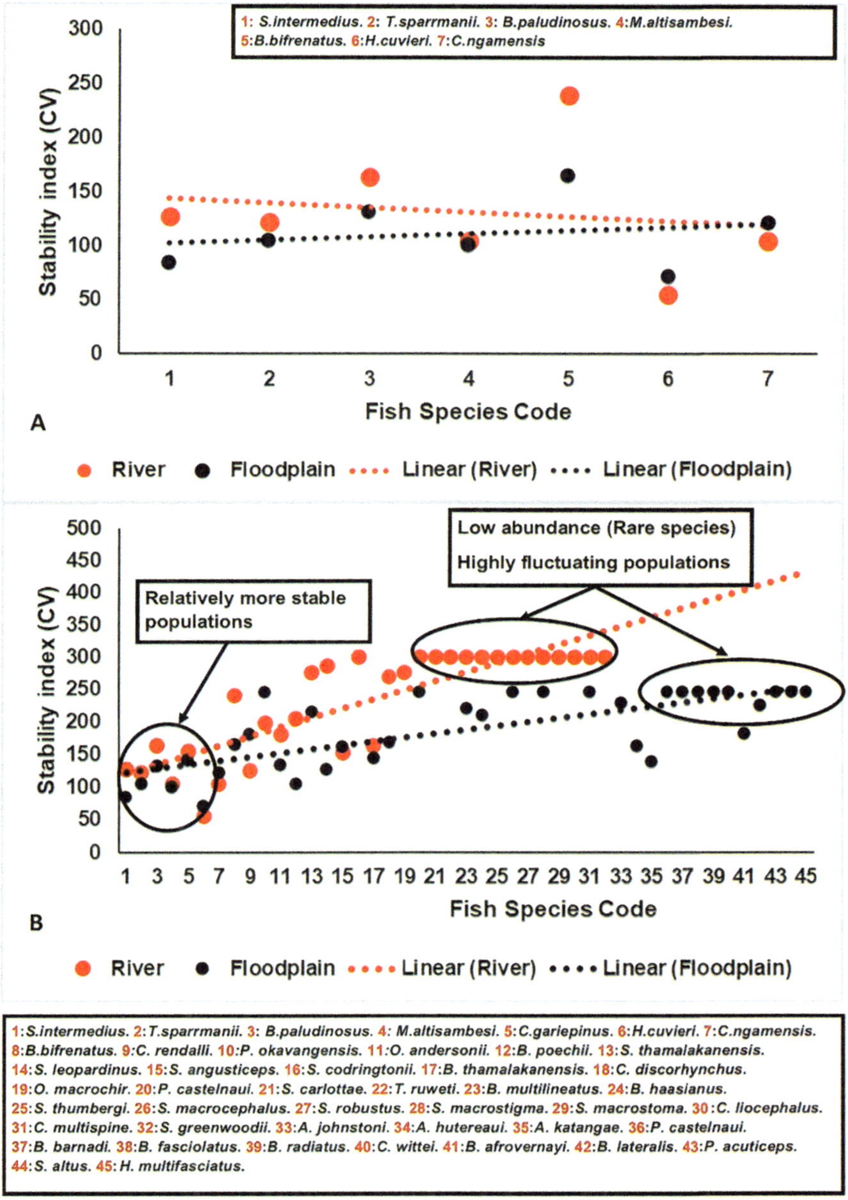

Fig. 7.5 Dynamic changes in the stability of a seasonal floodplain fish species assemblage at Nxaraga

slight exceptions were *Serranochromis angusticeps* and *Barbus thamalakanensis,* whose populations were categorized as rare (R) but did not fluctuate as much as the other species.

Similarly, some populations classified as abundant (A), such as *M. altisambesi* and *S. intermedius,* were classified as "moderately fluctuating" (CV = 67%) or "fluctuating" (CV = 89%), respectively. However, the *H. cuvieri* population, whose abundance was classified as common (C), was "moderately stable" (CV = 38%). Conversely, the *Barbus paludinosus*, *T. sparrmanii*, and *Barbus bifrenatus* populations, which were classified as abundant (A), had fluctuating (CV > 75) populations. Conversely, different observations were made for the floodplain habitat species assemblage, as summarized in Table 7.3. Apart from the *H. cuvieri* population, which was "moderately stable" (CV = 71), all the species populations were "fluctuating" (CV > 75). Generally, 5B illustrates that floodplain fish populations were more unstable than river fish populations ($\beta = -0.312$, 95% CI [−0.535, −0.089], $p = 0.007$, adjusted $R^2 = 0.173$). At the assemblage level, populations of smaller fish species were significantly unstable ($\beta = -0.274$, 95% CI [−0.545, −0.003], $p = 0.048$, adjusted $R^2 = 0.095$) compared to populations of larger fish species in both habitats (Fig. 7.6).

7.3.4 Environmental Variability and Stability

River Habitat

The first two axes (eigenvalue >1) were able to explain the relationship between environmental variability and the stability of the river habitat fish species assemblage (Table 7.4). These two axes cumulatively explained approximately 71% of the variance observed in the stability of the river fish species populations.

Overall, axis 1 of the PCA revealed a significant relationship ($p = 0.022$) between environmental variability and the observed variability in the river habitat fish species populations. In general, the two main axes identified by the PCA are the (i) conductivity and (ii) discharge:DO axes (Fig. 7.7). Nonetheless, there are some differences in the distributions of population stability for the different species along these two axes.

The stability of *C. ngamensis* is related to the conductivity gradient around the months of May and September. The stabilities of *C. gariepinus*, *H. cuvieri* and *B. paludinosus* are strongly correlated with the conductivity:discharge gradient in the months of June and November. The stabilities of *S. intermedius*, *M. altisambesi*, and *P. okavangensis* were correlated with the discharge:DO gradient in the months of October and August. The stability of the two cichlids, *T. sparrmanii* and *C. rendalli*, was neutral to any of the observed environmental or seasonal gradients. Three stable groups existed in the river habitat according to the cluster analysis (Fig. 7.8). These clusters include (i) the February, March and April group; (ii) the August and October group; and (iii) the May, September, November and June group.

Table 7.3 Summary of stability indices (CVs) and abundance categories for the same species in the two habitats of the study area

	River		Floodplain	
Species	CV (%)	Abundance	CV (%)	Abundance
Schilbe intermedius	89	A	84	A
Tilapia sparrmanii	146	A	104	A
Barbus paludinosus	123	A	131	A
Marcusenius altisambesi	67	A	100	A
Clarias gariepinus	115	C	141	R
Hepsetus cuvieri	38	C	71	C
Clarias ngamensis	105	C	121	C
Barbus bifrenatus	192	A	165	A
Coptodon rendalli	155	C	180	R
Petrocephalus okavangensis	155	C	245	R
Oreochromis andersonii	160	R	134	R
Barbus poechii	160	R	105	C
Synodontis thamalakanensis	224	R	214	R
Synodontis leopardinus	233	R	126	R
Serranochromis angusticeps	113	R	160	R
Sargochromis codringtonii	245	R	–	–
Barbus thamalakanensis	123	R	144	A
Cyphomyrus discorhynchus	218	R	167	R
Pollimyrus castelnaui	245	R	245	C
Sargochromis carlottae	245	R	–	–
Barbus multilineatus	245	R	218	C
Barbus haasianus	245	R	208	A
Serranochromis thumbergi	245	R	–	–
Serranochromis macrocephalus	–	–	245	R
Serranochromis robustus	245	R	–	–
Synodontis macrostigma	245	R	245	R
Synodontis macrostoma	245	R	–	–
Ctenopoma multispine	245	R	245	R
Sargochromis greenwoodii	245	R	–	–
Aplocheilichthys johnstoni	–	–	227	A
Aplocheilichthys hutereaui	–	–	162	A
Aplocheilichthys katangae	–	–	138	C
Pollimyrus castelnaui	–	–	245	C
Barbus barnadi	–	–	245	R
Barbus fasciolatus	–	–	245	R
Barbus radiatus	–	–	245	R
Coptostomabarbus wittei	–	–	245	R
Barbus afrovernayi	–	–	182	C
Brycinus lateralis	–	–	224	R
Pharyngochromis acuticeps	–	–	245	R
Serranochromis altus	–	–	245	R
Hemigrammocharax multifasciatus	–	–	245	R

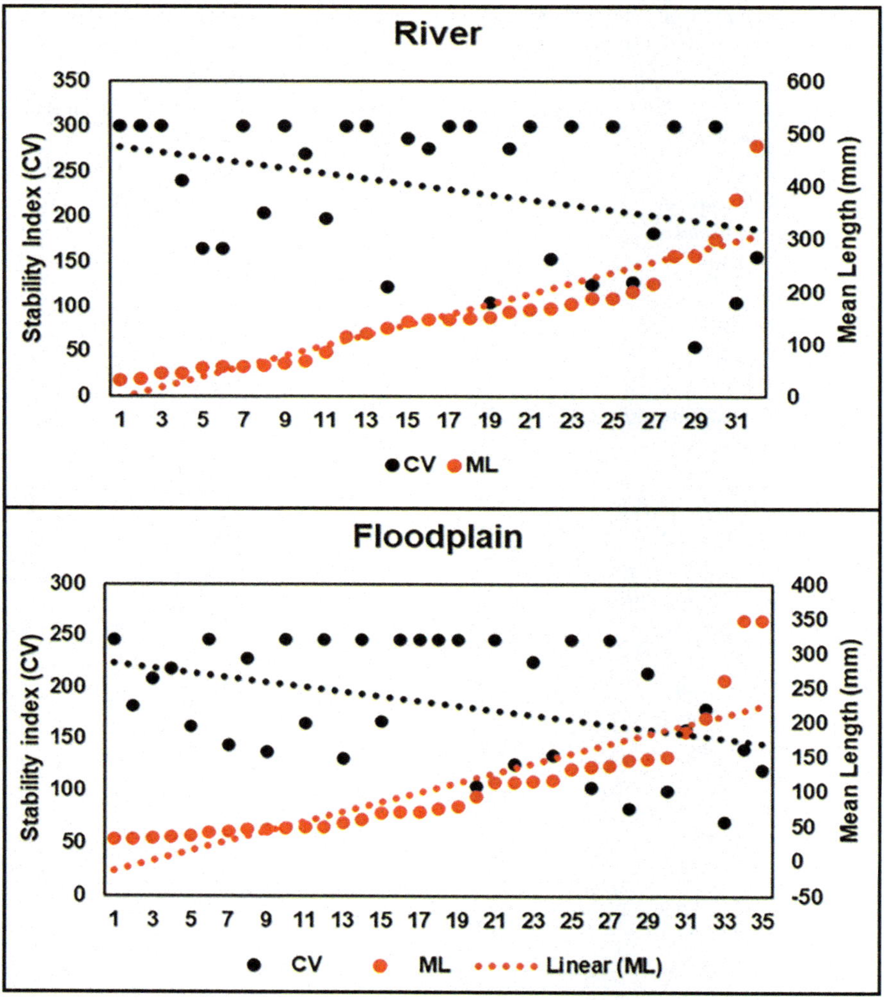

Fig. 7.6 Regression analysis illustrating the relationship between stability (CV) and mean fish size (mean length) within the two habitats in the study area

Table 7.4 Summary of results from PCA where *p* values are a result of Monte Carlo randomization

Axis	Eigenvalue	%Variance	Cumulative	*p*
1	4.455	44.451	44.451	0.022*
2	2.641	26.347	70.798	0.192

p values with an asterisk indicate a significant relationship at the 95% confidence level

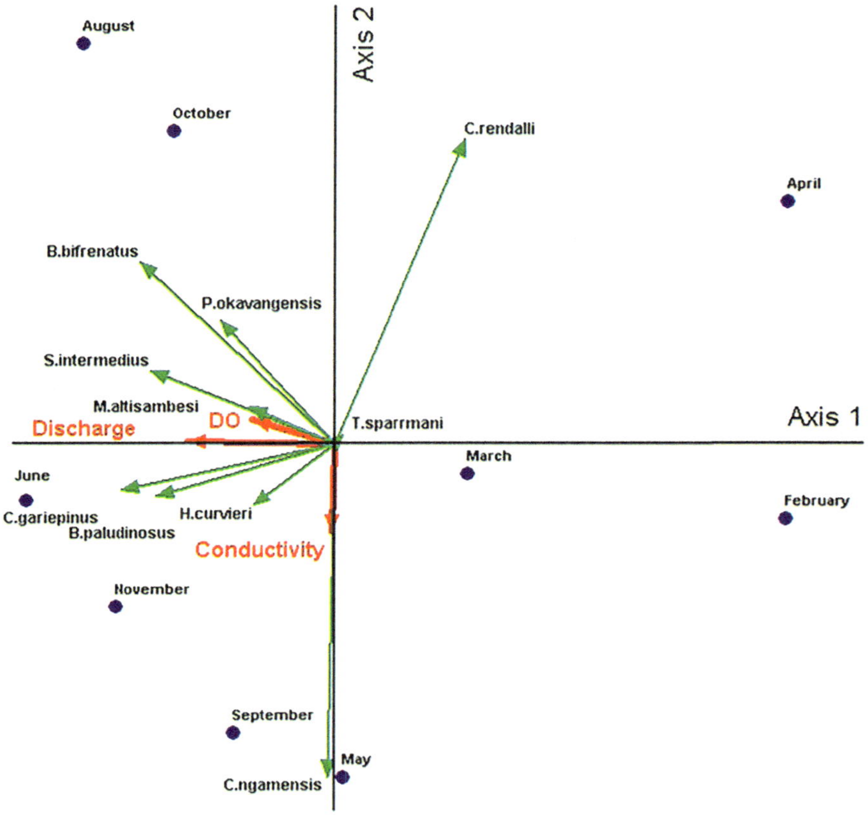

Fig. 7.7 Principal component analysis biplot illustrating the relationship between the stability of river habitat fish species and environmental variability at the study site

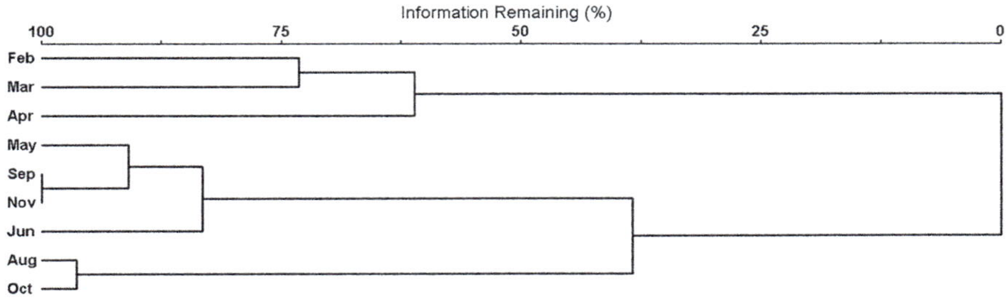

Fig. 7.8 Cluster analysis dendrogram illustrating stability groups in the river habitat fish species assemblage

Table 7.5 Summary of results from PCA where p values are a result of Monte Carlo randomization

Axis	Eigenvalue	%Variance	Cumulative	p
1	12.409	65.631	65.631	0.001*
2	3.657	19.344	84.976	0.992
3	1.412	7.468	92.444	1.000
4	1.247	6.594	99.038	0.994

p values with an asterisk indicate a significant relationship at the 95% confidence level

Floodplain Habitat

The first four axes (eigenvalue >1) cumulatively explained 99% of the variance observed in the stability dynamics of the different fish species populations in the floodplain habitat (Table 7.5). PCA revealed that there was a significant relationship ($p = 0.001$) between the environmental variability and the stability of the floodplain fish species assemblage.

The four axes are the (i) temperature, (ii) discharge, (iii) conductivity and (iv) DO gradients (Fig. 7.9). Generally, the hydrology (discharge) axis is associated with two topminnows (*A. johnstoni* and *A. hutereaui*) and two cyprinids (*B. afrovernayi* and *B. haasianus*) in May and June. The population stability of four other species, which included two cyprinids (*B. thamalakanensis* and *B. multilineatus*), a topminnow (*Aplocheilichthys katangae*) and a small mormyridy (*P. castelnaui*), was associated with a conductivity gradient in May and June. The population stability of two large predators (*S. intermedius* and *C. ngamensis*), three cyprinids (*Barbus poechii*, *B. paludinosus*, and *B. bifrenatus*), and one mormyridy (*M. altisambesi*) was associated with a DO gradient in August. The stabilities of the *H. cuvieri* and *T. sparrmanii* populations were weakly related to the temperature gradient, while Fig. 7.9 illustrates the strong correlation between the stabilities of these two populations.

Cluster analysis revealed stable seasonal partitioning, where three stable groups were identified in the floodplain habitat (Fig. 7.10). The three identified stable cluster groups are May–June, August–November, and September–October.

7.4 Discussion

7.4.1 General Observations

Our results showed that cichlids and cyprinids dominate both habitats (i.e. rivers and floodplains), which attests to their importance in river–floodplain fisheries. Cichlids are the backbone of an economically viable small-scale commercial fishery in the Okavango Delta (Mmopelwa et al. 2005). *O. andersonii*, *O. macrochir* and *C. rendalli* are the preferred species in Delta fisheries (Mosepele et al. 2007, 2022; Mosepele 2019). Therefore, cichlids constitute a key component of not only commercial fishing in the Delta (Mosepele et al. 2003) but also major subsistence species in riparian communities (Mosepele et al. 2006).

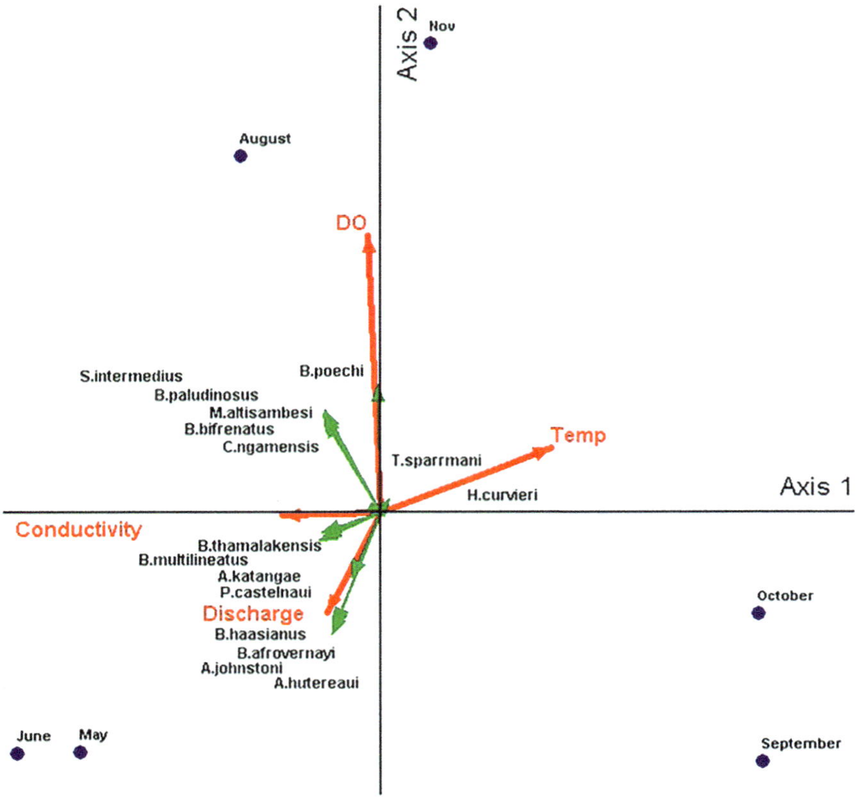

Fig. 7.9 Principal component analysis biplot illustrating the relationship between the stability of floodplain habitat fish species and environmental variability in the study area

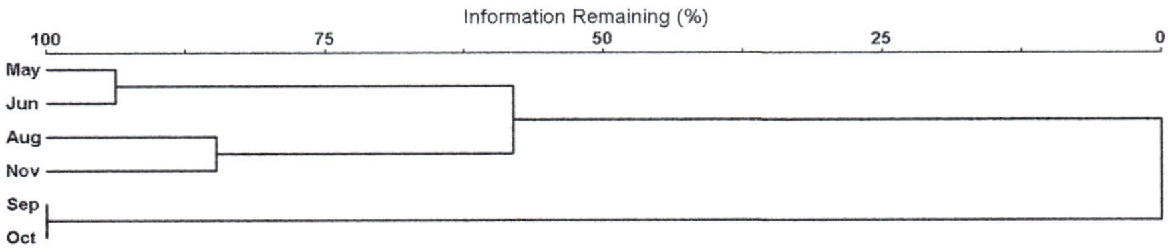

Fig. 7.10 Cluster analysis dendrogram illustrating stable groups in the floodplain habitat fish species assemblage

Cyprinids are key subsistence fishery species exploited by traditional fishing gear in the Delta (Mosepele et al. 2022). Young boys, using traditional hooks and lines, exploit larger cyprinids such as *B. poechii*; women, using "mosquito nets", exploit topminnows (*Aplocheilichthys* spp.) and smaller cyprinids such as *B. bifrenatus*; and women, using fishing baskets, exploit smaller cichlids such as

P. philander and smaller mormyrids such as *C. discorhynchus* (Mosepele et al. 2003, 2022; Mosepele 2019). Generally, smaller species have greater nutritional content than larger fish species. This makes them key sources of household nutrition in the Okavango Delta, particularly for socioeconomically marginalized households (Mosepele et al. 2006). Furthermore, Ngwenya and Mosepele (2008) showed that the delta's fishery is a source of nutrition for households with "chronically ill persons". The nutritional importance of small species has been highlighted in other African wetlands (FAO 2016) and other systems in Asia (Kongsbak et al. 2008; Roos et al. 2007).

7.4.2 *Fluxes in Fish Community Structure:* At the Microhabitat Scale

At the microhabitat level (river and floodplain habitats), the delta fish community undergoes dynamic seasonal variability. This finding is consistent with the observations of Mosepele et al. (2017) for the Okavango Delta fish community. Mosepele et al. (2017) also established this phenomenon at the macroscale in the Delta where fish dynamics are regulated by the seasonal flood pulse regime. Similarly, the effect of this flood pulse on the trophodynamics of the delta's fish community was observed by Mosepele et al. (2012). Temporally flooded floodplains act as nutrient pumps when they are flooded, which facilitates primary production for increased fish productivity (FAO 2016). This is what facilitates fish longitudinal migrations to these seasonally flooded habitats.

This study revealed temporal variation in the species dominance of the river channel fish community structure, whereby the fish community was dominated by different species at any given time. However, *Schilbe intermedius* is a dominant species in both river and floodplain habitats (Table 7.1 and Fig. 7.2). This observation is consistent with previous research showing that *S. intermedius* is ubiquitous in the delta region (Mosepele et al. 2017, 2022; Mosepele 2019). Furthermore, earlier studies in the Delta region have highlighted that *S. intermedius* is unutilized by commercial, recreational or subsistence fisheries. This might also attest to its dominance in the system because it is relatively unutilized. The principal target species in the Delta River fishery, *O. andersonii*, *O. macrochir*, and *C. rendalli* (Mosepele et al. 2022), are less important in the river channel fish community, which contradicts Merron and Bruton's observations (1988) for the delta. Generally, the community is dominated by predator fish species, and they completely dominate the fish community in September. Prey species (*B. paludinosus*) increase in dominance in October and November, while predator dominance decreases. The dominance of the river habitat by *C. gariepinus* in August could be due to Merron's (1993) annual catfish running in the delta. During this period, *C. gariepinus* undergoes longitudinal migration in the river channel and preys on *M. altisambesi,* which may backmigrate from floodplains (Merron 1993; Mosepele et al. 2009).

The floodplain fish community structure is dominated by *T. sparrmanii*, whose dominance increased over time, particularly from September until November, during receding floods. The presence of *C. rendalli* in September and October suggested that this species migrated into the floodplains for spawning immediately after the flood arrived. Its presence at this time could be a result of adults returning to the river habitat. Likewise, *T. sparrmanii* migrated into the floodplains at the onset of floods and until the end of the flood season. These results agree with those of Hoberg et al. (2002), who observed that these species and *C. gariepinus* dominated the floodplain species community in June 1998. The back migration of adults from floodplains after spawning has been previously observed (King et al. 2003). This clearly attests to the migrations of cichlids into floodplains for spawning, which agrees with the findings of Hoberg et al. (2002) and King et al. (2003), who observed that floodplain inundation enhances fish recruitment. Most likely, *C. gariepinus* and *C. ngamensis* migrated out into the floodplain to take advantage of prey species such as cichlids, mormyrids (especially *M. altisambesi*), and cyprinids (*Barbus* spp.). This is based on Merron's (1993a, b) observations that these catfish species feed predominantly on *M. altisambesi*. A similar relationship between seasonal variations in a predator and its prey has been previously observed by Lowe-McConnel (1979).

The relative importance of *H. cuvieri* in the floodplain habitat is consistent with the findings of Mosepele et al. (2009), who reported that this species prefers "slower-flowing, well-vegetated seasonal" swamps. Furthermore, while the key prey fish species for *H. cuvieri* is *B. thamalakanensis*, it also feeds on diverse prey fish species, which include topminnows (Mosepele et al. 2012). Therefore, intraspecific interactions could also contribute to the fish biodiversity dynamics observed in this area. Therefore, "predator–prey interactions" in floodplain fish dynamics are a key research area in the Okavango Delta.

7.4.3 Temporal and Spatial Fluctuations in Community Structure

There is a clear inverse relationship between fish species diversity in river and floodplain lagoon habitats. Generally, there is a gradual decrease in floodplain fish species diversity, which is related to an increase in river fish species diversity. However, the high diversity of floodplain species in May (Fig. 7.3) suggested the occurrence of optimum breeding and feeding conditions, which agrees with the findings of other studies (Junk 2000) on fish behaviour during flooding. This peak, with corresponding variability in the diversity of the river assemblage, attests to fish lateral migrations from the river habitat into the floodplain, which creates fluctuations in species diversity between the two habitats.

The fluctuations in river diversity suggest that species diversity is constantly expanding and contracting, perhaps in concert with abiotic (e.g. flood levels, water flow rates, water chemistry) and biotic (e.g. aquatic invertebrate densities) factors. This shows that species diversity is highly plastic, which is perhaps a perfect mirror of the seasonal contractions and expansions of environmental conditions in response to the flood regime. Mosepele et al. (2017) showed that these changes in fish diversity are driven by the delta hydrological regime. These fluxes in community structure are controlled by longitudinal and lateral migrations (Merron 1991). Longitudinal migrations occur within main river channels (Welcomme 1979). Therefore, a decrease in species diversity in the river channel would not necessarily suggest an increase in species diversity in the floodplains due to a species influx from the river channel. Fish might simply be migrating up and down the river channel, much in concordance with Junk's (2000) flood pulse concept for floodplain systems. Merron and Bruton (1984) also observed that fish undergo longitudinal migrations from Nxaraga downstream during the flooding season. This finding suggested that longitudinal migration might account for the dynamic interplay between species diversity and community structure development in river channels and lagoon habitats. Similarly, lateral migrations can account for the species diversity dynamics that exist between river channels and floodplain habitats.

This study revealed spatiotemporal variations in fish population stability in the Okavango Delta. These variations were observed within both the river and floodplain habitats (Fig. 7.3) and at temporal scales. It is possible that the observed dynamics in not only stability but also other community metrics could reflect Hubbell's "neutral theory" (Rosindell et al. 2011). Stability is driven by fish migrations within and between microhabitats in the delta. According to Ricklefs (2006), the "neutral theory" also concerns changes over time, particularly changes in abundance. According to Leigh Jr. et al. (2010), "neutral theory" focuses on species richness, relative abundance and spatiotemporal variations in species composition. Based on the basic framework of the river–floodplain complex in the delta, it can be argued that the river habitat is the repository of Hubbell's "meta-population". A "meta-population" is a source pool of organisms with the potential to produce offspring in the "local community" under study (Norris 2003; Rosindell et al. 2011). Therefore, the floodplain habitat, where fish species migrate for feeding and spawning during the flooding season, is the repository of a "local community". At the local community level, "birth, death and migration" are the key drivers of population dynamics (Norris 2003; Missa et al. 2016). Hence, the observed fluxes in fish biodiversity dynamics between floodplains and river habitats are an illustration of the "neutral theory". The greater instability of floodplain populations relative to river populations is driven by fish migrations onto floodplains and back to the river habitat. Other factors driving these dynamics include spawning (birth) and death (either by predation or by drying our pools) in seasonally flooded floodplains.

Our results revealed that floodplain fish populations are more unstable than river fish populations are (Table 7.2 and Fig. 7.3). This finding is consistent with the observations of Bruton et al. (1984) that fish communities are stable in river systems and highly unstable and mobile in delta floodplains. PCA revealed that there was a significant relationship between environmental variability and fish stability (in both river and floodplain habitats). Therefore, this makes sense if we assume that species migrate into floodplains primarily for breeding and feeding before floodplains dry again. This has been previously observed in other floodplain systems (Suarez et al. 2001). Hence, when floodplains are flooded, species diversity quickly increases, possibly characterized by frenzied feeding and spawning (birth), followed by rapid immigration to either the river channel or lagoon habitat. Predators such as *C. gariepinus* have developed specialized feeding ("death" for prey species) strategies (annual catfish run) that target key prey species (e.g. *M. altisambesi*) during this period (Merron 1993a, b). These ebbs and flows in population dynamics are illustrated by the steep slopes of the floodplain species diversity curve in Fig. 7.3. This observation agrees with that of Lowe-McConnell (1979), who highlighted that the high-water season in seasonal floodplains is the time for fish to feed and store fat before immigrating into river channels.

Inland river fisheries in Africa, Asia and Latin America are key sources of livelihood and sustenance for rural communities (UNEP 2010). Leveraging the value of inland fisheries is critical for the socioeconomic growth of developing economies. According to UNEP (2010), inland fisheries are a key component of diet and income for various communities around the world. Key riverine inland fisheries include those in systems such as the Amazon (Almeida et al. 2004), the Mekong River (Baran et al. 2007; Friend 2009), and the Bangweulu River (Imai 1998), which are major sources of livelihood for riparian communities. This is particularly important in Africa, where the African Union advocates that member states leverage the value of their natural resources for economic growth. Subsequently, SADC has also developed a blue economy strategy aimed at operationalizing this continental policy document at the regional level. The utilization of fisheries resources is also a realization of SDG 14, life under water (FAO 2015). However, the sustainable management of fisheries resources requires comprehensive knowledge of their dynamics.

References

Almeida OT, Lorenzen K, McGrath, DT (2004) The fisheries sector in the Amazon regional economy. In: Proceedings of the Second International Symposium on the Management of Large Rivers for Fisheries Volume II. Welcomme R, Petr T (eds) FAO Regional Office for Asia and the Pacific, Bangkok, Thailand. RAP Publication 2004/17, pp 15–24

Ashton PJ, Nordin L, Alonso LE (2003a) Introduction to the Okavango Delta and the AquaRAP Expedition. In: Alonso LE, Nordin L (eds) A rapid biological assessment of the aquatic ecosystems of the Okavango Delta, Botswana: High Water Survey. RAP Bulletin of Biological Assessment 27. Conservation International, Washington, DC, pp 29–37

Ashton PJ, Masundire H, Hart R, Prince-Nengu J, Botshelo O, Lekhuru M, Mehlomakulu M, Tylol I (2003b) Water quality of the Okavango Delta, Botswana. In: Alonso LE, Nordin L (eds) A rapid biological assessment of the aquatic ecosystems of the Okavango Delta, Botswana: High Water Survey. RAP Bulletin of Biological Assessment 27. Conservation International, Washington, DC, pp 38–53

Baran E, Jantunen T, Chong CK (2007) Values of inland fisheries in the Mekong River Basin. WorldFish Center, Phnom Penh

Begon M, Mortimer M (1986) Population ecology; a unified study of animals and plants, 2nd edn. Blackwell Scientific Publications, London

Begon M, Harper JL, Townsend CR (1986) Ecology: individuals, populations and communities. Blackwell Scientific Publications, Oxford, UK

Biggs RC (1976) The effects of the seasonal flood regime on the ecology of Chief's Island and the adjacent floodplain systems. In: Proceedings of the symposium on the Okavango Delta and its future utilization. National Museum, Gaborone, Botswana, pp 113–123

Bonyongo MC, Mubyana T (2004) Soil nutrient status in vegetation communities of the Okavango Delta floodplains. South African J Sci 100:337–340

Bruton MN, Skelton PH, Merron GS (1984) Brief synopsis of the main findings and objectives of the Okavango fish research programme performed by the JLB Smith Institute of Ichthyology. JLB Smith Institute of Ichthyology Investigational Report No. 17. Grahamstown, South Africa

Cantu NE, Winemiller KO (1997) Structure and habitat associations of Devil's River fish assemblages. Southwestern Naturalist 42:265–278

Cronberg G, Gieske A, Martins E, Prince-Nengu J, Stenstrom IM (1996) Major ion chemistry, plankton, and bacterial assemblages of the Jao/Boro River, Okavango Delta, Botswana: the swamps and floodplains. Arch Hydrobiol/Supp 107(3):335–407

Danet A, Mouchet M, Bonnaffé W, Thébault E, Fontaine C (2021) Species richness and food-web structure jointly drive community biomass and its temporal stability in fish communities. Ecol Lett 24(11):2364–2377. https://doi.org/10.1111/ele.13857.hal-03766980

FAO (2015) FAO and the 17 Sustainable Development Goals. Rome, Italy

FAO (2016) Fisheries in the drylands of sub-Saharan Africa – "Fish come with the rains". Building resilience for fisheries-dependent livelihoods to enhance food security and nutrition in the drylands. FAO Fisheries and Aquaculture Circular No. 1118. Rome, Italy

Friend RM (2009) Fishing for influence: fisheries science and evidence in water resources development in the Mekong basin. Water Altern 2(2):167–182

Grossman GD, Dowd JF, Crawford M (1990) Assemblage stability in stream fishes: a review. Environ Manag 14(5):661–671

Hart RK, Calver MC, Dickman CR (2002) The index of relative importance: an alternative approach to reducing bias in descriptive studies of animal diets. Wildl Res 29(5):415–421

Hilborn R, Walters CJ (1992) Quantitative fisheries stock assessment: choice, dynamics and uncertainty. Chapman and Hall Publishers, New York

Hoberg P, Lindholm M, Ramberg L, Hessen DO (2002) Aquatic food web dynamics in the Okavango Delta, Botswana. Hydrolobiologia 470:23–30

Imai I (1998) Sustainability of fishing in the Bangweulu Swamps, Zambia. Afr Study Monogr 19(2):69–86

Jetschke G (2002) The unified neutral theory of biodiversity and biogeography. Ecology 83(6): 1771–1772

Jolliffe IT (2002) Principal component analysis. Springer, New York

Junk WJ (2000) The Central Amazon river floodplain: concepts for the sustainable use of its resources. In: Junk WJ, Ohly JJ, Piedade MTF, Soares MGM (eds) The Central Amazon floodplain: actual use and options for a sustainable management. Backhuys Publishers, Leiden, pp 75–94

King AJ, Humphries P, Lake PS (2003) Fish recruitment on floodplains: the roles of patterns of flooding and life history characteristics. Can J Aquat Sci 60:773–786

Kolding J, Skaalevik A (2010) Pasgear II. Version 2.1. A database program for experimental or artisanal fishery data. University of Bergen, Norway

Kongsbak K, Thilsted SH, Wahed MA (2008) Effect of consumption of the nutrient-dense, freshwater small fish *Amblypharyngodon mola* on biochemical indicators of vitamin A status in Bangladeshi children: a randomised, controlled study of efficacy. Br J Nutr 99:581–597

Krah M, McCarthy TS, Annegarn H, Ramberg L (2004) Airborne dust deposition in the Okavango delta, Botswana, and its impact on landforms. Earth Surf Process Landf 29:565–577

Krah M, McCarthy TS, Huntsman-Mapila P, Wolski P, Annegarn H, Sethebe K (2006) Nutrient budget in the seasonal wetland of the Okavango Delta, Botswana. Wetl Ecol Manag 14:253–267

Leigh Jr EG, Rosindell J, Etienne RS (2010) Unified neutral theory of biodiversity and biogeography. Scholarpedia 5(11):8822

Leps J, Smilauer P (2003) Multivariate analysis of ecological data using CANOCO. Cambridge University Press, Cambridge, UK

Lindholm M (2002) Predator-induced cyclomorphosis of *Daphnia laevis* (Branchiopoda, Cladocera) in a tropical floodplain (Okavango Delta, Botswana). Crustaceana 75(6):803–814

Lowe-McConnell RH (1979) Ecological aspects of seasonality in fishes of tropical waters. Symp Zool Soc Lond 44:219–241

Lowe-McConnell RH (1987) Ecological studies in tropical fish communities. Cambridge University Press, Cambridge, UK

Lucas MC, Baras E (2000) Methods for studying spatial behavior of freshwater fishes. Fish Fish 1:283–316

McCarthy TS, Ellery WN (1995) Sedimentation on the distal reaches of the Okavango fan, Botswana, and its bearing on calcrete and silcrete (ganister) formation. J Sediment Res A65(1):77–90

Merron GS (1991) The ecology and management of the fishes of the Okavango delta, Botswana, with particular reference to the role of the seasonal floods. Rhodes University, Phd thesis. p 171

Merron GS (1993a) The diversity, distribution and abundance of the fishes in the Moremi Wildlife Reserve, Okavango Delta, Botswana. S Afr Wildf Res 23(4):115–122

Merron GS (1993b) Pack hunting in two species of catfish, *Clarias gariepinus* and *C. ngamensis*, in the Okavango Delta, Botswana. J Fish Biol 43:575–584

Merron GS, Bruton MN (1984) Report on the June – July 1984 expedition to the Okavango Delta, Botswana. JLB Smith Institute of Ichthyology Investigational Report No. 12. Grahamstown, South Africa

Merron GS, Bruton MN (1988) The ecology and management of the fishes of the Okavango delta, Botswana, with special reference to the role of the seasonal floods. J.L.B. Smith Institute of Ichthyology, Grahamstown, South Africa. Investigational Report No. 29

Missa O, Dytham C, Morlon H (2016) Understanding how biodiversity unfolds through time under neutral theory. Philos Trans R Soc B 371:20150226. https://doi.org/10.1098/rstb.2015.0226

Mladenov N, Huntsman-Mapila P, Wolski P, Masamba WRL, McKnight DM (2008) Dissolved organic matter accumulation, reactivity, and redox state in ground water of a recharge wetland. Wetlands 28(3):747–759

Mmopelwa G, Raletsatsi S, Mosepele K (2005) Cost Benefit Analysis of Commercial Fishing in Shakawe, Ngamiland, Botswana. Botswana Notes and Records 37:11–21

Mosepele K (2019) Dynamics of the Seasonal Floodplain Fishery of the Okavango Delta, Botswana. PhD Thesis, University of Bergen, Norway. (Skipnes Kommunikasjon/ University of Bergen: Bergen, Norway)

Mosepele K, Kolding J (2003) Fish Stock Assessment in the Okavango delta, Botswana – preliminary results from a length based analysis. In: Bernad T, Mosepele K, Ramberg L (eds) Proceedings of an "Environmental Monitoring of Tropical and Subtropical Wetlands" conference held at Maun Lodge, 4–6 December 2002, Maun, Botswana, pp 363–390

Mosepele K, Mmopelwa TG, Mosepele B (2003) Characterization and monitoring of the Okavango delta artisanal fishery. In: Bernard T, Mosepele K, Ramberg L (eds) Environmental Monitoring of Tropical and Subtropical Wetlands. University of Botswana, Maun and University of Florida, Gainesville, FLA, pp 391–413

Mosepele K, Ngwenya BN, Bernard T (2006) Artisanal fishing and food security in the Okavango Delta, Botswana. In: Ahmed A (ed) Global and local resources in achieving sustainable development, pp 159–168. (Interdescience Publishers: Geneva, Switzerland)

Mosepele K, Mmopelwa G, Mosepele B, Kgathi DL (2007) Indigenous knowledge and fish utilization in the Okavango Delta, Botswana: Implications for food security. In: Ahmed A (ed) Managing knowledge, technology and development in the era of information revolution, pp 292–302. (Griffith University: Australia)

Mosepele K, Moyle PB, Merron GS, Purkey D, Mosepele B (2009) Fish, floods and ecosystem engineers; interactions and conservation in the Okavango Delta, Botswana. Bioscience 59 (1):53–64

Mosepele K, Mosepele B, Wolski P, Kolding J (2012) Dynamics of the feeding ecology of selected fish species from the Okavango River delta, Botswana. Acta Ichthyologica et Piscatoria, 42 (4):271–289

Mosepele K, Kolding J, Bokhutlo T (2017) Fish community dynamics in an inland floodplain system of the Okavango Delta, Botswana. Ecohydrol Hydrobiol 17(2):89–102

Mosepele K, Kolding J, Bokhutlo T, Mosepele BQ, Molefe M (2022) The Okavango Delta: Fisheries in a fluctuating floodplain system. Front Environ Sci 10:854835. https://doi.org/10.3389/fenvs.2022.854835

Ngwenya B, Mosepele K (2008) Socio-economic survey of subsistence fishing in the Okavango Delta, Botswana. (Bay Publishing: Gaborone, Botswana)

Norris S (2003) Neutral theory: a new, unified model for ecology. Bioscience 53(2):124–129

Oberdorff T, Porcher J-P (1992) Fish assemblage structure in Brittany streams (France). Aquat Living Resour 5:215–223

Oliveira BF, Moore FC, Dong X (2022) Biodiversity mediates ecosystem sensitivity to climate variability. Commun Biol 5:628. https://doi.org/10.1038/s42003-022-03573-9 | www.nature.com/commsbio

Rader RB (1994) Macroinvertebrates of the northern Everglades: species composition and trophic structure. Florida Scientist 57:22–33

Ricklefs RE (1990) Ecology, 3rd edn. Freeman and Company, New York

Ricklefs RE (2006) The unified neutral theory of biodiversity: do the numbers add up? Ecology 87(6):1424–1431

Roos N, Wahab A, Hossain MAR, Thilsted SH (2007) Linking human nutrition and fisheries: incorporating micronutrient-dense, small indigenous fish species in carp polyculture production in Bangladesh. Food Nutr Bull 28(2):S280–S293

Rosindell J, Hubbell SP, Etienne RS (2011) The Unified neutral theory of biodiversity and biogeography at age ten. Trends Ecol Evol 26(7):340–348

Suarez YR, Petrere M, Catella AC (2001) Factors determining the structure of fish communities in Pantanal lagoons (MS, Brazil). Fish Manag Ecol 8:173–186

Ter Braak CJ, Smilauer P (2002) CANOCO reference manual and CanoDraw for Windows User's Guide: Software for Canonical Community Ordination (Version 4.5)

UNEP (2010) Blue harvest: inland fisheries as an ecosystem service. WorldFish Center, Penang

Welcomme RL (1979) The inland fisheries of Africa. CIFA Occasional Paper No. 7

Welcomme RL (1985) River fisheries. FAO fisheries technical paper, 262, Rome

B. Mosepele holds an MPhil in Fisheries Biology and Management from the University of Bergen, Norway. She is currently a PhD Candidate with the North-West University, in Potchefsroom, South Africa. I'm a Senior Lecturer at the Botswana University of Agriculture and Natural Resources (BUAN). Previously I worked as an Executive Secretary—Botswana Environmental Assessment Practitioners Association (BEAPA), Project Officer United Nations Development Program (UNDP) Ngamiland Sustainable Land Management (NSLM) Project, Ngamiland & Ghanzi District Environment Coordinator, Department of Environmental Affairs, Monitoring and Evaluation Specialist Southern Africa Regional Environmental Program (SAREP), and Biodiversity Officer—Conservation International (CI), Botswana Country Office. Her research interests include fisheries, aquaculture, aquatic biodiversity, freshwater macroinvertebrates ecology.

K. Mosepele (PhD) is the Vice Chancellor of the Botswana University of Agriculture and Natural Resources. He holds the degree of Doctor of Philosophy (PhD) in Fisheries Biology and Management from the University of Bergen, Norway. He also holds the degree of Master of Philosophy (MPhil) and a Postgraduate Diploma in Fisheries Biology and Management from the University of Bergen, Norway. Professor Mosepele obtained his degree of Bachelor of Science (BS) in Agriculture majoring in Fisheries Science from the New Mexico State University, USA. He previously served as a Professor of Fisheries Biology and Management and Deputy Director of the Okavango Research Institute, University of Botswana.

I. Mosie is a Senior Technician at the Okavango Research Institute, University of Botswana. He holds an MSc in Environmental Management from the University of the Orange Free State, Bloemfontein, South Africa. As a Senior Technician in the Monitoring and Field Support, he is dedicated to advancing sustainable fisheries management and conservation efforts in the Okavango Delta. With expertise in fisheries management and a passion for preserving aquatic biodiversity, he plays a pivotal role in conducting field research and implementing responsible practices. Through comprehensive surveys, he contributes to establishing baseline information, monitoring changes, and designing evidence-based strategies for sustainable harvesting.

K. Sianga is a Senior Lecturer at the Botswana University of Agriculture and Natural Resource. He holds a BSc in Conservation Biology and Wildlife Management and an MPhil and a PhD in Natural Resource Management. He worked as Project Assistant at the University of Botswana, Project Manager (Communities Living Among Wildlife Sustainably, CLAWS). His research interests include functional and vegetation heterogeneity and plant-wildlife interactions, fire ecology and management, human-wildlife interactions, spatial ecology. Dr Sianga has published four articles in peer reviewed journals and seven book chapters.

Chapter 8
Effectiveness of Community Participation in Natural Resources Management: Empirical Evidence from Magude District, Southern Mozambique

Tomás Adriano Sitoe and Diamantino Carlos Pinto

8.1 Introduction

Communities have been taking part in conservation since the 1960s. Some approaches for community participation emerged and have been gradually adopted (He et al. 2020). Community-based Natural Resource Management (CBNRM) as a term and approach emerged in the early 1990s, referring to the collective use and management of natural resources in rural areas by a group of people with a self-defined, distinct identity using communally owned facilities (Adams and Hulme 2001). The underlying idea behind CBNRM is that it represents a better management solution than fences, which exclude the population from using resources. A key principle underlying these community-based initiatives is to align long-term conservation with the short-term needs of local people, ensuring that community members gain some benefit from their participation in conservation efforts (Adams et al. 2004; Agrawal and Redford 2006). The earlier discourse of positioning local communities to an opposite side of conservation and of the necessity of top-down government compulsory management has also gradually shifted to the idea that effective public participation can ensure the consistency and continuity of information and promote the legitimacy and rationality of decision-making in conservation (Selin and Chavez 1995).

T. A. Sitoe (✉)
Agricultural Research Institute of Mozambique, Centre for Socioeconomic Studies, Maputo, Mozambique

D. C. Pinto
Ministry of Agriculture and Rural Development, Praça dos Heróis Moçambicanos, Maputo, Mozambique

W. L. Hambira et al. (eds.), *Environmental Change and Biodiversity Conservation in sub-Saharan Africa*, Advances in Global Change Research 75,
https://doi.org/10.1007/978-3-031-73136-5_8

A crucial aspect of CBNRM is the effective participation of institutions, local communities, and different organizations. Participation is based on the premise that the involvement of key actors in decision-making processes is fundamental to guarantee their sustainability (Robinson 1998; Matakala and Mushove 2001). However, the legitimacy of participation in decision-making processes, including participation in the context of Natural Resources Management Committees (CGRN abbreviation in Portuguese), is questioned because although theoretically these spaces are open to social participation in decision-making processes, only some actors would have access to this social participation, mainly social groups with greater economic resources and power. Therefore, there is a certain skepticism about the legitimacy of community participation in the CGRN, arising from conflicts of interest between the main actors in the management of natural resources.

In Mozambique, since the mid-1990s, important legal and political reforms have been registered that created conditions for the recognition of community rights over land, forest, and wildlife resources. This recognition has been translated into a movement to guarantee the rights of communities through the demarcation of community lands and through the creation of Natural Resources Management Committees. This movement has been encouraged by donors, NGOs, and the government. However, on a national scale, there are mixed results in relation to the implementation of CBNRM and the experiences of natural resource management by committees. There are some interesting initiatives, such as partnerships (joint-venture) in ecotourism projects that are fully aimed at promoting the transition to full community control, as well as experiences of community forest concessions where communities promote the development of their members and ensure the sustainability of forest resources. However, there are still many communities that adopt an "absentee landlord" attitude toward natural resources, and the internal accountability structures in the Community Management Committees for Natural Resources are weak, allowing excessive cutting of trees for charcoal production. Associated with this is the occurrence of forest fires and frequent poaching, as well as the fact that some community members use their position to steal funds for their own benefit.

In Mapulanguene, District of Magude, since 1997, a growing number of local communities exercise their rights and responsibilities for the control, rational use, conservation and sharing of benefits related to the management of natural resources in their areas of influence, in accordance with legislation and current sector policies. Despite this, thus far, few studies have been developed to prove the effectiveness of community participation in natural resources management communities.

The objective of the study is to analyze the effectiveness of community participation in the Natural Resources Management Commission (CGRN) in Mapulanguene, Magude District.

Specifically, the study aimed to:

- Analyze respondents' socioeconomic characteristics in the area of study.

- Analyze the deliberative process in the Natural Resources Management Committees, including the extent and reach of the results of public participation in the Natural Resources Management Committees.[1]

It is intended that this research contributes to building a knowledge base that can be useful in academia and support the development of appropriate policies for local development and Community-based Natural Resource Management.

8.2 Literature Review

Community-based natural resource management is fundamentally premised on institutional reforms that decentralize authority over natural resource management to local actors. In the absence of such reforms, the incentives for local groups of people to collectively invest in natural resource management are unlikely to exist or emerge (Nelson and Agrawal 2008). The theory behind Community-based natural resource management argues that the best way to manage natural resources is for local people to use their local knowledge and technologies (Blaikie 2006). The Natural Resources Management Committees are an attempt to devolve power from the State to local levels and often to local groups and institutions. The devolution of state power to community-based (local) forms of resource management allows "marginalized and neglected groups" (vulnerable and poor people) to have a greater role and participation in controlling natural resources. This devolution is only effective if the decision-making power in the co-management of natural resources through local social groups is not limited; that is, if decision-making power is not concentrated in certain groups and social sectors in the community.

The theory behind CBNRM is based on critical theory that is generally attributed to Jürgen Habermas, who focuses much on social and cultural functions that are served by deliberative spaces. Natural Resources Management Committees play a crucial role in generating ideas and information that can improve knowledge, improve understanding, and enhance the quality of decisions. The relevance of theories of deliberative democracy in CBNRM is justified by the fact that deliberative spaces become systematically distorted through manipulation, coercion, and misinformation (Parkins and Mitchell 2005) and the nature and complexity of the issues discussed, which are often highly contested. Deliberative democratic theory is concerned with "debate and discussion aimed at producing reasonable, well-informed opinion in which participants are willing to revise preferences in light of discussion, new information, and claims made by fellow participants" (Chambers 2003).

As a social process, deliberation or public debate relies on persuasion to induce participants to reflect on and change views (Dryzek 2000); it contrasts with other

[1] Deliberation refers to a form of reflection and discussion that involves weighing the pro and con views or viewpoints of some proposition and reaching some agreement. Discussion thus constitutes the critical element of deliberation.

communication approaches such as coercion, manipulation, or delusion (deception), which are achieved through ideological domination and the capture of interests by groups (Przeworski 1998). According to Abelson et al. (2003), the traditional view is that deliberation takes place within government as a feature of representative democracy. However, deliberation can also take place outside the governmental sphere as a process mediated by the mass media. According to these authors, outside the governmental sphere, deliberation can take the form of meetings between people.

According to Fearon (1998), discussing issues before making a decision provides the opportunity to (1) share views on a subject that voting does not allow "and associated activities such as the ability to communicate intensity of preferences and the relative weights of preferences"; (2) generate and consider a wider range of options or new alternatives that might not have been considered otherwise; (3) support or encourage more public-spirited proposals in contrast to those motivated by self-interest; (4) increase the legitimacy of the ultimate decision and to ease implementation or compliance with decision by giving everyone a say; and (5) improve the moral or intellectual qualities of the participants.

According to Abelson and Gauvin (2006), there are a plurality of evaluation approaches and methods, but it is possible to identify three main approaches: (1) *user-based,* which assumes that different participants have different goals and that evaluation must take these into account; (2) *theory-based,* which is driven by theories and models of public participation and applies normative evaluation universally to any public participation effort; and (3) *goal-free* evaluation, which is not constituted by any stated goals and is conducted in the absence of any theory. Most evaluation studies to date fall under the user-based category, and one distinguishing feature of public participation evaluation is the emphasis on either process or outcome evaluation (e.g., a decision or set of recommendations) (Abelson et al. 2003).

There is a strong association between the broad acceptance of the decision outcomes and "processes in which agencies are responsive, participants are motivated, the quality of deliberation is high, and the participants have at least a moderate degree of control over the process" (Beierle and Cayford 2002). The normative theory of the public participation model is based on the review of Habermas' concepts about ideal discourses and communicative competence, which identifies two major principles (meta principles): reasonableness (fairness) and competence, on which deliberative participation must be evaluated (Webler 1995). According to Abelson and Gauvin (2006), evaluation processes face a number of theoretical and practical challenges. First, the task of defining the end-point of a participation exercise for purposes of measuring effectiveness is often unclear. Second, the public participation process may be well run according to some criteria but not others. The third major challenge lies in the measurement criteria. Participant satisfaction is routinely used as a measure of success despite problems associated with its interpretation.

According to Ribot et al. (2006), the common outcome for CBNRM reform efforts in contexts of high resource value and low institutional accountability is that such reforms often occur only on paper, and the power to allocate resources continues to be monopolized by central actors with neopatrimonial motivations. Broadly speaking, participation means taking part in something; this can be "passive" when the subject just listens passively without contributing; "participation through consultation" when foreign experts, after gathering community opinions, analyse the information and decide what to do with the information; "functional participation" when people form groups to achieve predefined goals; and "interactive participation" when people participate in the joint analysis that leads to an action plan (Birou 1982). The approach is always from the point of view of active participation.

Communities can have different levels of control over forest resources, which can range from "complete" control (total responsibility) to "shared" control (shared responsibility) to very little control (advisory responsibility) (Matakala 1998). Intensive "problem-solving" processes are most effective for achieving a broad set of social goals, and participant motivation and agency responsiveness are key factors for success (Beierle and Cayford 2002). Armstein's ladder of participation provides a valuable framework for perceiving the degree of community participation in decision-making processes (Fig. 8.1 and Box 8.1).

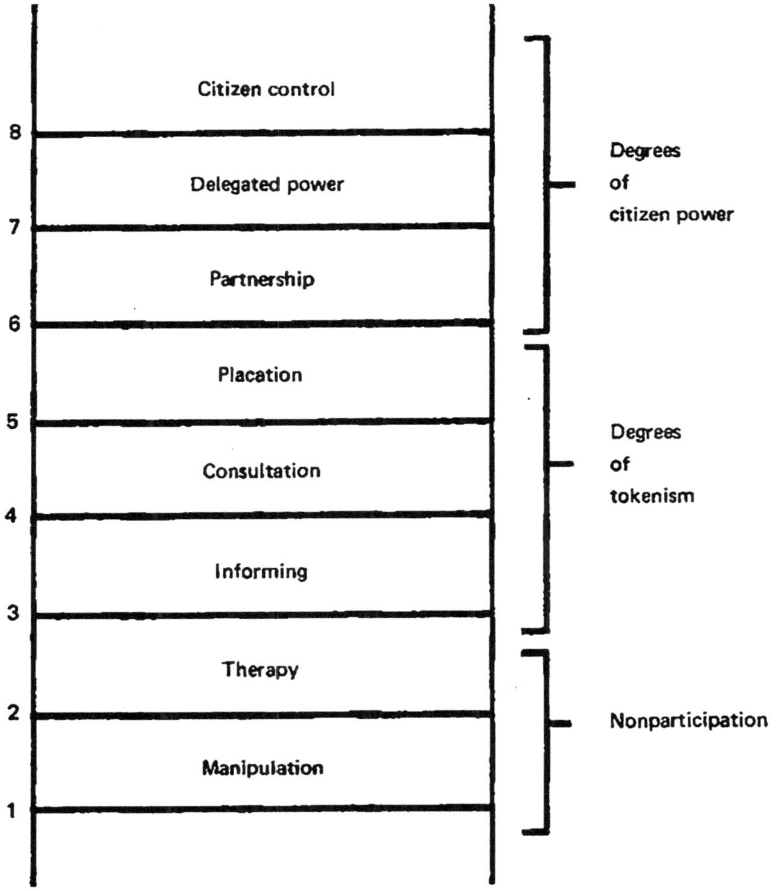

Fig. 8.1 Degrees of citizen participation. (Adapted from Armstein's ladder (1969))

Box 8.1 Degrees of Citizen Participation Description: Eight Rungs of the Ladder

1. *Manipulation* and 2. *Therapy*. These two rungs describe levels of "nonparticipation" that have been contrived by some to substitute for genuine participation. Their real objective is not to enable people to participate in planning or conducting programs but to enable powerholders to "educate" or "cure" the participants.
3. *Informing*. This is the most important first step to legitimate participation. However, more often than not, the emphasis is on a one-way flow of information. There is no channel for feedback.
4. *Consultation*. Arnstein notes that "inviting citizens' opinions, like informing them, can be a legitimate step toward their full participation." However, when consultation processes are "not combined with other modes of participation, this rung of the ladder is still a sham since it offers no assurance that citizen concerns and ideas will be taken into account. When power holders restrict the input of citizens' ideas solely to this level, participation remains just a window-dressing ritual. People are primarily perceived as statistical abstractions, and participation is measured by how many come to meetings, take brochures home, or answer a questionnaire.
5. *Placation (pacification, conciliation)*. Participation as placation occurs when citizens are granted a limited degree of influence in a process, but their participation is largely or entirely tokenistic: citizens are merely involved only to demonstrate that they were involved. In Arnstein's words, "An example of placation strategy is to place a few hand-picked 'worthy' poor on boards of Community Action Agencies or on public bodies like the board of education, police commission, or housing authority. If they are not accountable to a constituency in the community and if the traditional power elite holds the majority of seats, the have-nots can be easily outvoted and outfoxed."
6. *Partnership*. Power is redistributed through negotiation between citizens and power holders. Planning and decision-making responsibilities are shared, e.g., through joint committees.
7. *Delegated Power*. Participation as delegated power occurs when public institutions, officials, or administrators give up at least some degree of control, management, decision-making authority, or funding to citizens.
8. *Citizen Control*. Participation as citizen control occurs, in Arnstein's words, when "participants or residents can govern a program or an institution, be in full charge of policy and managerial aspects, and be able to negotiate the conditions under which 'outsiders' may change them."

8.3 Materials and Methods

8.3.1 Study Area

The District of Magude is in the northern part of Maputo Province, 150 km from the city of Maputo, between the parallel 26° 02′ 00″ south latitude and between 32° 17′ 00″ longitude (MAE 2014). This district is crossed by the Nkomati River, which favors agriculture and livestock practice, with Magude being considered the second largest producer of cattle in Maputo Province.

According to official data, the Magude surface is 6654 km^2, which corresponds to 26.4% of the total area of Maputo Province. The total population is estimated at 53,317 inhabitants according to the General Population and Housing Census carried out in 2007. It has four administrative posts (Mapulaguene, Mahel, Mutaze, and Ponjane), being the district that most suffered the effects of the war between 196 and 1992 when peace agreements were signed by the Mozambique Government and RENAMO (the Mozambican National Resistance Movement).

The creation of the natural resources management committee in Mapulanguene took place in 2007 and was preceded by a scenario in which there was overuse of forest resources, mainly by people from other provinces who went there for poaching. On the other hand, after the end of the armed conflict that devastated the country for 16 years (1976–1992), the populations returned to their areas of origin and without their own source of subsistence, as their cattle had been looted, and the sources of migratory employment were closed. The development of the large Limpopo cross-border park and ecotourism projects created an environment for the emergence of several investors in the region. There is still a conflict of interest between the main actors (local communities, outsiders, and the state). Skepticism regarding community participation is compounded by the fact that in the past, Mozambique had implemented various development approaches related to participation (people's farms, cooperatives, etc.); however, these approaches have created conflicts of interest and expropriation of communities (Fig. 8.2).

8.3.2 Data Collection

This is a case study using a qualitative research approach. Data and information were collected through individual interviews using a structured questionnaire. Individual interviews were carried out to collect data on the sociodemographic characteristics of the interviewees as well as information regarding the use and control of forest and wildlife resources by local communities. In total, 72 people were selected for interviews using a snowball technique. Among the selected people, 64 were members of the local Natural Resources Management Committee. Table 8.1 shows the sample composition.

Fig. 8.2 Geographic location of Magude. (Cartography: National Centre for Cartography and Teledetection (CENACARTA), Maputo)

Table 8.1 Number of people interviewed by category of origin

Category of origin	Number of people interviewed
Traditional authority	2
Government authority at local level	1
Local natural resources management officer	1
Resources Management Committee	64
Individuals	4
Total	72

In addition to individual interviews, we attended the Natural Resources Management Committee meeting that lasted 2 h. The Natural Resources Management Committee meeting was organized by the leaders of the Mapulanguene Natural Resources Management Committee. According to the Committee leaders, the meeting was organized with a minimum of 5 weeks of preparation. An "open" call for participation was made to members of the Natural Resources Management Committee, ensuring diversity and representativeness through inviting members from different villages and with different demographic characteristics. We adopted the deliberative assessment model suggested by Abelson et al. (2003) to evaluate the effectiveness of community participation in natural resources management committees. Abelson's model focuses on four evaluation components of the deliberative process: (i) the representation; (ii) the structure of the process or lines of procedure; (iii) the information used in the process; and (iv) the results and decisions resulting from the deliberative process. The four components can be grouped into elements: the process and outcomes.

8.4 Results and Discussion

8.4.1 Respondents' Socioeconomic Characteristics in the Study Area

Currently, the Natural Resources Management Committee in Mapulanguene is legalized and comprises 95 members, 50 men and 45 women. In general, members of the Natural Resources Management Committee in Mapulanguene belong to the association for charcoal production and sale. Eventually, this was the way they found to defend themselves from the exploitation of resources by outsiders and ensure biodiversity conservation and participation in the charcoal market, which was previously a business for people from other locations. The sociodemographic characteristics of the interviewees are summarized in Table 8.2.

Approximately 40% of respondents are female and mostly elderly. The fact that most of the respondents are male does not necessarily mean that women do not participate in biodiversity conservation programs or that women do not have control over natural resources. Women actively participate in the sale of charcoal produced

Table 8.2 Respondent's socioeconomic characteristics in Mapulanguene

Characteristic		($n = 72$)
Gender (= male, in %)		60
Age (years) (mean)		55
Education (%)	Illiterate	30
	Adult education	10
	Primary education	60
	Secondary education general	0

mainly by men. Agricultural production and livestock are also important economic activities that ensure household and community subsistence in Mapulanguene, and 90% of respondents practiced these activities. The fact that a significant number only have primary education represents a challenge for communities, especially for activities that require specific skills and abilities because scientific knowledge is necessary for good co-management of natural resources (Schusler et al. 2003).

8.4.2 *Legitimacy of Participation and Scope of Achievements of Deliberations on the Well-Being of Communities*

Socioeconomic demands raised by the community during the Natural Resource Management Committee meeting include absence of the on-site hospital, lack of water and school, and improved livelihoods. These concerns suggest an intrinsic link between the need for socioeconomic development and participation in biodiversity conservation programs. That is, while communities participate in biodiversity conservation programs, they are also concerned with seeing the socioeconomic needs that guarantee well-being resolved. A total of 98% of the Natural Resources Management Committee participants responded that the meeting format promoted adequate discussion; that is, it provided them with an equal opportunity to participate in the discussions, and 97.3% considered that the orientation of the meeting by the leaders was appropriate for the discussion of the topics.

According to the OECD (2020), time is one of the factors that distinguishes representative deliberative processes from most other types of stakeholders and citizen participation. Approximately 90% of the participants considered that they had enough time to fully discuss the issues and thought that the information and purpose of the meeting were clearly communicated (85.4% and 82%, respectively), suggesting the legitimacy of participation in the deliberative processes. From the demands listed above, there was greater satisfaction in relation to the fulfilment of water and school needs. The concern with the hospital had not been met satisfactorily, suggesting that despite some needs having been met, the fulfilment of the demands arising from participation in the deliberative process is not complete, which can affect the impact of programs for the preservation of biodiversity, which can be seen in the following result: one third of the participants think that the deliberation meetings are of little use because of the slowness in solving the problems posed. Therefore, the government's delay in fulfilling community demands can negatively affect natural resource co-management efforts.

The representativeness of the CGRN is not adequate to guarantee the sustainability of the natural resources in Mapulanguene since only the community is part of the CGRN, supported by the NGOs (ORAM, CARITAS), and the companies that have Land Use and Benefit Rights (DUAT) to carry out activities and exploitation of resources on site are not represented on the Natural Resources Management Committee. The nonparticipation of companies in natural resources management committees may be related to the fact that participation is voluntary, and the involvement of companies with communities is generally limited to the payment of social responsibility, whose value is stipulated by law at 20% of revenues from resource exploitation. The amount passed on by the companies to the community in 2012 was only 12,000 meticais, equivalent to USD 200.00, a value considered by the communities to be quite insignificant. On the other hand, licenses for the exploration of natural resources by companies are authorized by the government at the central level in Ministries with little involvement of the communities, which contributes to indifference toward the plundering of resources. The appropriate model for resource sustainability should include business involvement in local resource management committees.

8.5 Conclusion and Recommendations

Our study suggests that there is a link between participation in biodiversity conservation programs and solving socioeconomic needs that ensure the well-being of communities. The Natural Resources Management Committees have been a space for the deepening of democracy, allowing communities to discuss the problems that affect their development. There are important achievements arising from the participation of local populations in these decision-making spaces. However, delay in government response to participant concerns may affect the effectiveness of participation in these spaces, suggesting that deliberative spaces by themselves are unlikely to bring about the required caliber and speed of change. The government must be proactive and rapid in resolving the demands of the communities.

The forum participants stressed the importance of improving the quality of community consultation processes for the attribution of licenses to investors. Participants are aware of the need to preserve forest and wildlife resources, but the fact that exploration licenses are granted at the ministry level without adequate consultation with citizens makes them have a passive attitude toward conservation activities. A model of sustainable management of natural resources must involve all interested actors, defining the role of each actor and even the alternation of the presidency of the resource management committees. It is recommended that the government encourage the development of natural resource management models that involve a larger number of local actors (communities, private sector, and civil society). The government must establish mechanisms to ensure that private investors fulfil the commitments made during the community consultation processes for obtaining land use and enjoyment rights.

Mechanisms should be established to ensure that investors are obliged to establish partnerships that bring benefits to all parties involved, and during consultations, local authorities should explain to communities the advantages and disadvantages of projects to be undertaken by investors.

References

Abelson J, Gauvin F-P (2006) Assessing the impacts of public participation: concepts, evidence, and policy implications. Canadian Policy Research Networks. Research Report P06

Abelson J, Forest P, Eyles J, Smith P et al (2003) Deliberations about deliberation: issues in the design and evaluation of public consultation processes. Soc Sci Med 57:239–251

Adams WM, Hulme D (2001) If community conservation is the answer in Africa, what is the question? Oryx 35(3):193–200

Adams WM, Aveling R, Brockington D, Dickson B, Elliott J, Hutton J et al (2004) Biodiversity conservation and the eradication of poverty. Science 306:1146–1149. https://doi.org/10.1126/science.1097920. PMID: 15539593

Agrawal A, Redford K (2006) Poverty, development, and biodiversity conservation: shooting in the dark? Working Paper No. 26. Wildlife Conservation Society

Arnstein SR (1969) A ladder of citizen participation. J Am Plann Assoc 35(4):216–224

Beierle TC, Cayford J (2002) Democracy in practice: public participation in environmental decisions. Resources for the Future, Washington, DC

Birou A (1982) Dicionário das ciências sociais. Publicações Dom Quixote

Blaikie P (2006) Is small really beautiful? Community-based natural resource management in Malawi and Botswana. World Dev 34(11):1942–1957. https://doi.org/10.1016/j.worlddev.2005.11.023

Chambers S (2003) Deliberative democratic theory. Annu Rev Polit Sci 6:307–326

Dryzek JS (2000) Deliberative democracy and beyond: liberals, critics, contestations. Oxford University Press, New York

Fearon JD (1998) Deliberation as discussion. In: Elster J (ed) Deliberative democracy. Cambridge University Press, pp 44–68

He S, Yang L, Min Q (2020) Community participation in nature conservation: the Chinese experience and its implication to national park management. Sustainability 12(11):4760. https://doi.org/10.3390/su12114760

Matakala P (1998) Guião para trabalhadores de campo e investigadores em maneio florestal comunitário. Maputo. FAO

Matakala P, Mushove P (2001) Arranjos institucionais para o maneio comunitário dos recursos naturais (MCRN): perfis e análise de 42 iniciativas de MCRN em Moçambique. DNFFB, Maputo, p 142

MINISTÉRIO DA ADMINISTRAÇÃO ESTATAL (MAE) (2014). Perfil do distrito de Magude Província de Maputo. Maputo, Mozambique. Available at http://www.portaldogoverno.gov.mz. Retrieved on 2/09/2015

Nelson F, Agrawal A (2008) Patronage or participation? Community-based natural resource management reform in Sub-Saharan Africa. Dev Chang 39(4):557–585

OECD (2020) Innovative citizen participation and new democratic institutions: catching the deliberative wave. OECD Publishing, Paris. https://doi.org/10.1787/339306da-en

Parkins J, Mitchell R (2005) Public participation as public debate: a deliberative turn in natural resource management. Soc Nat Resour 18:529–540

Przeworski A (1998) Deliberation and ideological domination. In: Elster J (ed) Deliberative democracy. Cambridge University Press, Cambridge, pp 140–160

Ribot JC, Agrawal A, Larson AM (2006) Recentralizing while decentralizing: how national governments reappropriate forest resources. World Dev 34(11):1864–1886

Robinson M (1998) Democracy, participation, and public policy. In: Robinson M (ed) The democratic development state. Political and institutional design. Oxford University Press, Oxford, UK

Schusler TM, Decker DJ, Pfeffer MJ (2003) Social learning for collaborative natural resource management. Soc Nat Resour 15:309–326

Selin S, Chavez D (1995) Developing a collaborative model for environmental planning and management. Environ Manag 19:189–195

Webler T (1995) "Right" discourse in citizen participation: an evaluative yardstick. In: Renn O, Webler T, Wiedemann P (eds) Fairness and competence in citizen participation. Evaluating models for environmental discourse. Risk, governance and society, vol 10. Springer, Dordrecht. https://doi.org/10.1007/978-94-011-0131-8_3

Tomás Adriano Sitoe holds a doctorate in Rural Development from the Federal University of Rio Grande do Sul, Brazil and an MSc in Agricultural Development from UEM. Since 1990 he has been an employee of the Mozambique Ministry of Agriculture, where he directs the Centre for Socioeconomic Studies of the Mozambique Agricultural Research Institute. He has previously worked as a mathematics teacher between 1984 and 1989 at Francisco Manyanga Secondary School.

Diamantino Carlos Pinto holds a BSc with honours in Agriculture from the University Saint Thomas Mozambique. He works at the Mozambique Ministry of Agriculture in the agricultural extension services.

Chapter 9
Social Media Backlash and Tourism Boycott Demands After Botswana's Trophy Hunting Ban Lift: *Thematic and Sentiment Analysis*

Wame L. Hambira, Patricia K. Mogomotsi, Victor K. Muposhi, and Goemeone E. J. Mogomotsi

9.1 Introduction

The tourism industry is increasingly becoming inundated with threats of or actual boycotts of some destination. These more often result in tarnishing the image of those destinations and their associated tourist attractions. In the process, the image of businesses is tarnished and international travel to some destinations restricted, thus leading to the decline in tourism arrivals (Seyfi and Hall 2019; Shaheer et al. 2018). Further, the disruptions in financial investment and supply chain; decrease in tourist destinations' market shares and sales; and in some cases, substantial economic and personal hardship may be experienced (Seyfi and Hall 2019; Shaheer et al. 2018).

Pezzullo (2011:p.125), defines boycott campaign as a "… a concerted refusal to spend money-as well as to convince others to refuse to spend money-on a product or service in the hopes of changing specific condition(s) or practice(s) of an institution". Boycott sponsors may include various actors such as consumer groups, labour unions, organisations representing ethnic and racial minorities, religious groups, women's rights groups, environmental groups (Firedman 1999) and in the case of tourism in particular, tourists, non-governmental organisations, or governments (Luo and Zhai 2017). Nguyen et al. (2018) premise boycott types on consumer motivations and classify them as economic, religious, minority, ecological, social,

W. L. Hambira (✉) · V. K. Muposhi · G. E. J. Mogomotsi
Botswana University of Agriculture and Natural Resources, Gaborone, Botswana
e-mail: whambira@buan.ac.bw; vmuposhi@buan.ac.bw; gmogomotsi@buan.ac.bw

P. K. Mogomotsi
University of Botswana, Department of Economics, Gaborone, Botswana
e-mail: pmadigele@ub.ac.bw

W. L. Hambira et al. (eds.), *Environmental Change and Biodiversity Conservation in sub-Saharan Africa*, Advances in Global Change Research 75,
https://doi.org/10.1007/978-3-031-73136-5_9

and relational such that ultimately each is motive specific hence the boycotting characteristics will also differ.

A study by Nguyen et al. (2018) revealed five motivations influencing ecological boycott participation. These include environmentally sustainable consciousness, health consciousness, self-enhancement, meaning in life, and security. Ultimately, boycott campaign organisers and participants believe that their actions on the "wrong doer" will address not only economic conditions, but also ethical, moral, and/or political factors because of their interlinkages (Pezzullo 2011). Hence the boycotts have grown as tools for demanding behavioural and policy change and the fact that the international community may play centre stage is testament that the world is a global village. Some conservation activists have been advocating for the banning of hunting safaris or trophy hunting, a common tourism activity in Africa. Trophy hunting has, for years, been embraced as a one of the sustainable approaches to wildlife conservation in Africa but there are many dissenting views that have arisen in recent years as to whether indeed it can be viewed as a conservation tool (Baker 1997; International Institute for Environment and Development (IIED) 2020a; Lindsey et al. 2006; Mbaiwa 2017). Some schools of thought view restrictions associated with conservation policies as a ploy by conservationists and governments to deprive people of their land and resources (Kelly 2011).

Based on the definition of boycott by Peruzullo (2011), it can be deduced that tourism boycott is a rigorous campaign for refusal to spend money on tourist destinations, attractions or businesses in protest to an undesirable behaviour observed by the consumer of the tourist product. A research note by Shaheer, Insch, and Carr (2018) identified 146 tourist destination boycotts that took place between 1948 and 2015, about 90% of which emerged between 2003 and 2015. The note also revealed that the reasons for boycotting tourist destinations as well as attractions and services include violation of human or animal welfare and rights. Boycott organisers usually aim to promote what Penzullo (2011) refers to as "a public stigma" aimed at deterring consumers. Social media provides a platform and effective tool for perpetuating this stigma. Shaheer et al. (2018) argue that social media platforms such as Facebook and Twitter play a prominent role in online activism and consequently have changed the modern landscape of consumer boycotts. They argue that the potential negative impacts that such events can have on the tourism industry of a destination underscore the need to understand the intricacies involved. Hence scholars of recent have given attention to social media research due to among other reasons, production of user generated content on such platforms (Snelson 2016). Shaheer et al. (2018) further buttress that considerable growth of tourist destination boycotts occurred more especially ever since 2008 due possibly to social media. The increase in tourism destination boycotts especially with the advent of social media platforms such as Facebook therefore warrants research on how tourism boycotts occur on social media (Luo and Zhai 2017).

Botswana tourism has not been spared from boycott threats in social media platforms. Survival International, a UK-based global movement for tribal peoples' rights, campaigned for the boycott of Botswana tourism over alleged mistreatment of the "Bushmen" of the Central Kalahari Game Reserve (CKGR) by the Government of Botswana. The movement implored people to write to the Minister of Tourism in Botswana informing him they will not be coming to Botswana on account that, according to them, the Bushmen were being starved and intimidated off their ancestral land (Survival International 2020). The message was to be carried on email, Facebook, Twitter, and WhatsApp. A question posed on Tripadvisor on whether the call for the boycott had merit, revealed that some travellers did not think so as they were of the view that most of the time the people who suffer are not the ones whose influence was aimed at (Tripadvisor 2020). Another event that sparked a call for boycott of Botswana tourism as a destination of choice was the lifting of the hunting ban in Botswana in May 2019 after a 5-year moratorium. What started as a domestic governance issue on the part of the Botswana government led to an international outcry to boycott Botswana as a tourist destination. This was to be expected since according to Friedman (1999), international boycotts are growing due to social as well as environmental issues becoming more international and moreover environmental and animal rights movements are at the centre stage since the issues advocated for transcended national and regional boundaries.

The outcry against hunting in Botswana ignores the reality that neighbouring countries such as South Africa, Namibia, Zimbabwe, and Zambia still allow trophy hunting (Potgeiter 2019). Novelli et al. (2006), posits that in southern Africa, commercial use of wildlife involves wildlife safari, safari or trophy hunting tourism, game ranching, intensive ostrich and crocodile production and elephant culling among others. Consumptive wildlife products include game meat, hide and skins, ivory, and live sales. According to Potgeiter (2019), one of the travel agents that supported the call for boycott, still sold holidays to Botswana's neighbouring countries while refusing to make bookings for Botswana. The campaign to boycott Botswana tourism has raised questions as to whether the advocates are oblivious of the ecological and social negative impacts that have come as a result of the hunting ban. While tourists from developing countries may view travel as a means of reconnecting with nature considering that they come from highly urbanised areas, rural communities in the global south live side by side with the wildlife and therefore self-defence against dangerous wild animals is imperative, as well as consumptive use for subsistence and poverty alleviation (Novelli et al. 2006; Orams 2002).

In the view of Novelli et al. (2006), trophy hunting can be considered as a form of ecotourism aimed at maximising community benefits, as well as sustainable use of wildlife and conservation since it benefits both the environment and communities. It is also perceived that in some cases, trophy hunting has proven more beneficial than non-consumptive activities such as photographic tourism (Novelli and Humavindu 2005). Advocates against the hunting ban have been perceived to be using bullying tactics and twisting of facts against Botswana. Indeed, secondary crisis communication on social media can be viewed as a reflection of tourist's intentions to share and forward crisis information and leave negative words about the country on social media (see Luo and Zhai 2017).

Tourism destination boycott campaigns in the case of Botswana may be closely related to ecological boycotts. This is because tourism in Botswana is largely wildlife or nature based. This is important to point out since a tourism boycott may aim at a type of tourism activity or style of tourism development (see Luo and Zhai 2017). According to Nguyen et al. (2018), ecological boycott is triggered by company actions deemed harmful or abusive towards the environment. Consequently, this motivates individuals to express displeasure by refusing to buy or consume that company's products or services with the intention of improving the company's behaviour (Nguyen et al. 2018). The paper analyses user generated publicly available textual data on Facebook to determine the views of Facebook users on the lifting of the hunting ban in Botswana and resultant calls to boycott Botswana tourism.

9.2 Consumptive and Non-consumptive Tourism in Botswana: The Hunting Suspension and the Lifting Thereof

Historically, subsistence hunting was perceived as a customary right for many tribes in pre-colonial Botswana with some ethnic groups such as the San/Bushmen having it entrenched in their culture and way of life (LaRocco 2016; Madzwamuse 2010; Taylor 2001). Post-independence, the government of Botswana started regulating hunting through the introduction of the Unified Hunting Regulations of 1979 and associated Special Game Licences (Hitchcock 2001). According to LaRocco (2016), the special hunting licenses which conferred legitimacy to subsistence hunting were swapped out for Community Based Natural Resources Management (CBNRM) and a quota system from the year 1996. The CBNRM programme was introduced with the aim to empower local communities who live alongside natural resources, specifically wildlife, to benefit economically from what their local environments offered (Government of Botswana 2000). This positioned local communities to participate in tourism ventures and through the benefits thereof they are compelled to sustainably utilise the resources from which they derive their livelihoods (Hambira 2019).

Allowing local communities to benefit from the intrinsic value of the natural resources is only fair as they generally absorb most of the costs that come with their conservation due to their proximity to these natural resources (Botswana Government 2000). CBNRM activities mainly take place in areas designated as Wildlife Management Areas as well as Controlled Hunting Areas (CHAs) and according to Mbaiwa (2017) the main land use in these areas is wildlife utilisation and they are located between local community areas and national parks as well as game reserves. Some tourism activities occur in concession areas in the CHAs. CHAs are administrative land blocks used by the government to allocate hunting quotas and governance framework to provide for sustainable utilisation of wildlife resources (Blackie 2019). Mbaiwa (2017:p.52) states "data from Tawana Landboard indicate that a total of 68 concession areas are designated for wildlife-based tourism, of which 21 of the concession areas are reserved for citizen hunting, 32 for community-managed areas and 15 for private sport hunting concessions in which sport hunting by foreigners is permitted." It is therefore clear that hunting in Botswana is strictly controlled and can only occur in accordance with the terms and conditions of a licence issued under the Wildlife Conservation and National Parks Act of 1992.

The introduction of Community Trusts (CTs) or Community Based Organisations (CBOs) further buttressed efforts of controlled hunting so that community tourism takes place in an organised manner for the benefit of the community while at the same time upholding sustainable conservation principles. That is, CTs/CBOs represent communities or groups of communities adjacent to CHAs and carry out consumptive and non-consumptive tourism activities on behalf of communities. For example, through CTs/CBOs, communities may approach relevant Land Authorities to obtain 15-year Community Natural Resource Management Lease for the commercial use of natural resources; communities may also sublease or transfer any commercial natural resource user rights to joint venture partners with the help of CTs/CBOs and other relevant authorities (Government of Botswana 2000). In August 2013, the government of Botswana through the Ministry of Environment Wildlife and Tourism (MEWT, as it was called then), announced a temporary ban on hunting of wildlife in all CHAs effective January of 2014. The terms and conditions highlighted in the press release were as follows: (i) no quotas, licences or permits will be issued for hunting of part I and II Schedule of game animals stipulated in the Wildlife Conservation and National Parks Act; (ii) licences for game birds listed in part III of the same Act not affected unless the Department of Wildlife and National Parks finds reason to; (iii) the ban would not apply to hunting in registered game ranches (MEWT Press Release 2013).

The suspension of hunting was necessitated by declining trends of several wildlife species accruing to illegal offtake and habitat fragmentation or loss (Department of Wildlife and National Parks 2012; Chase 2011). It was argued that the ban was intended to be temporary in order to allow the MEWT (now Ministry of Environment and Tourism) to investigate the decline and come up with remedial measures to curb the situation. Given that there were communities already benefiting from trophy hunting through partnerships with trophy hunting operators as part of the CBNRM programme, this hunting suspension would not be without ramifications. It came with job losses, unavailability of free meat from trophy hunting, loss of income to Community Based Organisations/Trusts and associated social services, increased elephant populations and consequently increased human–wildlife conflicts (HWC); increased poaching; as well as reintroduction of negative attitudes towards wildlife conservation by local communities (Blakie 2019; IIED 2020b; Mbaiwa 2017).

Job losses were recorded due to the retrenchments from safari hunting companies. Further, there was scarcity of game meat since animal carcasses from hunts which were often given out for free were no longer available to communities as relish. Furthermore, there was a decline in income for traditional dancers who entertained hunter-tourists during hunting seasons thus resulting in the loss of household income. Corporate social responsibility initiatives offered by hunting safari companies and CBOs such as funeral schemes, housing provision for the vulnerable and scholarships were also negatively affected (Blakie 2019; Mbaiwa 2017). Even though one of the reasons for the hunting moratorium was decline in some wildlife species, the ban was applied to all animals, even those that were not in any danger of extinction such as elephants (Blakie 2019; Chase 2011). This led to an increase in human–wildlife conflicts, a significant reduction of the economic value of some of the more financially marginal concession areas, and degradation of the ecological landscape (Mogomotsi et al. 2020).

The imposition of the hunting moratorium by the Government of Botswana in 2014 meant that communities had to quickly adapt and find other means to compensate for the loss of income from trophy hunting. New strategies included seeking joint ventures/partnerships with companies operating photographic tourism and high-end accommodation facilities as well as engaging in agricultural economic activities (see IIED 2020b; Mbaiwa 2017). However, switching from trophy hunting operations to photographic safaris was not automatic as some of the areas were not suitable for the latter. This is bolstered by Mbaiwa (2017), who argues that peripheral areas used for hunting safaris are not necessarily ideal for photographic tourism. Moreover, earlier studies by Ecosurv (2012) also supported that the potential for photographic tourism in areas such as Boteti District were low. Potgeiter (2019) opines that since the hunting ban compelled hunting operators to switch to photographic tourism, should the Botswana photographic clientele buy in the boycott campaigns, partnerships with the hunting operators may become an option. In other words, the loyalty of the photographic tourists will render the ushering in of trophy hunting operations a non-urgent matter.

9.2.1 Theoretical Framework

The notion of sustainable development brought to the fore the importance of harmony between the developmental facets of the environment, economy, and society. This has led to the emergence of sustainable natural resources management approaches where the culture and livelihood needs of local communities must be explicitly considered in relation to the physical and ecological dynamics of natural resources in any given space and time. Similarly, successful implementation of tourism (nature-based tourism in particular) as a viable economic sector requires a more holistic and complex conceptualization that takes into cognizance human dimensions of value. In line with this perspective, this study has adopted the Sustainable Tourism Framework as its conceptual foundation. The framework borrows from the Sustainable Development discourse which is described as development that endeavours to strike a balance between societal, economic, and environmental aspects of the economy for the benefit of current and future generations (Brundtland 1987).

Similarly, sustainable tourism advocates for tourism development that is not at the expense of the environment and societal well-being (Liu 2003). Tourism in Botswana derives its value from the pristine wilderness and natural resources base more especially wildlife. Hence for the industry to continue contributing significantly to the country's GDP, the natural capital must be harnessed sustainably. However, tourism cannot rely on the natural capital at the expense of the societies therein. Tourism destinations rely on a well functional society, healthy environment and stable economy (Pyke et al. 2016). Therefore, tourism activities should not only benefit the environment and economy but the community as well by creating wealth generating opportunities (see Hambira 2019). Botswana conservation policies are premised on this understanding. In that manner, the Wildlife Policy of 2013 whose objective is to promote sustainable use of wildlife resources provides that in order to achieve its objective there should be a regulated wildlife-based industry guided by the principle of sustainable off-take, science-based quota setting, and efficient utilisation of wildlife resources (MEWT 2013: p. 18). Further, the CBNRM policy states that the natural resources have an embedded or intrinsic economic, environmental, and social value (Government of Botswana 2000).

Furthermore, sustainable utilisation of natural resources, specifically wildlife and veldt products, are considered a rural development strategy as identified in the Revised Policy on Rural Development of 2002 and the National Strategy for Poverty Reduction of 2003 (Government of Botswana 2000). Worth-noting also is the fact that it is a policy goal to ensure that the proceeds of the sale of natural resource concessions or hunting quotas accruing to communities have to be paid into a National Environmental Fund managed by the Ministry of Environment, Natural Resources Conservation and Tourism (Government of Botswana 2000). Striking a balance between the three dimensions of sustainable development has always eluded many governments and business entities. Some businesses tend to be profit oriented at the expense of the society and the environment while some conservationists are on the extreme side of preservation to the detriment of societies and economies. Similarly, there is no consensus on the social equity and economic viability of trophy hunting (Novelli et al. 2006). It is therefore imperative to understand the viewpoints of Botswana tourism destination boycott campaign "cheer leaders". Premised on the Sustainable Tourism Framework, the study endeavours to provide insights for a "win-win" solution.

9.3 Material and Methods

9.3.1 Study Area

This study focusses on Botswana, a sub-Saharan Africa with a population of 2,038,228 as of the last census of 2011 (Central Statistics Office 2011) with the 2020 projection standing at 2,351,627 (Worldometer 2020). The country, spread over an area of 581,730 square kilometres, is landlocked by South Africa, Namibia, Zambia, and Zimbabwe (Fig. 9.1).

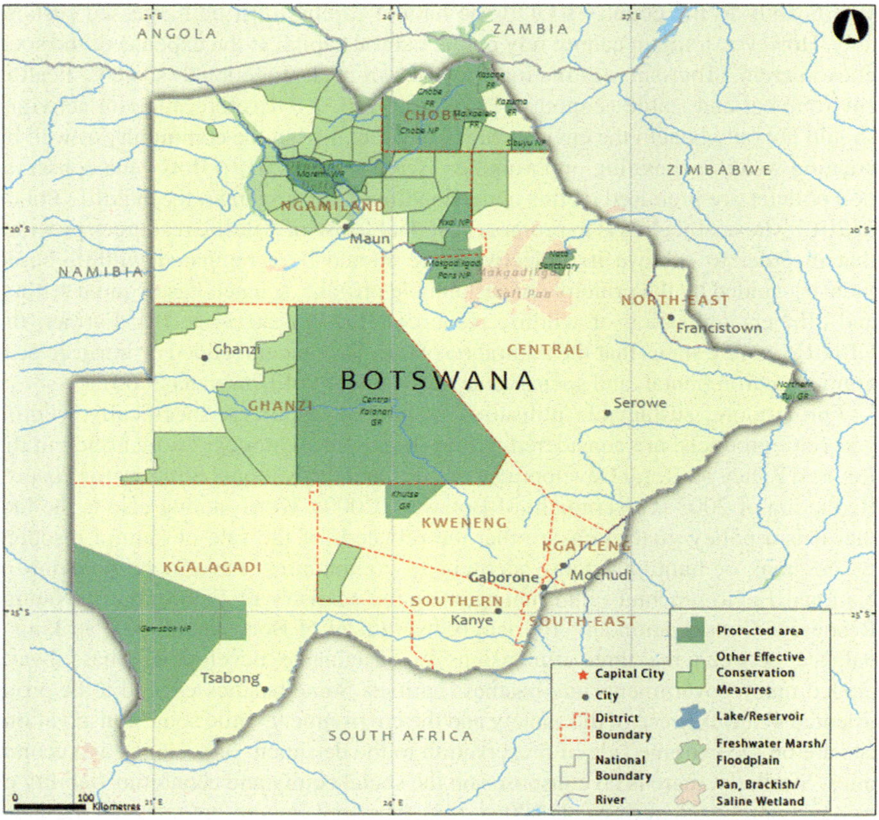

Fig. 9.1 Map showing the protected area and conservation network in Botswana. (Source: Government of Botswana (2022)

Botswana is classified as an upper middle-income country (The World Bank 2020). This income level ranking owes to the country's mineral sector (particularly the diamond industry) which contributes the most to GDP and accounts for 80% of export earnings (The World Bank 2020). The trickling down effects of the country's wealth, however, remains a challenge as income inequalities remain high and employment opportunities are limited (The World Bank 2020). Almost 40% of the country's land area is dedicated to conservation: 17% as national parks and reserves and 22% as Wildlife Management Areas (Department of Environmental Affairs, undated). This forms the basis of the tourism industry in Botswana making it the second most important economic sector contributing approximately US$ 575.30 million constituting about 3.2% GDP (Sunday Standard 2014).

9.3.2 Study Design and Framework

This study employs social media research where data derived from social media sources is used. Social media are described as web-based platforms where users generate and share content (Social Media Research Group 2016). Social media research is considerably new in academia and thus research paradigms to be applied are still under development (Gerges 2015; Van Osch and Coursaris 2014). Even though social media research is still developing, some social science research tools may be applied in social media research albeit the latter may require some manipulation for it to be useful (Social Media Research Group 2016).

According to Social Media Research Group (2016), social media research is twofold: 1) using social media as a research tool through, for example, surveys and 2) researching on the activity or content on the social media platform itself. The latter will be the focus of this study. A qualitative approach was therefore adopted due to the textual form of the data collected. This research engages a social media tool, known as Facebook, which falls under the Social Communities category (Van Looy 2016) also classified as a Social Networking Site (Social Media Research Group 2016). Facebook was preferred because according to Statista.com (2020), it is the most popular social media network worldwide with 2603 million users as of July 2020.

An online search was conducted on Facebook between July and September 2020 using the search terms "Facebook + hunting + ban + boycott + Botswana". The study focussed on user-generated, publicly available textual data, specifically posts and comments from May 2019 to December 2019. Collected data were meticulously reviewed for relevance, excluding posts unrelated to reactions to the lifting of the hunting ban. While such data can be used to extract frequencies, user profiles, and interactions (Social Media Research Group 2016), this study concentrated solely on the content of the posts, disregarding user demographics.

One of the primary challenges in analysing ethnographic observational data, including that sourced from social media platforms, is the assessment and mitigation of potential hidden biases (Faimau and Behrens 2019). To address this issue, this study utilized passive observation and thematic analysis to examine patterns in Facebook user interactions. Passive observation was specifically chosen to ensure that the researchers did not interact with individuals whose posts were analysed, thereby reducing the likelihood of introducing bias. Orben, Mutak, Dablander et al. (2018) posit that the passive consumption of information, wherein users do not interact with the content in any way, is becoming increasingly prevalent (see also Schlosser 2005; Burke et al. 2010; Burke et al. 2011; Gerson et al. 2017). This study concentrates on user-generated publicly available textual data, including posts and comments from individuals and tourism operators, such as booking agencies.

The posts selected for this research were identified through a Google search and were sourced from the following Facebook pages, listed in the order of their appearance:

1. The Botswana Gazette, on the topic "Botswana lifts hunting ban".
2. BWgovernment, on "Botswana controlled hunting program: a sustainable, science-based conservation strategy to benefit and empower communities."
3. BWgovernment, on "Government lifts hunting ban".
4. BWgovernment, on "Handover notes of the cabinet sub-committee on hunting ban social dialogue report by Hon Frans van Der Westhuizen—Minister of Local Government and Rural Development".

All comments deemed relevant were manually copied and pasted into a Word document for further analysis.

9.3.3 Thematic Analysis of Social Media Discourse

A thematic analysis was conducted to identify and analyse patterns within the data. The process involved several key steps:

1. *Familiarization:* The collected data were reviewed thoroughly to gain a comprehensive understanding. Data were first imported into RStudio, where they were cleaned and pre-processed using the tidyverse package. This involved removing duplicates, handling missing values, and standardizing text formats.
2. *Coding:* The cleaned data, with relevant excerpts were then coded manually to identify initial themes, with each segment of text being assigned a code that represented a specific concept or idea. The coded data were subsequently categorized into broader themes using a combination of inductive and deductive approaches.
3. *Categorization:* The codes were grouped into broader categories that represented larger themes. This categorization was iterative, involving repeated examination and refinement of the codes and categories. Inductive coding was used to allow themes to emerge directly from the data, while deductive coding was guided by existing theoretical frameworks relevant to the study. The dplyr package facilitated the categorization process by enabling efficient data manipulation and summarization.
4. *Theme identification:* To further refine and identify key themes, text mining techniques were applied using the tm and textclean R packages. The text mining process involved tokenization, stemming, and removal of stop words to focus on the most relevant terms. Frequency analysis and word cloud visualizations were used to highlight predominant themes. Additionally, co-occurrence analysis helped to identify relationships between different codes. To ensure the reliability and validity of the identified themes, intercoder reliability was assessed. The categories were further analysed to identify overarching themes. These themes were distilled into five major areas of focus:

- Governance and policy,
- Economic considerations,
- Conservation concerns,
- International reaction, and
- Alternative solutions.

9.3.4 Sentiment Analysis of Social Media Discourse

The corpus used for the sentiment analysis was derived from a document containing various opinions and comments regarding the lifting of the hunting ban in Botswana. The document included pro and anti-hunting sentiments expressed by individuals from different backgrounds and perspectives. The text was extracted from the document and cleaned to remove any non-alphabetic characters and convert all text to lowercase. This step ensured uniformity and eliminated any noise from the data. Common English stop words were removed to focus on the meaningful content.

Sentiment analysis was conducted using the TextBlob library, which provides tools for processing textual data and extracting sentiments. Each word in the cleaned corpus was analysed for sentiment polarity, which is a measure of how positive or negative a word is. Words with a sentiment polarity of zero (neutral words) were excluded from further analysis. To focus on the most impactful sentiments, the top 20 words with the highest absolute sentiment values (both positive and negative) were selected. These words were identified as having the most significant emotional impact on the discussion about the hunting ban. The selected top 20 sentiment words were visualized using a bar plot, created with the Seaborn library. The plot displays the sentiment polarity of each word on the x-axis and the words themselves on the y-axis. A vertical line at sentiment polarity zero was added for reference.

9.3.5 Ethical Considerations

Since data generated through social media was not meant for research in the first place, researchers are cautioned to be mindful of data representativeness, robustness, and ethics (Social Media Research Group 2016). With regard to representativeness, the researcher has to be cautious to avoid data skewness; with respect to large volumes of data likely to be generated should be manageable taking into account the method of analysis; while ethical consideration should encompass non-disclosure of identity. Effort has been taken to adhere to these set out principles. To avoid large volumes for data which may not be manageable in the end, other relevant posts were not considered as soon as no more new information was derived from reading the comments from the posts. Anonymity is also observed since no direct quotations or "voices" of the users were stated in the results. Only summaries of the textual data are referred to. This is important since passive analysis is being employed where publicly available data is used without first seeking consent from the social media users whose posts and comments were relevant to the study.

9.4 Results

9.4.1 Thematic Analysis of Discourse on Lifting Botswana's Trophy Hunting Ban

From the passive interaction with the data, it was deduced that the social media users comprised Botswana citizens and residents; foreigners who identified themselves as potential tourists and those who have been to Botswana as tourists; and to a lesser extent, tourist operators. The initial 20 codes were consolidated into five categories, focussing on boycott threats, destination image preservation, and animal welfare. Thematic analysis further distilled these into five major themes: Governance and Policy, Economic Considerations, Conservation Concerns, International Reaction, and Alternative Solutions, as shown in Table 9.1 and Fig. 9.2.

Table 9.1 Thematic analysis of the hunting ban lift in Botswana from local to global perspectives and solutions

Scale	Thematic aspect	Descriptors
Local	Governance and policy	Decision making: the decision is a political move Comments suggesting poor governance Learning from others: Botswana should learn from the experience of countries practicing trophy hunting
	Economic considerations	Tourism: tourism will go down Photographic safari is more profitable Beneficiation: comments on beneficiation will only benefit hunting operators
	Conservation Concerns	Species impact: elephants will become extinct; elephants have the right to exist Behavioural changes: elephants will become more aggressive Human–wildlife conflict: intrusion into elephant habitat by humans; find other means of compensating farmers terrorized by elephants
Global	International reaction	Condemnation: Western countries will not condone it Boycotts: tourists should boycott; boycott Botswana as a tourist destination Sanctions and image: call for sanctions; concern for the country's reputation; destination image
Overlapping	Alternative solutions	Alternatives to hunting: there are other alternatives to dealing with human–wildlife conflict (HWC) Economic alternatives: photographic safari is more profitable; find other means of compensating farmers

Fig. 9.2 Word cloud depicting thematic analysis of discourse on lifting Botswana's trophy hunting ban

1. *Governance and Policy:* The analysis revealed significant concerns regarding the political motivations behind the decision to lift the hunting ban. Critics highlighted that the move was largely seen as a political strategy rather than a well-considered conservation policy. The themes within this category suggested poor governance practices and emphasized the need for Botswana to learn from countries with experience in managing trophy hunting sustainably. However, this would be considered as mudslinging given the record and success of Botswana's conservation efforts, making it the leader globally when it comes to African elephant populations.
2. *Economic Considerations:* Economic impacts were a major theme, particularly regarding tourism. Concerns were raised that lifting the ban could negatively affect Botswana's tourism industry, which is heavily reliant on wildlife and photographic safaris. There was also scepticism about the purported economic benefits, with many asserting that the decision would primarily benefit hunting operators rather than the broader community.
3. *Conservation Concerns:* This theme encompassed worries about the ecological implications of lifting the ban. There were fears that the decision could lead to increased aggression in elephants, potentially resulting in more human–wildlife conflicts. Additionally, the potential for elephants to become extinct was a recurring concern, underscoring the perceived threat to biodiversity.
4. *International Reaction:* The global community's response was overwhelmingly negative. Many international stakeholders, including Western countries, expressed condemnation. Calls for boycotts and sanctions were prominent, reflecting the significant reputational risk for Botswana. This theme illustrated the global implications of the decision and the strong opposition from international conservation advocates.

5. *Alternative Solutions:* Despite the criticism, there were numerous suggestions for alternative approaches to managing human–wildlife conflict (HWC) and promoting economic benefits. Photographic safaris were frequently mentioned as a more profitable and sustainable alternative to trophy hunting. Additionally, there were calls for innovative compensation schemes for farmers affected by wildlife, highlighting the need for creative and community-focussed solutions.

9.4.2 *Sentiments on Trophy Hunting Ban Lift in Botswana*

The sentiment analysis of the comments regarding the lifting of the hunting ban in Botswana revealed a range of emotions and opinions, highlighting several key issues (see Fig. 9.3):

1. *Strong Negative Sentiments*
   - Words such as “horrible,” “heartbreaking,” and “shame” indicate intense negative emotions among individuals opposing the lifting of the hunting ban. These sentiments reflect concerns about the ethical implications of hunting and potential long-term negative impacts on wildlife conservation.
   - There are fears that the decision might damage Botswana’s international reputation and reduce tourism, which is a significant source of revenue.
2. *Positive Sentiments*
   - Conversely, words like “good,” “sustainable,” and “empower” reveal positive sentiments from supporters of the ban lift. These sentiments are associated with arguments for the benefits of controlled hunting for local communities and wildlife management.
   - Supporters suggest that lifting the ban could help manage the overpopulation of elephants and reduce human–wildlife conflicts, which have caused significant harm to human lives and livelihoods.

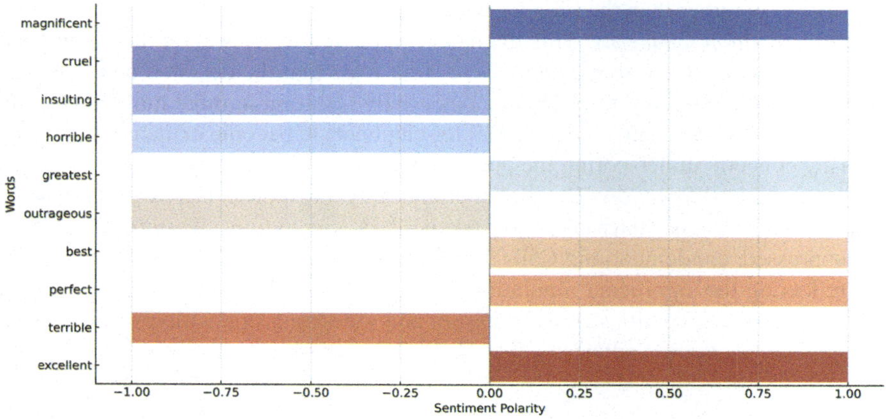

Fig. 9.3 Predominant sentiments in the discourse on lifting the trophy hunting ban in Botswana

3. *Ethical and Practical Considerations*
    - The sentiment analysis reveals a significant divide between ethical concerns about animal rights and practical considerations regarding human safety and economic benefits.
    - Comments reflect a struggle between preserving Botswana's wildlife heritage and addressing the immediate needs of its people, especially those in rural areas facing direct conflicts with wildlife.
4. *Policy and Public Opinion Impact*
    - The polarized sentiments suggest that any policy decision regarding wildlife management needs to carefully balance these conflicting viewpoints.
    - Public opinion is deeply split, indicating that any changes in policy will likely continue to provoke significant debate.

The sentiment analysis of the social media discourse provides a nuanced understanding of public reactions to the lifting of the hunting ban in Botswana. It highlights the complexity of balancing ethical considerations with practical needs in wildlife conservation policymaking. This information is crucial for stakeholders and policymakers to make informed decisions that address both conservation goals and the well-being of local communities.

9.5 Discussion

Boycott threats included comments about booking cancellations, calls for protests, action that will ensure the tourism sector collapses. Furthermore, a lot of frustration from such comments was displayed using words such as "what a shame", "heart breaking", "so sad", "horrifying", "disastrous", and "leave them alone". Boycotts have become popular especially for environmental movements (Penzullo 2011). Depending on their level of consumer support boycotts may have a direct impact on tourism through for example reduction in visitors (Pezzullo 2011; Yousaf 2019; Mkono and Holder 2019). According to Hahn and Albert (2017), consumers decide individually whether to partake in boycott and not all consumers are willing to sacrifice consumption of the product or service offered by the "misbehaving" entity depending on the costs and benefits. This was displayed through some comments which suggested their proponents might have had a change of heart. Such comments would for example be displayed after a constructive back and forth with proponents of pro-lifting of the hunting ban category. This means corporation by individuals is of paramount importance thus it remains to be seen how the tourists would buy in the call to boycott the Botswana tourist destination. Pezzullo (2011) argues that the economic incentives for institutions to implement change are minimal when the individuals engaging in boycotts are not original consumers of the targeted product or service. This suggests that for boycotts to be effective, they need to involve and impact the actual customer base of the institution, thereby creating a tangible economic pressure that compels the institution to respond and make changes.

We established from various comments that some proponents of the boycott have previously visited Botswana and vowed never to return, while others claimed to have cancelled their bookings following the decision to lift the hunting ban. Additionally, some identified themselves as potential visitors who were planning to come to Botswana but have since reconsidered. Potgeiter (2019) cautions southern African countries that still practice trophy hunting against implementing a ban, even as a short-term measure, to avoid undue pressure on their tourism industries. He argues that such a ban is a risky long-term strategy for conservation efforts, despite short-term praise from the international community.

Regarding the preservation of Botswana's destination image, some comments indicate that the country is perceived as a leader in conservation, and lifting the hunting ban could tarnish its reputation. This sentiment is reflected in threats to instigate negative press, warnings about losing respect as a global conservation leader, admonitions to lead by example, and accusations of selling the country to wealthy hunters. Research has shown that a negative image can indirectly and adversely impact a destination, attraction, or tourism service provider (Mkono and Holder 2019; Pezzullo 2011; Yousaf 2019). Furthermore, the social identity of citizens in negatively perceived countries is also threatened (Yousaf and Li 2015). If attempts to portray a country negatively are successful, the underlying issues leading to this perception should be addressed (Yousaf 2019).

Research indicates that social media users from non-elephant range countries dominate discussions, focussing on elephant welfare and trophy hunting, issues not critical to elephant conservation (Hammond et al. 2022). In contrast, individuals in African elephant range countries predominantly address human-elephant conflict, coexistence, poaching, and advocate for elephant tourism (Muposhi et al. 2016). This disparity highlights the underrepresentation of African voices on social media, particularly those experiencing the direct impacts of living alongside wildlife (Fink et al. 2020). Consequently, it is essential to ensure that social media discourse does not disproportionately influence policymakers, given the insufficient representation of local stakeholders and the limited attention to key conservation threats (Barrios-O'Neill 2021; Dickman et al. 2020).

A nuanced perspective is that for years, Botswana has been recognized for its exemplary leadership in sustainable tourism, earning international accolades from bodies such as the World Travel and Tourism Council. Notably, the Tourism for Tomorrow Destination award was attributed to the country's hunting ban—a decision that has since been reversed, potentially jeopardizing its reputation for sustainability (IIED 2020b). This reversal garnered praise from animal welfare organizations and anti-trophy hunting groups. Should boycott threats materialize and tarnish Botswana's image, it would be essential to develop strategies to mitigate this impact, possibly through re-branding efforts.

The theme of animal welfare was underscored by concerns about the potential extinction of elephants due to the lifting of the hunting ban, which could increase poaching opportunities. Additional concerns included human intrusion on elephant habitats, human–wildlife conflict (HWC), disregard for nature, and perceptions that such decisions aim to eradicate the remaining elephant populations, emphasizing that elephants have a right to live. Hoffer (2002) presents a nuanced discussion on the polarized perspectives surrounding consumptive tourism. Proponents argue that trophy hunting can be a tool for population control and maintaining ecosystem balance, suggesting that it contributes to conservation funding and local community benefits. Opponents, however, vehemently oppose the practice, condemning the killing of animals for personal satisfaction and questioning the ethical and ecological justifications provided by its supporters. This debate underscores the complex interplay between conservation strategies and ethical considerations in wildlife management.

Violations of animal welfare and inappropriate environmental actions are some of the bases for advocating for destination boycotts (Shaheer et al. 2019; Nguyen et al. 2018). This is seen in some quarters as disregard of the great conservation efforts upheld by countries such as Botswana and is testimony that colonialism is still hovering around as African presidents are expected to please the developed world's expectation at the expense of the needs of their people (Potgeiter 2019). This has created a north–south divide with regard to perception, consumption, and conservation of wildlife (Smith and Duffy 2003). Furthermore, boycotts are often triggered by undesirable behaviour (Hahn and Albert 2017) and socially irresponsible actions (Klein et al. 2004), raising critical questions about who determines what constitutes desirable and socially responsible behaviour.

Therefore, addressing these complex conservation issues necessitates systems thinking and a global perspective. Countries like Botswana must balance global conservation goals from the Kunming–Montreal Global Biodiversity Framework with local sustainability and accountability. The framework promotes sustainable use, positioning people within nature's limits, and urges globally informed, locally tailored policies. Botswana should leverage its rich biodiversity and conservation ethos to implement strategies aligning with global standards while addressing local needs. By integrating systems thinking, Botswana can develop holistic approaches that consider ecological, social, and economic dimensions, promoting community involvement, ensuring equitable resource distribution, and fostering sustainable livelihoods without compromising ecological integrity. Achieving conservation goals requires blending global frameworks with local realities. Botswana's commitment to sustainability and accountability can model inclusive and adaptive strategies for other countries. This balance is crucial for long-term biodiversity conservation and human well-being, ensuring both nature and people thrive.

9.6 Conclusion

Social media, particularly Facebook and X (formerly Twitter), has become a powerful platform for shaping public and social discourse, including environmental and conservation issues. Activists utilise social media platforms to promote "pro-environmental" and "pro-sustainable" behaviours, including opposing tourism to destinations with differing environmental practices. This study highlights ongoing debates over optimal strategies for sustainable wildlife conservation in Africa, emphasizing the need to consider local socio-economic contexts of individual countries. Botswana, a leader in wildlife conservation, has leveraged its positive image to develop its tourism sector, the second highest contributor to its GDP. Negative campaigns and social media backlashing could harm both the tourism sector and the welfare of communities dependent on it by overlooking the complex balance between conservation and rural livelihoods. Assessing the impact of negative media framing, backlash, and the slow recovery of countries and conservation and tourism stakeholders post-COVID-19 is challenging due to the complexities introduced by the global pandemic. This underscores the necessity for strategic, data-driven interventions to ensure sustainability in these sectors.

References

Baker JE (1997) Development of a model system for touristic hunting revenue collection and allocation. Tour Manag 18:273–286

Barrios-O'Neill D (2021) Focus and social contagion of environmental organization advocacy on Twitter. Conserv Biol 35(1):307–315

Blakie I (2019) The impact of wildlife hunting prohibition on the rural livelihoods of local communities in Ngamiland and Chobe District areas, Botswana. Cog Soc Sci 5(1):1558716

Brundtland GH (1987) Our common future: Report of the world commission on environment and development. Geneva, UN-Dokument A/42/427

Burke M, Marlow C, Lento T (2010) Social network activity and social well-being. In: Proceedings of the SIGCHI conference on human factorsin computing systems, New York, NY, pp 1909–1912

Burke M, Kraut R, Marlow C (2011) Social capital on Facebook: differentiating uses and users. In: CHI 2011 proceedings of the SIGCHI conference on human factors in computing systems, Vancouver, pp 571–580. https://doi.org/10.1145/1978942.1979023

Central Statistics Office (2011) 2011 Population and housing census: preliminary brief. Retrieved https://ecastats.uneca.org/aicmd/Portals/0/Census%202011%20Preliminary%20%20Brief%20Sept%2029%202011.pdf

Chase M (2011) Dry season fixed wing aerial survey of elephants and wildlife in Northern Botswana: elephant without boarders

Department of Environmental Affairs (undated). Action Plan for Implementing the Convention on Biological Diversity'sProgramme of Work on Protected Areas. Botswana report submitted to the Secretariat of the Convention on Biological Diversity. https://www.cbd.int/doc/world/bw/bw-nbsap-powpa-en.pdf

Department of Wildlife and National Parks (2012) Aerial census of animals in Botswana: 2012 dry season. Government Printer, Gaborone, Botswana

Dickman A, Johnson PJ, Coals P, Harrington LA, Tyrrell P, Somerville K et al (2020) Wars over wildlife: green militarisation and just war theory. Conserv Soc 18(3):293–297

Ecosurv. (2012) Strategic environmental assessment for the Okavango Delta Ramsar Site (ODRS). Maun. Unpublished report

Faimau, G., & Behrens, C. (2019). The Use of the Qualitative Approach and Digital Ethnography in the Study of Religion and New Media: Reflections from Research Fieldwork in Botswana. Botswana Notes and Records 51: 64–77

Fink C, Hausmann A, Di Minin E (2020) Online sentiment towards iconic species. Biol Conserv 241:108289

Friedman M (1999) Consumer boycotts: effecting change through the marketplace and media. Routledge, New York

Gerges M (2015) Paradigms in social media studies. Master Thesis in Master Program in Social Sciences Digital Media and Society. Uppsala University. Retrieved https://www.diva-portal.org/smash/get/diva2:857296/FULLTEXT01.pdf

Gerson J, Plagnol AC, Corr PJ (2017) Passive and active Facebook use measure (PAUM): validation and relationship to the reinforcement sensitivity theory. Pers Individ Diff 117:81–90

GOB (2022) "Botswana State of Environment and Outlook Report 2022", Government of Botswana, Gaborone.

Government of Botswana (2000) Community-based natural resource management policy. Government Printer, Gaborone

Hahn T, Albert N (2017) Strong reciprocity in consumer boycotts. J Bus Ethics 145(3):509–524

Hambira WL (2019) A review of community social upliftment practices by tourism multinational companies in Botswana. In: Stone MT, Lenao M, Moswete N (eds) Natural resources, tourism, and community livelihoods in Southern Africa: challenges of sustainable development. Routledge, London. https://doi.org/10.4324/9780429289422

Hammond NL, Dickman A, Biggs D (2022) Examining attention given to threats to elephant conservation on social media. Conserv Sci Pract 4(10):e12785

Hitchcock RK (2001) 'Hunting is our heritage': the struggle for hunting and gathering rights among the San of Southern Africa. In: Anderson DG, Ikeya K (eds) Parks, property, and power: managing hunting practice and identity within state policy regimes. National Museum of Ethnology, Osaka, pp 139–156

Hofer D (2002) The Lion's share of the hunt. Trophy hunting and conservation: a review of the Eurasian tourist hunting market and trophy trade under CITES. TRAFFICEurope Regional Report, Brussels

International Institute of Environment and Development (IIED) (2020a) Impacts of trophy hunting-lessons from the hunting ban in Botswana. Retrieved from https://www.iied.org/impacts-trophy-hunting-lessons-hunting-ban-botswana

International Institute of Environment and Development (IIED) (2020b) A tale of two villages, community perspectives on the impact of Botswana's trophy hunting ban. Retrieved from https://www.iied.org/tale-two-villages-community-perspectives-impact-botswanas-trophy-hunting-ban

Kelly AB (2011) Conservation practice as primitive accumulation. J Peasant Stud 38(4):683–701

Klein JG, Smith NC, John A (2004) Why we boycott: consumer motivations for boycott participation. J Mark 68(3):92–109

LaRocco A (2016) The comprehensive hunting ban: strengthening the state through participatory conservation in contemporary Botswana. In: Ramutsindela M, Miesche G, Boehi M (eds) The politics of nature and science in Southern Africa. Basler Afrika Bibliographien, pp 179–207. Retrieved https://pdfs.semanticscholar.org/14c1/1d0a67ee72011728efb77d9fff1b359a9ade.pdf

Lindsey PA, Alexander R, Frank LG, Mathieson A, Romanach SS (2006) Potential of trophy hunting to create incentives for wildlife conservation in Africa where alternative wildlife-based land uses may not be viable. Anim Conserv 9:283–298

Liu Z (2003) Sustainable tourism development: A critique. J Sustain Tour 11(6):459–475 https://doi.org/10.1080/09669580308667216

Luo Q, Zhai X (2017) "I will never go to Hong Kong again!" how the secondary crisis communication of "Occupy Central" on Weibo shifted to a tourism boycott? Tour Manag 62:159–172

Madzwamuse M (2010) Adaptive or anachronistic? Maintaining indigenous natural resource governance systems in Northern Botswana. In: Nelson F (ed) Community rights, conservation and contested land: the politics of natural resource governance in Africa. Earthscan, London, pp 241–268

Mbaiwa JE (2017) Poverty or riches: Who benefits from booming tourism industry in Botswana? J Contemp Afr Stud 35(1):93–122

Ministry of Environment Wildlife and Tourism (2013) Press release: Hunting ban in Botswana-Message from Permanent Secretary. Retrieved from https://www.facebook.com/148228411926492/posts/press-release-hunting-ban-in-botswana-message-from-permanent-secretarythe-minist/500849569997706/

Mkono M, Holder A (2019) The future of animals in tourism recreation: Social media as spaces of collectivemoral reflexivity. Tour Manag Perspect 29:1–8. https://doi.org/10.1016/j.tmp.2018.10.002
Mogomotsi PK, Mogomotsi PK, Dipogiso K, Phonchi-Tshekiso ND, Stone LS, Badimo D (2020) An analysis of communities' attitudes toward wildlife and implications for wildlife sustainability. Trop Conserv Sci 13:194008292091560. https://doi.org/10.1177/1940082920915603
Muposhi VK, Gandiwa E, Bartels P, Makuza SM (2016) Trophy hunting, conservation, and rural development in Zimbabwe: issues, options, and implications. Int J Biodivers 2016:1
Nguyen T, Ngo H, Ngo P, Kang GD (2018) Understanding the motivations influencing ecological boycott participation: an exploratory study in Viet Nam. Sustain For 10(12):4786
Novelli M, Humavindu MN (2005) Wildlife tourism. Wildlife use vs local gain: trophy hunting in Namibia. In: Novelli M (ed) Niche tourism: current issues, trendand cases. Elsevier, Oxford, pp 171–182
Novelli M, Barnes JI, Humavindu M (2006) The other side of the ecotourism coin: consumptive tourism in southern Africa. J Ecotour 5(1–2):62–79. https://doi.org/10.1080/14724040608668447
Orams MB (2002) Feeding wildlife as a tourist attraction: a review of issues and impacts. Tour Manag 23:281–293
Orben AC, Mutak A, Dablander F, Hecht M, Krawiec JM, Valkoviˇcová Nand Kos̄ıte D (2018) FromFace-to-face to Facebook: probing the effects of passive consumptionon interpersonal attraction. Front Psychol 9:1163
Pezzullo PC (2011) Contextualizing boycotts and buycotts: the impure politics of consumer-based advocacy in an age of global ecological crises. Commun Crit/Cult Stud 8(2):124–145
Potgeiter GC (2019) Why calling for a tourist boycott of Botswana over hunting is a bad idea. https://africasustainableconservation.com/2019/07/04/why-calling-for-a-tourist-boycott-of-botswana-over-hunting-is-a-bad-ideaj/
Pyke S, Hartwell H, Black A, Hemingway A (2016) Exploring well-being as a tourism product resource. Tour Manag 55:94–105
Schlosser AE (2005) Posting versus lurking: communicating in a multiple audience context. J Consum Res 32:260–265
Seyfi S, Hall CM (2019) Sanctions and tourism: effects, complexities and research. *Tourism Geographies*, 22(4):1–19
Shaheer I, Insch A, Carr N (2018) Tourism destination boycotts–are they becoming a standard practise? Tour Recreat Res 43(1):129–132
Shaheer I, Carr N, Insch A (2019) What are the reasons behind tourism boycotts. Anatolia 30(2):294–296
Smith M, Duffy R (2003) The ethics of tourism development. London:Routledge
Snelson C (2016) Qualitative and mixed-methods social media research: A review of the literature. Int J Qual Methods 15(1):1–15. https://doi.org/10.1177/160940691562457
Social Media Research Group (2016) Using social media for social research: an introduction. https://assets.publishing.service.gov.uk/government/uploads/system/uploads/attachment_data/file/524750/GSR_Social_Media_Research_Guidance_-_Using_social_media_for_social_research.pdf accessed 10 September 2020
Statistica (2020) Most popular social networks as of July 2020, ranked by number of active users. Retrieved https://www.statista.com/statistics/272014/global-social-networks-ranked-by-number-of-users/
Sunday Standard (2014) Tourism contributes 3.2% to GDP. https://www.sundaystandard.info/tourism-contributes-3-2-percent-to-gdp/
Survival International (2020) Boycott Botswana Tourism. https://www.survivalinternational.org/emails/boycott-botswana
Taylor MJ (2001) Narratives of identity and assertions of legitimacy: Basarwa in northern Botswana. In: Anderson DG, Ikeya K (eds) Parks, property, and power: managing hunting

practice and identity within state policy regimes. National Museum of Ethnology, Osaka, pp 157–181

The World Bank (2020) Botswana, retrieved https://data.worldbank.org/country/botswana?view=chart

Trip advisor (2020) Does the call for tourists to boycott Botswana have merit? https://www.tripadvisor.com/ShowTopic-g293766-i9284-k6946352-Does_the_call_for_Tourists_to_boycott_Botswana_have_merit-Botswana.html

Van Looy A (2016) Social media management: technologies and strategies for creating business value. Springer International Publishing, New York

van Osch W, Coursaris CK (2014) Social media research: an assessment of the Domain's productivity and intellectual evolution. Commun Monogr 81(3):285–309. https://doi.org/10.1080/03637751.2014.921720

Worldometer (2020) Botswana demographics. Retrieved https://www.worldometers.info/demographics/botswana-demographics/

Yousaf S, Li H (2015) Social identity, collective self esteem and country reputation: the case of Pakistan. J Prod Brand Manag 24(4):399–411

Yousaf S (2019) Lost destination marketing opportunities due to international sporting boycott. *Ann Leis Res* 22(4):415–423

Wame L. Hambira (PhD) is an Associate Professor of Environmental Management at the Botswana University of Agriculture and Natural Resources (BUAN). Prof Hambira has served the *Intergovernmental Science-Policy Platform for Biodiversity and Ecosystem Services (IPBES) as an* expert in the scoping of the *nexus assessment of the interlinkages among biodiversity, water, food and health in the context of climate change as well as a* scoping expert for the global assessment of the second biodiversity and ecosystem assessment. Prof Hambira has also served as a United Nations Environment Program (UNEP) Review Editor for the Solution pathways for transformation of environmental systems chapter for the seventh Global Environment Outlook.

Patricia K. Mogomotsi (PhD) is an Associate Professor of Economics in the Department of Economics, Faculty of Social Science, University of Botswana. Prior to that, she served as research Associate Professor (Natural Resources Economics) at the Okavango Research Institute, University of Botswana. She holds a PhD in Economic and Management Sciences from the Northwest University. She also holds Master of Economics from the Rhodes University, Master of Financial Management from Amity University, a postgraduate Bcom (Hons) in Economics from Stellenbosch University, and BA (Economics) from the University of Botswana. Professor Mogomotsi has published extensively on environmental and natural resources economics, tourism management, water institutions and regulation, waste management, tourism economics, and climate change adaptation.

Victor K. Muposhi is a Conservation Scientist and Associate Professor of Wildlife Conservation in the Department of Wildlife and Aquatic Resources at the Botswana University of Agriculture and Natural Resources. His research interests include biodiversity, ecosystems dynamics, spatial ecology, socio-ecological systems, human-wildlife coexistence, and sustainability science.

Goemeone E. J. Mogomotsi (PhD) is currently the Director, Legal Services at the Botswana University of Agriculture and Natural Resources. Further, he is an Associate Professor (International Environmental Law and Policy) at the University of Botswana. He holds PhD in Law from the University of the Witwatersrand; LLM in Commercial Law from the University of South Africa; LLM in International Trade, Business & Investment Law, University of the Western Cape; Master of Arts, Master of Public Administration, and Bachelor of Laws from the University of Botswana. He also holds Bachelor of Financial and Investment Analysis and Master of Financial Management from Amity University. He has published widely on indigenous peoples' rights, international water law and diplomacy, and natural resources use law and policy.

Part II
Sustainable Range and Forests Management

Chapter 10
Forest Resources Utilisation and Rural Livelihoods in Sub-Saharan Africa: Exploring e-Extension Model for Sustainable Forest Management

Dixon Olutade Torimiro, Iris Mosipuri, Oluwagbenga Titus Alao, Nelson Motlapele Tselaesele, and Tsheko Baruti

10.1 Introduction

Forests, generally, are very essential in contributing to the livelihoods of the rural populace. In sub-Saharan Africa, Kaimowitz (2003), Edusah (2011), and Agrawal et al. (2013) identified forest goods and services as very important for rural livelihoods through the provision of food, medicine, shelter, fuel, and cash income. Many rural farm families, in most agrarian sub-Saharan Africa, collect forest resources and utilise them in diverse ways, either for direct consumption or for economic survival. According to Oksanen and Mersmann (2003), more than 15 million people in the region are estimated to earn their cash income from forest-related enterprises such as fuelwood and charcoal sales, small-scale saw-milling, commercial hunting, and handicraft excluding about 200,000–300,000 people are directly employed in the commercial timber industry. According to the Bolt et al. (2005), for example, more than half of the Congo Basin Forest population depends on forests for a variety of reasons, including food, fuel, livestock grazing, and medicine. This was supported

D. O. Torimiro (✉) · N. M. Tselaesele · T. Baruti
Department of Agricultural Education and Agricultural Extension and Rural Development, Botswana University of Agriculture and Natural Resources, Gaborone, Botswana
e-mail: dtorimiro@buan.ac.bw; ntselaes@buan.ac.bw; tzbaruti@buan.ac.bw

I. Mosipuri
Department of Biometry and Mathematics-ICT Unit, Botswana University of Agriculture and Natural Resources, Gaborone, Botswana
e-mail: imosipuri@buan.ac.bw

O. T. Alao
Department of Agricultural Extension and Rural Development, Osun State University, Ejigbo Campus, Osogbo, Nigeria
e-mail: oluwagbenga.alao@uniosun.edu.ng

W. L. Hambira et al. (eds.), *Environmental Change and Biodiversity Conservation in sub-Saharan Africa*, Advances in Global Change Research 75,
https://doi.org/10.1007/978-3-031-73136-5_10

by Tieguhong et al. (2012), Tieguhong and Nkamgnia (2012), and Mendako et al. (2022), who alluded that forestry plays an important role in the livelihood improvement of poor people in rural areas as their day-to-day needs depend on it which includes most of their protein, medicines, energy, materials, and income. Forestry benefits include significant economic outputs such as income and employment generated by forest industries, recreational purposes, as well as the broader contribution that forests make indirectly to other sectors such as tourism (Agrawal et al. 2013).

FAO (2003) estimated the value of net exports of various wood-based products from countries in sub-Saharan Africa between 1993 and 2002 to be more than US$2 billion. Furthermore, of late, there is a serious concern about the indiscriminate destruction of tropical forests throughout the World. According to Laurance and Bierregaard (1997) and Whitmore (1997), It is estimated that between 1981 and 1990 an average of 15.4 million hectares of tropical forests were destroyed. Every year, total forest conservation of 21 million hectares, that is, about 1.2% of the world's tropical forests are being destroyed or degraded, especially in Amazonia, parts of South-East Asia, and Africa (Whitmore 1997). Furthermore, the forests are bedevilled with forest fires, climate change, overutilisation of forest products, illegal cutting of trees, low employment, low income and poverty among the forest communities, inadequate legal policy frameworks, and overpopulation (Solomon and Uzoma 2014; Shivani et al. 2020; Ekott 2020).

It is against this background that this chapter, using desk research, expounded on forest resources utilisation and rural livelihoods in sub-Saharan Africa with a view to exploring the e-extension model for sustainable forest management information system (FMIS). Specifically, each of the sections discussed the challenges in forest management; forest management information system (FMIS), which covers data generation, data distribution with the flavour of e-extension or cyber extension and data utilisation capturing diverse forest stakeholders; and finally, exposition is offered on how FMIS impacts on the rural livelihoods.

10.2 Challenges in Forest Management

This section covers the various challenges plaguing forest management. For a better understanding of the forest situation, a comparative analysis of time series data (2000–2021) on net change in tree cover, tree cover loss, the percentage of tree cover loss due to deforestation, the equivalent percentage decrease in tree cover, carbon dioxide emissions, and gained tree cover in relation to percent global total were carried out from 12 selected sub-Saharan African countries, that is, Nigeria, Togo, Burkina Faso, and Cote d'Ivoire (West African region); Rwanda, Tanzania, Kenya, Malawi, and Ethiopia (East African region); South Africa and Botswana (Southern African region); and Democratic Republic of Congo (Central African region) (Table 10.1).

Data on the net change in tree cover in Table 10.1 show that the countries in the East, Southern, and Central African regions experienced the highest percentage

Table 10.1 Distribution of some selected sub-Saharan African countries by the net change in tree cover, tree cover loss, and CO_2 emissions (2000–2021)

Country	The net change in tree cover (2000–2020)	Tree cover loss (2001–2021)	CO_2 emissions
Nigeria	−1.47 Mha (−6.1%)	0.000114 ha	604 Mt
Togo	−133 kha (−5.9%)	67,300 ha	31.2 Mt
Burkina Faso	−9.17 kha (−0.36%)	131 ha	0.577 Mt
Cote d'Ivoire	−1.65 Mha (−6.9%)	0.000346 ha	1720 Mt
Rwanda	−17.0 kha (−2.9%)	40,600 ha	25.8 Mt
Tanzania	−3.82 Mha (−11%)	0.000286 ha	970 Mt
Kenya	−285 kha (−6.0%)	368,000 ha	179 Mt
Malawi	−222 kha (−7.9%)	20,900 ha	84.3 Mt
Ethiopia	−738 kha (−3.6%)	448,000 ha	204 Mt
South Africa	−142 kha (−1.9%)	0.000153 ha	895 Mt
Botswana	−1.27 kha (−0.71%)	508 ha	0.090 Mkt
Democratic Republic of Congo	−6.00 Mha (−3.6%)	0.00171 ha	10,500 Mt

Source: Generated from Global Forest Watch. https://www.globalforestwatch.org/dashboards

increase in forest cover than the countries in the West African region. The highest carbon dioxide emissions were experienced by the Democratic Republic of Congo with 10,500 Mt followed by Cote d'Ivoire with 1720 Mt Tanzania and the lowest was Botswana with 0.090 Mt.

Based on data from Global Forest Watch, Fig. 10.1 illustrates the percentage of tree cover loss due to deforestation in 12 sub-Saharan African countries. Nigeria and Rwanda experienced notably higher tree cover loss compared to the other countries. In contrast, some countries experienced very low or no deforestation, demonstrating effective forest management. This is crucial as forest loss negatively impacts human life and the various animal species that rely on forests for shelter and food. In addition, people are less able to benefit from the natural resources that these ecosystems supply when forests are lost. This may result in more poverty, and in certain situations, people may need to relocate in search of a forest where they can survive. Deforestation must, therefore, be prevented at all costs since it contributes to around 10% of global warming, and there is no possible way to combat the climate issue if deforestation continues (Skutsch et al. 2007).

Burkina Faso had the highest percentage decrease in tree cover as compared to the Democratic Republic of Congo, Botswana, and Rwanda due to deforestation (Fig. 10.2). This implies that the forest dependants in those countries might experience threats to forest products that they rely on for their day-to-day survival. In addition, a decrease in tree cover may lead to a more extreme temperature that could be inimical to plants and animals on which the forest dwellers are further dependent on.

In the sub-Saharan African region, Nigeria topped the list of the gained tree with 928 kha, which amounted to 0.17% of the global total, followed by Cote d'Ivoire with 882 kha, which amounted to 0.17% showing that gained tree cover is the highest in West African region as compared to other African regions (Fig. 10.3).

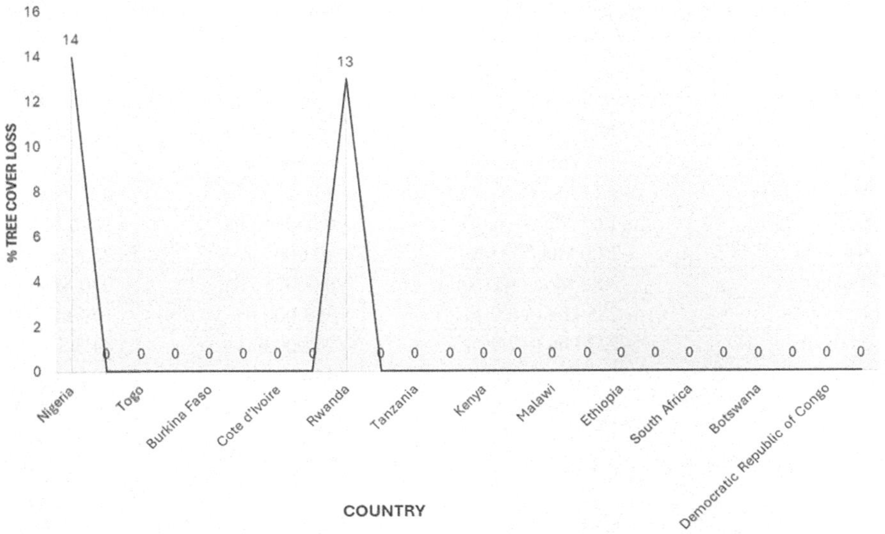

Fig. 10.1 Tree cover loss due to deforestation by selected sub-Saharan African Countries. (Source: Generated from Global Forest Watch. https://www.globalforestwatch.org/dashboards)

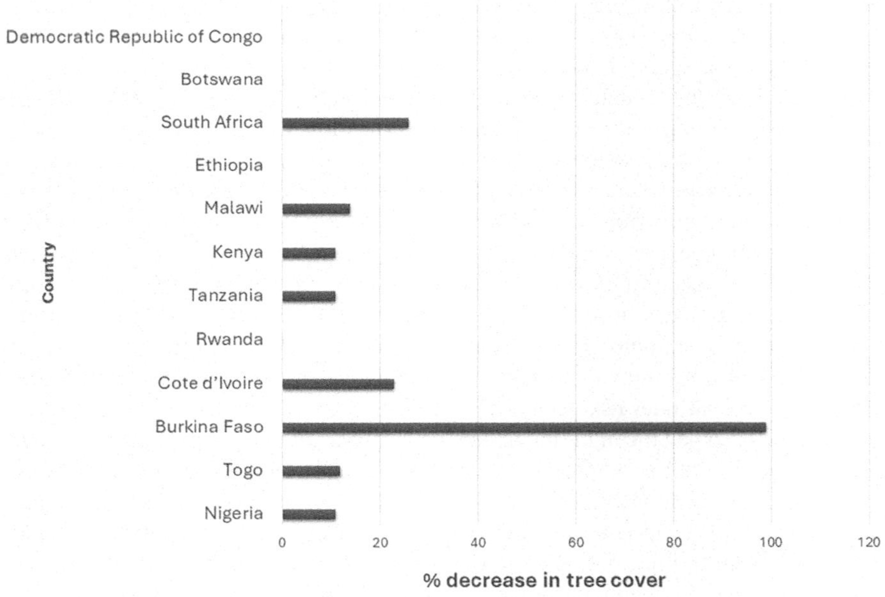

Fig. 10.2 Equivalent percentage decrease in tree cover in some selected sub-Saharan African countries. (Source: Generated from Global Forest Watch. https://www.globalforestwatch.org/dashboards)

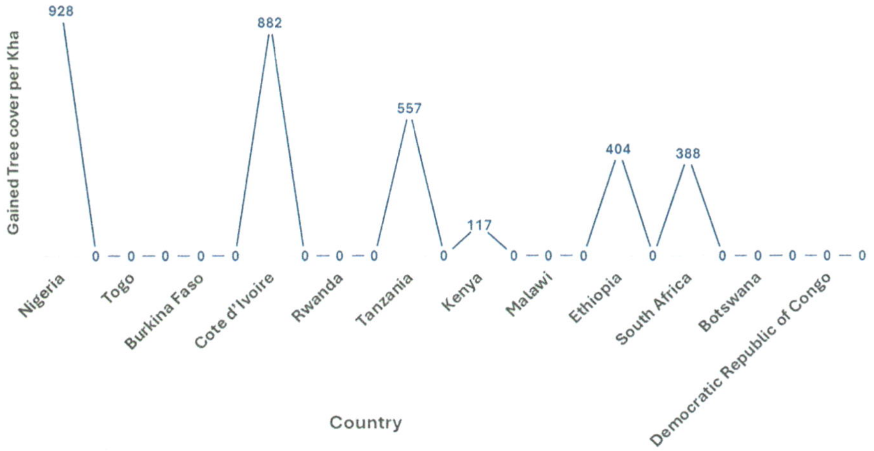

Fig. 10.3 Gained tree cover in relation to % global total by selected sub-Saharan African Countries. (Source: Generated from Global Forest Watch. https://www.globalforestwatch.org/dashboards)

Some of the specific examples of forest management challenges are depicted in Fig. 10.4 and further discussed as follows:

Inadequate legal or policy frameworks Nigeria's forests cover about 96,043 square km—that is about 10% of the country's landmass. But the policy and presence of authorities in these sanctuaries to implement this policy is either non-existent or, at best, sporadic. This has led to Nigeria's forest areas being poorly managed, which in turn has led to them being exploited by criminals (Olaniyan 2018).

Overpopulation The exponential growth of the human population over the last few centuries has caused encroachment in the forest and consequent destruction of forest and forest products. The growth rate of the world population was approximately 2% per annum from 1960 to 2000, which indicated a potential population doubling every 35 years and thus could cause ecological unsustainability (Roser et al. 2013).

Soil erosion FAO and ITPS (2015) classified Africa among the five of seven regions of the world mostly threatened by soil erosion (FAO 2019). Soil erosion, though, natural process of wearing away topsoil, but human activities have been identified as accelerating the process due to the removal of vegetation, or any activity that renders the ground dry.

Climate change Climate change is a product of human-made activities. For instance, human activities such as cutting forest trees, and burning fuels to produce energy, have released large amounts of carbon dioxide and other gases into the atmosphere (FAO 2015). In many poor rural landscapes, where access to inputs like fertiliser is limited, farming communities have met the food demands of growing

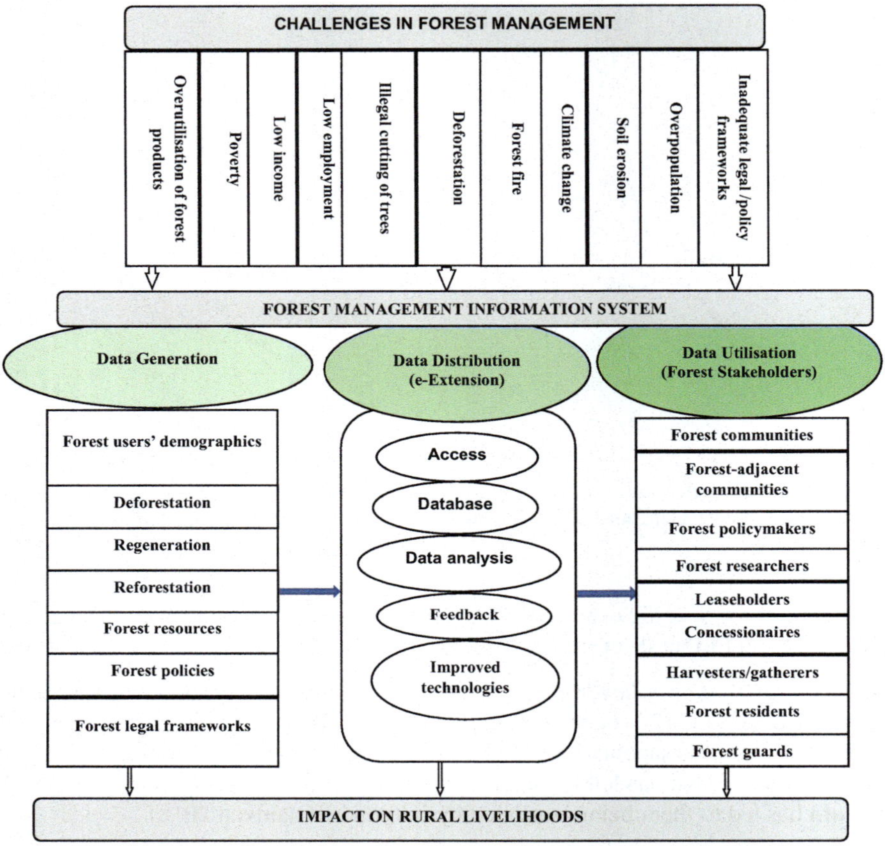

Fig. 10.4 e-Extension model for sustainable forest management information. (Source: Developed by authors)

populations through the extensification of agriculture rather than through intensification. Increased inter-annual rainfall variability and the frequency of extreme events due to climate change have been found to accelerate the rates of degradation of soil and water resources upon which farming communities depend for their livelihoods (Dube 2003; Akinyemi 2017).

Forest fire Wildfire, also called forest, bush, or vegetation fire, can be described as any uncontrolled and non-prescribed combustion or burning of plants in a natural setting such as a forest, grassland, brushland, or tundra, which consumes the natural fuels and spreads based on environmental conditions (e.g. wind, topography). On an annual basis, human acts of carelessness such as leaving campfires unattended and negligent discarding of cigarette butts, hunting, and burning of grasses to rejuvenate by pastoralists have resulted in wildfire disasters. In addition, accidents, deliberate acts of arson, burning of debris, and fireworks are as well other contributors to wildfires. Consequently, lead to a deterioration of the air quality, and loss of property, crops, resources, animals, and people. Furthermore, while wildfires damage the habitat of animals, causing them to wander in cities, many of them die in the fires unable to escape. These fires destroy the vegetation, soil quality, and overall flora and fauna; kill or injure unintended plants or trees.

Deforestation Deforestation in turn leads to a reduced ability to capture CO_2, thus exasperating the greenhouse gas problem. Deforestation is also strongly associated with loss of habitat and extinction. Agriculture is responsible for nearly 80% of global deforestation. Another 14% is attributed to logging, 5% to firewood collection, and the balance results from other causes. Human population increase is related to all these deforestation pressures. More people mean we need more food, more wood products, and more firewood.

Illegal cutting of trees The primary factors that lead to current forest conditions include logging large trees, fire suppression, and livestock grazing. Since the beginning of the twentieth century, these three factors have been available in western Nigerian forests, and they continue to play a role today. Between 2002 and 2021, Nigeria lost 153 kha of humid primary forest, making up 14% of its total tree cover loss in the same period. The total area of humid primary forest in Nigeria decreased by 8.1% in this period (Ogunyale 2022).

Low employment Due to the low employment rate, the youths search for non-available white-collar jobs in the cities and return to the rural area to farm in the reserved forest that is still fertile, thereby creating jobs and absorbing those whose livelihoods had always come from the forest.

Poverty The poverty rate in Nigeria, already high, only appears to be rising. According to the World Bank (2022) four in ten Nigerians—about 80 million people—were living in poverty in 2019, with the COVID-19 pandemic pushing another five million people below the poverty line by 2022. In many of these forest reserves, there are very few things people (surrounding communities) can do to survive except hunting, farming, or logging in the protected forests. So many communities in Nigeria are heavily dependent on the forest. It is, thus, creating serious problems for conservation not just in Nigeria but in most parts of Africa.

Overutilisation of forest products Due to poverty in the rural communities and low employment in the cities some of the able-bodied youths are into logging, farming in the conserved forest, and hunting wild animals. As the population increases forest product utilisation increases. Since forest products are not the elastic effect of an increase leads to the reduction of forest products.

Low income The peasant farmers in the rural community earn low income from their farm produce but depend on the forest products to augment through hunting, logging, and other forest-diminishing activities.

10.3 Sustainable Forest Management Information System: e-Extension Model

The use of information and communication technology (ICT) in forestry management is becoming more important, most especially, in the field of agricultural extension. Generally, agricultural extension is a process through which improved practices, technologies, skills, and knowledge are transferred from research to farm families to improve their farm production, income, and standard of living. However, e-extension includes a wide variety of information and communication technologies (ICTs) that can communicate, create, store, disseminate, and manage agricultural information. It can also be referred to as cyber extension, which is the application of online networks power, computer communications, and digital interactive multimedia to facilitate the distribution of improved agricultural technologies. In the context of forestry, e-extension is presumed to be a digital approach to forest management information system (FMIS) through information generation, distribution, and utilisation by the end users (Fig. 10.4).

To prevent our forest and forest products from diminishing and invariably going to extinction, a functioning forest management information system must be put in place. This section, therefore, offers an exposition on the e-extension model for operationalising a sustainable forest management information system. It focusses on forest management data generation, forest management data distribution, and forest management data utilisation.

10.3.1 Forest Management Data Generation (FMDG)

For a proper management information system to be functional data must be generated, distributed, and (e-extension) data utilised by all the forest stakeholders. In this component, data generated will include data on forest users' demographics, timber harvests (deforestation), re-growth or re-planting (regeneration) practices, weeding and/or timber, stand improvement practices, reforestation, and tree planting, among others.

10.3.2 Forest Users' Demographics

Forest management entails strong participation from both the forest users' communities and forest managers. The participation differs from location to location due to the diversity of forest resources, landscape, climate, vegetation, and the goods and services provided in that forest area. According to Tadesse et al. (2017), the demographic factor plays an important role in determining the forest user's participation in forest management activities. These demographic factors include users' age, gender, educational background, income status, and population. For example, age and education were found to significantly influence household reliance on forests with decreasing forest dependency as age and educational level increased (Garekae et al. 2017). In addition, gender as a factor can also contribute to determining the forest user's participation in forest management activities. For instance, women have been found to play a more crucial role in managing forest resources because they mostly used forest products in their daily domestic activities (Mwangi et al. 2011; Dhruba Bijaya et al. 2016). Mather and Needle (2000) established inverse relationships between national trends in forest areas and population; however, the relationships are found to be very complex to predict (Meyerson 2004). Also, several authors (Yemiru et al. 2010; Biland et al. 2021) have established income from forests has alleviated poverty among rural people. Hence, proper records of the forest users must be collected and kept in the data bank and updated from time to time.

10.3.3 Deforestation, Forest Regeneration, and Reforestation Practices

Generating data on forest management is critical for understanding the situation of the forest for planning sustainable forest management and restoration, and conservation activities. This section elucidates the possible data to be generated in modelling e-extension for sustainable forest management. This entails data on deforestation, forest regeneration, and reforestation practices. This could serve as a reference data set that could be used in presenting the situation of the forest ecosystems and in investigating the value of forests for species, ecosystems, and their services. Geographical Information System (GIS) and a drone could be deployed in data collection, which enables professionals to do advanced mapping and spatial analytics for forest vigilance purposes. Advanced mapping and spatial analytics includes monitoring of species, human–animal conflict, management of protected areas, etc.

Ways of deforestation through tree felling, bushfires, erosion, among others, as well as strategies for reforestation through natural regeneration, natural dispersal, and artificial afforestation are direct processes through which supposedly imbalanced forest ecosystems are being maintained. Data generation on individual, communal, and government initiatives revolving around deforestation and reforestation are presumed to be captured in the model.

Some of the questions that may be explored in generating data on deforestation are what hectarage of the forest is lost and at what rate? Which region is most affected? What is the quantity of carbon dioxide released due to deforestation? To what extent do indigenous people as well as animal and plant species are threatened by deforestation? Possible data that may be explored on forest regeneration include density (that is, a measure of the number of trees per unit area), stocking (that is, the number of well-spaced trees relative to some reference density), distribution (including distance measurement between trees, the coefficient of variation, or spatial pattern statistics), acceptable crop species, field measurement of competition levels, acceptable growth or productivity by the regenerating forest and scale (Brand et al. 1991). So, data on deforestation must be properly kept in controlling the rate of deforestation.

Forest resources Forest resources are benefits derived from forest land, including forest products such as timber for wood products, soil productivity, water and air, fisheries, wildlife habitats, recreation opportunities, and aesthetic or other traditional values of the forest land. In addition, forest resources consist of so many other segments such as wild fruits, game animals, medicinal herbs, roots, firewood, edible insects, and other lower animals on which data could be generated.

Forest policies Comprehensive and well-coordinated forest policy should be enacted and made available to guide the misuse of forest resources.

Forest legal frameworks There should be a legal framework to guide forest activities.

10.4 Forest Management Data Distribution (FMDD)

10.4.1 e-Extension

The main task of forestry e-extension according to Igbokwe et al. (2021) is to disseminate necessary information to the clients to change their attitudes towards environmental and developmental problems and adopt suitable innovations that can enhance the conservation of the forest. This can be termed as technology transfer which includes aspects of the movement of knowledge and networking which needs communication between the producer organisations and recipients. The forest management data distribution through the vehicle of the e-extension model involves data access, database, data analysis, feedback, and improved technologies, which are consequently utilised by the various forest stakeholders. In other words, all the stakeholders in forest management must have unreserved/free access to the information data bank through extension agents using the concept of social media for easy access; all the data generated must be kept in the bank for easy access; data collected must be analysed from time to time for prediction and explanation of events in the forest; adequate feedback must be enforced to ensure that the forest user received the correct message on the forest utilisation for effective management of the forest; and innovations on improved technologies in forest management could be made to reach forest users through e-extension (Fig. 10.5).

Fig. 10.5 Pictorial representation of e-extension for sustainable forest management information system

10.5 Forest Management Data Utilisation (FMDU)

(a) *Forest communities:* forest communities survive if the forest survives because they depend on the forest produce and products to survive hence need to utilise information on forest management.
(b) *Forest adjacent communities:* Communities living adjacent to forest reserves can partner with others to co-manage the forest resources through e-extension services.
(c) *Forest policymakers:* forest policy makers should review the old policy and make new ones that will conserve forest produce and products. The gap between policy and implementation should be closed for effective forest conservation and management.
(d) *Forest researchers:* forest researchers must measure and analyse tree and forest growth, develop new trees, and breed commercial trees that grow fast and are ready for logging within a few years.
(e) *Leaseholder:* forest leaseholders should be properly monitored, and their activities should be checked from time to time to prevent mismanagement of the forest, forest produce, and products.
(f) *Concessionaires/harvesters/gatherers:* Forest concessionaires, harvesters/gatherers' activities must be put in check to promote the sustainability of the forest produce and products.
(g) *Forest residents:* forest residents should use the forest resources wisely and obey the set rules for the sustainability of the wild animals, forest produce, and products.
(h) *Forest guards:* professional forest guards properly trained and retrained must be employed to reduce the menace of poachers and prevent the misuse of forest produce and products.

If a functional forest management information system is properly explored by deploying an e-extension model in the sub-Saharan African region, it will positively impact the rural livelihoods and diverse forest management stakeholders.

10.6 Conclusion

Responsible use of forest resources is very imperative for the survival of forest dependants and indeed all the forest management stakeholders in sub-Saharan Africa. This must be done through a workable forest management approach that is sustainably adaptable and practicable, and readily embraced by the diverse forest management stakeholders within the African continent. The proposed e-extension model has great potential in meeting this aspiration in providing an appropriate sustainable forest management system based on data generation, distribution, and utilisation if given a trial. It is, therefore, recommended that the e-extension model is trial tested for its efficacy to be realised.

References

Agrawal A, Cashore B, Hardin R, Shepherd G, Benson C, Miller D (2013) Economic contributions of forests [online], pp 1–132. Available at: https://www.un.org/esa/forests/wp-content/uploads/2015/12/EcoContrForests.pd

Akinyemi FO (2017) Climate change and variability in Semiarid Palapye, Eastern Botswana: an assessment from smallholder farmers' perspective. Am Meteorol Soc 9(3):349–365

Biland M, Zeb A, Ullah A, Kaechele H (2021) Why do households depend on the forest for income? Analysis of factors influencing households' decision-making behaviors. Sustainability 13:9419. https://doi.org/10.3390/su13169419

Bolt K, Hamilton KE, Lange G-M, Markandya A, Ordoubadi MS, Pedroso-Galinato S, Ruta G, Silva P, Tajibaeva L (2005) Where is the wealth of nations : measuring capital for the 21st century (English). Washington, D.C. : World Bank Group. http://documents.worldbank.org/curated/en/287171468323724180/Where-isthe-wealth-of-nations-measuring-capital-for-the-21st-century

Brand DG, Leckie DG, Cloney EE (1991) Forest regeneration surveys: design, data collection, and analysis. For Chron 67(6):249–253

Dhruba Bijaya GC, Cheng S, Xu Z, Bhandari J, Wang L, Liu X (2016) Community forestry and livelihood in Nepal: a review. J Anim Plant Sci 26(1):1–12

Dube OP (2003) Impact of climate change, vulnerability, and adaptation options: exploring the case for Botswana through Southern Africa: a review. Botswana Notes Rec 35(2003):147, 22 pages–168

Edusah SE (2011) The impact of forest reserves on livelihoods of fringe communities in Ghana. J Sci Technol 31(1):10

Ekott N (2020) Four million hectares of forests lost globally in 2019 – Report. Nigeria: Premium Times, June 4, 2020. https://www.premiumtimesng.com/news/top-news/396016-four-million-hectares-of-forests-lost-globally-in-2019-report.html?tztc=1

FAO (2003) Forestry outlook study for Africa. Regional report – opportunities and challenges towards 2020. FAO Forestry Paper 141, Rome

FAO (2015) Climate change and food security: risks and responses. Food and Agriculture Organization of the United Nations, Rome
FAO (2019) Soil erosion: the greatest challenge to sustainable soil management, Rome. 100 pp. Licence: CC BY-NC-SA 3.0 IGO
FAO & ITPS (2015) Status of the world's soil resources – main report. Food and Agriculture Organization of the United Nations and Intergovernmental Technical Panel on Soils, Rome. 649 pp (also available at http://www.fao.org/3/a-i5199e.pdf)
Garekae H, Thakadu OT, Lepetu J (2017) Socio-economic factors influencing household forest dependency in Chobe Enclave, Botswana. Ecol Process 6:40. https://doi.org/10.1186/s13717-017-0107-3
Global Forest Watch. https://www.globalforestwatch.org/dashboards/country/NGA
Igbokwe M, Oluwasegun O, Agube E (2021) Role of extension forest officers in forest conservation in Cross River State, Nigeria. J Agric Ext 25(3):69–80
Kaimowitz D (2003) Not by bread alone. Forests and rural livelihoods in sub-Saharan Africa. In: Oksanen T, Pajari B, Tuomasjukka T (eds) Forests in poverty reduction strategies: capturing the potential. EFI Proceedings No. 47. European Forest Institute, Joensuu, Finland, pp 45–64. Available online: http://www.efi.fi/attachment/f5d80ba3c1b89242106f2f97ae8e3894/241e80d8e1b2b0919426d5a82060db7e/Proc_47.pdf
Laurance WF, Bierregaard RO (1997) Tropical forest remnants: ecology, management, and conservation of fragmented communities. University of Chicago Press, Chicago
Mather AS, Needle CL (2000) The relationships of population and forest trends. Geogr J 166:2–13. https://doi.org/10.1111/j.1475-4959.2000.tb00002.x
Mendako RK, Tian G, Ullah S, Sagali HL, Kipute DD (2022) Assessing the economic contribution of forest use to rural livelihoods in the Rubi-Tele Hunting Domain, DR Congo. Forests [online] 13(1):130. https://doi.org/10.3390/f13010130
Meyerson FAB (2004) Population growth and deforestation: a critical and complex relationship. PRB. https://www.prb.org/resources/population-growth-and-deforestation-a-critical-and-complex-relationship/
Mwangi E, Meinzen-Dick R, Sun Y (2011) Gender and sustainable forest management in East Africa and Latin America. Ecol Soc 16(1). http://www.jstor.org/stable/26268834
Ogunyale K (2022) "We depend on the forest": logging worsens climate impacts for Ogun people in Nigeria. Twentyten Daily. http://twentytendaily.com
Oksanen T, Mersmann C (2003) Forests in poverty reduction strategies: an assessment of PRSP processes in sub-Saharan Africa. In: Forests in poverty reduction strategies: capturing the potential. EFI Proceedings No. 47. European Forest Institute, Finland, pp 121–158
Olaniyan A (2018) How poor management of Nigerian forests led to exploitation by criminals. https://phys.org/news/2018-12-poor-nigerian-forests-exploitation-criminals.html
Roser M, Ritchie H, Ortiz-Ospina E, Rodés-Guirao L (2013) World population growth. Published online at OurWorldInData.org. Retrieved from: https://ourworldindata.org/world-population-growth [Online Resource]
Shivani U, Rashmi P, Bhumija K, Sharma RK (2020) Human overpopulation: impact on environment. In: Megacities and rapid urbanization: breakthroughs in research and practice. IGI Global, Engineering Science Reference
Skutsch M, Bird N, Trines E, Dutschke M, Frumhoff P, de Jong BHJ, van Laake P, Masera O, Murdiyarso D (2007) Clearing the way for reducing emissions from tropical deforestation. Environ Sci Pol 10(4):322–334
Solomon LL, Uzoma DC (2014) Challenges of forest management and security issues in Nigeria. https://www.researchgate.net/publication/268216469_Challenges_of_forest_management_and_security_issues_in_Nigeria
Tadesse S, Woldetsadik M, Senbeta F (2017) Forest users level of participation in a participatory forest management program in Southwestern Ethiopia. For Sci Technol 13(4):164–173. https://doi.org/10.1080/21580103.2017.1387613
Tieguhong JC, Nkamgnia EM (2012) Household dependence on forests around Lobeke National Park, Cameroon. Int For Rev 14(2):196–212

Tieguhong JC, Ndoye O, Grouwels S, Mala WA, Betti JL (2012) Rural enterprise development for poverty alleviation based on non-wood forest products in Central Africa. Int For Rev 14(3):363–379. https://doi.org/10.1505/146554812802646701. https://www.forestasyst.org/managementplan.html

Whitmore TC (1997) Tropical forest disturbance, disappearance and species loss. In: Laurance WF, Bierregaard RO (eds) Tropical forest remnants: ecology, management, and conservation of fragmented communities. The University of Chicago Press, Chicago, pp 3–12

World Bank (2022) Nigeria poverty assessment 2022: a better future for all Nigerians. https://www.worldbank.org/en/news/infographic/2022/03/21/afw-nigeria-poverty-assessment-2022-a-better-future-for-all-nigerians

Yemiru T, Roos A, Campbell BM, Bohlin F (2010) Forest incomes and poverty alleviation under participatory forest management in the Bale Highlands, Southern Ethiopia. Int For Rev 12(1):66

Dixon Olutade Torimiro, PhD, is Professor of Agricultural Extension and Rural Sociology and Current Head, Department of Agricultural Education and Agricultural Extension and Rural Development at the Botswana University of Agriculture and Natural Resources, Gaborone. He was ECOWAS-AAU Fellow at the Northwest University, Mafikeng, South Africa; International Extension Scholar- in-Residence at the Ohio State University, Columbus, USA; and European Commission expert evaluator, Brussels, Belgium. He has won a good number of research and travel grants, including CODESRIA, AAU/DFID, IDRC, and ESRC. He is well published in the areas of Agricultural Extension and Rural Community Development.

Iris Mosipuri is a Senior Instructor in the Department of Biometry and Mathematics in the Botswana University of Agriculture and Natural Resources, Gaborone. She holds a Bachelor of Science degree in Information Technology from the Limkokwing University of Creative Technology, Botswana, and MSc in Computer Systems Management from Botho University, Botswana. Her research interests include computer technology and information systems.

Oluwagbenga Titus Alao, PhD, is an Associate Professor of Agricultural Extension and Rural Sociology, and is currently the Acting Dean, Faculty of Agricultural Production and Management, College of Agriculture, Ejigbo Campus, Osun State University, Osogbo, Nigeria. He has won IDRC and AAU/DFID research and travel grants. He has attended many learned conferences locally and internationally. He has published articles and chapters in books related to rural development, extension and gender studies.

Nelson Motlapele Tselaesele is a Senior Lecturer in Rural Sociology and Agricultural Extension. He holds a PhD in Agribusiness from the University of Queensland (Australia), MSc in Management of Agricultural Knowledge System (Wageningen University), and Bachelor of Arts in Social Science (University of Botswana). His research interests include rural development and management of natural resources.

Tsheko Baruti holds BSc in Agricultural Extension and Rural Development from the Fort Hare University, South Africa, and is currently working as Senior Instructor in the Department of Agricultural Education and Agricultural Extension and Rural Development at the Botswana University of Agriculture and Natural Resources. He has carried out a few research projects including study on the effects of mulching on tomatoes, study on milling companies in Botswana, and study on land use planning in Tukulu farm in South Africa.

Chapter 11
Anthropological Impact on Natural Forests in the Lower Okavango Delta, Botswana

Reniko Gondo, Oluwatoyin Dare Kolawole, and Chiedza Ngonidzashe Mutanga

11.1 Introduction

While there is no universally agreed definition of a forest, Bera et al. (2020) define it as a "large expansion or territory that is mainly dominated by various types of trees and shrubs". Whereas the definition of Bera et al. (2020) does not specify how big *a large expansion or territory* is, Citaristi (2022) offers a more precise meaning of the term. A forest is, therefore, conceived as a piece of land of at least 0.5 hectares (ha) with trees and shrubs and with a minimum tree height of five metres along with a 10% canopy (Citaristi 2022). Also, the Forest Survey of India (FSI) (2021) used a set of guidelines to define a forest as a land area of a minimum of 1 ha size, and which has a tree canopy cover of more than 10%. Due to human stress and various developmental activities like infrastructure and land use and land cover change (LULCC), among others, vast forest land has been converted into a sparse or isolated forest. The continuous human encroachment along with developmental activities over forest lands aggravates the high magnitude of forest degradation.

It is noteworthy that about one-fifth of the world's forest is found in Africa (Bromhead 2012) and the continent is home to the second largest rainforest—the Congo Basin, after the Amazon Forest in South America (Zhongming et al. 2019). The central African forests are estimated to provide subsistence to over 60 million people (De Wasseige et al. 2015). Forest landscapes are fundamental to sustaining life on earth (Canton, 2021) and are of relevance to the African continent where most of the population is rural (Köhl et al. 2015; Lufumpa 2005). Forests provide ecosystem services in the form of water regulation (Creed and Van Noordwijk 2018), carbon, nitrogen cycles, and pollination (Le Quéré et al. 2009). Besides their

R. Gondo (✉) · O. D. Kolawole · C. N. Mutanga
Okavango Research Institute, University of Botswana, Maun, Botswana
e-mail: rgondo@ub.ac.bw; tkolawole@ub.ac.bw; mutangac@ub.ac.bw

W. L. Hambira et al. (eds.), *Environmental Change and Biodiversity Conservation in sub-Saharan Africa*, Advances in Global Change Research 75,
https://doi.org/10.1007/978-3-031-73136-5_11

contribution to soil formation (Canton, 2021), forests also help in combating soil erosion (Bennett et al. 2009). The importance of forests includes the provision of habitat for wildlife, a form of recreation and spirituality for millions of people in Africa and perhaps elsewhere. As noted by Allendorf et al. (2014), forests are sacred to many cultures. The forest landscapes are replete with products that are essential to everyday life and the livelihood of most rural people. Thus, the list of forest products includes food, building materials, medicines, and energy (Nybakk et al. 2009). Specifically, more than 62% of the African population depends on goods and services from the ecosystem (Huge et al. 2020), many of which are forest-related and 90% of rural communities depend on firewood and charcoal for energy, especially for cooking (Bamwesigye et al. 2020). The forest is also a source of medicinal products for about 80% of the world's population, and a total of 5400 medicinal plants were documented across the globe (Otieno et al. 2015).

Lately, there has been a constant decline in global forest cover. Statistics show that a total of 4128 million hectares (ha) were covered by forest in the year 1990 (Giesen 2015) and this has declined to about 3999 million ha in 2015 (Keenan et al. 2015). Thus, there is a worldwide decline in forest cover. This deterioration is attributed to anthropogenic activities especially LULCC due to industrial development, commercial agriculture, and urbanisation, among others (Olorunfemi et al. 2020; Forest Conservation Botswana 2013). Botswana, like other developing countries, is experiencing forest depletion, and this is mainly attributed to urbanisation, particularly in major population centres and settlements having tourism-based economy (Forest Conservation Botswana 2013). That Maun is the tourism hub of Botswana is not an overstatement; the alluring effect of the Okavango Delta has engendered the expansion and urbanisation of the settlement in the last several years.

One of the environmental impacts of urbanisation is the conversion of land, especially forest land. Studies that used different methodologies on forest and land conversion have been widely conducted (Drummond and Loveland 2010; Nazir and Ahmad 2018; Twongyirwe et al. 2011) while studies on the relationship between urbanisation and forest land loss especially at the local level in Botswana are still lacking. This chapter, therefore, aims at analysing the impact of urbanisation on forest degradation in the Maun settlement in the lower Okavango Delta, Botswana. Specifically, this chapter determines the extent of forest conversion as a result of urbanisation. Of late, the use of remote sensing (RS) data has been of fundamental importance in monitoring the dynamics of forest cover patterns. The use of RS and geographic information systems (GIS) provides one of the most accurate means of measuring the extent and patterns of changes in forest cover conditions over time.

11.2 Material and Methods

11.2.1 Study Area

Maun is the tourism capital of Botswana and the administrative centre of the Ngamiland District (Fig. 11.1). The village is the headquarters of several Safari and Air Charter operations that engage in tourism offerings in the Okavango Delta (Manwa et al. 2017). The settlement was established in 1915 at the capital of the BaTswana people. The name Maun is a derivative of the SeYei word "Maung" which is translated to "The place of river reeds" (Saum 2010). Before the arrival of the BaTswana, Maun was a small Yei Village. The village became the capital of the Tawana people in 1915 (Saum 2010). The capital was transferred from Toteng. In a broader sense, Maun is the gateway for exploring much of northern Botswana (Mbaiwa et al. 2007); it is the natural hub for international visitors to explore much of the Tsodilo hills and the Makgadikgadi pans (Stone 2014). At the centre of the village is the Thamalakane River (Fig. 11.1).

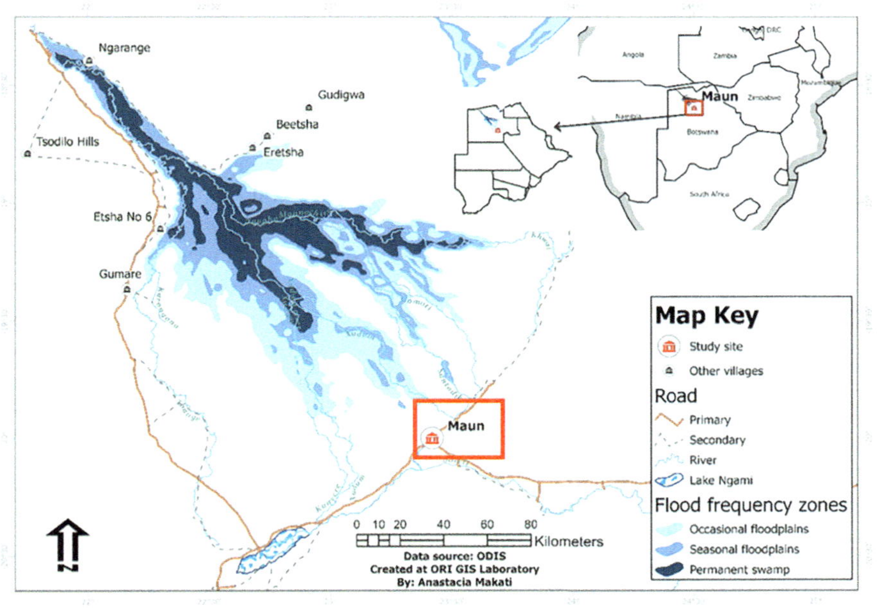

Fig. 11.1 Map of the study area. (Source: Okavango Research Institute GIS LAB (2023))

11.2.2 Remote Sensing Data Processing

This study used remotely sensed data to analyse forest cover distribution and assess the impact of human activities on forest cover in Maun from 1992 to 2022. The pre-processing, for noise removal (radiometric correction), was performed following Lin et al. (2015) and Bijaya et al. (2016), and this was carried out using ArcGIS 10.8 software. For cover classification, the Landsat MSS images of 1992, Landsat ETM+ images of 2002, 2013 and Landsat 8 OLI images of 2022 were used (Table 11.1). Landsat imagery pre-processing and supervised classification using a random forest classifier technique was performed for forest cover distribution following Churches et al. (2014) and Wu et al. (2019), which involves the use of multi-temporal datasets to differentiate areas of land cover change between dates of imaging (Garai and Narayana 2018). Images were classified into five LULC classes, namely, built-up, dense/thick forest, medium forest, light forest, and water bodies (Table 11.2). As recommended by Svatonova (2016), the recognition and differentiation of different categories of forest cover relied upon a visual interpretation of satellite images involving the observation of attributes such as shape, size, pattern, tone and texture, among others.

11.2.3 Mapping Forest Cover Distribution

The spatial distribution of forest cover maps for the years 1992, 2002, 2013, and 2022 were prepared using the random forest classification method. While bands 1–7 (Landsat 5) and 1–9 (Landsat 8) and thermal bands (6 and 10) for Landsat 5 and 8, respectively, were used to derive LST, they were not used in mapping forest cover distribution. Five classes were derived, and these are bare land, dense/thick forest, medium forest, light forest, and water bodies. To measure the accuracy of the mapping, an accuracy assessment was conducted by computing a confusion matrix where 450 random points were generated over the study area. Accuracy measures, namely, overall producer and user accuracies as well as kappa coefficient were then calculated.

Table 11.1 Characteristics of satellite imageries

Satellite (sensor)	Path/row	Acquisition date	Cloud cover	Spatial resolution	Year	Source
Landsat 5 (TM)	174/074	22 Sept	0%	30 m	1992	USGS
Landsat ETM+	174/074	20 Sept	0%	30 m	2002	USGS
Landsat ETM+	174/074	17 Sept	0%	30 m	2013	USGS
Landsat 8 (OLI)	174/074	14 Sept	0%	30 and 15 m	2022	USGS

Table 11.2 Forest cover categories in the study area

Forest classification	Description
Built-up	Areas devoid of vegetation, with buildings, pavements
Dense/thick forest	Areas with over 80% tree cover
Medium forest	All lands with a tree cover of canopy density between 40% and 80%
Light forest	Areas with scattered trees and more grasslands tree density of less than 40%
Water bodies	Areas covering the surface of channelised water

11.2.4 Mapping Normalised Difference Vegetation Index

NDVI is an index used to quantify vegetation cover. This is the most commonly used vegetation index utilised in monitoring the disturbance of forests by human activities (Nath and Acharjee 2013). Based on Bijaya et al. (2016), on the one hand, NDVI values that tend towards −1 signify built-up or barren land. Thus, all negative values denote non-vegetated surfaces and water bodies. On the other hand, those NDVI values that tend towards 1, designate vegetation cover. To compute NDVI, Eq. (11.1) was employed (Gamon et al. 2005; Bijaya al. 2016). In this study, NDVI was derived from Landsat 5 and 8 for the years 1992, 2002, 2012, and 2022. To derive NDVI, the red (band 3) and near-infrared (band 4) for Landsat 5 and bands 4 and 5 for Landsat 8 were used. In this chapter, we adopted the formula used by Guha et al. (2018) in their study of Florence and Naples city in Italy.

$$\begin{aligned} \mathrm{NDVI} &= \frac{\mathrm{Band\,4} - \mathrm{Band\,3}}{\mathrm{Band\,4} + \mathrm{Band\,3}} \text{ for Landsat 5} \\ &\text{and } \frac{\mathrm{Band\,5} - \mathrm{Band\,4}}{\mathrm{Band\,5} + \mathrm{Band\,4}} \text{ for Landsat 8 for the period.} \end{aligned} \quad (11.1)$$

11.2.5 Mapping Land Surface Temperature

To derive LST, we employed a three-step computation procedure used for Landsat 5 and a six-step method used to estimate LST from Landsat 8. For Landsat 5 and 8, LST is estimated using bands 6 and 10, respectively. We avoided band 11 for Landsat 8 because the band is highly affected by split radiation and using it to derive LST gives inaccurate results. However, other studies recommended the use of both bands 10 and 11 in deriving LST (Mao et al. 2005; Sobrino et al. 1994; Erdős et al. 2014). To make the images usable, the acquired satellite images were pre-processed. The pre-processing stage comprises radiometric and geometric corrections using constants derived from the metadata file downloaded together with the images. In this chapter, steps 1–3 were used to derive LST from Landsat 5, and steps 1–7 were adopted to derive LST from Landsat 8 data.

Step 1: conversion of the digital number to spectral radiance

$$L_\lambda = L_{\text{MIN}} + \left(L_{\text{MAX}} - L_{\text{MIN}}\right) \times \text{DN} / 255 \tag{11.2}$$

where L_λ stands for the spectral radiance, L_{MIN} is 1.238, L_{MAX} is 15.303 and DN is the digital number. In stage 2, we converted spectral radiance to temperature in Kelvin using formula (11.2).

Step 2: conversion of spectral radiance to temperature in Kelvin:

$$T_{\text{B}} = \frac{K_2}{\text{Ln} / \left(\frac{K_1}{L_\lambda} + 1\right)} \tag{11.3}$$

where K_1 and K_2 are calibration constants 607.76 and 1260.56, respectively, for the thermal band and these are derived from the metadata. The T_{B} in the formula stands for brightness temperature, which is measured in Kelvin. The final stage in the computation of LST entails the conversion of temperature in Kelvin to temperature in degrees Celsius using formula (11.3).

Step 3:

$$\text{LST} = T_{\text{B}} - 273.15 \tag{11.4}$$

where LST stands for land surface temperature in degrees Celsius and T_{B} stands for brightness temperature in Kelvin.

For Landsat 8, Operational Land Imager/Thermal Infrared Sensor (OLI/TIRS band 10 was used and formula (11.4) was used to convert digital numbers (DN) into spectral radiance (USGS 2019).

Step 4:

$$L_\lambda = M_{\text{L}} \times Q_{\text{cal}} + A_{\text{L}} \tag{11.5}$$

Whereas L_λ denotes the top of atmosphere (TOA) spectral radiance, M_{L} stands for the band-specific multiplicative rescaling factor and A_{L} stands for the band-specific additive rescaling factor and Q_{cal} is the quantised and calibrated standard product pixel values found in the metadata file downloaded together with the images (Table 11.3).

Table 11.3 Multiplicative and additive rescaling factors for Landsat 8 data

		RADIANCE_MAXIMUM_ BAND_X		QUANTIZE_CAL_MAX_ BAND_x	
Landsat	Band	Minimum	Minimum	Minimum	Maximum
8	10	0.10033	22.00180	1	65,535

Table 11.4 Thermal conversion constants values (K_1 and K_2) for Landsat 8

Landsat	Band	K_1	K_2
8	10	774.8853	480.8883

In the second state, brightness temperature, which is defined as electromagnetic radiation (EMR) travelling in the air from the Earth's top atmosphere (Abd Latif et al. 2018), was calculated by transforming TOA radiance by multiplying TIRS data by the thermal constant using an Eq. (11.2) (USGS 2019).

Step 5:

$$T_{\mathrm{B}} \frac{K_2}{\ln\left(\frac{K_1}{L_\lambda}+1\right)} \tag{11.6}$$

In the equation, the thermal constants K_1 and K_2 (see Table 11.4) were obtained from the metadata file. While T_B stands for the TOA brightness temperature in Kelvin, L_λ stands for the TOA spectral radiance. The K_1 and K_2 are band-specific thermal conversion constants obtained from the metadata file of Landsat 8 (Table 11.4).

Based on Dissanayake et al. (2019), computing the LST using Landsat 8 data requires the evaluation of the land surface emissivity, and this was done following Eq. (11.6) (Li et al. 2017; Mallick 2012).

Step 6:

$$\varepsilon = m\mathrm{Pv} + n \tag{11.7}$$

In the equation, the symbol ε represents land surface emissivity (LSE), n is 0.986 and Pv stands for the proportion of vegetation. Pv was calculated using Eq. (11.4) (Neinavaz et al. 2020; Samarasinghe et al. 2022).

Step 7:

$$\mathrm{Pv} = \left(\frac{\mathrm{NDVI} - \mathrm{NDVI}_{\min}}{\mathrm{NDVI}_{\max} - \mathrm{NDVI}_{\min}}\right)^2 \tag{11.8}$$

11.3 Results

11.3.1 Population Growth Trends

While Botswana has had a decennial housing and population census since 1971, the latest census was conducted after 11 years because of the delay caused by COVID-19, which disrupted development activities throughout the world (Leslie 2020).

Findings from the study of patterns and trends of urbanisation derived from the decennial census show an increase in the population of the study area. Thus, the population of the village increased by about 38.9% between 1992 and 2002 and by 33.4% between 2001 and 2011 (Statistics Botswana 2022). Findings show that the population living in the village within the period 2011–2022 increased by 23%. On average, the population of Maun increased by approximately 31.4% between 1991 and 2022 (Statistics Botswana 2022). The increase in population also implies an increased demand for accommodation and related services, and infrastructure. The increase in the built-up area, which was a response to housing demand, rapidly transformed the various LULCs of Maun, which consequently reduced forest cover.

11.3.2 Forest Cover Analysis

The statistics and forest cover map for the year 1992 are shown in Table 11.5 and Fig. 11.2. Findings show that while in 1992 built-up covered 7.62 km^2 (18.7%), dense/thick forest covered 2.81 km^2 (6.9%) of the land area. The largest (20.87 km^2) area constituting 51.1% of the study area was covered by light forest. Whereas medium forest covered 8.47 km^2 (20.7%), water covered 1.05 km^2 (2.6%) of the Maun settlement. The statistics and the spatial distribution of forest cover for the year 2002 are also shown in Table 11.5 and Fig. 11.2. Built-up increased from 7.62 km^2 (18.7%) in 1992 to 9.66 (23.4%) in 2002. Thus, built-up increased by 4.7% as forests were cut down due to developmental activities. Dense/thick forest cover increased to 4.44 km^2 and the medium forest also increased to 13.45 km^2 (32.9%). There was a decline in the spatial extent of light forests from 20.87 km^2 (51.1%) in 1992 to 8.75 km^2 (21.4%) in 2002.

11.3.3 Accuracy Assessment

The accuracy assessment performed using a confusion matrix (Tables 11.6, 11.7, 11.8, and 11.9) involves evaluating the performance and reliability of the classification model in correctly categorising pixels within satellite imagery (Stehman and

Table 11.5 Forest classification in 1992

Forest classification	Area in km^2 (1992)	%	Area in km^2 (2002)	%
Water bodies	1.05	2.6	4.62	11.4
Medium forest	8.47	20.7	13.45	32.9
Dense/thick forest	2.81	6.9	4.44	10.9
Light forest	20.87	51.1	8.75	21.4
Built-up	7.62	18.7	9.66	23.4
Total	**40.82**	**100**	**40.82**	**100**

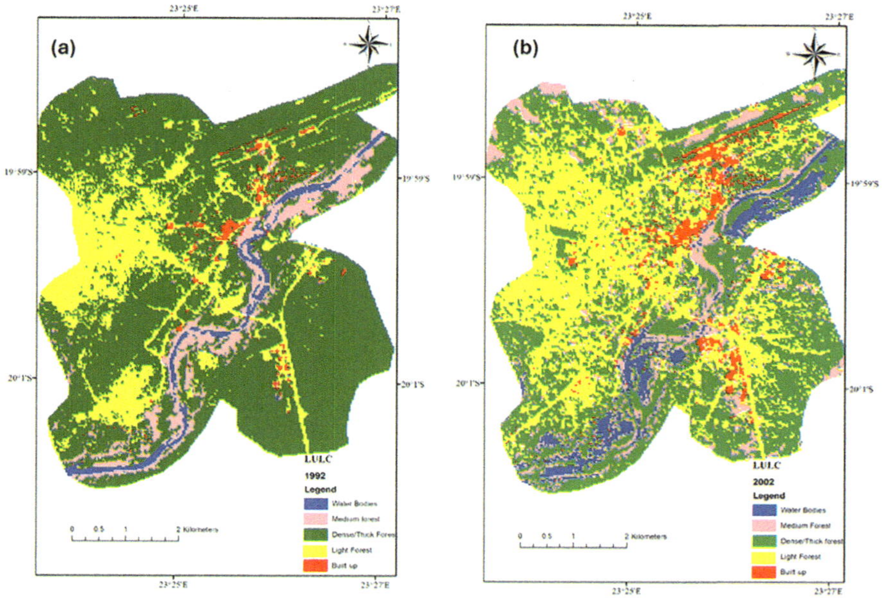

Fig. 11.2 Map of forest classification of (**a**) year 1992 and (**b**) year 2002

Table 11.6 Accuracy assessment 1992

	Built-up	Dense/thick forest	Light forest	Medium forest	Water bodies	Total
Built-up	18	1	0	0	1	20
Dense/thick forest	0	60	0	1	0	61
Light forest	0	1	55	0	0	56
Medium forest	1	0	1	43	3	48
Water bodies	2	1	0	0	60	63
Total	21	63	56	44	64	248
User accuracy	0.86	0.95	0.98	0.98	0.94	**0.95**
Producer accuracy	0.90	0.98	0.98	0.90	0.95	

Foody 2009). It provides a quantitative measure of how well the classification model is performing. It allows for the gauging of the level of accuracy achieved by the model in differentiating various land cover or land use classes (Congalton 2001). Without such an evaluation, it is not possible to ascertain the reliability of the classification results, making it challenging to trust the outputs for any subsequent analysis or decision-making (Stehman and Foody 2009). Moreover, the accuracy assessment helps in identifying the strengths and weaknesses of the classification algorithm used (Stehman and Foody 2009). While, on the one hand, user accuracy

Table 11.7 Accuracy assessment 2002

	Medium forest	Dense/thick forest	Light forest	Built-up	Water	
Medium forest	30	0	1	0	0	31
Dense/thick forest	1	48	0	1	0	50
Light forest	0	3	55	1	1	60
Built-up	0	0	0	58	1	59
Water	1	1	2	1	43	48
Total	32	52	58	61	45	248
User accuracy	0.94	0.92	0.95	0.95	0.96	**0.94**
Producer accuracy	0.97	0.96	0.92	0.98	0.90	

Table 11.8 Accuracy assessment 2013

	Medium Forest	Dense/thick forest	Light forest	Built-up	Water	
Medium forest	40	1	0	0	0	41
Dense/thick forest	1	48	1	1	0	51
Light forest	1	0	50	0	1	52
Built-up	0	0	9	50	1	60
Water	1	1	0	8	34	44
Total	43	50	60	59	36	248
User accuracy	0.93	0.96	0.83	0.85	0.94	**0.90**
Producer accuracy	0.98	0.94	0.96	0.83	0.77	

denotes the proportion of correctly classified pixels or observations in the remote sensing data compared to the reference data or ground truth, the producer accuracy, on the other hand, is calculated by dividing the total number of classified points that agree with the reference data by the total number of reference points for that class. For most applications, a minimum user accuracy and producer accuracy of 85% to 90% is often considered acceptable (Stehman and Foody 2009). In the case of this study, the producer, user, and overall accuracy were within the recommended threshold.

The spatial statistics and variations of forest cover for the years 2013 and 2022 are shown in Table 11.10 and Fig. 11.3. While in 2013 built-up covered 12.9 km^2 (31.6%) in 2022 it covered 20.53 km^2 (50.3%). On the one hand, dense/thick forests continued to decline from 9.5 km^2 (23.3%) in 2013 to 6 km^2 (14.7%) in 2022. While the light forest was 14.3 km^2 (35%) in 2013, it declined in 2022 to 3.0 km^2 (7.3%).

11.3.4 Analysis of the Spatial Pattern of NDVI

The change in vegetation cover within a given area was assessed by analysing NDVI. On the one hand, healthy and undisturbed green vegetation has high visible light absorption and high near-infra-red reflectance capacity, which consequently leads to higher NDVI values in healthy and dense vegetation than in dry vegetation, bare land and water bodies. On the other hand, dry vegetation, bare land and water

Table 11.9 Accuracy assessment 2022

	Medium forest	Dense/thick forest	Light forest	Built-up	Water	
Medium forest	59	2	0	0	0	60
Dense/thick forest	1	48	1	1	0	51
Light forest	1	0	34	0	2	36
Built-up	0	0	9	56	1	66
Water	1	1	0	8	23	33
Total	62	51	44	65	26	248
User accuracy	0.95	0.94	0.77	0.86	0.88	**0.89**
Producer accuracy	0.98	0.94	0.94	0.84	0.69	

Table 11.10 Forest classification in 2013 and 2022

Forest classification	Area in km^2 (2013)	%	Area in km^2 (2022)	%
Water bodies	1.22	3	2.08	5.1
Medium forest	2.9	7.1	9.21	22.6
Dense/thick forest	9.5	23.3	6	14.7
Light forest	14.3	35	3	7.3
Built-up	12.9	31.6	20.53	50.3
Total	40.82	100	40.82	100

absorb more near-infrared and reflect more visible light, consequently leading to lower NDVI values in areas covered by dry, built and water. The minimum and maximum NDVI for 1992 and 2002 were −0.20, 0.02 and 0.13 and 0.43, respectively (Fig. 11.4). While the NDVI ranged from 0.01 to 0.05 in the western side of Maun in 1992, the NDVI for the same area in 2002 ranged from 0.05 to 0.08. The northern, eastern, and southern parts of the study area had an NDVI ranging from 0.06 to 0.08 and 0.09 to 0.14. The area along the central part and the banks of the Thamalakane River had an NDVI ranging from 0.13 to 0.43 in 1992 and from 0.25 to 0.66 in 2002.

The distribution of NDVI for the years 2013 and 2022 is shown in Fig. 11.5. The minimum NDVI for 2013 and 2022 ranged from −0.09 to 0.03 and −0.07 to 0.05, respectively, while the maximum NDVI for the same period ranged from 0.05 to 0.13 and 0.20 to 0.37, respectively. In 2013, the NDVI values in the western, northern, and eastern sides of the study area ranged from 0.01 to 0.03. Incredibly low (−0.04–0.04) values of NDVI were registered along the Thamalakane riverbanks.

11.3.5 Analysis of the Spatial Pattern of LST and Land Cover Type

This section of the chapter presents the findings of the LST distributions across five different land cover types from 1992 to 2002 (Table 11.11 and Fig. 11.6). The results show that while dense/thick forests had an LST of 23.25–27.88 °C and 13.31–17.01

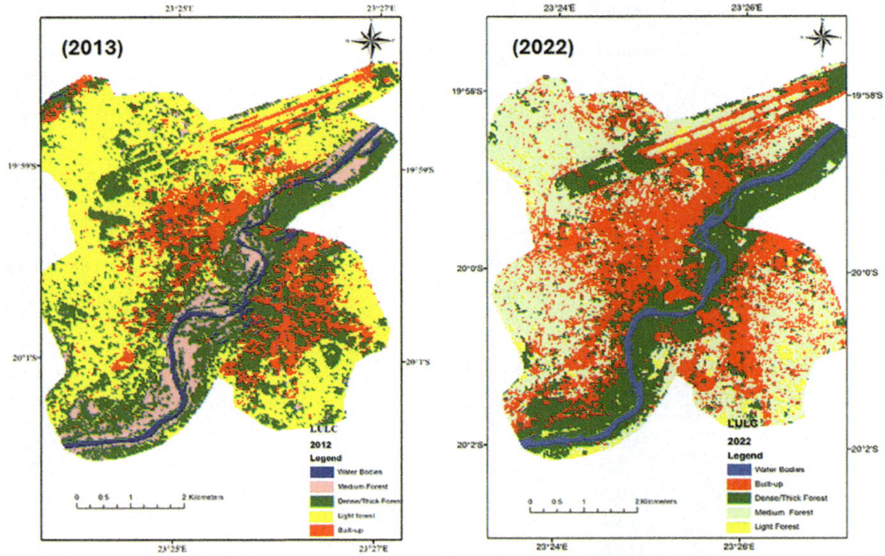

Fig. 11.3 Map of forest classification, 2013 and 2022

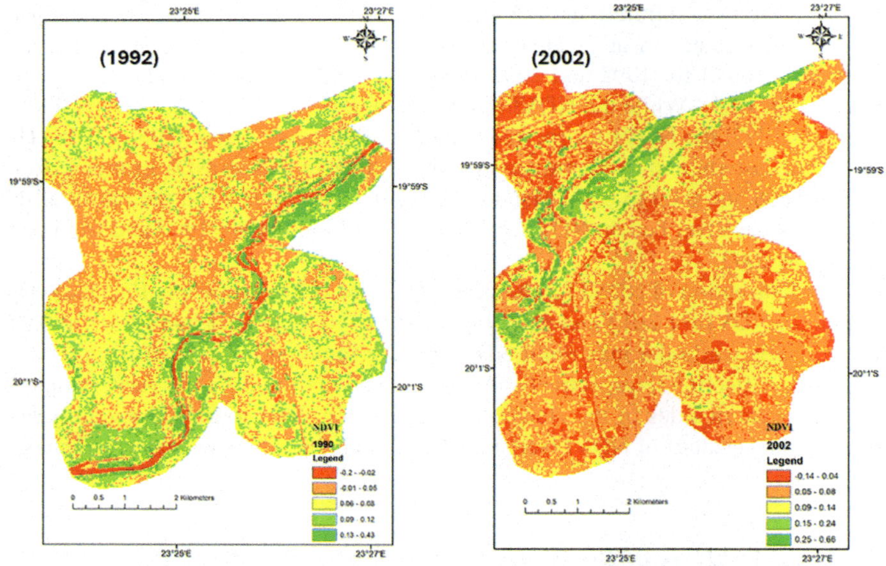

Fig. 11.4 Vegetation distribution based on NDVI in 1992 and 2002

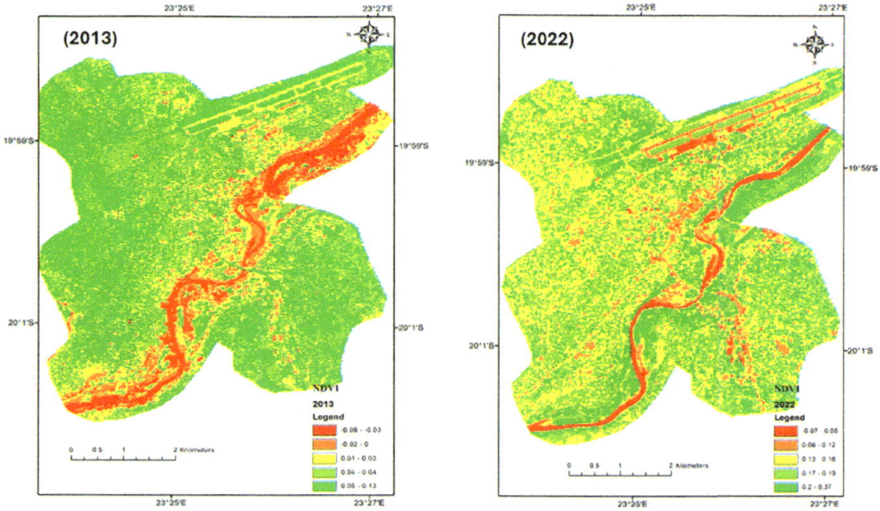

Fig. 11.5 Vegetation distribution based on NDVI in 2013 and 2022

Table 11.11 LST distribution between 1992 and 2002

LULC type	LST (°C), 1992	LST (°C), 2002
Dense/thick forest	23.25–27.88	13.31–17.01
Medium forest	27.89–31.23	17.02–20.16
Light forest	31.24–33.66	20.17–22.4
Built-up	33.67–35.22	22.41–24.1
Water bodies	35.23–38	24.11–28.77

°C in 1992 and 2002, respectively, the medium forest had an LST of 27.89–31.23°C and 17.02–20.16 °C in the same period, respectively. Light forests had an LST ranging from 31.24 to 33.66 °C and from 20.17 to 22.4 °C in 1992 and 2002, respectively. The highest (35.23–38.0 °C and 24.11–28.77 °C) LST was recorded in bare land for both 1992 and 2002. It is worth noting that water bodies and dense/thick forests had the same effect of lowering LST.

The findings on the distribution of LST in different land cover types for 2013 and 2022 are shown in Table 11.12 and Fig. 11.7. The least (21.06–22.79 °C and 21.79–24.74 °C) LST for 2013 and 2022 were recorded on water bodies and dense/thick forests and the highest (26.69–30.42 °C and 21.06–22.79 °C) LST were recorded in built-up area. While the medium forest had an LST ranging from 20.2 to 22.0 °C and from 15.5 to 24.1 °C in 2012 and 2022, respectively, the light forest had an LST of 23.3–24.6 °C and 30.2–34.5 °C within the same period. While high temperatures were recorded in tarred roads, buildings, and areas with bare soils among other human-modified environments, lower LST were observed in areas with water bodies and those covered by dense forest.

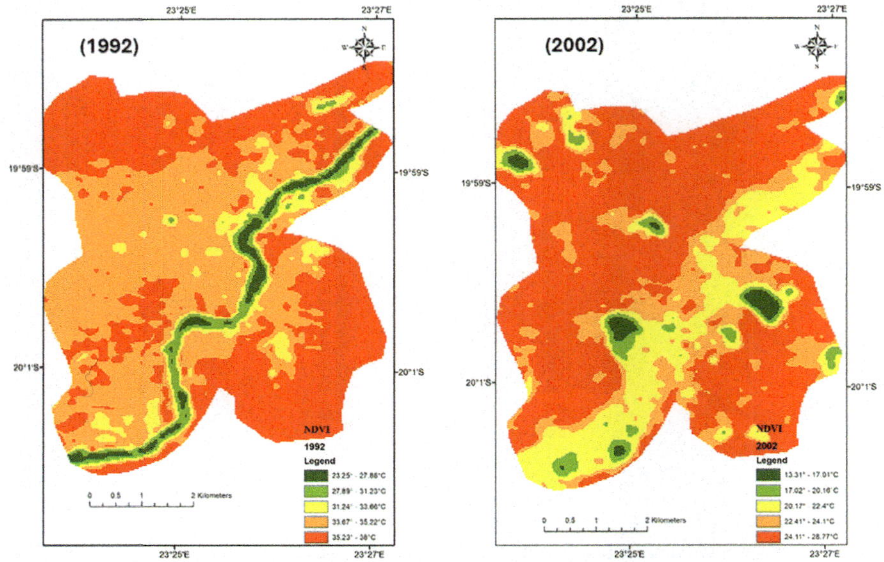

Fig. 11.6 Land surface temperature distribution, 1992 and 2002

Table 11.12 LST distribution between 2013 and 2022

LULC type	LST (°C), 2013	LST (°C), 2022
Dense/thick forest	21.06–22.79	21.79–24.74
Medium forest	22.80–24.55	24.75–26.52
Light forest	26.56–27.68	30.2–34.5
Built-up	25.84–26.68	34.6–40.7
Water bodies	26.69.2–30.42	13.0–15.7

11.3.6 Relationship Between LST and NDVI

To better understand the impact of humans on land surface temperature and vegetation cover, a simple regression analysis was conducted to test the relationship between LST and NDVI within a three-decadal period of 1992–2022 (Figs. 11.8, 11.9, 11.10, and 11.11). The findings revealed a strong but negative relationship between LST and NDVI ($t = -0.410$, $p \leq 0.01$) for the first 10-year period (1992–2002). However, there was no significant relationship between LST and NDVI ($t = -0.26$, $p = 0.798$) in 2002–2012. The result for the period 2012–2022 shows a significant but negative relationship between LST and NDVI ($t = -0.57$, $p \leq 0.00$). Considering the relationship between the two variables over 30 years (1992–2022), the result reveals a significant but negative relationship between LST and NDVI ($t = -0.410$, $p \leq 0.00$). This implies that an increase in LST leads to a decrease in vegetation cover and it is equally the same when LST is low because

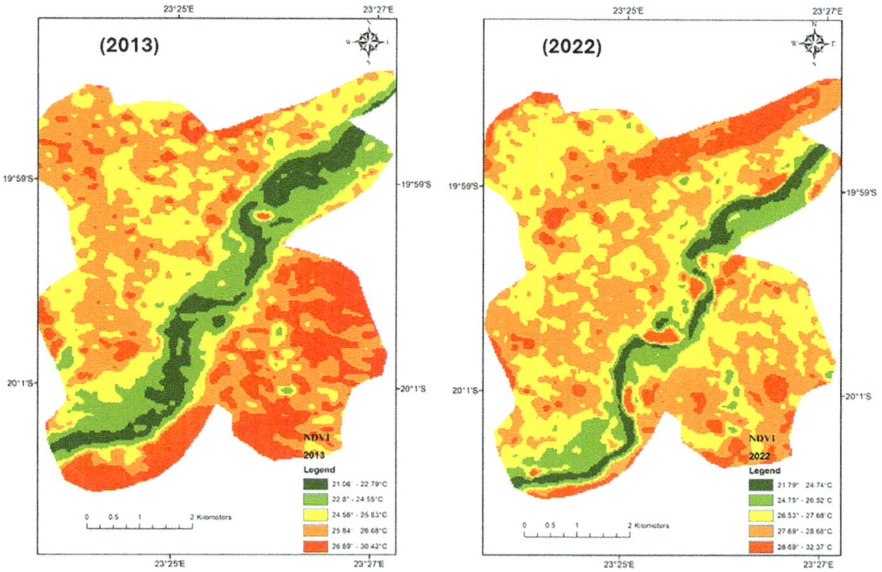

Fig. 11.7 Land surface temperature distribution, 2013 and 2022

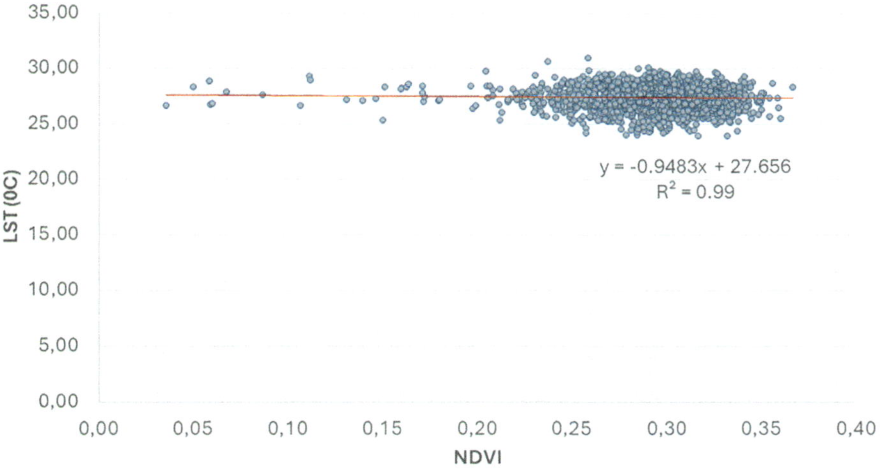

Fig. 11.8 Relationship between LST and NDVI in 1992–2002

lower ground temperatures result in dense vegetation cover (Olorunfemi et al. 2020). This was noted in the previous section where dense vegetation was found along the river channels and in other locations, whose vegetations are still relatively untouched.

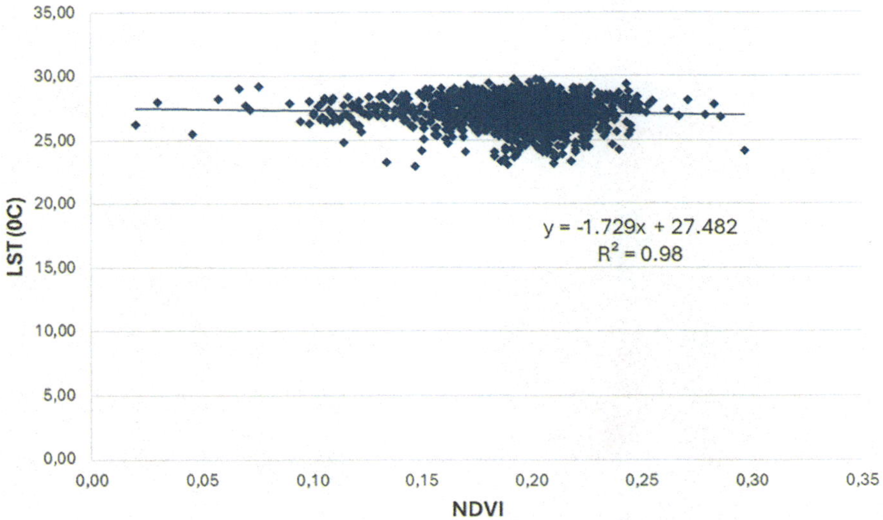

Fig. 11.9 Relationship between LST and NDVI in 2002–2012

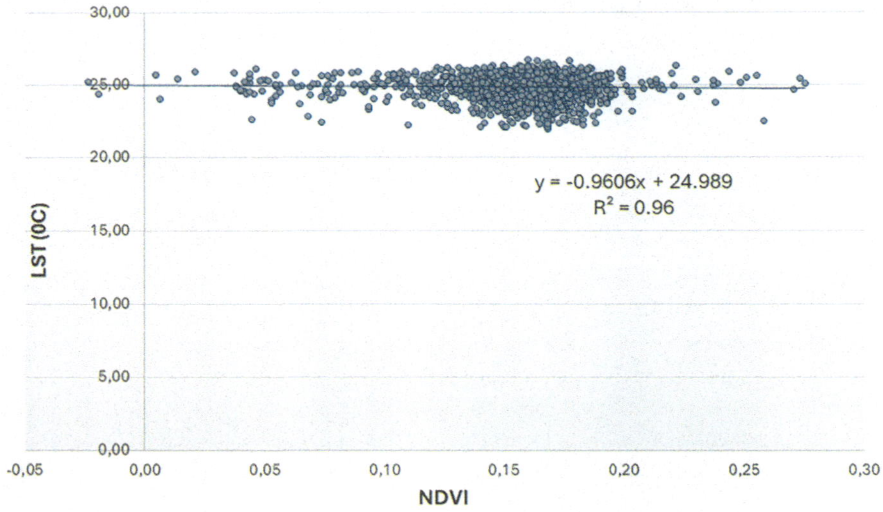

Fig. 11.10 Relationship between LST and NDVI in 2012–2022

11.4 Discussion

The general objective of this study was to analyse the impact of human activities on the forest cover in Maun between 1992 and 2022. The specific objectives were to analyse the spatiotemporal variations in LST and NDVI in the study area and assess the relationship between LST and NDVI over 30 years. Findings showed that dense/

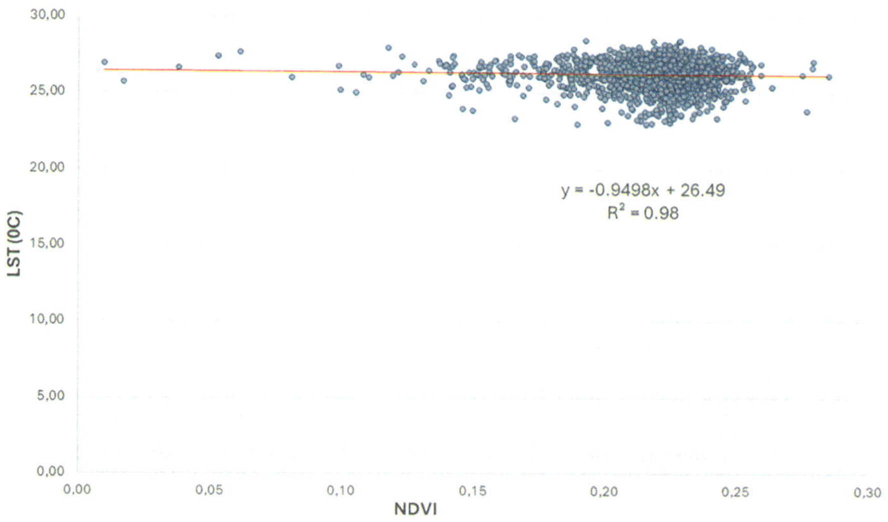

Fig. 11.11 Relationship between LST and NDVI in 1992–2022

thick forests thrived in cool moist areas like riverbanks and where people had had a minimum disturbance. Light vegetation and bare land result in high temperatures; bare land and built-up areas had no vegetation at all.

Population increases lead to a change in land use and turn, affect forest cover. In the current study, the population of Maun increased from 60,263 in 2011 to 85,293 in 2022 indicating a growth rate of 3.3% over 11 years. Literature has shown that there is an association between population increase and forest cover decline (Foster and Rosenzweig 2003; Mather and Needle 2000). Here population increase plays a significant role in LULCC resulting in biophysical environmental changes.

The spatial–temporal distribution of forest cover in the study area was a function of human activities; as the human population increased in number and density, more space was needed for accommodation, industrial development, and related infrastructures. Garai et al. (2022) and Ullah et al. (2023) observed similar findings in Shangala, Pakistan, where the pattern of forest cover deteriorated with an increase in the construction of tarmac, concrete surface, and industrial sites. In general, Nikkala et al. (2022) and Andronis et al. (2022) observe that human activities alter forest cover, which might result in the depletion of vegetation if sound environmental management is not given significant consideration.

Several studies (Liu et al. 2022; Ghouri et al. 2022; Leta et al. 2022; Zadan and Javed 2022) noted a strong and negative correlation between NDVI and LST. The trend was that NDVI was higher in wet seasons and moist areas than in dry seasons and areas. This study found that bare land comprising built-up areas and open spaces increased in spatial extent between 1992 and 2022. Similar results were reported by Adegoke et al. (2003) who analysed the impact of changes in land use in the settlement of Zhuang in China to assess the effect of LULCC and thermal signature association. Similarly, our results are similar to those of Garai et al. (2022) who found that bare soil and built-up areas have a higher LST than water bodies and forested areas. In the current study, findings revealed that the LST increased with the conversion of forest cover surfaces into bare land and concrete pavements. Also, the findings showed that LST increased in bare land areas over the study period probably due to the effect of global warming and anthropogenic land surface modifications.

11.5 Conclusion

This chapter mapped and analysed forest cover variations over 30 years in the lower Okavango Delta. Forest cover changes were analysed in terms of NDVI and the influence of human activities on LST. Satellite data retrieved from USGS were analysed in GIS 10.8. A supervised classification method using a random forest classifier was used to assess the spatial variations in forest cover changes as a result of human activities. Findings revealed that an increase in population resulted in the expansion of built-up areas and related infrastructure, which also led to a reduction in the spatial extent of forest cover. The increase in LST over three decades in the study area may have impacted human well-being and the environment at large. All things considered, these changes may have impacted settlers' livelihoods, human and environmental health in the long run.

11.6 Recommendations and Policy Implications

This chapter affirmed that an increase in population has a strong relationship with a decrease in vegetation cover induced by modernisation and expansion of human settlements. All things considered, future generations will suffer dire consequences associated with climate change and variability if the trend in vegetation cover loss is sustained. It is, therefore, imperative that the local authority and perhaps the national government would need to adopt conservation measures to address the challenges associated with vegetation loss. This can be achieved through the implementation of reafforestation and afforestation initiatives at the local level. The information on forest cover dynamics provided in this chapter would serve as baseline data for further spatial–temporal analysis of natural resource loss within the study area and the entire Okavango Delta in general. Also, policies geared towards local-level human capacity development in forest conservation are needed to stop forest depletion within the entire Okavango Delta region. The maintenance of the ecological integrity of the study area would require policies that emphasise impactful forestry research and forest conservation efforts.

References

Adegoke JO, Pielke RA Sr, Eastman J, Mahmood R, Hubbard KG (2003) Impact of irrigation on midsummer surface fluxes and temperature under dry synoptic conditions: A regional atmospheric model study of the US High Plains. Mon Weather Rev 131(3):556–564

Abd Latif M, Wan Rasidah K, Ahmad Zuhaidi Y (2018) Is plantation forestry a wise investment? A perspective from Malaysia's initiatives. J Trop For Sci 30(5):461–467

Allendorf TD, Brandt JS, Yang JM (2014) Local perceptions of Tibetan village sacred forests in northwest Yunnan. Biol Conserv 169:303–310

Andronis V, Karathanassi V, Tsalapati V, Kolokoussis P, Miltiadou M, Danezis C (2022) Time series analysis of Landsat data for investigating the relationship between land surface temperature and forest changes in Paphos Forest, Cyprus. Remote Sens 14(4):1010

Bamwesigye D, Kupec P, Chekuimo G, Pavlis J, Asamoah O, Darkwah SA, Hlaváčková P (2020) Charcoal and wood biomass utilization in Uganda: the socioeconomic and environmental dynamics and implications. Sustainability 12(20):8337

Bennett EM, Peterson GD, Gordon LJ (2009) Understanding relationships among multiple ecosystem services. Ecol Lett 12(12):1394–1404

Bera B, Saha S, Bhattacharjee S (2020) Forest cover dynamics (1998 to 2019) and prediction of deforestation probability using binary logistic regression (BLR) model of Silabati watershed, India. Trees For People 2:100034

Bromhead MA (2012) Forest, trees, and woodlands in Africa: an action plan for World Bank engagement. World Bank, Washington, DC

Bijaya GC, Cheng S, Xu Z, Bhandari J, Wang L, Liu X (2016) Community forestry and livelihood in Nepal: A review. J Anim Plant Sci 26(1)

Canton H (2021) Food and Agriculture Organization of the United Nations—FAO. In The Europa directory of International Organizations 2021 (pp. 297–305). Routledge.

Churches CE, Wampler PJ, Sun W, Smith AJ (2014) Evaluation of forest cover estimates for Haiti using supervised classification of Landsat data. Int J Appl Earth Obs Geoinf 30:203–216

Citaristi, I. (2022). Specialized agencies and related organizations within the un system: Food and agriculture organisation of the United Nations—FAO. In The Europa Directory of International Organizations 2022 (pp. 307–315). Routledge.

Congalton RG (2001) Accuracy assessment and validation of remotely sensed and other spatial information. Int J Wildland Fire 10(4):321–328

Creed IF, van Noordwijk M (2018) Forest and water on a changing planet: vulnerability, adaptation and governance opportunities. A global assessment report. IUFRO World Series, 38

De Wasseige C, Tadoum M, Atyi EA, Doumenge C (2015) The forests of the Congo Basin-Forests and climate change. Weyrich, Neufchateau

Dissanayake DMSLB, Morimoto T, Ranagalage M, Murayama Y (2019) Land-use/land-cover changes and their impact on surface urban heat islands: Case study of Kandy City, Sri Lanka. Climate 7(8):99

Drummond MA, Loveland TR (2010) Land-use pressure and a transition to forest-cover loss in the eastern United States. BioScience 60(4):286–298

Erdős L, Bátori Z, Tölgyesi C, Körmöczi L (2014) The moving split window (MSW) analysis in vegetation science–an overview. Appl Ecol Environ Res 12(3):787–805

Food and Agriculture Organization of the United Nations. New partnership for Africa's development. https://reliefweb.int/organization/fao

Forest Conservation Botswana (2013) Forest management and use in Botswana: brief situation analysis and options for the Forest Conservation Strategy. Paper prepared by the Centre for Applied Research for the Forest Conservation Board Botswana, Centre for Applied Research

Foster AD, Rosenzweig MR (2003) Economic growth and the rise of forests. Q J Econ 118(2):601–637

Garai S, Khatun M, Singh R, Sharma J, Pradhan M, Ranjan A, Rahaman SM, Khan ML, Tiwari S (2022) Assessing the correlation between Rainfall, normalized difference Vegetation Index (NDVI) and land surface temperature (LST) in Eastern India. Saf Extreme Environ 4(2):119–127

Garai D, Narayana AC (2018) Land use/land cover changes in the mining area of Godavari coal fields of southern India. The Egyptian Journal of Remote Sensing and Space Science, 21(3):375–381

Gamon JA, Kitajima K, Mulkey SS, Serrano L, Wright SJ (2005) Diverse optical and photosynthetic properties in a neotropical dry forest during the dry season: implications for remote estimation of photosynthesis 1. Biotropica: Biol Conserv 37(4):547–560

Ghouri AY, Khan A, Rasheed F (2022) Monitoring the land surface temperature and its correlation with NDVI of Chiniot by using GIS technology and remote sensing. Earth Environ Sci Res Rev 5(2):01, 9

Giesen W (2015) Utilising non-timber forest products to conserve Indonesia's peat swamp forests and reduce carbon emissions. J Indones Nat Hist 3(2):17–26

Guha S, Govil H, Dey A, Gill N (2018) Analytical study of land surface temperature with NDVI and NDBI using Landsat 8 OLI and TIRS data in Florence and Naples city, Italy. Eur J Remote Sens 51(1):667–678

Huge J, Rochette AJ, de Béthune S, Paitan CP, Vanderhaegen K, Vandervelden T, Van Passel S, Vanhove MPM, Verbist B, Verheyen D, Waas T (2020) Ecosystem services assessment tools for African Biosphere Reserves: A review and user-informed classification. Ecosyst Serv 42:101079

Keenan RJ, Reams GA, Achard F, de Freitas JV, Grainger A, Lindquist E (2015) Dynamics of global forest area: results from the FAO Global Forest Resources Assessment 2015. For Ecol Manag 352:9–20

Köhl M, Lasco R, Cifuentes M, Jonsson Ö, Korhonen KT, Mundhenk P, Stinson G (2015) Changes in forest production, biomass, and carbon: Results from the 2015 UN FAO Global Forest Resource Assessment. For Ecol Manage 352:21–34

Le Quéré C, Raupach MR, Canadell JG, Marland G et al (2009) Trends in the sources and sinks of carbon dioxide. Nat Geosci 2:831–836

Leslie D (2020) Tackling COVID-19 through responsible AI innovation: five steps in the right direction. Harvard Data Sci Rev 10

Leta T, Wondu H, Chala D, Yadeta D (2022) Forest vegetation dynamics and their response to land surface temperature in West Arsi Zone Southwest Ethiopia. Forest 8(12):01–08

Lin D, Lai J, Yang B, Song P, Li N, Ren H, Ma K (2015) Forest biomass recovery after different anthropogenic disturbances: relative importance of changes in stand structure and wood density. Eur J For Res 134:769–780

Li L, Hao T, Chi T (2017) Evaluation on China's forestry resources efficiency based on big data. J Clean Prod 142:513–523

Liu W, Jia B, Li T, Zhang Q, Ma J (2022) Correlation analysis between urban green space and land surface temperature from the perspective of spatial heterogeneity: a case study within the sixth ring road of Beijing. Sustainability 14(20):13492

Lufumpa CL (2005) The poverty–environment nexus in Africa. Afr Dev Rev 17(3):366–381

Manwa H, Saarinen J, Atlhopheng JR, Hambira WL (2017) Sustainability management and tourism impacts on communities: residents' attitudes in Maun and Tshabong, Botswana. Afr J Hosp Tour Leis 6(3):1–5

Mallick A (2012) Encroachment on the rights of the Adivasis: Colonial Forest Policy in 19th century Chota Nagpur and Santal Parganas. In Proceedings of the Indian History Congress (Vol. 73, pp. 747–755). Indian History Congress

Mao K, Qin Z, Shi J, Gong P (2005) A practical split-window algorithm for retrieving land-surface temperature from MODIS data. Int J Remote Sens 26(15):3181–3204

Mather AS, Needle CL (2000) The relationships of population and forest trends. Geogr J 166(1):2–13

Mbaiwa JE, Toteng EN, Moswete N (2007) Problems and prospects for the development of urban tourism in Gaborone and Maun, Botswana. Dev South Afr 24(5):725–740

Ministry of Environment, Forest and Climate Change. (2021). India State of the Forest Report 2021: Uttarakhand (ISBN 978-81-950007-3-1). Ministry of Environment, Forest and Climate Change.

Nath B, Acharjee S (2013) Forest cover change detection using normalized difference vegetation index (NDVI): a study of Reingkhyongkine lake's adjoining areas, Rangamati, Bangladesh. Indian Cartogr 33(2):348–403

Nazir N, Ahmad S (2018) Forest land conversion dynamics: a case of Pakistan. Environ Dev Sustain 20:389–405

Neinavaz E, Skidmore AK, Darvishzadeh R (2020) Effects of prediction accuracy of the proportion of vegetation cover on land surface emissivity and temperature using the NDVI threshold method. Int J Appl Earth Obs 85:101984

Nikkala S, Peddada JR, Neredimelli R (2022) Correlation analysis of land surface temperature on Landsat-8 data of Visakhapatnam Urban Area, Andhra Pradesh, India. Earth Sci Inf 15(3):1963–1975

Nybakk E, Crespell P, Hansen E, Lunnan A (2009) Antecedents to forest owner innovativeness: an investigation of the non-timber forest products and services sector. For Ecol Manag 257(2):608–618

Olorunfemi IE, Fasinmirin JT, Olufayo AA, Komolafe AA (2020) GIS and remote sensing-based analysis of the impacts of land use/land cover change (LULCC) on the environmental sustainability of Ekiti State, southwestern Nigeria. Environ Dev Sustain 22:661–692

Otieno J, Abihudi S, Veldman S, Nahashon M, van Andel T, de Boer HJ (2015) Vernacular dominance in folk taxonomy: a case study of ethnospecies in medicinal plant trade in Tanzania. J Ethnobiol Ethnomed 11(1):1–7

Samarasinghe JT, Gunathilake MB, Makubura RK, Arachchi SM, Rathnayake U (2022) Impact of climate change and variability on spatiotemporal variation of forest cover; World Heritage Sinharaja Rainforest, Sri Lanka. Forest and Society 6(1):355–377

Saum R (2010) Chapter Eight. Promise and reality of community based natural resource management in Botswana: common-pool resource use and institutional change in Ikoga, Okavango Delta (panhandle). In: Disputing the floodplains. Brill, pp 361–412

Sobrino JA, Li ZL, Stoll MP, Becker F (1994) Improvements in the split-window technique for land surface temperature determination. IEEE Trans Geosci Remote Sens 32(2):243–253

Statistics Botswana (2022) 2022 Population and housing census preliminary results V2. Government Printers, Goborone

Stehman SV, Foody GM (2009) Accuracy assessment. In: The SAGE handbook of remote sensing. Sage, London, pp 297–309

Stone LS (2014) Perceptions of nature-based tourism, travel preferences, promotions and the disparity between domestic and international tourists: the case of Botswana. Arizona State University

Svatonova H (2016) Analysis of visual interpretation of satellite data. Int Arch Photogramm Remote Sens Spat Inf Sci 41:675–681

Twongyirwe R, Majaliwa JGM, Ebanyat P, Tenywa MM, Sheil D, Heist MV, Oluka M, Kumar L (2011) Dynamics of forest cover conversion in and around Bwindi impenetrable forest, Southwestern Uganda. J Appl Sci Environ Manag 15(1)

Ullah W, Ahmad K, Ullah S, Tahir AA, Javed MF, Nazir A, Abbasi AM, Aziz M, Mohamed A (2023) Analysis of the relationship among land surface temperature (LST), land use land cover (LULC), and normalized difference vegetation index (NDVI) with topographic elements in the lower Himalayan region. Heliyon 9(2):e13322

United States Geological Survey (2019) Landsat 8 (L8) Data Users Handbook, LSDS-1574 Version 5.0

Wu Z, Dai E, Wu Z, Lin M (2019) Future forest dynamics under climate change, land use change, and harvest in subtropical forests in Southern China. Landsc Ecol 34:843–863

Zadan KHAN, Javed A (2022) Correlation between land surface temperature (LST) and normalized difference vegetation index (NDVI) in Wardha Valley Coalfield, Maharashtra, Central India. Nova Geod 2(3):53–53

Zhongming Z, Linong L, Xiaona Y, Wangqiang Z, Wei L (2019) Congo Basin deforestation threatens food and water supplies throughout Africa. World Resource Institute

Reniko Gondo holds a PhD degree from the University of Botswana. He earned his MSc and BSc degrees from the Bindura University of Science Education and Zimbabwe Open University (Zimbabwe), respectively. Currently, he serves as a Senior Research Fellow at the Okavango Research Institute (ORI), University of Botswana (UB). Gondo worked as a geography teacher for ten years in Zimbabwe. Additionally, he served as a part-time lecturer for 3 years at the Bindura University of Science Education. Gondo joined the Okavango Research Institute, University of Botswana, in February 2016. He authored over 20 publications focusing on the interface between Indigenous and Western knowledge in the governance of natural resources. Gondo is a reviewer for *Water Resources Management*, *Scientific African*, *Sustainable Water Resources Management*, and *Geography and Sustainability*.

Oluwatoyin Dare Kolawole is Professor of Rural Development at the Okavango Research Institute (ORI), University of Botswana, Maun, Botswana. He is Research Fellow at the Centre for Gender and Africa Studies, University of the Free State, South Africa. He is also an Adjunct Faculty at the College of Health and Social Sciences in the Eastern University, St. Davids, Pennsylvania. He works at the interface of science, policy, and agriculture with major interests in politics of knowledge, sustainable agriculture, natural resources conservation, and community development. He was the University of Canterbury Visiting Canterbury Fellow in 2014 and a CODESRIA laureate in 1994 and 2006.

Chiedza Ngonidzashe Mutanga is a Senior Research Fellow in Sustainable Tourism at the Okavango Research Institute, University of Botswana. She holds a DPhil in Tourism and Wildlife Management from Chinhoyi University of Technology, Zimbabwe. Her research interests include sustainable tourism development, with a focus on wildlife tourism, nature-based tourism, community-based tourism and community livelihoods, ecotourism, protected area tourism, community-based natural resource management, and wildlife resource conservation.

Chapter 12
The Ecology and Management of the Coffee Forests of Ethiopia

Feyera Senbeta

12.1 Introduction

Dense natural forests once covered a considerable area of the Ethiopian highlands (Logan 1946; IUCN 1990; McCann 1997; Darbyshire et al. 2003). The current mosaic-like landscapes with isolated trees and small fragments of primary and secondary forests are also proof of how much of Ethiopia's highlands were once covered by forests. However, a substantial portion of these highlands are today bereft of forest vegetation and are largely covered by farmlands and settlements. The southwest and southeast parts of the country currently have few scattered forest fragments. Some of these forest fragments may be regrowth after the abandonment of agricultural fields and home gardens (McCann 1997). These forest dynamics probably influenced today's forest structure, species diversity and composition. Many of these remnant forest fragments are primarily moist-evergreen Afromontane forests with the presence of some dry Afromontane forests. A portion of the moist-evergreen Afromontane forest has been acknowledged as the center of diversity and origin of *Coffea arabica* L. (arabica coffee) populations (Strenge 1956; Meyer 1965; Aga et al. 2003; Tesfaye 2006). This forest belt with an occurrence of naturally grown arabica coffee plants is usually called a "coffee forest." Coffee forests exist in fragments that are geographically separated from one another by agriculture and settlements. They are mostly found in the southwest, with only one patch occurring in the southeastern part of the country. These coffee forest fragments currently contain the major global wild arabica coffee genetic resources (Anthony et al. 2001; Aga et al. 2003; Silvarolla et al. 2004; Tesfaye 2006).

F. Senbeta (✉)
Department of Biological Sciences, Faculty of Sciences, Botswana University of Agriculture and Natural Resources, Gaborone, Botswana
e-mail: fwakjira@buan.ac.bw

W. L. Hambira et al. (eds.), *Environmental Change and Biodiversity Conservation in sub-Saharan Africa*, Advances in Global Change Research 75,
https://doi.org/10.1007/978-3-031-73136-5_12

Beyond its center of wild arabica coffee genetic pools, coffee forests offer enormous ecological and social benefits to the local community and beyond (Mittermeier et al. 2011). An important habitat for the conservation of forest bird species (EWNHS 1996; Rodrigues et al. 2018), the diversity of plants (Senbeta 2006; Schmitt et al. 2009), and the diversity of wild arabica coffee populations (Tesfaye 2006; Kotecha 2008). These forest fragments also provide critical habitat for several wild animals (Rodrigues et al. 2019) and support high carbon stocks (Sanchez et al. 2021). They also provide significant socioeconomic benefits to nearby communities (Senbeta et al. 2013; Ango et al. 2014; Shumi et al. 2019a; Schultner et al. 2021). The local populations dwelling in and nearby the coffee forests heavily depend on various forest products for their livelihoods, for example, honey, coffee, timber, water, and spices. They are further important for managing watersheds, serving as water catchments, and acting as barriers to erosion as they are situated on mountains.

Despite their importance, however, land use pressure and climate change have already put coffee forest fragments in danger (Senbeta 2006; Tadesse et al. 2008; Davis et al. 2012; Tadesse et al. 2014; Shumi et al. 2018; Sanchez et al. 2021). Overexploitation, ruthless destruction, as well as degradation, have made coffee forests the most vulnerable and threatened ecosystem (Woldemariam et al. 2002; Hylander et al. 2013; Tadesse et al. 2014; Ango 2018; Ango et al. 2020). Threats to coffee forests may also have been exacerbated by unclear forest (property) rights and regulations, and population growth (Shumi et al. 2019a). Although the outcomes are not satisfactory, conservation efforts have been ongoing for the past three decades to address these threats (Labouisse et al. 2008; Tadesse et al. 2016). These initiatives have included the creation of biosphere reserves, participatory forest management programs, and genetic reserves for the coffee forests. In fact, an array of sustainable management and conservation initiatives are required to meet the needs and aspirations of the current generation without endangering the forest's ability to survive in the future.

This review paper provided new insights into the ecology, ecosystem services, threats, and management status of the Ethiopian coffee forests. To achieve this, I have consulted a wide range of scholarly perspectives, including works on the location and distribution of coffee forests, ecology, diversity, socioeconomic significance, the value of habitat for fauna and flora, and conservation management. Finally, suggestions are forwarded for conserving the last remaining coffee forests.

12.2 Physical Settings

12.2.1 Location and Distribution of Coffee Forests

Coffee forests are located in the southwest and southeast parts of Ethiopia (Fig. 12.1), at an altitudinal range of 1000–2000 m above sea level. The potential range of coffee forests is thought to cover an area of around 900,000 hectares (ECTDMA 2015),

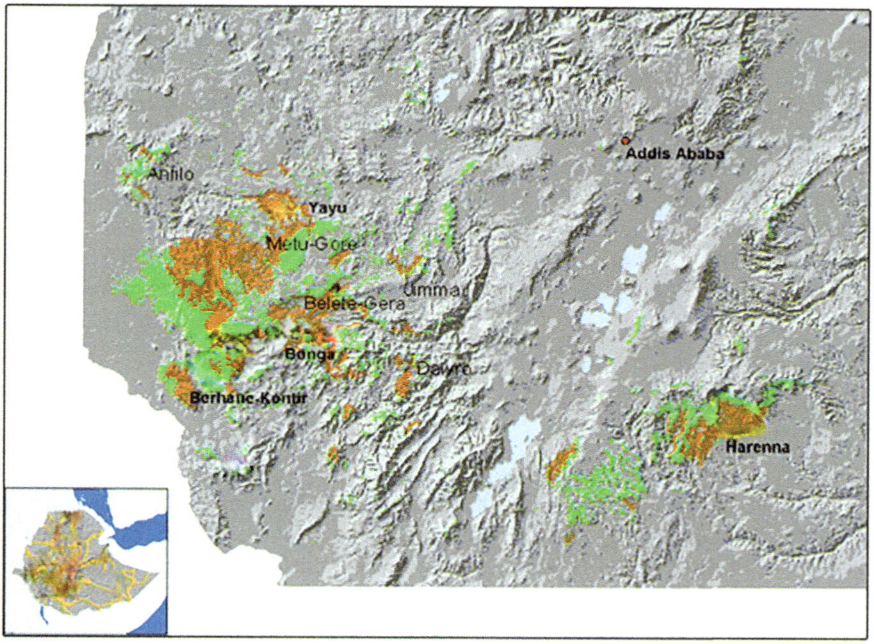

Fig. 12.1 A topographic map of south and southwest Ethiopia displays the distribution of forests in green and brown colours (brown colour shows coffee forests distribution) (COCE 2004)

but due to land use pressure, these forests have already been severely degraded, fragmented, and partly turned into non-forest land uses (Senbeta 2006). The coffee forests occur in areas with diverse topography, soil types, and climatic conditions, such as in undulating terrain, at valley bottoms, and along incised rivers that flow into valleys. The diverse vegetation units found in coffee forests are home to a wide variety of plant and animal species. Patches of diverse grasslands and wetlands, each with its own distinctive faunistic and floristic connections, are interspersed across the coffee forests (Senbeta 2006; Rodrigues et al. 2018). Currently, the country's southwest area, specifically the Belete-Gera, Sigmo-Gatira, Yayu, Boginda, Gergeda, Bonga, Shaka, Shako, Majang, and Maji, contains the largest coffee forests, as well as in the Harenna Forest in the southeast.

12.2.2 Geology and Soils

The country's western region is made up of alkali olivine basalt and tuffs volcanic rock, while the southeast's coffee forest region was formed by outpourings of lava during the Oligocene and Miocene geological eras (Mohr 1965). This was probably created between 40 and 25 million years ago and was formed before the Great Rift

Valley System (Mohr 1971; Umer and Bonnefille 1998). In coffee forests, the soils are often of medium texture, brown or reddish-brown, deep, and readily permeable (Murphy 1968). These soils are typically described as reddish-brown silt-clay that is abundant in simple exchangeable cations. With depth, the color may be redder, and the clay content may rise. Soil analysis of coffee forests uncovered various soil types that include Cambisols, Acrisols, Regosols, and Nitosols with the soil's pH ranged from 4.2 to 6.6 (Senbeta 2006). These soils have low levels of readily available phosphorus and are acidic to slightly acidic (e.g., Murphy 1968; Willson 1985). It appears that the phosphorus needs of the coffee plants growing in these soils are met by substances generated during the breakdown of organic debris or weathering.

12.2.3 Climate

Coffee forests have adapted to a variety of climate patterns, including those with total annual precipitation between 1000 and over 2000 mm, monthly minimum temperatures between 11 and 19 °C, and monthly maximum temperatures between 23 and 29 °C (Asfaw 2003). However, the southwest coffee forests exhibit unimodal patterns of rainfall, with some areas receiving rain for much of the year and the highest rainfall occurring primarily from June to September. The southeast coffee forest area has bimodal patterns of precipitation, with the main rainy season occurring from March to May (Asfaw 2003; Senbeta 2006). As the country's coffee-growing areas receive plenty of rainfall, numerous rivers flow in and out of the coffee forests.

12.3 Forest Structure and Composition

12.3.1 Forest Structure

The attribute of a forest known as "forest structure" may represent the vertical and horizontal forest structures, for example, vertical stratification, forest type, density, and size. Forest structure serves as a template for biodiversity and ecosystem functions as well as a by-product of forest dynamics and biophysical processes (Spies 1998). Natural forces including wind, fire, succession, and forest management all have an impact on how a forest develops and changes its structure. The past, present, and future of a forest ecosystem can all be better understood by having a solid understanding of forest structure. Understanding the patterns of change in forest structure across time and space can form the basis of forest management plans that aim to maintain forest functions and services (Spies 1997). Density and vertical stratification are presented in this paper to describe the structural characteristics of coffee forests.

Density refers to the number of individuals per unit area. In this paper, density was measured using the number of woody plants per plot (0.04 hectares). Table 12.1 displays some density characteristics of studied coffee forest (i.e., Harenna, Maji, Sheko, Bonga, and Yayu) fragments (Woldemariam 2003; Senbeta 2006). Additionally, Table 12.1 displays the median number of coffee plants per plot and how frequently they appear in the studied forests. The density of woody plants varied from 9309 to 69,130 per hectare in five studied coffee forest fragments of Ethiopia. These density variations could be explained by the interplay between or among various historic elements at each site. For instance, the stage of the forest's succession may be to be blamed for the high plant densities in some sites, for example, in the Yayu coffee forest. A forest's succession processes may be impacted by a variety of natural disturbances, such as fire. Likely there have been some disturbances to the Yayu Forest in the past. Pioneer species may flourish densely together at the beginning of successional development up until they reach the peak stage (climax stage) before many of them are scrubbed because of the Competitive Exclusive Principle (Ewel 1983). On the contrary, the low woody plant density in the Harenna coffee forest demonstrates the persistence of human disturbance at the time of the vegetation survey.

The term "stratification" describes the vertical layering of vegetation. Figure 12.2 displays the general vertical stratifications of coffee forest from southwest Ethiopia. According to a study by Senbeta (2006), most coffee forests consist of three tree strata: the emergent (>30 m tall), the middle tree (15–30 m tall), and the small trees and shrubs layer (2–15 m tall). The emergent tree layer usually possesses a few trees raised above the middle tree layer. Different tree species occupy the upper canopy layer of coffee forests, which is likely due to the variations in edaphic, climatic, or historical aspects of each coffee forest. The middle canopy layer can be either continuous or discontinuous and possesses a narrow canopy. Usually, the bottom small tree layer creates a dense canopy. Forest ferns and grasses make up much of the scanty herb layer. There are lots of lianas and strangling epiphytes. According to Thiel et al. (2021), forest communities commonly possess six vertical layers: the soil, the forest floor, the herbaceous, the shrubby, the canopy, and the emergent

Table 12.1 Density of woody plants (≥0.5 m in height) within a plot size of (20 m × 20 m) in five coffee forests in Ethiopia

	Name of coffee forests				
Characteristics[a]	Harenna	Maji	Sheko	Bonga	Yayu
No. of plots	24	10	37	28	48
Min woody plant density/plot	89	432	303	169	955
Max woody plant density/plot	1980	1209	1756	1459	7684
Median of woody plant density/plot	178	665	570	777	2642
Woody plant density/ha	9309	18,183	18,981	19,232	69,130
% frequency of coffee plants	71	100	91	93	100
Median of coffee density/plot	18	385	46	51	901

[a]Senbeta (2006)

Fig. 12.2 An example of the coffee forest layers, which are made up of different tree species, Gesha forest site in southwest Ethiopia. (Photograph by Feyera Senbeta)

layers. Human influences can alter a forest's vertical stratification in addition to the physical environment. In most of Ethiopia's coffee forests, logging and human settlement have long been there. Given that many of these forest tree species have been used for timber, logging must have played a significant role in reducing the emergent canopy trees, for example, *Olea welwitschii and Pouteria adolfi-friederici.*

12.3.2 Floristic Composition

Many scholars have studied the floristic composition of coffee forests (e.g., Friis 1992; Woldemariam 2003; Schmitt 2006; Senbeta 2006; Hundera et al. 2013). A study by Senbeta (2006) was the most detailed and covered five coffee forest fragments—Bonga, Sheko, Yayu, Harenna, and Maji—and recorded over 700 vascular plant species. According to the author, 649 angiosperm species, 2 gymnosperm species, 50 pteridophyte species, and 118 families were recorded in total. Only 40% of the 118 families in total were found in all five forests, and 21% of the recorded species were found in only one forest. This implies that many of these coffee forest species are rare and have a small geographic range. Rare taxa are those with narrow ranges or low abundance (Gaston 1994). Any divergence of biological, physical, or both could limit the species' ability to spread geographically or in terms of abundance (Cowling 1990; Goldblatt 1997). It appears that conservation areas must be duplicated across the coffee forests to fully capture the variability through the forest gradients while exercising conservation efforts.

Broad-leaved species and coniferous species (*Podocarpus falcatus*) typically coexist in the canopy layers of coffee forests. It is noteworthy that *Pouteria adolfi-friederici* becomes more common in the southwest, but *Podocarpus falcatus* progressively gets scarce in the southwest and more prevalent in the southeast. Apart from a few forest tree species, for example, *Filicium decipiens,* coffee forests are floristically similar. Table 12.2 displays the coffee forests' most prevalent canopy tree, small trees, shrubs, herbs, lianas, and epiphytic species.

12.4 Coffee Forest Ecosystem Services

Forests provide a variety of ecosystem functions and services. Similarly, the Ethiopian coffee forests provide various ecosystem services at local, regional, and global levels. An overview of some of the important ecosystem services that coffee forests offer is described below.

Biodiversity Coffee forests harbor diverse biological resources. These forests are situated on mountainous terrain, allowing for extensive ecological gradients along the altitudes. Along these gradients, there are distinct vegetation units that sustain a variety of flora and fauna. As an example, around 10% of the 7600 plant species that make up the Eastern Afromontane Biodiversity hotspot were found in the coffee forests (Birdlife International 2012). The coffee forests support over 700 vascular plants at least 5% endemics (Senbeta 2006), 50 mammals (Agebo and Tekalign 2022), over 200 birds (EWCA 2013), 40 amphibian species (Mengistu et al. 2013), and many reptiles. The previous section of this chapter provides a list of characteristic plant species found in coffee forests. Coffee forests also support numerous endemic species of fish, reptiles, and amphibians, although complete data recording is lacking. Table 12.3 displays some of the animal species (Rodrigues et al. 2019; Agebo and Tekalign 2022), and Table 12.4 portrays some of the characteristic bird species recorded in the coffee forests (EWCA 2013; Exotic Birding 2023), respectively.

Diversity of arabica coffee Many scholars concur that arabica coffee's greatest diversity may be found in the montane forests of Ethiopia, where it is thought have originated. Various studies (e.g., Strenge 1956; Meyer 1965; Aga et al. 2003; Tesfaye 2006) have confirmed the presence of diverse arabica coffee genetic resources in the coffee forests of Ethiopia. According to Kotecha (2008), Ethiopia is assumed to have 99.8% of global arabica coffee genetic pools. The most significant gene pool for the crop is found in the coffee forests. According to Tesfaye (2006), considerable genetic variation exists among and between several forest-dwelling wild coffee

Table 12.2 The most characteristic tree, shrub, herb, liana, and epiphyte species in the coffee forests

Canopy tree species	Small tree species	Shrub species	Herb species	Liana species	Epiphyte species
Albizia gummifera	*Allophylus abyssinicus*	*Acalypha acrogyna*	*Adenostemma mauritianum*	*Clematis longicauda*	*Aerangis luteoalba*
Albizia schimperiana	*Bersema abyssinica*	*Acanthus eminens*	*Aframomum corrorima*	*Combretum paniculatum*	*Arthropteris monocarpa*
Apodytes dimidiata	*Blighia unijugata*	*Allophylus macrobotrys*	*Aneilema beniniense*	*Gouania longispicata*	*Asplenium aethiopicum*
Celtis africana	*Bridelia micrantha*	*Crossopteryx febrifuga*	*Asystasia schimperi*	*Hippocratea africana*	*Asplenium sandersonii*
Cordia africana	*Cassipourea malosana*	*Erythrococca abyssinica*	*Corymborkis corymbi*	*Hippocratea goetzei*	*Bulbophyllum intertextum*
Croton macrostachyus	*Chionanthus mildbraedii*	*Galiniera saxifraga*	*Desmodium hirtum*	*Jasminum abyssinicum*	*Drynaria volkensii*
Ekebergia capensis	*Coffea arabica*	*Halleria lucida*	*Hypoestes aristata*	*Landolphia buchananii*	*Loxogramme abyssinica*
Ilex mitis	*Dracaena afromontana*	*Maytenus arbutifolia*	*Impatiens hochstetteri*	*Oncinotis tenuiloba*	*Lycopodium clavatum*
Millettia ferruginea	*Ehretia cymosa*	*Ocimum urticifolium*	*Oxalis corniculata*	*Rubus apetalus*	*Peperomia tetraphylla*
Mimusops kummel	*Elaeodendron buchananii*	*Psychotria orophila*	*Pentas schimperiana*	*Rubus steudneri*	*Polystachya bennettiana*
Olea welwitschii	*Lepidotrichilia volkensii*	*Psychotria orophila*	*Peperomia abyssinica*	*Saba comorensis*	*Polystachya rivae*
Polyscias fulva	*Maesa lanceolata*	*Psydrax parviflora*	*Pilea tetraphylla*	*Sicyos polycanthus*	*Pteris catoptera*
Prunus africana	*Nuxia congesta*	*Rytigynia neglecta*	*Sanicula elata*	*Tiliacora troupinii*	*Tectaria gemmifera*
Sapium ellipticum	*Oxyanthus speciosus*	*Sida rhombifolia*	*Satureja simensis*	*Toddalia asiatica*	
Schefflera abyssinica	*Rothmannia urcelliformis*	*Vangueria apiculata*			
Trichilia dregeana	*Schrebera alata*				
	Teclea nobilis				
	Vepris dainellii				

populations. The author further argued that in contrast to the domesticated types or landraces grown in the world, wild arabica coffee populations are genetically different and diverse. Natural coffee populations in the forest have such great genetic diversity in part because of the region's extensive ecological heterogeneity in elevation and highly dissected and undulating topography.

Yet the narrow genetic material that was moved to Yemen from Ethiopia some centuries ago was what granted arabica coffee to expand throughout the world (Purseglove 1968; Ferwerda 1976). Although the exact date is unknown, it was

Table 12.3 Common mammal species of coffee forests

Common name	Scientific name
Aardvark	*Orycteropus afer*
Abyssinian genet	*Genetta abyssinica*
African buffalo	*Syncerus caffer*
African civet	*Civettictis civetta*
Baboon	*Papio anubis daguera*
Blue monkey	*Cercopithecus mitis*
Bohor reedbuck	*Redunca redunca*
Bushbuck	*Tragelaphus scriptus*
Bushpig	*Potamochoerus larvatus*
Chlorocebus	*Cercopithecus aethiops*
Colobus monkey	*Colobus guereza*
Common duiker	*Sylvicapra grimmia*
De Brazza's monkey	*Cercopithecus neglectus*
Ethiopian hare	*Lepus fagani*
Giant forest hog	*Hylochoerus meinertzhageni*
Greater kudu	*Tragelaphus strepsiceros*
Leopard	*Panthera pardus*
Lion	*Panthera leo*
Mountain nyala	*Tragelaphus buxtoni*
Porcupine	*Hystrix cristata*
Serval	*Leptailurus serval*
Shelmitmat	*Viverrdae sanguineus*
Warthog	*Phacochoerus aethiopicus*
Wild hunting dog	*Lycaon pictus*
Bats	

assumed/estimated that the first coffee bean arrived in Yemen around 575 AD. According to Ferwerda (1976), the coffee plant then moved to Java from Yemen in the late seventeenth century, then moved to the Amsterdam Botanical Garden around 1706, and then spread to Latin America in the early eighteenth century. Thus, the current global arabica coffee production system is said to be based on narrow genetic pools (Tesfaye 2006). Hence, we can confidentially argue that the Ethiopian wild arabica coffee populations offer a great opportunity for future coffee breeding and improvement programs (Meyer 1965; Woldemariam et al. 2002). The wild coffee populations in the coffee forests are therefore critical for international coffee breeding programs that strive to discover low caffeine content, drought tolerant, and disease resistant, and increase productivity, and other desirable traits (Wellman 1961; Schoen and Brown 1993; Nevo 1998; Silvarolla et al. 2004). For instance, the assessment of three breeding programs—low caffeine content, enhanced yields, and resistance to coffee berry disease and coffee rust—was used to determine the wild coffee genetic resources values. According to this valuation, coffee genetic resources have a resultant commercial value of between US$ 0.5 and

Table 12.4 Common bird species of coffee forests

Common name	Scientific name
Abyssinian black-headed oriole	*Oriolus monacha*
Abyssinian cat bird	*Parophasma galinieri*
Abyssinian ground hornbill	*Bucorvus abyssinicus*
Abyssinian ground thrush	*Turdus piaggiae*
African hill babbler	*Alcippe abyssinica*
Banded barbet	*Lybius guifsobalito*
Black-winged lovebird	*Agapornis taranta*
Blue winged goose	*Cyanochen cyanoptera*
Chestnut-naped francolin	*Francolinus leucoscepus*
Crested francolin	*Ortygornis sephaena*
Namaqua dove	*Oena capensis*
Olive-bellied sunbird	*Nectarinia olivacea*
Red-chested cuckoo	*Cuculus solitarius*
Rouget's rail	*Rougetius rougetii*
Scarce swift	*Apus myioptilus*
Sharpe's starling	*Cinnyricinclus sharpei*
Snowy-headed robin chat	*Cossypha niveicapilla*
Stuhlmann's starling	*Poeoptera stuhlmanni*
Tambourine dove	*Turtur tympanistria*
Thick-billed raven	*Caruvus crassirostris*
Wattled ibis	*Bostrychia carunculata*
White backed black tit	*Parus leuconotus*
White-billed starling	*Onychognathus albirostris*
White-checked turaco	*Tauraco ruspolii*
White-cheeked turaco	*Tauraco leucotis*
Yellow-billed duck	*Anas undulata*
Yellow-fronted parrot	*Poicephalus flavifrons*

US$ 1.5 billion (Hein and Gatzweiler 2006). This shows how economically valuable Ethiopia's coffee genetic resources are.

Wood and non-wood forest products Coffee forests offer economic, social, and environmental benefits to the adjacent communities and beyond. It is a source of livelihood for many people who dwell in or nearby the coffee forests (Ango et al. 2014; Shumi et al. 2019a; Schultner et al. 2021). It provides diverse wood and non-wood forest products that include wood, fuelwood, honey, coffee, spices, foods, medicine, fresh water, genetic resources, and others (see, e.g., Fig. 12.3). According to Seyoum (2010), non-timber forest products such as spices, coffee, and honey offer over 87% of the households' cash income, with coffee accounting for over 70% of this total revenue in some areas. Concurrently, tree and shrub species also provide a range of use categories, including fuelwood, building materials, household utensils, agricultural implements, poles and wood, animal fodder, and coffee shade, which are crucial for the livelihood of rural farmers in southwest Ethiopia (Shumi et al. 2019b).

Fig. 12.3 Two important spice species in the southwest coffee forests are *Aframomum corrorima* (left), *Piper capense* (middle), and *Coffea arabica* (right). (Photograph by Feyera Senbeta)

Coffee forests support various plant species that have medicinal value. Around 50 plant species from 31 families were recorded by the local community as therapeutic plants (Senbeta et al. 2013). These species have historically been used to treat a variety of human and animal illnesses, including rabies, viral diseases, headaches, stomachaches, and wounds. Some of these plant species include *Argomuellera macrophylla*, *Elaeodendron buchananii*, *Filicium decipiens*, *Landolphia buchananii*, *Lippia adoënsis*, *Macaranga capensis*, *Maesa lanceolata*, *Mimusops kummel*, *Ocotea kenyensis*, *Paullinia pinnata*, *Ricinus communis*, *Ritchiea albersii*, *Sapium ellipticum*, *Strychnos mitis*, *Trema orientalis*, *Trichilia prieuriana*, and *Warburgia ugandensis*.

Ecotourism Coffee forests have huge potential for ecotourism development, leisure activities, spiritual enlightenment, and aesthetic enjoyment. Ecotourism is a non-extractive type of tourism that is becoming more popular around the world. If ecotourism is properly planned and managed, it could help coffee forest conservation by generating finance. The natural and cultural elements of the coffee forests are an excellent resource to boost ecotourism. The coffee forests have excellent natural and cultural features for promoting ecotourism. Recently tourists, among others, are eager to observe wild coffee plants in their natural habitat (forest) and want to taste and drink coffee where it originated. Some of the main tourist attraction features of coffee forests, other than wild coffee plants, are the different coffee production systems, distinctive forest and agricultural landscapes, waterfalls, cultural/historical sites, and caves, as well as distinctive forest birds and other animals. Thus, guided cultural and nature tours have the potential to draw tourists into the coffee forest localities of Ethiopia.

Regulating and supporting services Coffee forests also preserve watersheds, uphold water quality, combat climate change, store carbon, and discharge and/or recharge streams. Coffee forests play a fundamental role in the protection of watersheds, helping to capture and transmit water and keep soil from eroding. The stabilization of water quality and preservation of the streams' and rivers' natural flow patterns are crucial functions of coffee forests (Umencdeo et al. 1993). According to available data, coffee forests provide a unique watershed management function.

This value is related to the presence and occurrence of high precipitation regimes in the area where coffee forests are found. Because of the presence of trees, net precipitation can increase absolutely. Although the amount of groundwater and stream flow may increase as a result, it is challenging to pinpoint exactly how this would affect the hydrological cycle. The impact will in any event vary from location to location based on elements including the frequency of wind, wind speed, size and direction of mountains, height, climatic variables, and vegetation types (Umencdeo et al. 1993; Stadtmuller 1987).

Tropical forests maintain 40–50% of the carbon produced by terrestrial vegetation (Erb et al. 2018). Likewise, coffee forests store and sequester carbon from the atmosphere and regulate air quality. According to Sanchez et al. (2021), montane forests in Africa including Ethiopian coffee forests, store 149.4 megagrams of above-ground carbon per hectare, which is more than many other forest types.

12.5 Coffee Management and Production Systems

The historical account of coffee domestication and use in Ethiopia is uncertain and requires investigation. Local sources, however, revealed that roasted coffee mixed with butter (locally known as *Buna Qalaa*) was an essential long-expedition nourishment since ancient times for the Oromos, the largest ethnic group in Ethiopia. Long before coffee became the beverage of choice for many, it was known that the Oromo used *Buna Qalaa* as a preservative for the food boluses when they were traveling on military expeditions during ancient times (Woldemariam 2015; Bulitta and Duguma 2021). Still, this practice is very common among the diverse Oromo communities in Ethiopia. The Oromo's emotional attachment to the plant and their long-standing and renowned tradition of drinking and/or directly consuming roasted coffee beans point to an earlier domestication and use of the plant by Oromos than other people (Bulitta and Duguma 2021). Coffee continues to hold a unique place in the Oromo people's daily lives during a spectrum of sociopolitical events, such as conflict resolution, social interaction, politics, festivals, and tragedy.

Coffee production and cultivation are thought to have taken on a traditional form before becoming a global phenomenon. Traditional coffee cultivation mimics the conditions found in coffee's original habitat in which people use their indigenous knowledge systems to cultivate the crop (Bulitta and Duguma 2021). The conventional method makes coffee a component of a complex agroforestry network system. Because the traditional growing approach does not focus on maximizing harvests for markets, production is typically low.

Today, coffee is grown in several geographical and agro-ecological zones of Ethiopia via different coffee production systems. When you travel across coffee-growing areas of the country, you can notice forest-based coffee production systems, mostly managed by smallholder farmers. These coffee production systems include (1) unmanaged coffee forests (coffee forest), (2) semi-managed coffee forests, (3) garden coffee, and (4) plantation coffee. Generally, Ethiopia's coffee production systems greatly support biodiversity and are environmentally friendly (Woldemariam 2003; Senbeta 2006; Tesfaye 2006; Labouisse et al. 2008; Rodrigues et al. 2018; Shumi et al. 2019b) as compared to global coffee production systems. A brief description of each coffee production is provided below.

Coffee Forest

In this system, no management techniques are employed to increase coffee yields; rather, coffee berries are harvested from naturally grown coffee plants (see Fig. 12.4). Species composition, diversity, and structure of the forest are less influenced by human activity. In a forest's stratification, which frequently consists of a variety of trees, coffee is one of the understorey plants. Wild coffee plants are distributed locally unevenly in the forests due to a variety of environmental factors, including habitat patchiness, substrate, dispersion restrictions, competition, and the extent of human intervention (Vuilleumier and Simberloff 1980). About 10% of the nation's total coffee production is thought to be covered by this system (ECTDMA 2015).

Semi-Managed Coffee Forest

In this system, coffee is planted or maintained from naturally grown coffee, but understory trees, shrubs, and lianas are routinely cut down to increase the yield of coffee (see Fig. 12.5). It employs a management system that transforms the coffee forests into a semi-managed coffee forest by removing undesired plants to make

Fig. 12.4 Bird's eye view of a typical coffee forest from Southwest Ethiopia. (Photograph by Feyera Senbeta)

Fig. 12.5 Semi-managed coffee forest system. (Photograph by Feyera Senbeta)

space for coffee plants which often reduce species diversity and composition (Woldemariam 2003). Due to the frequent removal of non-coffee smaller to medium plants, this system frequently obstructs regeneration, reduces density, and accelerates the extinction of some non-coffee species. Consequently, both vertical and horizontal structures of the semi-managed coffee forest system are dominated by coffee plants, unlike the coffee forest system. The density of shade trees is significantly diminished to limited plants per hectare. About 35% of the nation's total coffee is produced from this system (ECTDMA 2015).

Garden Coffee
This method of producing coffee involves managing and planting coffee plants in the vicinity of the farmers' homes (see Fig. 12.6). Since this system is well-managed, there is a dramatic decrease in both the diversity and density of non-coffee plants. Some farmers may grow improved coffee varieties in garden coffee system. Planting trees usually results in shade trees. Less number of shade trees are usually present per hectare. This system is responsible for roughly 15% of the nation's total coffee production (ECTDMA 2015).

Coffee Plantation
Intensively managed coffee production systems have also been expanding in the country in recent years (see Fig. 12.7). Plantation coffee is grown by private companies or government enterprises on a large-scale (Woldemariam 2015). Plantation coffee growers use improved coffee varieties and fertilizers (Tadesse et al. 2014). It

Fig. 12.6 Garden coffee system. (Photograph by Feyera Senbeta)

Fig. 12.7 Coffee plantation in southwest Ethiopia. (Photograph by Tadesse Woldemariam)

typically has good management and makes up 15% of the overall coffee production in the country (ECTDMA 2015). This production system still supports a high diversity of shade trees either planted or naturally grown leftover trees.

12.6 Threats to Coffee Forests

Over the past century, due to human activities, Ethiopian forests have experienced severe rates of degradation and deforestation including coffee forests (Logan 1946; McCann 1997; Tadesse et al. 2008; Hylander et al. 2013; FAO 2020). Human activities such as habitat loss and fragmentation (land-use change), over-exploitation, and variables connected to climate change have been and remain the primary causes of forest loss in Ethiopia. The loss of forests has significantly disrupted ecosystem services and severely exacerbated soil erosion, hydrological imbalance, loss of biodiversity, and accelerated poverty in many parts of the country (Woldemariam et al. 2002; Davis et al. 2012; Shumi et al. 2018; FAO 2020; Sanchez et al. 2021). Likewise, threats to coffee forests owing to logging, climate change effects, human settlement, and the expansion of agriculture to forest lands have accelerated in recent years. The following paragraphs provide a summary of the major causes of the destruction and clearing of coffee forests in Ethiopia.

Agricultural expansion The expansion of smallholder farms and commercial coffee and tea plantations has been continuing to be the principal threats to coffee forests (see Fig. 12.8). For instance, in southwest Ethiopia since the early 1970s, approximately 100,000 ha of natural coffee forests have been cleared for commercial tea and/or coffee plantations expansion (Woldemariam et al. 2002; Senbeta 2006). According to Logan (1946), a sizable block of forest encompassed a significant portion of southwestern Ethiopia throughout the 1940s and 1950s. After

Fig. 12.8 An example of commercial tea plantation and smallholder farm expansion to the coffee forest, in southwest Ethiopia. (Photograph by Feyera Senbeta)

60–70 years, the forest area has now been reduced to patchy relics. Owing to their proximity to the agricultural frontiers, several of these fragments are currently undergoing additional fragmentation. According to a study by Tadesse et al. (2008), the forest cover around Mizan Teferi town decreased from 71% to 48% between the years 1973 and 2005 owing to political, social, or economic factors. Many farmers transformed the forest into a settlement, agriculture, and coffee agroforestry system. The future of coffee forests has also been dubious because of the recent expansion of land appropriation for commercial tea and coffee plantations (Ango 2018).

In addition to the coffee forest appropriation for other uses, the ongoing intensification of coffee management by smallholder coffee farmers has also contributed to a homogenization of the coffee forest biodiversity (Senbeta 2006; Hylander et al. 2013; Tadesse et al. 2014). Recent studies (e.g., De Beenhouwer et al. 2015; Senbeta 2018; Zewdie et al. 2022) have shown a decline in forest biodiversity as coffee management intensity increased in southwest Ethiopia. However, the rate of deforestation was relatively lower in the altitudinal belt where coffee grows best compared to the forest outside of the coffee-growing zone (Hylander et al. 2013).

Settlement At least four settlement initiatives that were executed in Ethiopia over the past seven decades have significantly altered the land use patterns and forest ecosystem. With the change of government in 1974, all land was first nationalized alongside new land reforms, and peasant associations were created to distribute land among their members in Ethiopia. Due to the change in land use rights and the resulting first-time access to land for peasants, this proclamation resulted in a migration movement from the northern highlands and other parts of the country to the western highlands where forest was abundant (McCann 1995). The Ethiopian Government established a relocation program that allowed people to move from drought-prone to drought-unaffected areas because of the 1984–1985 drought. Following that, in 1985/86, the Ethiopian Government unveiled a brand-new settlement plan known as villagization. The goal of the villagization was to create small towns out of isolated homesteads so that locals could more easily access resources like clean water, medical care, and power (McCann 1995). Again, due to drought-related issues, a sizable population moved to non-drought-prone areas in the late 1990s and early 2000s. All these policies, initiatives, and modifications were impromptu and often ill-planned. These were all followed by forest degradation and deforestation, particularly the coffee forest because of its conducive environment for living. As a result, many people moved into the coffee forest, resided there, and transformed it into non-forested land.

Fig. 12.9 Illegal logging in the southwest coffee forest. (Photograph by Feyera Senbeta)

Logging Likewise, logging has also devastated a significant portion of the coffee forests, in particular illegal logging (see Fig. 12.9). In Ethiopia, logging was one of the primary drivers of forest loss. Along with logging, access to the forest is made possible by the roads built for logging by landless people. Farmers removed the trees from lands that had once been cut by saw millers in several areas of the southwest coffee forest.

Climate change and forest fire On top of land use pressure, recent climate change has put immense pressure on coffee forests and their associated biodiversity. Different models showed the susceptibility of coffee forests to climate change (Davis et al. 2012). According to the present wild arabica coffee occurrences, ecological tolerances, and bioclimatic unsuitability studies, it was predicted that due to climate change and associated stress, wild arabica coffee populations will have a high probability of extinction soon in Ethiopia. Continuous deforestation and forest degradation are also expected to increase greenhouse gas emissions (IPCC 2007). According to Geeraert et al. (2019), forest loss can contribute 30% of global greenhouse gas emissions annually. In addition to agricultural landscape expansion, wildfires, and prescribed burning can have considerable detrimental effects on coffee forests. Recently, some of the coffee forest areas are experiencing frequent drought and this has aided the repeated occurrence of forest fires. Frequent drought conditions in some coffee forest areas have increased the risk of forest fires. Figure 12.10 shows an example of a forest fire incidence in the Gesha coffee forest, southwest Ethiopia.

Fig. 12.10 Forest fire in the Gesha coffee forest in southwest Ethiopia. (Photograph by Feyera Senbeta)

12.7 Sustainable Management and Use

Ethiopia has been working to conserve coffee forests and other forest ecosystems since the 1980s when the Department of Forestry designated 58 Natural Forest Priority Areas (NFPAs), which were thought to cover around 2.5 million hectares of natural forests (EFAP 1994). The NFPA designation aimed to improve the nation's forest protection, production, and conservation. The NFPA included many coffee forest fragments, such as Harenna, Setema, Yayu, Belete-Gera, Bonga, and Sheko forests. However, the designation of NFPAs never improved conservation efforts (Labouisse et al. 2008; Tadesse et al. 2008). Local populations living in and around the forests were hardly involved in the process; and the approach was command-and-control type with due emphasis on "protection," which led to major forest degradation and deforestation in many of the designated forests, including coffee forests.

Back in 1998, three coffee forest fragments in southwest Ethiopia, namely, Berhan-Kontir, Boginda-Yeba, and Geba-Doggi, were selected for the conservation of coffee forest genetic resources with financial support from the European Union.

Due to ineffective administration and coordination among the implementing partner institutions, the initiative hardly succeeded after 10 years of execution (Labouisse et al. 2008). Again, in the late 1990s and early 2000s, the Participatory Forest Management (PFM) scheme was introduced as a prominent method of forest management in Ethiopia. In contrast to the early command-and-control type of management, PFM was introduced to stop the problem of deforestation and to improve the livelihoods of forest-dependent communities. Since then, numerous PFM projects have been initiated by various NGOs in coordination with the Department of Forestry in many parts of the country (Gobeze et al. 2009; Lemenih et al. 2015). In this context, several studies have indicated a favorable impact of PFM on forest conditions (Ameha et al. 2014; Tadesse et al. 2016), whereas other studies have reported the opposite (Mohamed 2006). To ascertain the true impact of PFM interventions on the preservation of forest ecosystems and, consequently, on the protection of coffee forests, more research is required.

Back in 2002, a collaborative research project entitled "Conservation and use of the wild populations of *Coffea arabica* in the montane rainforests of Ethiopia" was launched by the Centre for Development Research (ZEF), University of Bonn, the Institute of Biodiversity Conservation, the Addis Ababa University, and the Ethiopian Institute of Agricultural Research as main partners. Phase I of this project concentrated on and covered the following six study themes: forest biodiversity, coffee genetics and genomics, coffee ecophysiology, coffee pathology, socioeconomic and institutional aspects of coffee forests (ZEF 2006). The project's final report suggested the UNESCO Man and the Biosphere Programme as a viable option for the conservation of coffee forests.

Subsequently, the Kaffa and Yayu Coffee Forest Biosphere Reserves were established in 2010 as a mechanism for sustainable development and preservation of the coffee forests. Two more UNESCO biosphere reserves have recently been nominated in the coffee forests. This strategy was expected to promote local development through market linkage, local product certification, and value addition and enhance in situ conservation. Yet there are no guarantees that coffee forests and their genetic resources would be conserved because of the establishment of biosphere reserves (Aerts et al. 2015).

Finally, many actions are required to enhance management while also enhancing the local communities' ability to support themselves in and around the forest. To encourage the sustainable management and use of coffee forests, the following issues must be considered: (1) participatory development and promoting the potential wealth of the coffee genetic resource into actual benefits for the local communities living in the area through developing environmental marketing schemes is required (e.g., eco-labeling, certification of wild coffee, water trading, and ecotourism); and (2) promoting financial incentives for conservation of wild coffee and the forests through the development of international fund-raising approaches for financing the conservation of unique coffee forest ecosystem; and designing conservation strategy that empowers local people and also provide safeguards against overuse of coffee forests is quite essential.

12.8 Conclusion

Coffee forests have proven to be a significant ecosystem for storing biodiversity and the only refugia for populations of wild arabica coffee. Since coffee forests are the final resort for maintaining wild arabica coffee populations, they must be preserved as forest ecosystems to protect a variety of species that occupy various forest niches including coffee. Coffee forests also have a wider range of other commercially valuable plant species that the local population depends on for their survival. Despite their significance for the environment and the economy, coffee forests are in danger of disappearing because of land use pressures such as habitat destruction, commercial coffee and tea plantations, and agricultural expansion. It is projected that climate change could pose greater threats to the range of wild coffee populations and their ecosystem. Conservation activities have been continuing for the past three decades to address these challenges with limited success. The long-term survival of the coffee forests, therefore, requires additional and substantial conservation initiatives. Reducing greenhouse gas emissions and the preservation of forest biodiversity both require reduction of deforestation and forest degradation as well as sustainable forest management and use.

References

Aerts A, Berecha G, Honnay O (2015) Protecting coffee from intensification. Science 347(6218):139

Aga E, Bryngelsson T, Bekele E, Salamon B (2003) Genetic diversity of forest coffee (*Coffea arabica* L.) in Ethiopia as revealed by random amplified polymorphic DNA (RAPD) analysis. Hereditas 138:36–46

Agebo A, Tekalign W (2022) Terrestrial medium and large-sized mammalian species diversity in Michole Community Protected Forest, Southern Ethiopia. BMC Zool 7:20

Ameha A, Larsen O, Lemenih M (2014) Participatory forest management in Ethiopia: learning from pilot projects. Environ Manag 53:838–854

Ango TG (2018) "Medium-scale" forestland grabbing in the southwestern highlands of Ethiopia: impacts on local livelihoods and forest conservation. Land 7(1):24

Ango TG, Börjeson L, Senbeta F et al (2014) Balancing ecosystem services and disservices: smallholder farmers' use and management of forest and trees in an agricultural landscape in southwestern Ethiopia. Ecol Soc 19(1):30

Ango TG, Hylander K, Börjeson L (2020) Processes of forest cover change since 1958 in the coffee-producing areas of Southwest Ethiopia. Land 9:278

Anthony F, Bertrand B, Quiros O et al (2001) Genetic diversity of wild coffee (*Coffea arabica* L.) using molecular markers. Euphytica 118:53–65

Asfaw G (2003) Breaking the cycle of recurrent famine in Ethiopia: natural resources management and drought-related famine prevention, Addis Ababa

Birdlife International (2012) Ecosystem profile: eastern Afromontane biodiversity hotspot. Birdlife International

Bulitta BJ, Duguma LA (2021) The unexplored socio-cultural benefits of coffee plants: implications for the sustainable management of Ethiopia's coffee forests. Sustain For 13(7):3912

COCE (2004) Conservation and use of wild populations of *Coffea arabica* in the montane rainforests of Ethiopia (COCE) 2002–2006: an Ethiopian-German collaborative research project. Hosted by ZEF, University Bonn

Cowling RM (1990) Diversity components in a species-rich area of the Cape Floristic Region. J Veg Sci 1:699–710

Darbyshire I, Lamb H, Umer M (2003) Forest clearance and regrowth in northern Ethiopia during the last 3000 years. The Holocene 13(4):537–546

Davis AP, Woldemariam T, Baena S et al (2012) The impact of climate change on indigenous Arabica coffee (*Coffea arabica*): predicting future trends and identifying priorities. PLoS One 7(11):e47981

De Beenhouwer M, Aerts R, Hundera K et al (2015) Management intensification in Ethiopian coffee forests is associated with crown habitat contraction and loss of specialized epiphytic orchid species. Basic Appl Ecol 16:592–600

ECTDMA (2015) Ethiopian coffee & tea development and marketing authority, Addis Ababa

EFAP (1994) Ethiopian forestry action program. Final report, vol. II – the challenge for development. Transitional Government of Ethiopia, Ministry of Natural Resources Development and Environmental Protection, Addis Ababa

Erb K, Kastner T, Plutzar C et al (2018) Unexpectedly large impact of forest management and grazing on global vegetation biomass. Nature 553:73–76

EWCA (2013) Bale Mountains National Park: a traveller's guidebook. The Ethiopian wildlife conservation authority (EWCA), Addis Ababa

Ewel J (1983) Succession. In: Golley GB (ed) Tropical rainforest ecosystems: structure and function. Elsevier, Amsterdam, pp 9–26

EWNHS (1996) Important bird areas of Ethiopia. Ethiopian Wildlife and Natural History Society (EWNHS), Addis Ababa

Exotic Birding (2023) Bird checklist for Harenna Forest. http://www.exoticbirding.com/ethiopia/harenna-forest/checklist.html. Accessed 1 Mar 2023

FAO (2020) Global forest resources assessment 2020: main report. Food and Agriculture Organization of the United Nations, Rome

Ferwerda FP (1976) Coffees Coffea spp. (Rubiaceae). In: Simmonds NW (ed) Evolution of crop plants. Longman, London, pp 257–260

Friis I (1992) Forests and forest trees of northern east tropical Africa. Kew Bull Additional Series, London

Gaston KJ (1994) Rarity. Chapman and Hall, London

Geeraert L, Hulsmans E, Helsen K et al (2019) Rapid diversity and structure degradation over time through continued coffee cultivation in remnant Ethiopian Afromontane forests. Biol Conserv 236:8–16

Gobeze T, Bekele M, Lemenih M et al (2009) Participatory forest management and its impacts on livelihoods and forest status: the case of Bonga forest in Ethiopia. Int For Rev 11(3):346–358

Goldblatt P (1997) Floristic diversity in the Cape flora of South Africa. Biodivers Conserv 6:359–377

Hein L, Gatzweiler F (2006) The economic value of coffee (Coffea arabica) genetic resources. Ecol Econ 60:176–185

Hundera K, Aerts R, Fontaine A et al (2013) Effects of coffee management intensity on composition, structure and regeneration status of Ethiopian moist evergreen Afromontane forests. Environ Manag 51:801–809

Hylander K, Nemomissa S, Delrue J et al (2013) Effects of coffee management on deforestation rates and forest integrity. Conserv Biol 27(5):1031–1040

IPCC (2007) Climate change 2007: the fourth assessment report of the intergovernmental panel on climate change. IPCC, Geneva

IUCN (1990) Ethiopia national conservation strategy phase one. March 1990, Gland

Kotecha S (2008) Arabica's from the Garden of Eden-Coffea Aethiopica. Available via DIALOG. https://revistacafeicultura.com.br/index.php?mat=24556. Accessed 20 Nov 2022

Labouisse JP, Bellachew B, Kotecha S et al (2008) Current status of coffee (*Coffea arabica* L.) genetic resources in Ethiopia: implications for conservation. Genet Resour Crop Evol 55:1079–1093
Lemenih M, Allan C, Biot Y (2015) Making forest conservation benefit local communities: participatory forest management in Ethiopia. Farm Africa
Logan WEM (1946) An introduction to the forest of central and Southern Ethiopia. Imp For Inst Paper 24:1–58
McCann JC (1995) People of the plow: an agricultural history of Ethiopia, 1800–1990. University of Wisconsin Press, Madison
McCann JC (1997) The plow and the forest: narratives of deforestation in Ethiopia 1840–1992. Envi Hist 2(2):138–159
Mengistu AA, Nagel P, Getahun A et al (2013) Review article updated review of amphibian diversity, distribution, and conservation in Ethiopia. Ethiop J Biol Sci 12(1):81–116
Meyer FG (1965) Notes on wild Coffea arabica from Southwestern Ethiopia, with some historical considerations. Econ Bot 19:136–151
Mittermeier RA, Turner WR, Larsen FW et al (2011) Global biodiversity conservation: the critical role of hotspots. In: Zachos FE, Habel JC (eds) Biodiversity hotspots: distribution and protection of conservation priority areas. Springer, Berlin, pp 3–22
Mohamed BS (2006) Impact of joint forest management on Handeni hill forest reserve and adjacent communities. Sokoine University of Agriculture, Morogoro
Mohr PA (1965) Reclassification of the Ethiopian Cenozoic volcanic succession. Nature 208:71–78
Mohr PA (1971) The geology of Ethiopia, 2nd edn. University College of Addis Ababa Press, Addis Ababa
Murphy HE (1968) A report on the fertility status and other data on some soils of Ethiopia. College of Agriculture, Haile Sellassie I University, Dire Dawa
Nevo E (1998) Genetic diversity in wild cereals: regional and local studies and their bearing on conservation ex-situ and in situ. Genet Resour Crop Evol 45:355–370
Purseglove JW (1968) Tropical crops. In: Dicotyledons, 2nd edn. Longmans, Green and Co. Ltd, London
Rodrigues P, Shumi G, Dorresteijn I et al (2018) Coffee management and the conservation of forest bird diversity in south-western Ethiopia. Biol Conserv 217:131–139
Rodrigues P, Dorresteijn I, Senbeta FJ (2019) Living on the edge: rapid assessment of the mammal community in a coffee forest in southwestern Ethiopia. Afr J Ecol 57(2):279–285
Sanchez C, Sullivan MJP, Platts PJ et al (2021) High aboveground carbon stock of African tropical montane forests. Nature 596:536–542
Schmitt C (2006) Montane rainforest with wild *Coffea arabica* in the Bonga region (SW Ethiopia): plant diversity, wild coffee management and implications for conservation. Cuvillier Verlag, Göttingen
Schmitt CB, Senbeta F, Denich M et al (2009) Wild coffee management and plant diversity in the montane rainforest of southwestern Ethiopia. Afr J Ecol 48:78–86
Schoen DJ, Brown AHD (1993) Conservation of allelic richness in wild crop relatives is aided by assessment of genetic markers. Proc Natl Acad Sci USA 90:10623–10627
Schultner J, Dorresteijn I, Manlosa AO et al (2021) Ecosystem services from forest and farmland: present and past access separates beneficiaries in rural Ethiopia. Ecosyst Serv 48:101263
Senbeta F (2006) Biodiversity and ecology of Afromontane rainforests with wild *Coffea arabica* L. populations in Ethiopia. Cuvillier Verlag, Göttingen
Senbeta F (2018) Diversity, threats, and conservation status of coffee forest in Ethiopia. Ethiop J Biol Sci 17(2):169–188
Senbeta F, Woldemariam T, Kelbessa E et al (2013) Diversity of useful plants in the coffee forest of Ethiopia. J Ethnobiol Res Appl 11:049–069
Seyoum A (2010) Microeconomics of wild coffee genetic resources conservation in southwestern Ethiopia: Forest zoning and economic incentives for conservation. Shaker Verlag, Germany

Shumi G, Schultner J, Hanspach J et al (2018) Land use legacy effects on woody vegetation in agricultural landscapes of SW Ethiopia. Divers Distrib 24:1136–1148

Shumi G, Dorresteijna I, Schultner J et al (2019a) Woody plant use and management in relation to property rights: a social-ecological case study from southwestern Ethiopia. Ecosyst People 15(1):303–316

Shumi G, Rodrigues P, Schultner J et al (2019b) Conservation value of moist evergreen Afromontane Forest sites with different management and history in southwestern Ethiopia. Biol Conserv 232:117–126

Silvarolla MB, Mazzafera P, Fazuoli LC (2004) A naturally decaffeinated arabica coffee. Nature 429:826

Spies TA (1997) Stand structure, function, and composition. In: Kohm KA, Franklin JF (eds) Creating a forestry for the 21st century: the science of ecosystem management. Oxford University Press, pp 11–30

Spies TA (1998) Forest structure: a key to the ecosystem. Northwest Sci 72(2):34–39

Stadtmuller T (1987) Cloud forests in the humid tropics: a bibliographic review. United Nations University, Tokyo

Strenge H (1956) Wild coffee in Kaffa Province of Ethiopia. Trop Agr 33:297–301

Tadesse D, Woldemariam T, Senbeta F (2008) Deforestation of Afromontane rainforests in Ethiopia. In: Sanchez IB, Alonso CL (eds) Deforestation research progress. Nova Science Publishers, pp 19–39

Tadesse G, Zavaleta E, Shennan C (2014) Coffee landscapes as refugia for native woody biodiversity as forest loss continues in southwest Ethiopia. Biol Conserv 169:384–391

Tadesse S, Woldetsadik M, Senbeta F (2016) Impacts of participatory forest management on forest conditions: evidence from Gebradima Forest, southwest Ethiopia. J Sustain For 35(8):604–622

Tesfaye K (2006) Genetic diversity of wild *Coffea arabica* populations in Ethiopia as a contribution to conservation and use planning. Cuvillier Verlag, Göttingen

Thiel S, Tschapka M, Heymann EW et al (2021) Vertical stratification of seed-dispersing vertebrate communities and their interactions with plants in tropical forests. Biol Rev 96:454–469

Umencdeo C, Ilmouk N, Pereza M, Lockhus N (1993) Tropical montane cloud forests: conservation status and management issues. In: Hamilton LS, Juvik JO, Scatena FN (eds) Tropical montane cloud forests. East-West Center, Program on Environment, pp 17–24

Umer M, Bonnefille R (1998) A late glacial/late Holocene pollen record from highland peat at Tamsaa, Bale Mountains, South Ethiopia. Glob Planet Change 16/17:121–129

Vuilleumier F, Simberloff D (1980) Ecology versus history as determinants of patchy and insular distribution in high Andean birds. Evol Boil 12:235–379

Wellman FL (1961) Coffee: botany, cultivation and utilization. Leonard Hill Books, London

Willson KC (1985) Climate and soil. In: Clifford MN, Willson KC (eds) Coffee: botany, biochemistry and production of beans and beverage. Croom Helm, London, Sydney, pp 97–107

Woldemariam T (2003) Vegetation of the Yayu Forest in SW Ethiopia: impacts of human use and implications for in situ conservation of wild *Coffea arabica* L. populations. Cuvillier Verlag, Göttingen

Woldemariam T (2015) Coffee production systems in Ethiopia. Ethiopian Coffee Forest Forum, Addis Ababa

Woldemariam T, Denich M, Teketay D et al (2002) Human impacts on *Coffea arabica* L. genetic pools in Ethiopia and the need for its in-situ conservation. In: Engels JMM, Rao VR, Brown AHD, Jackson MT (eds) Managing plant genetic diversity. IPGRI, Italy, pp 237–247

ZEF (2006) ZEF annual report 2005/2006: knowledge and development. Centre for Development Research, University of Bonn

Zewdie B, Tack AJM, Ayalew B et al (2022) Plant biodiversity declines with increasing coffee yield in Ethiopia's coffee agroforests. J Appl Ecol 00:1–11

Feyera Senbeta received his BSc in Forestry from the Alemaya University of Agriculture in 1994, MSc in Natural Forest Management from the Swedish University of Agricultural Sciences in 1998, and PhD in Natural Sciences from the University of Bonn in 2006. He was appointed at Addis Ababa University in 2008 as an Assistant Professor and was promoted to Associate Professor and Professor in 2014 and 2021, respectively. Since 2022, he has been working at the Botswana University of Agriculture and Natural Resources. He has authored or co-authored over 120 journal articles, edited books, book chapters, and conference proceedings. He is on Editorial Boards of the *Journal of Forestry and Natural Resources* and the *Ethiopian Journal of Development Research*.

Chapter 13
Impacts of Forest Fires in Botswana: Implications for Forest Resources Conservation and Management

Keoikantse Sianga, Witness Mojeremane, Ayana Angassa Abdeta, Kamogelo Makgobota, and Summer Mabula

13.1 Introduction

The incidence of fire has been an essential process affecting the terrestrial ecosystem for over 350 million years (Santín and Doerr 2016). According to Santín and Doerr (2016), humans have always experienced fire in fire-prone ecosystems, and its use can be noted as the first tool utilized to induce ecosystem dynamism. The literature demonstrates that globally, forest fires are predominantly caused by lightning (i.e., natural fires) (Komarek 1971; Pyne and Laven 1996; Scott et al. 2013; Moris et al. 2020). For example, fires common in the boreal forest of North America (Nash and Johnson 1996; Kasischke et al. 2002; Veraverbeke et al. 2017; Hanes et al. 2019) and in the Alps (Müller et al. 2013; Vacchiano et al. 2018) are induced by lightning. Both in the literature and media, the effects of uncontrolled fire on ecosystems have been considered a problem, with commonly held views of increasing fire occurrence, intensity, hazard and resulting losses (Santín and Doerr 2016).

Throughout the world, vast coverage of forestlands has been burned and destroyed by uncontrolled fires over centuries, causing serious environmental and economic problems (Rowell and Moore 2000; Lemessa and Perault 2001). Considerable losses of forest resources and biological diversity have resulted in

K. Sianga (✉)
Botswana University of Agriculture and Natural Resources, Department of Wildlife and Aquatic Resources, Gaborone, Botswana
e-mail: ksianga@buan.ac.bw

W. Mojeremane · A. A. Abdeta · K. Makgobota · S. Mabula
Botswana University of Agriculture and Natural Resources, Department of Range and Forest Resources, Gaborone, Botswana
e-mail: wmojerem@buan.ac.bw; aabdeta@buan.ac.bw; kmakgobota@buan.ac.bw; smabula@buan.ac.bw

W. L. Hambira et al. (eds.), *Environmental Change and Biodiversity Conservation in sub-Saharan Africa*, Advances in Global Change Research 75,
https://doi.org/10.1007/978-3-031-73136-5_13

environmental changes and the disruption of ecosystem services (Rowell and Moore 2000). Furthermore, the challenges and increasing patterns of forest and grassland fires are becoming serious concerns around the world (Rowell and Moore 2000; van Wees et al. 2021). For example, certain ecosystems (e.g., the Amazon Rainforest in Brazil and the cloud forest of Chiapas in Mexico) that sustained considerable damage from fires have recently been vulnerable and subjected to fires due to the impacts of climate change (Rowell and Moore 2000; van Wees et al. 2021). The damage caused by widespread fires on various continents is evident. According to Giglio et al. (2013), forest fires comprised the largest proportion of the area burned in Equatorial Asia (62%), followed by North America (48%) and Asia (36%). Grassland fires were the most prevalent in Central Asia and South America, while cropland fires were dominant in Europe and the Middle East. Shrubland burning comprised a relatively small proportion of the area burned in all regions except Australia, where 37% of the area burned belonged to the shrubland category. Above all, the savanna of Africa is considered the largest proportion (71%) of the area burned on the global scale, and the major fire of 2000 that occurred in Ethiopia is a good example (Lemessa and Perault 2001).

Fire is considered one of the key drivers in the dynamics of vegetation and environmental changes (Scott et al. 2000) and ecological interactions. Some studies have demonstrated that fire can destroy approximately 80% of ecosystem components, including forests and related resources (Bond and Van Wilgen 1996). On the other hand, a recent study reported an important role of fire in shaping the evolution of terrestrial plants (Bond and Scott 2010) and hence a major driver of trait innovations among terrestrial biodiversity (Crisp et al. 2011; He et al. 2011, 2012; Lamont et al. 2013). The magnitude of vegetation losses to fire fluctuates among plant communities and species. Some species of plants are more susceptible to fire than others with flammable leaves, shoots, and litter (Bond and Van Wilgen 1996).

13.2 Methodology

This chapter aims to review the effects of fire on forest resources in Botswana to inform sustainable policies for possible management scenarios. This chapter presents historical perspectives about fire and its impacts on ecosystem dynamics and plant communities and policy directions to minimize fire hazards for a more sustainable coexistence with the incidence of a forest fire.

13.3 Results and Discussion

13.3.1 Causes of Forest Fire in Botswana

Botswana is one of the sub-Saharan countries that is affected by large wildfires annually (Fig. 13.1) (Botswana Environment Statistics 2017; Maabong and Mphale 2021). Forest fires across districts in Botswana occur in two ways: those ignited by lightning (i.e., natural) and those ignited by man (Dube 2013; Fox et al. 2017; Maabong and Mphale 2021). Most forest fires are attributed to human activities such as using fire to clear land and remove dead biomass for improved cultivation, grazing and fishing (Dwyer et al. 2000; Andreae and Merlet 2001; Dube 2013; Chinamatira et al. 2016; Cassidy et al. 2022), eliminating disease-bearing insects and snakes, smoking out beehives, making charcoal, cooking, and warming (Dube 2013; Chinamatira et al. 2016).

Some forest fires have been occurring due to anti-social behaviors such as careless disposal of cigarettes, arson, poachers flushing out the game, and inappropriate

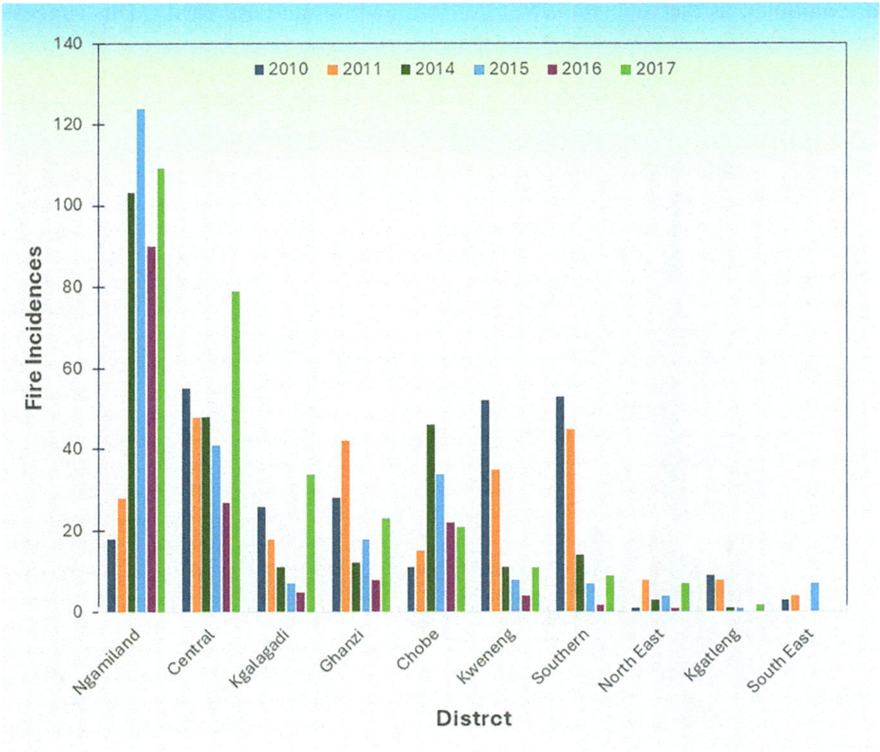

Fig. 13.1 Recorded number of fire incidents by districts between 2010 and 2017. (Data sources: Botswana Environment Statistics 2017)

fire handling by children (Dube 2013; Department of Forestry and Range Resources (DFRR) 2016). In addition, the DFRR (2016) noted that tourists and truck drivers in the Chobe region (Botswana) often make fires in open areas with a heavy load of dead wood, herbaceous biomass, and litter to cook and scare wild animals at night and thereafter do not completely extinguish them. These unextinguished fires end up escaping and igniting adjacent dry light biomass, resulting in large fires. Similarly, the Chobe region experiences cross-border fires, probably because it shares international borders with Namibia, Zambia, and Zimbabwe, and as such, cross-border fires are evident (Fox et al. 2017). It is also possible that fires in the Chobe region are set by illegal hunters to attract wild animals into burned areas. In areas around the Okavango Delta, fire is set illegally by hunters or photographic safari hunters' staff often prior to the arrival of the seasonal flood to draw in game closer to their safari camps (Cassidy 2009).

A review of the literature demonstrated that between 2006 and 2010 and 2012 and 2017 (Fig. 13.2), variable amounts of hectares in Botswana were consumed by fires caused by various means (Dube 2013; Botswana Environment Statistics 2017). Natural fires caused by lightning are rare and not considered a serious problem in Botswana because they usually occur during the rainy seasons when their impacts are minimal, as the vegetation will be fresh and do not burn easily. This suggests

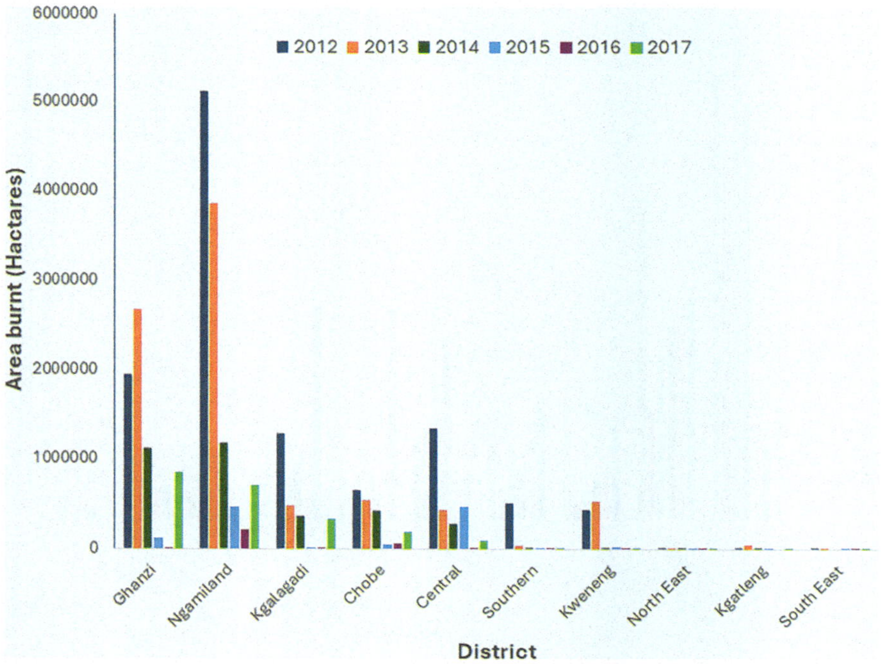

Fig. 13.2 Estimated area burnt (hectares) per district between 2012 and 2017. (Data sources: Botswana Environment Statistics 2017)

that most fire ignitions in Botswana are from anthropogenic sources (Fox et al. 2017; Maabong and Mphale 2021).

In addition, spontaneous natural peat fires reported in some parts of the Okavango Delta (Botswana) during very hot summers in dry years (Ellery et al. 1989; Gumbricht et al. 2002; Dube 2013) are uncommon and infrequent. Fires in Botswana usually start between April and June, but they are not as intense as fires experienced in the late season from August to October (Mojeremane 2004). Late season fires are uncontrollable because they often occur when the vegetation is dry and due to heat and strong winds (Maabong and Mphale 2021).

13.3.2 The Impacts of Fires on Biodiversity in Botswana

Botswana experiences patches of forest fires during the dry season (Heinl et al. 2006). These fires are often frequent and intensive in regions with higher rainfall, such as the gradient between the Chobe Forest region and the Okavango Delta in the Ngamiland region. This is because the accumulation of biomass due to adequate precipitation and higher soil moisture around wetlands enhances grass growth in wet and dry seasons, providing adequate fuel load for forest fires. These fires result in the loss of habitats and forage and hence negatively impact wildlife. Often, these fires are associated with ecosystem degradation and ultimately the loss of wildlife species that depend upon these resources (Dube 2007, 2009; Batisani and Yarnal 2010; DFRR 2011). Some studies conducted in the Chobe Forest Reserve (Botswana) demonstrated that fires that frequent this region, in combination with elephant herbivory (pyric herbivory), have elevated the mortalities of *Baikiaea plurijuga* and *Pterocarpus angolensis* (Ben-Shahar 1996, 1998; Mmolotsi et al. 2012; Fox et al. 2017), reducing plant diversity in the region. Fire is a factor that restricts the diversity and abundance of tree species, preventing the formation of canopies (Higgins et al. 2000; Sankaran et al. 2004). Large trees of these species can also be killed by fire (Van der Walt and le Riche 1984; Seymour and Huyser 2008). Seymour and Huyser (2008) reported that well-established *Vachellia erioloba*, which are an important keystone species in the region, were killed by frequent fires. This suggests that an increase in fire frequency could have implications for biodiversity loss.

Fires in the region have decreased the survivorship of year-old seedlings of *B. plurijuga* and *P. angolensis* and others, limiting the regeneration capabilities of these forest ecosystems in northern Botswana (Ben-Shahar 1996, 1998; Rietbergen-McCracken and Abaza 2014) and thus compromising the foraging and browsing patterns of large herbivores such as eland (*Taurotragus oryx*), roan (*Hippotragus equinus*), sable (*Hippotragus niger*), and kudu (*Tragelaphus strepsciceros*). However, it must be noted that in heterogeneous savanna ecosystems where there is a higher density of herbivores, such as in northern Botswana, it is difficult to isolate the impacts of fires from pyric herbivory (Ben-Shahar 1998; Georgiadis and McNaughton 1990; Mmolotsi et al. 2012). Similarly, these forest fires in Botswana negatively impact the livelihoods of local communities. For example, products such as thatching grass, food and medicinal plants, which are seasonal and valuable to communities, are always consumed by fires, thus affecting their livelihoods (Parry and Campbell 1990; Lepetu et al. 2009). Similarly, MODIS satellite imagery (2008) demonstrated that a mega fire in the Ghanzi (Botswana) region burned approximately 3.6 million hectares and extended into the inner parts of the Central Kalahari Game Reserve (CKGR) (Williams et al. 2011).

The damage caused by this mega fire was estimated at $U239 000, a cumulative estimate of the loss of grazing and tourism resources (DFRR 2007). However, some studies conducted in Botswana demonstrated positive impacts of fire on the ecosystems. For example, in the Koanaka Hills (northwestern Botswana), the occurrence of fires was associated with an increase in the species richness of herpetofauna (Kennedy et al. 2012). Smaller species invade burned areas to forage on seeds and seedlings, thus attracting a diversity of insects (Blaum et al. 2007; Thies and Lewis 2015). In the Okavango Delta, floodplains and grasslands that experience fire annually are hotspots for biodiversity relative to unburned patches (Tacheba et al. 2009; Fynn et al. 2014). Higher water tables in these grasslands allow regrowth immediately after a fire during the dry season and are favored by large populations of herbivores such as the plains zebra (*Equus burchellii*), sable, roan, warthog (*Phacochoerus africanus*), impala (*Aepyceros melampus*), and others (Fynn et al. 2014). These habitats offer bridging resources that are essential before the end of the dry season (Owen-Smith 2002; Fynn et al. 2014; Sianga and Fynn 2017; Fynn and Provenza 2023) and thus buffer wildlife from starvation, which could result in dry season mortalities (Fynn and Provenza 2023). This adaptive strategy of herbivores favoring floodplains and grasslands after the fire was also observed in other ecosystems: the Rukwa Valley floodplains in Tanzania (Vesey-FitzGerald 1960), Yaere floodplains in Cameroon (Pamo 1998), and Kgaswane Mountain Reserve, South Africa (Parrini and Owen-Smith 2010).

Fire and herbivory are believed to have evolved together, modifying the plant communities into the diversity of habitat patches observed in the landscapes today (Salvatori et al. 2001; Fuhlendorf and Engle 2001; Hassan 2007; Fuhlendorf et al. 2009). Grazing maintains a relatively low fuel load (Knapp and Seastedt 1986). This determines the frequency and intensity of forest fires in some ecosystems (Knapp and Seastedt 1986; Savage and Swetnam 1990; Fuhlendorf and Engle 2004; Fuhlendorf et al. 2009). Pyric herbivory is important in maintaining grass-tree savannas in the form we see today. For instance, the elimination of rinderpest in the early 1960s in the Serengeti ecosystem (Tanzania) led to the explosion of the Serengeti wildebeest population, elevating the grazing pressure, minimizing fire occurrences and grass competition, and promoting the growth of small trees (Sinclair 1979). Fire encourages herbivores to focus their grazing in certain patches relative to others and hence aids in the maintenance of grazing lawns in ecosystems (McNaughton 1984; Anderson et al. 2006). These grazed patches will eventually develop into mosaics of disturbed patches across landscapes (Fuhlendorf and Engle 2001; Hassan 2007), which are important in promoting greater forage and vegetation structural diversity (Fuhlendorf and Engle 2004). Therefore, pyric herbivory promotes habitat heterogeneity, biological diversity and ecosystem functioning

(Knapp and Seastedt 1986; Savage and Swetnam 1990; Fuhlendorf and Engle 2004; Fuhlendorf et al. 2009). In this respect, some studies have demonstrated that primary productivity in some ecosystems would have probably declined if pyric herbivory was excluded in the protocols of ecosystem management in North America (Boerner 1982), Serengeti grasslands in east Africa (McNaughton 1979, 1984), and the Taiga forest, Alaska (Van Cleve et al. 1983).

However, the impact of fires in savanna ecosystems is variable, depending on their frequency and season of ignition (Bond and Van Wilgen 2012). Frequent fires can be detrimental to ecosystems: large amounts of dry-weight herbage burned contribute to a significant loss of total nitrogen content in forage through volatilization (Ojima et al. 1994; Blair 1997; Fynn et al. 2003). This results in the reduction of plant diversity and forage, which determines the foraging patterns of several herbivores. For example, an adaptation of the intermediate disturbance hypothesis demonstrates that infrequent levels of disturbance by fires will keep species diversity at low levels because there will be an accumulation of low-quality dry forage that will limit sunlight reaching the new shoots and hence will not benefit herbivores. However, as the frequency of fires increases, diversity will increase due to a reduction in competitive plants and will attract a diversity of wildlife species to the burnt patches until an optimum is reached; thereafter, as the frequency continues to increase, the diversity will eventually decline (Connell 1978), as plants will be burned more frequently than they can recover (Gough et al. 1994; Gough and Grace 1998). This demonstrates that the impacts of forest fires are variable due to their frequencies and intensities.

13.3.3 Fire Prevention and Management in Botswana

Great efforts have been made in Botswana to reduce the impacts of fires, including fuel-load reduction and modification, prescribed burning, and firebreak construction. Firebreaks can effectively reduce the fire spread rate (Fernández et al. 2019), fire size, and fire severity (Cochrane et al. 2012). Firebreaks generally refer to a belt or patch of land cleared of flammable fuels to prevent the occurrence and spread of a fire (Ascoli et al. 2018). Firebreaks comprise both natural barriers, such as rivers (Price et al. 2007), and artificial barriers, including roads (Laschi et al. 2019), areas cleared of vegetation (Knapp et al. 2017) and prescribed burning areas (Suffling and Perera 2008). Roads can directly prevent the spread of fire and serve as control lines for prescribed burning (Narayanaraj and Wimberly 2011, 2012) and during firefighting. Botswana has created firebreaks that are spread throughout the country (~10,000 km long), and some degree of maintenance is carried out (~P30 million annually) prior to the fire season to keep them free of fuel. Although firebreaks have been used in fire management in Botswana for a long time, their effectiveness in preventing fire spread and severity has not been evaluated. Uncontrolled fires continue to burn and spread in Botswana despite the presence and maintenance of firebreaks annually.

Historically, fire prevention and management in Botswana were duties of the Ministry of Environment and Tourism and the Ministry of Agriculture. Fires in conservation areas were managed by the Ministry of Environment and Tourism, particularly through the Department of Wildlife and National Parks and the DFRR (Dube 2013), whereas those outside conservation areas were the responsibility of the Ministry of Agriculture, guided by the Agricultural Resources Conservation Act of 1974 (ARB 1976; Dube and Mafoko 2009; Magole 2009). However, the legislation informing the management of fires in Botswana is managed by the DFRR through the Herbage Preservation Act of 1978 and Forest Act of 1968 (Dube and Mafoko 2009). These legal instruments are used to punish criminal acts causing fire outbreaks in Botswana. However, due to the lack of reporting of fire igniters, frequent fires are still visible, and challenges in identifying culprits remain. As a result, the causes of most fires in Botswana are documented as unknown (Dube 2013). An information campaign aimed at educating the public, particularly youth, is organized annually. During the fire season, messages are broadcast through national radio and television, and informative leaflets to raise public awareness are distributed throughout the country. In addition, regular patrols are organized, mostly along the boundaries of areas where most fires start. The patrols are aimed at providing public information, enforcement of the law and detection and timely intervention in case of fire.

13.3.4 Challenges of Fire Prevention and Management in Botswana

The managers of fires in Botswana, the DFRR, have limited capacity in the application and usage of appropriate technology in detecting fires. Fire detection technology that uses MODIS satellite imagery in the country is centralized in Gaborone, and hence, there is a need for this technology to be deployed in districts across the country. This will probably assist in the detection of fires earlier and enhance immediate communication of early warnings to the respective communities (Dube 2013). More skilled personnel are therefore required to operate this technology to enhance its effectiveness in managing fires. However, despite limitations in financial resources and skilled personnel, the predicted increase in forest fires due to climate change requires the adoption and scale up of this fire detection and early warning technology to effectively manage fires in Botswana (Rowell and Moore 2000). Applications of modest technology such as the MODIS sensor are mandatory for tracking and detecting the possibilities of mega fire outbreaks in Botswana (Marumbwa et al. 2011; Dube 2013).

13.4 Conclusion

This chapter demonstrated that the impacts of forest fires in Botswana are variable with season and plant communities. The magnitude of vegetation losses to fire fluctuates greatly among plant communities and species. Some species with flammable leaves, shoots, and litter are more susceptible to fire than others. However, with the impacts of climate change on ecosystems, fires are predicted to become frequent and will challenge the management of forest resources. As such, fire policies and management protocols should be adaptive and promote the adoption and enrollment of effective technology utilized in the management of forest fires. This will assist in the management and prevention of mega fires in the country.

References

Anderson MJ, Ellingsen KE, McArdle BH (2006) Multivariate dispersion as a measure of beta diversity. Ecol Lett 9:683–693

Andreae MO, Merlet P (2001) Emission of trace gases and aerosols from biomass burning. Glob Biogeochem 15:955–966

ARB (1976) Agricultural Resources Board (ARB) annual report. Ministry of Agriculture, Republic of Botswana, Gaborone

Ascoli D, Russo L, Giannino F et al (2018) Firebreak and fuelbreak. In: Manzello S (ed) Encyclopedia of wildfires and wildland-urban interface (WUI) fires. Springer

Batisani N, Yarnal B (2010) Rainfall variability and trends in semiarid Botswana: implications for climate change adaptation policy. Appl Geogr 30:483–489

Ben-Shahar R (1996) Woodland dynamics under the influence of elephants and fire in northern Botswana. Vegetatio 123:153–163

Ben-Shahar R (1998) Changes in structure of savanna woodlands in northern Botswana following the impacts of elephants and fire. J Plant Ecol 136:189

Blair JM (1997) Fire, navailability, and plant response in grasslands: a test of the transient maxima hypothesis. Ecology 78:2359–2368

Blaum N, Rossmanith E, Jeltsch F (2007) Land use affects rodent communities in Kalahari savannah rangelands. Afr J Ecol 45:189–195

Boerner REJ (1982) Fire and nutrient cycling in temperate ecosystems. Bioscience 32:187–192

Bond WJ, Scott AC (2010) Fire and the spread of flowering plants in the Cretaceous. New Phytol 188:1137–1150

Bond WJ, Van Wilgen BW (1996) Why and how do ecosystems burn? In: Fire and plants. Springer

Bond WJ, Van Wilgen BW (2012) Fire and plants. Springer Science & Business Media

Botswana Environmet Statistics (2017) Natural disasters digest report. Statistics Botswana

Cassidy L (2009) Fire, wealth and access to wetland resources in the panhandle of Botswana's Okavango Delta. 2009 IEEE international geoscience and remote sensing symposium, Cape Town

Cassidy L, Perkins JS, Bradley J (2022) Too much, too late: fires and reactive wildfire management in northern Botswana's forests and woodland savannas. Afr J Range Forage Sci 39:160–174

Chinamatira L, Mtetwa S, Nyamadzawo G (2016) Causes of wildland fires, associated socioeconomic impacts and challenges with policing, in Chakari resettlement area, Kadoma, Zimbabwe. Fire Sci Rev 5:1–11

Cochrane MA, Moran CJ, Wimberly MC et al (2012) Estimation of wildfire size and risk changes due to fuels treatments. Int J Wildland Fire 21:357–367

Connell JH (1978) Diversity in tropical rainforests and coral reefs. Science 199:1302–1310
Crisp MD, Burrows GE, Cook LG et al (2011) Flammable biomes dominated by eucalypts originated at the Cretaceous–Palaeogene boundary. Nat Commun 2:1–8
DFRR (2007) Fire season report 2006 — summary report on the 2006 Botswana Fire Season, Department of Forest and Range Resources, Division of Conservation and Management, Fire Management Section, Ministry of Environment Wildlife and Tourism, Gaborone
DFRR (2011) Fire season preparedness, Division of Conservation and Management, Fire Management Section, Department of Ministry Management, Ministryof Environment Wildlife and Tourism
DFRR (2016) Chobe district bush fire risk management plan. Kasane
Dube OP (2007) Fire weather and land degradation. In: Climate and land degradation. Springer
Dube OP (2009) Linking fire and climate: interactions with land use, vegetation, and soil. Curr Opin Environ Sustain 1:161–169
Dube OP (2013) Challenges of wildland fire management in Botswana: towards a community inclusive fire management approach. Weather Clim Extrem 1:26–41
Dube OP, Mafoko JG (2009) Botswana case study. SAFNet, Africa environment outlook case studies: impacts of fires on the environment, SARDC, Thomas Theis Nielsen, Nairobi
Dwyer E, Pereira JM, Grégoire JM et al (2000) Characterization of the spatio-temporal patterns of global fire activity using satellite imagery for the period April 1992 to March 1993. J Biogeogr 27:57–69
Ellery W, Ellery K, McCarthy TS et al (1989) A peat fire in the Okavango Delta, Botswana, and its importance as an ecosystem process. Afr J Ecol 27:7–21
Fernández P, Rodríguez A, Gutiérrez D et al (2019) Firebreaks as a barrier to movement: the case of a butterfly in a Mediterranean landscape. J Insect Conserv 23:843–856
Fox JT, Vandewalle ME, Alexander KA (2017) Land cover change in northern Botswana: the influence of climate, fire, and elephants on semiarid savanna woodlands. Land 6:73
Fuhlendorf SD, Engle DM (2001) Restoring heterogeneity on rangelands: ecosystem management based on evolutionary grazing patterns: we propose a paradigm that enhances heterogeneity instead of homogeneity to promote biological diversity and wildlife habitat on rangelands grazed by lives. Bioscience 51:625–632
Fuhlendorf SD, Engle DM (2004) Application of the fire–grazing interaction to restore a shifting mosaic on tallgrass prairie. J Appl Ecol 41:604–614
Fuhlendorf SD, Engle DM, Kerby JA et al (2009) Pyric herbivory: rewilding landscapes through the recoupling of fire and grazing. Conserv Biol 23:588–598
Fynn RWS, Provenza FD (2023) Functional adaptive resources for large herbivores in African savannas: an ecological-gradient based framework. Front Conserv Sci 4:1133329
Fynn RWS, Haynes RJ, O'connor TG (2003) Burning causes long-term changes in soil organic matter content of a South African grassland. Soil Biol Biochem 35:677–687
Fynn RWS, Chase M, Röder A (2014) Functional habitat heterogeneity and large herbivore seasonal habitat selection in northern Botswana. S Afr J Wildl 44:1–15
Georgiadis NJ, Mcnaughton SJ (1990) Elemental and fibre contents of savanna grasses: variation with grazing, soil type, season and species. J Appl Ecol 1:623–634
Giglio L, Randerson JT, Van Der Werf GR (2013) Analysis of daily, monthly, and annual burned area using the fourth-generation global fire emissions database (GFED4). J Geophys Res Biogeosci 118:317–328
Gough L, Grace JB (1998) Herbivore effects on plant species density at varying productivity levels. Ecology 79:1586–1594
Gough L, Grace JB, Taylor KL (1994) The relationship between species richness and community biomass: the importance of environmental variables. Oikos 70:271
Gumbricht T, McCarthy TS, McCarthy J et al (2002) Remote sensing to detect subsurface peat fires and peat fire scars in the Okavango Delta, Botswana. S Afr J Sci 98:351–358
Hanes CC, Wang X, Jain P et al (2019) Fire-regime changes in Canada over the last half century. Can J For Res 49:256–269

Hassan SN (2007) Effects of fire on large herbivores and their forage resources in Serengeti, Tanzania. PhD Thesis, Norwegian University of Science and Technology

He T, Lamont BB, Downes KS (2011) Banksia born to burn. New Phytol 191:184–196

He T, Pausas JG, Belcher CM et al (2012) Fire-adapted traits of Pinus arose in the fiery Cretaceous. New Phytol 194:751–759

Heinl M, Neuenschwander A, Sliva J et al (2006) Interactions between fire and flooding in a southern African floodplain system (Okavango Delta, Botswana). Landsc Ecol 21:699–709

Higgins SI, Bond WJ, Trollope WS (2000) Fire, resprouting and variability: a recipe for grass-tree coexistence in savanna. J Ecol 88:213–229

Kasischke ES, Williams D, Barry D (2002) Analysis of the patterns of large fires in the boreal forest region of Alaska. Int J Wildland Fire 11:131–144

Kennedy AM, Marais J, Bauer AM et al (2012) Effect of fire on the herpetofauna of the Koanaka Hills, Ngamiland, Botswana. Check List 8:666–674

Knapp AK, Seastedt TR (1986) Detritus accumulation limits productivity of tallgrass prairie. Bioscience 36:662–668

Knapp EE, Lydersen JM, North MP et al (2017) Efficacy of variable density thinning and prescribed fire for restoring forest heterogeneity to mixed-conifer forest in the central Sierra Nevada, CA. For Ecol Manag 406:228–241

Komarek EV (1971) Lightning and fire ecology in Africa. In Tall timbers fire ecology conference, vol 11, pp 473–509

Lamont BB, He T, Downes KS (2013) Adaptive responses to directional trait selection in the Miocene enabled Cape proteas to colonize the savanna grasslands. Evol Ecol 27:1099–1115

Laschi A, Foderi C, Fabiano F et al (2019) Forest road planning, construction, and maintenance to improve forest fire fighting: a review. Croat J For Eng 40:207–219

Lemessa D, Perault M (2001) Forest fires in Ethiopia: reflections on socioeconomic and environmental effects of the fires in 2000. Assessment study June–September. UNDP Emergencies Unit for Ethiopia and Brown University, Addis Ababa

Lepetu J, Alavalapati J, Nair PK (2009) Forest dependency and its implication for protected areas management: a case study from Kasane Forest Reserve, Botswana. Int J Environ Res 3(4):525–536

Maabong K, Mphale K (2021) Wildfires in Botswana and their frequency of occurrence. Clim Atmos Sci 11:689–696

Magole L (2009) The "shrinking commons" in the Lake Ngami grasslands, Botswana: the impact of national rangeland policy. Dev South Afr 26:611–626

Marumbwa FM, Borstlap G, Kusane I (2011) AMESD SADC Thema services user guide—deliverable R181/91

McNaughton SJ (1979) Grazing as an optimization process: grass-ungulate relationships in the Serengeti. Am Nat 113:691–703

McNaughton SJ (1984) Grazing lawns: animals in herds, plant form, and coevolution. Am Nat 124:863–886

Mmolotsi RM, Obopile M, Kwerepe BC et al (2012) Studies on Mukwa (*Pterocarpus angolensis*, DC) dieback in Chobe Forest Reserves in Botswana. J Plant Stud 1:154

Mojeremane W (2004) Wildland fires in Botswana. Fire Manag Today 64:34–36

Moris JV, Conedera M, Nisi L et al (2020) Lightning-caused fires in the Alps: identifying the igniting strokes. Agric For Meteorol 290:107990

Müller MM, Vacik H, Diendorfer G et al (2013) Analysis of lightning-induced forest fires in Austria. Theor Appl Climatol 111:183–193

Narayanaraj G, Wimberly M (2011) Influences of forest roads on the spatial pattern of wildfire boundaries. Int J Wildland Fire 20:792–803

Narayanaraj G, Wimberly M (2012) Influences of forest roads on the spatial patterns of human- and lightning-caused wildfire ignitions. Appl Geogr 32:878–888

Nash CH, Johnson EA (1996) Synoptic climatology of lightning-caused forest fires in subalpine and boreal forests. Can J For Res 26:1859–1874

Ojima DS, Schimel DS, Parton WJ et al (1994) Long-and short-term effects of fire on nitrogen cycling in tallgrass prairie. Biogeochemistry 24:67–84
Owen-Smith RN (2002) Adaptive herbivore ecology: from resources to populations in variable environments. Cambridge University Press
Pamo ET (1998) Herders and wildgame behaviour as a strategy against desertification in northern Cameroon. J Arid Environ 39:179–190
Parrini F, Owen-Smith N (2010) The importance of post-fire regrowth for sable antelope in a Southern African savanna. Afr J Ecol 48:526–534
Parry DC, Campbell BM (1990) Wildlife management areas of Botswana. Botsw Notes Rec 22:65–77
Price O, Edwards A, Russell-Smith J (2007) Efficacy of permanent firebreaks and aerial prescribed burning in western Arnhem Land, Northern Territory, Australia. Int J Wildland Fire 16:295–305
Pyne SJ, Laven RD (1996) Introduction to wildland fire, 2nd edn. John Wiley and Sons Inc, New York
Rietbergen-McCracken J, Abaza H (2014) Economic instruments for environmental management: a worldwide compendium of case studies. Routledge
Rowell A, Moore PF (2000) Global review of forest fires. Forests for Life Programme Unit. WWF International
Salvatori V, Egunyu F, Skidmore AK et al (2001) The effects of fire and grazing pressure on vegetation cover and small mammal populations in the Maasai Mara National Reserve. Afr J Ecol 39:200–204
Sankaran MJ, Ratnam J, Hanan NP (2004) Tree-grass coexixtence in savannas revisted-insights from examination of assumptions and mechanisms invoked in existring models. Ecol Lett 7:480–490
Santín C, Doerr SH (2016) Fire effects on soils: the human dimension. Philos Trans R Soc Lond Ser B Biol Sci 371:20150171
Savage M, Swetnam TW (1990) Early 19th-century fire decline following sheep pasturing in a Navajo ponderosa pine forest. Ecology 71:2374–2378
Scott AC, Cripps JA, Collinson ME et al (2000) The taphonomy of charcoal following a recent heathland fire and some implications for the interpretation of fossil charcoal deposits. Palaeogeogr Palaeoclimatol Palaeoecol 164:1–31
Scott AC, Bowman DM, Bond WJ et al (2013) Fire on earth: an introduction. John Wiley & Sons
Seymour CL, Huyser O (2008) Fire and the demography of camelthorn (*Acacia eriobola* Meyer) in the southern Kalahari-evidence for a bonfire effect? Afr J Ecol 46:594–601
Sianga K, Fynn R (2017) The vegetation and wildlife habitats of the Savuti-Mababe-Linyanti ecosystem, northern Botswana. Koedoe 59:1–6
Sinclair ARE (1979) The eruption of the ruminants. In: Sinclair ARE, Norton-Griffiths M (eds) Serengeti: dynamics of an ecosystem. Chicago University Press, Chicago, pp 82–103
Suffling R, Perera A (2008) Chapter 4. Characterizing Natural Forest Disturbance Regimes: Concepts and Approaches. In A. Perera, L. Buse & M. Weber (Ed.), *Emulating Natural Forest Landscape Disturbances: Concepts and Applications* (pp. 43–54). New York Chichester, West Sussex: Columbia University Press. https://doi.org/10.7312/pere12916-007
Tacheba B, Segosebe E, Vanderpost C et al (2009) Assessing the impacts of fire on the vegetation resources that are available to the local communities of the seasonal wetlands of the Okavango, Botswana, in the context of different land uses and key government policies. Afr J Ecol 47:71–77
Thies ML, Lewis PJ (2015) Effects of fire on small mammals of the Koanaka Hills, Northwestern Botswana. Museum of Texas Tech University
Vacchiano G, Foderi C, Berretti R et al (2018) Modelling anthropogenic and natural fire ignitions in an inner-alpine valley. Nat Hazards Earth Syst Sci 18:935–948
Van Cleve K, Dyrness CT, Viereck LA et al (1983) Taiga ecosystems in interior Alaska. Bioscience 33:39–44
Van der Walt PT, le Riche EN (1984) The influence of veld fire on an Acacia erioloba community in the Kalahari Gemsbok National Park. Koedoe 27:103–106

van Wees D, van Der Werf GR, Randerson JT et al (2021) The role of fire in global forest loss dynamics. Glob Chang Biol 27:2377–2391

Veraverbeke S, Rogers BM, Goulden ML et al (2017) Lightning as a major driver of recent large fire years in North American boreal forests. Nat Clim Chang 7:529–534

Vesey-FitzGerald DF (1960) Grazing succession among East African game animals. J Mammal 41:161–172

Williams J, Albright D, Hoffmann AA et al. (2011) Findings and implications from a coarse-scale global assessment of recent selected mega-fires. In: Proceedings of the 5th international wildland fire conference, Suncity, p 19

Dr Keoikantse Sianga is a Senior Lecturer at the Botswana University of Agriculture and Natural Resource. He holds a BSc in Conservation Biology and Wildlife Management and an MPhil and a PhD in Natural Resource Management. He worked as Project Assistant at the University of Botswana, Project Manager (Communities Living Among Wildlife Sustainably, CLAWS). His research interests include functional and vegetation heterogeneity and plant-wildlife interactions, fire ecology and management, human-wildlife interactions, and spatial ecology. Dr Sianga has published four articles in peer reviewed journals and seven book chapters.

Witness Mojeremane is a Professor in the Department of Range and Forest Resources at the Botswana University of Agriculture and Natural Resources (BUAN). He holds a Diploma in Forestry, a Diploma in Human Resources Management, a Bachelor of Science (Honours) in Forestry, a Master of Science in Tropical Environmental Forestry, and a PhD in Ecology and Resource Management. He joined BUAN as an instructor in 1994. His research interests include forest sciences and has been engaged in several consultancy and research Projects. Professor Mojeremane has published 52 articles in refereed journals, 16 book chapters, and 3 articles in refereed conference proceedings.

Professor Ayana Angassa Abdeta is a Professor in the Department of Range and Forest Resources—Botswana University of Agriculture and Natural Resources. He received BSc and MSc from the Alemaya University of Agriculture, Ethiopia, PhD from the Norwegian University of Life Sciences, Department of International Environment and Development Studies; Postdoctoral Training from both the University of Hohenheim (Food Security Centre) and University of Cologne (Botanical Institute) in Germany, Certificate (2015) in Disaster Risk Management from the Humboldt University of Berlin, Certificate (2020) in Project Management Foundations from the University of Cape Town, and Certificate (2022) in Agroforestry Systems from the University of Freiburg, Germany.

Mr. Kamogelo Makgobota is a technician in the Department of Range and Forest Resources, Faculty of Natural Resources, Botswana University of Agriculture and Natural Resources. He joined BUAN in 2017 as an intern and was involved in several consulting projects. He is currently pursuing an MSc in Crop Science studying the effectiveness of beneficial rhizobia on indigenous wild legume establishments with poor natural regeneration. Mr Makgobota has coauthored six research papers. His research interest is to discover and assess natural-based solutions that can address societal challenges and provide ecological sustainability, which benefits people and the natural flora.

Ms Summer Mabula is an Ecologist with an experience in and enthusiasm for teaching and carrying out field work. She has expertise in training and conducting range assessments, plant identification, environmental monitoring as well as writing and presenting scientific reports. She is passionate about ecology and possesses an in-depth knowledge about ecosystems. She holds a Bachelor of Science in Range Science from the Botswana University of Agriculture and Natural Resources

Chapter 14
Woody Plant Encroachment as a Process of Landscape Change in Arid Rangelands: Effective Control Methods and Enhancing Herbaceous Species Richness: A Review

Ayana Angassa Abdeta, Keoikantse Sianga, Summer Mabula, and Witness Mojeremane

14.1 Introduction

Rangelands are one of the most important types of global ecosystems, occupying over 41% of the world's terrestrial area (Cao et al. 2019; Liao et al. 2018; Suttie et al. 2005). Extensive areas of global rangelands experience various phases of deterioration in the context of climate change and increasing human impacts, such as grazing pressure and suppression of range fires (White et al. 2000). Among these deterioration phases, woody plant encroachment in rangelands has attracted global attention (Archer et al. 1995; Eldridge et al. 2011; Naito and Cairns 2011; Oba et al. 2000). Woody plant encroachment has been a common problem worldwide, including in the rangelands of Africa (Levick and Rogers 2011), North America (Goslee et al. 2003; McGlynn and Okin 2006), Latin America (Cipriotti et al. 2012), Australia (Robinson et al. 2008), Mediterranean regions (Maestre et al. 2009), Inner Mongolia in China (Zhou 1990; Tong et al. 2001; Xiong et al. 2003; Peng et al. 2013; Zhang et al. 2017), and alpine grasslands of the Tibetan Plateau (Zhang et al. 2006; Wang et al. 2013).

Existing evidence suggests that arid and semiarid rangelands worldwide are being altered by the phenomenon known as "woody plant encroachment" (Adamoli et al. 1990; Archer et al. 1995; Moleele et al. 2002). Encroachment by woody plant

A. A. Abdeta (✉) · S. Mabula · W. Mojeremane
Botswana University of Agriculture and Natural Resources, Department of Range and Forest Resources, Gaborone, Botswana
e-mail: aabdeta@buan.ac.bw; smabula@buan.ac.bw; wmojerem@buan.ac.bw

K. Sianga
Botswana University of Agriculture and Natural Resources, Department of Wildlife and Aquatic Resources, Gaborone, Botswana
e-mail: ksianga@buan.ac.bw

W. L. Hambira et al. (eds.), *Environmental Change and Biodiversity Conservation in sub-Saharan Africa*, Advances in Global Change Research 75,
https://doi.org/10.1007/978-3-031-73136-5_14

species has changed the balance in the proportion of trees and shrubs relative to grasses and herbs, which have been classified as a form of land-use degradation. Woody plant encroachment in African rangelands can be explained in terms of the proliferation in the density and cover of indigenous invasive woody plants (Tews et al. 2004; Van Auken 2009), which significantly influences rangeland ecosystems. Woody plant encroachment causes the alteration of arid rangelands to a bushy environment, along with modifications in the natural structure and functionality of the rangeland ecosystem, which includes primary productivity, conservation of rangeland biodiversity, and carbon sequestration (Briggs et al. 2005; Eldridge et al. 2011; Peng et al. 2013). The expansion of woody plant encroachment in rangelands may modify the conditions of the rangeland's physical environment, such as soil structure and its properties (Angassa et al. 2012; Archer et al. 1995; Zhang et al. 2001; Li and Gao 2012), watershed management and hydrological processes, such as groundwater storage (Shiferaw et al. 2021; Peng et al. 2014). Such divergences in climate episodes may facilitate the process of woody plant encroachment toward a progressive increase in the density of bush encroachment and its irreversible impact (Archer et al. 1995; Scheffer and Carpenter 2003).

Woody plant encroachment reduces the abundance and yield of perennial grasses and other palatable herbaceous species (Hagos and Smit 2005; Ward 2005), thereby decreasing the capacity of rangelands for ungulates (Ward 2005; Wiegand et al. 2006), which in turn affects the livelihoods of pastoralists and the tourism sector in general (Naito and Cairns 2011). Furthermore, the impact of woody plant encroachment on the capacity of rangelands to accommodate ungulates threatens biodiversity and livestock production, particularly grazers (Kraaij and Ward 2006; Angassa and Oba 2008) and the livelihood of pastoral communities (Kgosikoma et al. 2012) in arid environments. Therefore, studies on woody plant encroachment in arid rangelands are decisive in terms of advancing knowledge on the response and resilience of rangeland ecosystems to climate change and human impacts. A literature review focusing on a body of literature could help accurately identify the ecological impacts of woody plant encroachment and will assist in designing suitable management options for the control of invasive woody plants in rangelands.

The literature has identified the key factors that cause the spread of woody plant encroachment in rangelands, for instance, global warming (Tietjen et al. 2010), increasing CO_2 concentration (Mooney and Cleland 2001; Stevens et al. 2016), fire suppression (Oba et al. 2000; Roques et al. 2001; Van Langevelde et al. 2003; Lohmann et al. 2014), increasing grazing pressure (Van Vegten 1984; Skarpe 1990), dissemination of seeds of invasive species by grazing animals (Chuong et al. 2016), local denudation, cultivation, advancement of commerce and transportation (Van Auken 2009), and reduced browsing pressure (Angassa and Oba 2008). Increased rainfall (Joubert et al. 2008) and soil characteristics (Wilson and Bowman 1994; Mourik et al. 2007; Sankaran et al. 2008) are other possible causes of bush encroachment.

Rangeland ecosystem responses to such driving forces can impact the processes of woody plant encroachment in rangelands. However, discussions remain about what triggers woody plant encroachment in rangelands and the interactions of the various root causes (D'Odorico et al. 2010, 2012). Research on the mechanism of woody plant encroachment entails long-standing and incessant scrutinizing of the process of woody plant encroachment (Roques et al. 2001; Van Auken 2009). Continuing detection techniques include field observations within a specific location (Cipriotti et al. 2012) and the dynamic understanding of past aerial photos and remote sensing images (Goslee et al. 2003; Cao et al. 2018) or indigenous knowledge systems (Angassa and Oba 2008). Various models and approaches are proposed as useful sources for examining the dynamics and processes of woody plant encroachment in rangelands. These include the dynamics of rangeland vegetation indicators (Xu et al. 2010), spectral mixture analysis (Sohn and McCoy 1997), time series data analysis (Wilfong et al. 2009; Zhou et al. 2013), and state-and-transition models (Westoby et al. 1989), which are established techniques for studying and monitoring the dynamics and processes of woody plant encroachment in rangelands.

However, the use of certain techniques, such as satellite images and spectral mixture analysis, in evaluating the spread of bush encroachment might be distorted due to the challenges attributed to the high resemblance between the bands of herbaceous and woody layers. The absence of actual techniques and adequate data for scrutinizing woody plant encroachment at the regional level limits the investigation of the processes of woody plant encroachment. According to Cao et al. (2019), the development of bush control methods and monitoring techniques, accompanied by an extensive geographical range of rangeland woody plant cover dynamic facts and figures, can assist in examining the effects of various driving forces at a broad range of natural events. Several methods can help to model the mechanisms and processes of bush encroachment in rangelands.

The following review of the literature highlights that the invasion of woody plant encroachment causes serious problems that go beyond mere deterioration of rangelands, discusses general and specific solutions, and concludes that appropriate control measures and public awareness are needed to mitigate the adverse effects of invasive woody plant encroachment in rangelands. We describe ongoing developments in research on the processes of bush encroachment in African rangelands, mechanisms of bush encroachment using state-and-transition models, options for suppressing invasive woody plants, and enhancing herbaceous forage productivity in the rangelands. Finally, some conclusions are drawn based on the literature review.

14.2 Bush Encroachment in Arid Rangelands

Woody plant encroachment is a serious problem in global rangelands. Woody plant encroachment is an impediment to livestock production and wildlife management in arid and semiarid rangelands worldwide due to a decline in grass cover (e.g., Oba et al. 2000; Roques et al. 2001). This alteration in plant community composition from grass to a shrub-dominated environment may be related to landscape heterogeneity, suggesting that this situation is probably sustained by unique rangeland ecological sites due to resource availability or disturbance regimes such as fire and grazing (D'Odorico et al. 2012). The authors present specific management interventions that accord with specific natural events, including prescribed fires, for the control of woody plant encroachment. They further address the use of modeling as a management tool to determine the times and places where interventions would have the highest effect.

14.3 Processes and Mechanisms of Bush Encroachment

The dynamics of woody plant encroachment (Van Vegten 1984) are frequently explained in terms of equilibrium between rangeland vegetation dynamics and grazing pressure. The equilibrium dynamic assumes a linear relationship between woody plant encroachment and livestock grazing pressure (e.g., Coppock 1994). The mechanisms of woody plant encroachment might, however, be more effectively examined using the state-and-transition model (Westoby et al. 1989; Angassa and Oba 2008). States are patch mosaics of vegetation composition that are usually described by the species composition of the dominant vegetation, which could account for nonlinear changes in woody plant encroachment (Westoby et al. 1989). The variations between vegetation states are shown by the level of grazing pressure and the age of enclosures against open-grazed areas (Angassa and Oba 2008).

Allen-Diaz and Bartolome (1998) argue that the pattern of vegetation dynamics along with the increased age of enclosures in the state-and-transition model is nonlinear, demonstrating a significant time dependency that can be described using long-term research and observations within at least 5- to 10-year-old enclosures. Knowledge about forces that trigger shifts between vegetation states would be useful in understanding the mechanisms of bush encroachment (Belayneh and Tessema 2017) in terms of transitions between vegetation states. Additionally, the state-and-transition model can be used to describe threats (i.e., factors that promote woody plant encroachment in the rangelands) and opportunities (i.e., fodder production for grazing animals) (Angassa and Oba 2008).

In the savanna ecosystems of Africa, the processes of bush encroachment in traditional range enclosures according to different age chronosequences of vegetation states are sparsely reported (Angassa and Oba 2008). Traditional range enclosures involving semiprivate grazing lands are widely used by pastoralists in the Horn of Africa for seasonal grazing by calves and sick or weak animals (Oba et al. 2000; Angassa and Oba 2008). Range enclosures of different ages may reflect different vegetation states that can be spatially arranged along an age chronosequence within the landscape conditions. For example, earlier studies (Oba et al. 2000) evaluated the potential effects of enclosures on vegetation dynamics by comparing woody plant encroachment density within enclosures against open-grazed areas. Others have mentioned the effects of age chronosequence on vegetation states to describe the dynamics of rangeland vegetation. Thus, the current review chapter considers the responses of invasive woody encroachment to the effects of grazing pressure and the age or time of protection of rangelands from livestock grazing.

A study by Angassa and Oba (2008) compared the influences of grazing pressure using enclosures against adjacent open-grazed rangeland along with different ages of enclosures and vegetation states. In doing so, the authors investigated woody vegetation density in terms of mature trees, saplings and seedlings, bush density, species composition, and the invasive woody species ratio (IWSR). They further defined IWSR as the density of invasive woody encroachment to that of noninvasive woody species (i.e., a ratio of 1.0 represents a high invasive threat, while <1.0 represents a lesser threat of encroachment). The study by Angassa and Oba (2008) further considered the hypothetical scenario for understanding the possible trajectory of individual woody species (vegetation states and the densities of invasive species) in the long term by doubling the age of enclosures.

14.3.1 Impacts of Grazing Pressure

The effects of grazing pressure have long been recognized as a major cause of woody plant encroachment influencing the dynamics of rangeland vegetation based on data generated from enclosures against open-grazed rangeland areas. Previous studies (Smart et al. 1985) stated that reduced grazing pressure increased the densities of woody plant encroachment in enclosures compared to open-grazed areas. The study by Angassa and Oba (2008) suggests that woody plant encroachment is a potential problem inside enclosures compared to open-grazed rangelands. Similar evidence from other parts of the world (Liu et al. 2020) suggested that the exclusion of grazing is an effective way to promote plant productivity and diversity in alpine areas. In northern Ethiopia, Abebe et al. (2006) reported increased woody species density in area enclosures. A study by Oba et al. (2000) from northern Kenya indicated that the exclusion of grazing promotes woody plant encroachment. Others (e.g., Mwalyosi 2000) reported increased densities of woody plant encroachment in enclosures than in open-grazed rangelands. A study by Smart et al. (1985) in Uganda indicated that long-term exclusion of grazing increased woody plant densities, while a study by Hayashi (1996) from Eastern Kenya reported a higher density of woody plant encroachment in enclosures than in open-grazed areas. Comparable evidence from other regions showed that the absence of livestock grazing results in a marked regeneration of woody species (Eccard et al. 2000). A study by Mwalyosi (2000) in Tanzania indicated increased densities of woody plant encroachment inside enclosures when compared to open-grazed areas. According to Eccard et al. (2000), the density of shrub encroachment was higher inside enclosures than in adjacent open-grazed areas. Contrary to the findings of Eccard et al. (2000), Trodd and Dougill (1998) did not confirm increased grazing pressure as a factor for the expansion of woody plant encroachment. The current literature review suggests insufficient evidence to prove the claim that a single factor contributes to woody plant encroachment under communal rangelands against protected areas (Angassa and Oba 2008). Angassa and Oba (2010) stated that enclosures perhaps served as a suitable environment for the regeneration of invasive woody plants as a result of reduced disturbance compared to open-grazed areas. On the other hand, frequent disturbances by livestock trampling could reduce soil moisture regimes. According to Oba et al. (2000), the causes of woody plant encroachment in arid rangelands are not fully understood, and the lack of difference between protected and unprotected areas showed that the proliferation of woody plant encroachment is widespread and that the causes of bush encroachment may be beyond heavy grazing (Devine et al. 2017).

14.3.2 Vegetation States and Transitions

The dynamics of woody plant encroachment in the various vegetation states can be described using the state-and-transition model (Franklin et al. 1987; Westoby et al. 1989). The different vegetation states explained in terms of enclosures, open-grazed areas, and different ages of enclosures (i.e., long-term versus short-term protection from grazing) suggest that the ecological mechanisms of bush encroachment may be linked to the suppression of fire, soil type, interannual rainfall variability, exclusion of browsers, competition, and applications of other methods of invasive woody plant control (Angassa and Oba 2008; Belayneh and Tessema 2017). The same authors reported differences in the density of woody plant encroachment between various vegetation states, suggesting that long-term protection from disturbance (grazing and fire) and differences in the age of enclosures are likely to facilitate the expansion of woody plant encroachment. Other scholars (O'Connor et al. 2014; Belayneh and Tessema 2017; Devine et al. 2017; Venter et al. 2018) stated the complex interactions among rainfall, grassland productivity, grazing and browsing intensities, fire, CO_2 concentration in the atmosphere, and soil properties as major drivers of woody plant encroachment. The proliferation of woody plant encroachment is a serious problem for the productivity of herbaceous plants, creating a scarcity of feeds for animals (Angassa et al. 2012). In another study, Angassa and Oba (2008) indicated that short-term protection from grazing enhanced the regeneration of invasive woody plant encroachment against long-term protection from grazing. The possible explanation provided by the authors was that under long-term protection, the vegetation state was probably harvested/removed by the local community, while in short-term protection, improved growth conditions most likely promoted the regeneration of increased woody plant encroachment from the soil seed bank. Existing studies (Angassa and Oba 2008) have shown that the invasive woody plant species ratio declined in the medium-age vegetation states. The decline might be attributed to self-thinning mortality among individual plant species, as well as other disturbance factors. According to Franklin et al. (1987), tree mortality increases between 10 and 20 years because of competition within or between plant species. Zhang et al. (2005) reported a decline in plant species composition in the medium-age vegetation states compared to the short-term and long-term vegetation states resulting from the exclusion of grazing. Generally, fluctuations in terms of the densities of woody plant encroachment linked to the age chronosequence of enclosures suggest that shifts in vegetation structure and mechanisms of bush encroachment seem unpredictable. From the current review, it seems that the influences of climate variability and soil seed bank supply combined with other factors are probably more important in regulating the proliferation of woody plant encroachment in rangelands. Angassa and Oba (2008) have shown that doubling the age of protection from grazing negatively affected most invasive woody species, suggesting that the decrease in the density is possibly caused by the dieback of individual species (Franklin et al. 1987).

Bush encroachment is a worldwide phenomenon that has created a serious problem for the pastoral industry (Wiegand et al. 2006). Although the causes of bush encroachment are diverse and complex, the proliferation of bush encroachment is triggered by the suppression of fire (e.g., Oba et al. 2000), heavy grazing pressure (e.g., Skarpe 1990), increasing atmospheric CO_2 (O'Connor et al. 2014; Stevens et al. 2016), rainfall (Devine et al. 2017) and climate variability. Reduced fuel load also reduces fire frequency, making bush encroachment control difficult (e.g., Archer et al. 1995).

14.4 Interventions for Controlling Woody Plant Encroachment

To control the adverse effects of bush encroachment in savanna grazing lands, various bush encroachment control methods are widely applied, including fire, tree cutting and combinations thereof. While many earlier studies emphasized the effects of woody plant control in terms of reducing woody cover, the short- and long-term responses of individual species to different bush encroachment control methods are poorly documented. The effect of fire on woody plants (e.g., Ruthven et al. 2003) suppresses the growth of woody plant encroachment by killing or reducing the abundance of invasive woody encroachers. The collective effects of fire and grazing (Salvatori et al. 2001) and cutting inhibit the recovery of woody plant encroachment (Sawadogo et al. 2002; Ruthven et al. 2003), while grazing could play a major role in shaping the dynamics of rangeland vegetation. According to Oba (1990), woody plant encroachment control may either facilitate the regeneration of invasive woody plants by breaking seed dormancy or prevent seedling recruitment by damaging tree seedlings (Kraaij and Ward 2006) depending on the intensity of disturbance and types of woody plant species. A study by Angassa and Oba (2008) demonstrated the potential role of bush encroachment control methods focusing on the responses of individual woody plant encroachment in terms of damaging the stump, coppicing individual species, and regenerating woody seedlings. The authors highlighted that different invasive woody plant species respond differently to various bush encroachment control methods. Previous studies have also shown that understanding the effects of various bush encroachment control methods and the responses of individual woody species is essential for the management and improvement of rangeland ecosystems. Understanding and identifying the reactions of individual species is essential in developing effective bush encroachment control strategies to address the challenges for the expansion of woody plant encroachment. Therefore, bush encroachment control methods can be used to create awareness for the management and improvement of degraded rangelands.

14.4.1 Responses of Individual Woody Species to Experimental Treatments

A study conducted by Angassa and Oba (2008) recorded the mortality of a few individual woody species in response to the control treatment (i.e., protected areas from disturbances such as fire and grazing). The authors further found that invasive woody species that experienced high mortality in response to the various treatments developed a strategy of investing in greater seedling recruitment posttreatment. For example, Pauchard et al. (2003) argued that protection from disturbance factors promotes bush encroachment in rangelands. A study conducted in the rangelands of southern Ethiopia by Angassa and Oba (2008) suggested that stump deaths following cutting and fire treatment were species-specific and woody species classes (Oba 1990). Greater than 50% ($n = 21$) of the invasive woody plant species considered in the study by Angassa and Oba (2008) were highly susceptible to fire, suggesting that fire could be one of the best management tools for suppressing invasive woody plants. According to Fulbright (1996), woody plant stump mortalities in response to disturbance factors might suggest changes in management strategy that can greatly reduce populations of invasive woody plants. Sawadogo et al. (2002), in West Africa, observed that the presence of grazing and browsing animals increased mortality among coppicing woody plants postcutting and fire treatment. Similarly, Ruthven et al. (2003) argued that the impacts of disturbances usually result in a long-term reduction in the density of woody plants. The current review of the literature suggests the use of integrated control techniques for the effective management/suppression of invasive woody plant species. A study conducted in the Maasai Mara National Reserve in Kenya by Salvatori et al. (2001) confirmed that fire and cutting effectively control bush encroachment and facilitate shifts in vegetation from shrubland to grassland. In a similar study, Ruthven et al. (2003) confirmed a postfire decline in many woody plant species.

14.5 Enhancing Herbaceous Plants and Diversity

The threat of woody plant encroachment in savanna rangelands is understandable (Oba et al. 2000). This is particularly true because it reduces perennial grass production and leads to the loss of biodiversity (Hagos and Smit 2005). The main reason for the control of bush encroachment is to enhance herbaceous forage productivity and diversity to increase societal benefits from the rangelands through improved livestock production. In *The Ecology of Natural Disturbance and Patch Dynamics* by Pickett and White (1985), disturbances are distinct events that interrupt the plant community structure through transformations of biotic environments and habitat conditions in which colonization of the disturbed microhabitat takes place.

The shifts in plant community structure toward herbaceous species dominance stimulate plant succession processes where species that tolerate the impacts of bush encroachment could be substituted by vulnerable species. Earlier literature (Fulbright 1996; Grime 1977) has proven that disturbances of natural vegetation can shift rangeland vegetation from dominance by woody plant encroachment toward dominance by herbaceous plants. A study by Grime (1977) confirmed that disturbances of grassland systems can also produce changes in the life history strategies of individual species in response to the intensities of driving forces. On the other hand, Vetaas (1992) argued that the reaction of herbaceous species differs against various bush encroachment control methods and microclimate conditions. Different bush encroachment control methods, including hand removal of trees, are commonly used to increase herbaceous production and species diversity. For example, a study conducted by Bates (2005) suggested that tree cutting, and prescribed fire followed by rest from grazing enhanced herbaceous plant diversity. Others reported the importance of prescribed fire and grazing in the restoration and conservation of savanna rangeland changes in herbaceous species richness. According to Bates et al. (2005), herbaceous species biomass productivity and diversity were effectively increased following the thinning of invasive woody plants. According to Fulbright (1996), bush encroachment control methods are management systems for promoting herbaceous species composition and diversity. Although the effects of bush encroachment control methods in promoting herbaceous species productivity, richness, and diversity have been well documented (e.g., Sawadogo et al. 2002), information regarding the potential role of such techniques in influencing policies and development strategies is limited. Angassa and Oba (2008) also reported the positive effects of different combinations of bush encroachment control methods, including tree cutting and fire, tree cutting, fire and grazing, and tree cutting alone. Herbaceous species richness and diversity were greatly improved in response to the above treatments. This indicates the possible benefits of bush encroachment control methods for enhancing herbaceous species production, richness, and diversity.

The control of bush encroachment using tree cutting and fire considerably increased herbaceous species richness posttreatment (Angassa and Oba 2008). This implies that the novel herbaceous species following tree cutting and fire treatments are probably those that are suppressed by the effects of bush encroachment. The study by Angassa and Oba (2008) suggests that a management strategy needs to be planned so that such treatments are followed by proper resting from grazing for at least a year before grazing by livestock is resumed. On the other hand, the authors stated that certain herbaceous species increased while others declined in terms of abundance. Other species might decline due to a lack of tolerance to disturbance. The reintroduction of grazing after resting might allow a rapid build-up of herbaceous biomass. However, this method is unlikely to be suited to the wider rangelands that are currently suffering from bush encroachment due to the cost of land rehabilitation. This review, therefore, emphasizes the importance of suppressing invasive woody plants for forage management at a small scale, which could easily be integrated into the existing systems of rangeland management.

Generally, this section broadly aims to understand the effectiveness of bush encroachment control methods for enhancing forage productivity in the rangelands of Africa. The current review suggests that the adoption of these methods would be dependent on the outcomes, both in terms of meeting the goals of ecosystem restoration and conservation of biodiversity as well as overall improvement of rangeland resource management. This review has made suggestions as to how the various bush encroachment control methods could be tailored for the purposes of rangeland resource management. The suggested methods and technologies are easier to adopt if they are incorporated into the existing community practices of rangeland management. To meet the challenge of broad acceptance, this section summarizes the effects of different bush control methods on herbaceous vegetation to present various management and policy scenarios. It seems that bush encroachment control methods meaningfully improved the recovery of herbaceous vegetation composition and productivity. Angassa and Oba (2008) suggested that invasive woody plant cutting followed by prescribed fire showed superior results to other treatments in terms of promoting herbaceous biomass. This section further highlights that the presumed advantage of excluding fire and grazing from rangelands to improve rangelands' productivity was untested. In contrast, a combination of cutting, fire, and grazing substantially increased herbaceous biomass and diversity.

14.6 Conclusion

The present review of the literature synthesized how enclosures of different age categories could elucidate temporal variation in the density of bush encroachment. The review shows that woody plant encroachment is globally expanding and attracting the attention of several researchers with increased publications since the 1970s (Fig. 14.1), suggesting key strategies for intervention to suppress woody plant encroachment in rangelands.

The regeneration of invasive woody plant species in the enclosures exceeded that in the open-grazed rangelands, suggesting that the effects of grazing by livestock alone had a minimal effect in driving woody plant encroachment in the rangelands. Conversely, long-term exclusion of grazing did not reduce the problems of woody plant encroachment compared to year-round grazing. It seems that the mechanisms of woody plant encroachment are more regulated by stochastic events, such as climate variability, than by grazing by livestock. As reviewed in section four, bush encroachment control methods considerably reduced the regeneration of invasive woody plant species with varied levels of effects on individual species. The evidence suggests that the impact of bush encroachment control methods on invasive woody plants seems to have a short-term effect and calls for sustained follow-up at appropriate intervals. Species-specific reactions to the various bush encroachment

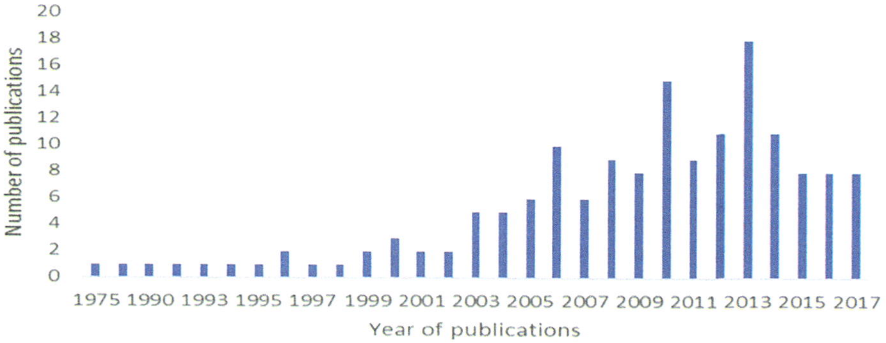

Fig. 14.1 Trends of various numbers of studies on bush encroachment by their year of publication. (Source: Directeur de recherche GONGSHEBIEKE Ayiben GRELLIER Séraphine LIU Xinyao 2017–2018)

control methods have important implications for the management, conservation, and sustainable utilization of rangelands in Africa. The current review of the literature also considered the reactions of herbaceous productivity, species richness, and diversity against the different bush encroachment control methods. Bush encroachment control methods such as tree cutting followed by prescribed fire, tree cutting followed by fire and grazing, and fire and grazing treatments improved herbaceous biomass and diversity, while continuous grazing greatly reduced herbaceous biomass. The outcomes in terms of herbaceous biomass and restoration of herbaceous plant biodiversity in response to the various bush encroachment control treatments have important implications for policy formulation, proper management, and sustainable utilization of rangeland resources. The use of prescribed fire and grazing seems to be more competent in promoting forage production for livestock and restoring herbaceous plant biodiversity. Defining the comparative advantage of different bush encroachment control methods in terms of forage for livestock and ease of implementation has immediate policy implications. The present review suggests that there is ample scope to inform policy where resource users need to improve the management of rangeland ecosystems by focusing on more effective bush encroachment control interventions.

References

Abebe MH, Oba G, Angassa A, Weladji RB (2006) The role of area enclosures and fallow age in the restoration of plant diversity in northern Ethiopia. Afr J Ecol 44: 507–514. https://doi.org/10.1111/j.1365-2028.2006.00664.x

Adamoli J, Sennhauser E, Acer JM, Rescia A (1990) Stress and disturbance: vegetation dynamics in the dry Chaco region of Argentina. J Biogeogr 17(4/5):491–500. https://doi.org/10.2307/2845381

Allen-Diaz B, Bartolome JW (1998) Sagebrush-grass vegetation dynamics: Comparing classical and state-transition models. Ecol Appl 8:795–804. https://doi.org/10.2307/2641267. https://www.jstor.org/stable/2641267

Angassa A, Oba G (2008) Effects of management and time on mechanisms of bush encroachment in southern Ethiopia. Afr J Ecol 46(2):186–196. https://doi.org/10.1111/j.1365-2028.2007.00832.x

Angassa A, Oba G (2010) Effects of grazing pressure, age of enclosures and seasonality on bush cover dynamics and vegetation composition in southern Ethiopia. J Arid Environ 74(1):111–120. https://doi.org/10.1016/j.jaridenv.2009.07.015

Angassa A, Oba G, Telera A (2012) Bush encroachment control demonstrations and management implications on herbaceous species in savannas of southern Ethiopia. Trop Subtrop Agroecosyst 15(1):173–185

Archer S, Schimel DS, Holland EA (1995) Mechanisms of shrubland expansion: land use, climate, or CO_2? Clim Chang 29(1):91–99. https://doi.org/10.1007/BF01091640

Bates JD (2005) Herbaceous response to cattle grazing following juniper cutting in Oregon. Range Ecology and Management 58:225–233

Bates JD, Miller RF, Svejcar T J (2005) Long-term successional trends following western juniper cutting. Range Ecology and Management 58:533–541

Belayneh A, Tessema ZK (2017) Mechanisms of bush encroachment and its interconnection with rangeland degradation in semiarid African ecosystems: a review. J Arid Land 9:299–312

Briggs JM, Blair JM, Mccarron J (2005) An ecosystem in transition: cause and consequences of the conversion of mesic grassland to shrubland. BioScience 55(7):561–572. https://doi.org/10.1641/0006-3568(2005)055

Cao X, Liu Y, Liu Q, Cui X, Chen X, Chen J (2018) Estimating population structure of encroaching shrubs in arid/semiarid grassland using high-spatial-resolution remotely sensed imagery. Remote Sens Environ 216:572–585. https://doi.org/10.1016/j.rse.2018.07.025

Cao X, Liu Y, Cui X, Chen J, Chen X (2019) Mechanisms, monitoring, and modelling of shrub encroachment into grassland: a review. Int J Digit Earth 12(6):625–641. https://doi.org/10.1080/17538947.2018.1478004

Chuong J, Huxley J, Spotswood EN, Nichols L, Mariottet P, Suding KN (2016) Cattle as dispersal vectors of invasive and introduced plants in a California annual grassland. Rangel Ecol Manag 69:52–58

Cipriotti PA, Aguiar MR, Wiegand T, Paruelo JM (2012) Understanding the long-term spatial dynamics of a semiarid grass-shrub steppe through inverse parameterization for simulation models. Oikos 121:848–861. https://doi.org/10.1111/j.1600-0706.2012.20317.x

D'Odorico P, Fuentes JD, Pockman WT, Collins SL, He Y, Medeiros JS, Dewekker S, Litvak ME (2010) Positive feedback between microclimate and shrub encroachment in the northern Chihuahuan Desert. Ecosphere 1(6):1. https://doi.org/10.1890/ES10-00073.1

D'Odorico P, Okin GS, Bestelmeyer BT (2012) A synthetic review of feedbacks and drivers of shrub encroachment in arid grasslands. Ecohydrology 5(5):520–530. https://doi.org/10.1002/eco.259

Devine AP, McDonald RA, Quaife T, Maclean IMD (2017) Determinants of woody encroachment and cover in African savannas. Oecologia 2017(183):939–951. https://doi.org/10.1007/s00442-017-3807-6

Eccard JA, Walther RB, Milton SJ (2000) How livestock grazing affects vegetation structures and small mammal distribution in the semi-arid Karoo. J Arid Environ 46:103–106. https://doi.org/10.1006/jare.2000.0659

Eldridge DJ, Bowker MA, Maestre FT, Roger E, Reynolds JF, Whitford WG (2011) Impacts of shrub encroachment on ecosystem structure and functioning: towards a global synthesis. Ecol Lett 14(7):709–722. https://doi.org/10.1111/j.1461-0248.2011.01630.x

Franklin JF, Shugart HH, Harmon ME (1987) Tree death as an ecological process. Bioscience 37:550–556

Fulbright TE (1996) Viewpoint: A theoretical basis for planning woody plant control to maintain species diversity. Journal of Range Management 49:554–559

Goslee SC, Havstad KM, Peters DPC, Rango A, Schlesinger WH (2003) High-resolution images reveal rate and pattern of shrub encroachment over six decades in New Mexico, U.S.A. J Arid Environ 54(4):755–767. https://doi.org/10.1006/jare.2002.1103
Grime JP (1977) Evidence for the existence of three primary strategies in plants and its relevance to ecological and evolutionary theory. American Naturalist 111:1169–1194
Hagos MG, Smit GN (2005) Soil enrichment by *Acacia mellifera* subsp detinens on nutrient poor sandy soil in a semiarid southern African savanna. J Arid Environ 61(1):47–59. https://doi.org/10.1016/j.jaridenv.2004.08.003
Hayashi I (1996) Five years experiment on vegetation recovery of drought deciduous woodland in Kitui, Kenya. J Arid Environ 34:351–361. https://doi.org/10.1006/jare.1996.0115
Joubert DF, Rothauge A, Smit GN (2008) A conceptual model of vegetation dynamics in the semi-arid Highland savanna of Namibia, with reference to bush thickening by *Acacia mellifera*. J Arid Environ 72(12):2201–2210. https://doi.org/10.1016/j.jaridenv.2008.07.004
Kgosikoma EO, Harvie BA, Mojeremane W (2012) Bush encroachment in relation to rangeland management systems and environmental conditions in the Kalahari ecosystem of Botswana. Afr J Agric Res 7(15):2312–2319
Kraaij T, Ward D (2006) Effects of rain, nitrogen, fire, and grazing on tree recruitment and early survival in bush-encroached savanna, South Africa. Plant Ecol 186:235–246. https://doi.org/10.1007/s11258-006-9125-4
Levick SR, Rogers KH (2011) Context-dependent vegetation dynamics in an African savanna. Landsc Ecol 26(4):515–528. https://doi.org/10.1007/s10980-011-9578-2
Li X, Gao Y (2012) Effects of shrub encroachment in desert grassland on runoff and the induced nitrogen loss in southeast fringe of Tengger Desert. Acta Ecol Sin 32(24):7828–7835. https://doi.org/10.5846/stxb201204010458
Liao C, Clark PE, DeGloria SD (2018) Bush encroachment dynamics and rangeland management implications in southern Ethiopia. Ecol Evol 8:11694–11703
Liu X, Ma Z, Huang X, Li L (2020) How does grazing exclusion influence plant productivity and community structure in the alpine grasslands of the Qinghai Tibetan Plateau? Global Ecol Conserv\ 23:e01066
Lohmann D, Tietjen B, Blaum N, Joubert DF, Jeltsch F (2014) Prescribed fire as a tool for managing shrub encroachment in semi-arid savanna rangelands. J Arid Environ 107:49–56. https://doi.org/10.1016/j.jaridenv.2014.04.003
Maestre FT, Bowker MA, Puche MD, Belén Hinojosa M, Martínez I, García-Palacios P, Castillo AP et al (2009) Shrub encroachment can reverse desertification in semiarid Mediterranean grasslands. Ecol Lett 12(9):930–941. https://doi.org/10.1111/j.1461-0248.2009.01352.x
McGlynn IO, Okin GS (2006) Characterization of shrub distribution using high spatial resolution remote sensing: ecosystem implications for a former Chihuahuan Desert grassland. Remote Sens Environ 101:554–566
Moleele NM, Ringrose S, Matheson W, Vanderpost C (2002) More woody plants? The status of bush encroachment in Botswana's grazing areas. J Environ Manag 64(1):3–11. https://doi.org/10.1006/jema.2001.0486
Mooney HA, Cleland EE (2001) The evolutionary impact of invasive species. Proc Natl Acad Sci USA 98(10):5446–5451. https://doi.org/10.1073/pnas.091093398
Mourik AA, Van Langevelde F, Van Tellingen E, Heitkonig IMA, Gaigher I (2007) Stability of wooded patches in a South African nutrient-poor grassland: do nutrients, fire or herbivores limit their expansion? J Trop Ecol 23(5):529–537. http://www.jstor.org/stable/4499130
Mwalyosi RBB (2000) Vegetation changes following land reclamation in the Kondoa Eroded Area, Central Tanzania. Afr J Ecol 38:265–268. https://doi.org/10.1046/j.1365-2028.2000.00257.x
Naito AT, Cairns DM (2011) Patterns and processes of global shrub expansion. Prog Phys Geogr 35(4):423–442. https://doi.org/10.1177/0309133311403538
O'Connor TG, Puttick JR, Hoffman MT (2014) Bush encroachment in southern Africa: changes and causes. Afr J Range Forage Sci 31(2):67–88

Oba G (1990) Effects of wildfire on a semi-desert riparian woodland along the Turkwel River, Kenya, and management implication for Turkana pastoralists. Land Degradation and Rehabilitation 2:247–259

Oba G, Post E, Syvertsen PO, Stenseth NC (2000) Bush cover and range condition assessments in relation to landscape and grazing in southern Ethiopia. Landsc Ecol 15:535–546. https://doi.org/10.1023/A:1008106625096

Pauchard A, Alaback PB, Edlund EG (2003) Plant invasions in protected areas at multiple scales: Linaria vulgaris (Scrophulariaceae) in the west Yellowstone area. Western North American Naturalist 63:416–428

Peng H, Li X, Tong S (2013) Effects of shrub encroachment on biomass and biodiversity in the typical steppe of Inner Mongolia. Acta Ecol Sin 33:7221–7229

Peng H, Li X, Tong S (2014) Effects of shrub (*Caragana* Lam.) encroachment on water redistribution and utilization in the typical steppe of Inner Mongolia. Acta Ecol Sin 34(9):2256–2265

Pickett STA, White PS (1985) The Ecology of natural disturbance and patch dynamics. Academic Press, New York

Robinson TP, van Klinken RD, Metternicht G (2008) Spatial and temporal rates and patterns of mesquite (*Prosopis* species) invasion in Western Australia. J Arid Environ 72(3):175–188. https://doi.org/10.1016/j.jaridenv.2007.05.011

Roques KG, O'Connor TG, Watkinson AR (2001) Dynamics of shrub encroachment in an African savanna: relative influences of fire, herbivory, rainfall, and density dependence. J Appl Ecol 38(2):268–280. https://doi.org/10.1046/j.1365-2664.2001.00567.x

Ruthven DC, Braden AW, Knutson HJ, Gallagher JF, Synatzske DR (2003) Woody vegetation responds to various burning regimes in South Texas. Journal of Range Management 56:159–166

Salvatori V, Egunyu F, Skidmore AK et al (2001) The effects of fire and grazing pressure on vegetation cover and small mammal populations in the Maasai Mara National Reserve. Afr J Ecol 39:200–204

Sankaran M, Ratnam J, Hanan N (2008) Woody cover in African savannas: the role of resources, fire, and herbivory. Glob Ecol Biogeogr 17(2):236–245. https://doi.org/10.1111/j.1466-8238.2007.00360.x

Sawadogo L, Nygard R, Pallo F (2002) Effects of livestock and prescribed fire on coppice growth after selective cutting of Sudanian savannah in Burkina Faso. Annals of Forest Science 59:185–195

Scheffer M, Carpenter SR (2003) Catastrophic regime shifts in ecosystems: linking theory to observation. Trends Ecol Evol 18(12):648–656. https://doi.org/10.1016/j.tree.2003.09.002

Shiferaw H, Alamirew T, Dzikiti S, Bewket W, Zeleke G, Schafner U (2021) Water use of *Prosopis juliflora* and its impacts on catchment water budget and rural livelihoods in Afar Region, Ethiopia. Sci Rep 11:2688. https://doi.org/10.1038/s41598-021-81776-6

Skarpe C (1990) Shrub layer dynamics under different herbivore densities in an arid savanna, Botswana. J Appl Ecol 27(3):873–885. https://doi.org/10.2307/2404383

Smart NOE, Hatton JC, Spence DHN (1985) The effects of long-term exclusion of large herbivores on vegetation in Murchison Falls National Park, Uganda. Biol Conserv 33:229–245. https://doi.org/10.1016/0006-3207(85)90015-1

Sohn Y, McCoy RM (1997) Mapping desert shrub rangeland using spectral unmixing and modelling spectral mixtures with TM data. Photogramm Eng Remote Sens 63(6):707–716

Stevens N, Erasmus BFN, Archibald S, Bond WJ (2016) Woody encroachment over 70 years in South African savannahs: overgrazing, global change or extinction aftershock? Philos Trans R Soc B 371:20150437. https://doi.org/10.1098/rstb.2015.0437

Suttie JM, Reynolds SG, Batello C (2005) Grasslands of the world. Food and Agriculture Organization of the United Nations, Roma

Tews J, Schurr F, Jeltsch F (2004) Seed dispersal by cattle may cause shrub encroachment of *Grewia* flava on southern Kalahari rangelands. Appl Veg Sci 7:89–102. https://doi.org/10.1111/j.1654-109X.2004.tb00599.x

Tietjen B, Jeltsch F, Zehe E, Classen N, Groengroeft A, Schiffers K, Oldeland J (2010) Effects of climate change on the coupled dynamics of water and vegetation in drylands. Ecohydrology 3(2):226–237. https://doi.org/10.1002/eco.70

Tong C, Yong W, Wu Y, Zhao L, Jiang C, Yong S (2001) Change in the spatial structure of grassland vegetation in the Xilin River Basin. Acta Sci Natur Univ Neimonggol 32:562–566

Trodd NM, Dougill AJ (1998) Monitoring vegetation dynamics in semi-arid African rangelands: Use and limitations of Earth observation data to characterize vegetation structure. Appl Geogr 18:315–330

Van Auken OW (2009) Causes and consequences of woody plant encroachment into western North American grasslands. J Environ Manag 90(10):2931–2942. https://doi.org/10.1016/j.jenvman.2009.04.023

Van Langevelde F, Van de Vijver C, Kumar L, Van de Koppel J, De Ridder N, Van Andel J, Skidmore AK, Hearne JW, Stroosnijder L, Bond WJ, Prins HHT, Rietkerk M (2003) Effects of fire and herbivory on the stability of savanna ecosystems. Ecology 84(2):337–350

Van Vegten JA (1984) Thornbush invasion in a savanna ecosystem in eastern Botswana. Vegetation 56:3–7. https://doi.org/10.1007/BF00036129

Venter ZS, Cramer MD, Hawkins HJ (2018) Drivers of woody plant encroachment over Africa. Nat Commun 9:2272. https://doi.org/10.1038/s41467-018-04616-8

Vetaas OR (1992) Micro-site effects of trees and shrubs in dry savannas. Journal of Vegetation Science 3:337–344

Wang J, Cai Y, An Z, Liu J (2013) The distribution changes of alpine shrub and its climatic driving forces in Junhe sub-watershed based on remote sensing. Resour Sci 35(6):1300–1309

Ward D (2005) Do we understand the causes of bush encroachment in African savannas? Afr J Range Forage Sci 22(2):101–105. https://doi.org/10.2989/10220110509485867

Westoby M, Walker B, Noy-Meir I (1989) Opportunistic management for rangelands not at equilibrium. J Range Manag 42(4):266–274

White RN, Murray S, Rohweder M (2000) Pilot analysis of global ecosystems: grassland ecosystems. World Resources Institute, Washington, DC. http://www.wri.org/wr2000

Wiegand K, Saltz D, Ward D (2006) A patch-dynamics approach to savanna dynamics and woody plant encroachment-insights from an arid savanna. Perspect Plant Ecol Evol Syst 7(4):229–242. https://doi.org/10.1016/j.ppees.2005.10.001

Wilfong BN, Gorchov DL, Henry MC (2009) Detecting an invasive shrub in deciduous forest understories using remote sensing. Weed Sci 57(5):512–520. https://doi.org/10.1614/WS-09-012.1

Wilson BA, Bowman DMJS (1994) Factors influencing tree growth in tropical savanna: studies of an abrupt eucalyptus boundary at Yapilika, Melville Island, Northern Australia. J Trop Ecol 10(1):103–120. http://www.jstor.org/stable/2559235

Xiong X, Han X, Bai Y, Pan Q (2003) Increased distribution of *Caragana microphylla* in rangelands and its causes and consequences in Xilin River Basin. Acta Pratacul Sin 12:57–62

Xu M, Cao C, Tong Q, Li Z, Zhang H, He Q, Gao M et al (2010) Remote sensing based shrub above-ground biomass and carbon storage mapping in Mu Us Desert, China. Sci China Technol Sci 53(s1):176–183. https://doi.org/10.1007/s11431-010-3231-1

Zhang H, Shi P, Zheng Q (2001) Research progress in relationship between shrub invasion and soil heterogeneity natural semi-arid grassland. Acta Phytoecol Sin 25(3):366–370

Zhang J, Zhao H, Zhang T, Zhao X, Drake S (2005) Community succession along a chronosequence of vegetation restoration on sand dunes in Horqin Sandy Land. J. Arid Environ. 62:555–566

Zhang W, Zhang Y, Wang Z, Ding M, Yang X, Lin X, Yan Y (2006) Analysis of vegetation change in Mt. Qomolangma natural reserve. Prog Geogr 25:12–21

Zhang P, Qing H, Zhang L, Xu Y, Mu L, Ye R, Qiu X, Chang H, Shen H, Yang J (2017) Population structure and spatial pattern of *Caragana tibetica* communities in Nei Mongol shrub-encroached grassland. Chin J Plant Ecol 41(2):165–174. https://doi.org/10.17521/cjpe.2015.0448

Zhou D (1990) Encroached-grassland of *Caragana* Lam. in Inner Mongolia. Grassland Pratac 3:17–19

Zhou Y, Chen J, Chen X, Cao X, Zhu X (2013) Two important indicators with potential to identify *Caragana microphylla* in xilin gol grassland from temporal MODIS data. Ecol Indic 34:520–527. https://doi.org/10.1016/j.ecolind.2013.06.014

Professor Ayana Angassa Abdeta is a Professor in the Department of Range and Forest Resources—Botswana University of Agriculture and Natural Resources. He received BSc and MSc from the Alemaya University of Agriculture, Ethiopia, PhD from the Norwegian University of Life Sciences, Department of International Environment and Development Studies, Postdoctoral Training from both the University of Hohenheim (Food Security Centre) and University of Cologne (Botanical Institute) in Germany, Certificate (2015) in Disaster Risk Management from the Humboldt University of Berlin, Certificate (2020) in Project Management Foundations from the University of Cape Town, and Certificate (2022) in Agroforestry Systems from the University of Freiburg, Germany.

Dr. Keoikantse Sianga is a Senior Lecturer at the Botswana University of Agriculture and Natural Resource. He holds a BSc in Conservation Biology and Wildlife Management, and an MPhil and a PhD in Natural Resource Management, both from the University of Botswana. His research interests include functional and vegetation heterogeneity and plant-wildlife interactions, fire ecology and management, human-wildlife interactions, and spatial ecology. Dr Sianga has published four articles in peer reviewed journals and seven book chapters.

Ms. Summer Mabula is an Ecologist with an experience in and enthusiasm for teaching and carrying out field work. She has expertise in training and conducting range assessments, plant identification, environmental monitoring as well as writing and presenting scientific reports. She is passionate about ecology and possesses an in-depth knowledge about ecosystems. She holds a Bachelor of Science in Range Science from the Botswana University of Agriculture and Natural Resources.

Witness Mojeremane is a Professor in Forest Sciences at the Botswana University of Agriculture and Natural Resources (BUAN). He holds a Diploma in Forestry, a Diploma in Human Resources Management, a Bachelor of Science (Honours) in Forestry, a Master of Science in Tropical Environmental Forestry, and a PhD in Ecology and Resource Management. He joined BUAN as an instructor in 1994. His research interests include forest sciences and has been engaged in several consultancy and research Projects. Professor Mojeremane has published 52 articles in refereed journals, 16 book chapters, and 3 articles in refereed conference proceedings.

Chapter 15
Evaluating Effects of Pre-sowing Treatments on Germination of *Brachystegia boehmii* and *Piliostigma thonningii*

Witness Mojeremane, Ronnie Matlanaga Mmolotsi, Melusi Rampart, Demel Teketay, Obusitswe Kakambi, Topoyame Isaac Makoi, Ishmael Kopong, Kamogelo Makgobota, Thembinkosi Mathowa, Gosiame Neo-Mahupeleng, and Keoikantse Sianga

15.1 Introduction

Forests and woodlands are critical components of the ecosystem, providing a variety of goods and environmental services to society (CSO 2004; Rampart et al. 2021a). In Botswana and other places, forests and woodlands are valuable natural resources. They supply fuelwood, charcoal, wood for construction and domestic utensils, fruits, traditional medicine, and animal feed to adjacent communities (Bekele 1994; Cavendish 2000; Shackleton et al. 2007; O'Brien and Morris 2014). They also prevent soil erosion (Zhaoa et al. 2009) and desertification (Carucci 2000;

W. Mojeremane (✉) · R. M. Mmolotsi · M. Rampart · D. Teketay · T. I. Makoi · I. Kopong · K. Makgobota
Department of Range and Forest Resources, Faculty of Natural Resources, Botswana University of Agriculture and Natural Resources, Gaborone, Botswana
e-mail: wmojerem@buan.ac.bw; rmmolots@buan.ac.bw; mrampart@buan.ac.bw; dteketay@buan.ac.bw; tmakoi@buan.ac.bw; ikopong@buan.ac.bw; kmakgobota@buan.ac.bw

O. Kakambi
Department of Forestry and Range Resources, Ministry of Environment and Tourism, Kasane, Botswana

T. Mathowa
Department of Crop and Soil Sciences, Faculty of Agriculture, Botswana University of Agriculture and Natural Resources, Gaborone, Botswana
e-mail: tmathowa@buan.ac.bw

G. Neo-Mahupeleng · K. Sianga
Department of Wildlife and Aquatic Resources, Faculty of Natural Resources, Botswana University of Agriculture and Natural Resources, Gaborone, Botswana
e-mail: gmahupeleng@buan.ac.bw; ksianga@buan.ac.bw

W. L. Hambira et al. (eds.), *Environmental Change and Biodiversity Conservation in sub-Saharan Africa*, Advances in Global Change Research 75,
https://doi.org/10.1007/978-3-031-73136-5_15

Malagnoux et al. 2008) protect watersheds (Malagnoux et al. 2008; Portillo-Quintero et al. 2015), conserve biodiversity (Wu et al. 2010), and sequester carbon (Sedjo 2001).

Globally, forests and woodlands are overexploited which jeopardizes the benefits they provide. The degradation of forests and woodlands due to overexploitation is well documented in Botswana (Sebego et al. 2019; Teketay et al. 2018; Akinyemi and Kgomo 2019) and elsewhere (Pouliot et al. 2012; Rudel 2013; Fox et al. 2017). Forest and woodland ecosystems are threatened by anthropogenic activities such as conversion to farmland, infrastructure development, extraction for firewood and construction material, seasonal forest fires, livestock and wild animal grazing, and climate change. Despite threats to forests and woodlands, knowledge of tree and shrub propagation for reforestation and afforestation programmes in Botswana is scanty.

Direct sowing of seeds in the field is an efficient and cost-effective method for restoring degraded forests or abandoned fields around the world (King and Keeland 1999; Engel and Parrotta 2001; Hooper et al. 2002; Sampaio et al. 2007). Seed germination has recently become an important tool for restoring and conserving many threatened plant species around the world (Vodouhè et al. 2011; Iralu and Upadhaya 2018). Seeds have been used to propagate seedlings for mass tree planting programmes (Mbailwa et al. 2023). However, propagating seedlings in arid and semi-arid environments is difficult. Seeds of many woody plants are difficult to germinate in the field (Luna et al. 2011) even when planted under optimal conditions (Botsheleng et al. 2014; Rampart et al. 2021a). This may be due to undeveloped embryos (Amoakoh et al. 2017) and seed coat impermeability that blocks entrance of water, gases, and other external elements (Smýkal et al. 2014). Seed coat impermeability is a mechanism that prevents seeds of many legume plants from germinating (Holmes et al. 1987; Pan et al. 2023) because it prevents water uptake (Owens et al. 1995). Studies conducted in Botswana (Mojeremane et al. 2017, 2018, 2021; Koobonye et al. 2018; Odirile et al. 2019; Botumile et al. 2020; Mmolutsi et al. 2020; Latiwa et al. 2023) and elsewhere (Amri 2010; Azad et al. 2011; Sinhababu and Banerjee 2013) reported seed coat impermeability in many legume trees.

Seeds with impermeable coats may need the application of various scarification techniques to break dormancy to achieve rapid, uniform, and high rate of germination (Teketay 1996a; Gebre and Karam 2004; Zincirkiran et al. 2010; Müller et al. 2017; Taghizadeh and Sajadi 2023). Many researchers have successfully utilized cold stratification, nicking, acid, boiling water to scarify seeds into order to break dormancy and accelerate germination (Teketay 1996b, 1998; Mojeremane et al. 2020, 2021; Latiwa et al. 2023; Marfoua et al. 2023). Scarification techniques have shown to break tough seed coats, thereby allowing water and oxygen uptake by the seeds. Each pre-sowing treatment technique has advantages and disadvantages that vary according to plant species (Mojeremane et al. 2021).

Brachystegia boehmii Taub. is a forest tree species that belongs to the Leguminosae family and is native to eastern and southern Africa (Palgrave 2002). It occurs at elevations ranging from 700 to 1600 m. It is a key component of the miombo woodlands, which is found in Tanzania, Zimbabwe, Zambia, the Democratic Republic of

Congo, Malawi, Mozambique, Angola, and northern Botswana (Bingham et al. 2023). According to Storrs (1995) and Moura et al. (2017), this species frequently occurs in poorly drained soils. It is a semi-deciduous tree (Storrs 1995) reaching a height of 5–16 m high (Palgrave 2002) with a flat-crowned shape and leaves that droop like a feather. *Brachystegia boehmii* has grey to grey-black bark with rectangular scales and deep vertical fissures on mature trees and purple-grey on branches. Leaves are alternate compound paripinnate with 11–30 pairs of leaflets on each leaf. Leaflets are oblong to oblong triangular with a slightly wider base than apex, subsessile, and crowded on the ochreous-rusty pubescent rachis. Flowers have velvety brown sepals and pedicels, and a greenish-white corolla. Fruits are a flattened rhomboid pod that are green when young, orange-brown when maturing and become woody and dehiscing explosively with age. Seeds are shiny dark brown and disc-like.

The wood is reddish brown, heavy, tough, and strong with coarse interlocking grains. It is used for mining timber, domestic tools, canoes, construction material, joinery, firewood, and charcoal (Storrs 1995). The timber is susceptible to borers if used untreated (Palmer and Pitman 1972; Bruschi et al. 2014). The bark contains 3% tannin; an extract has been used as the final dressing in the tanning process of hides (Palgrave 2002). The bark is fibrous and used for making baskets, fishnets (Storrs 1995; Moura et al. 2017), and beds and ropes (Moura et al. 2017). The bark is peeled and used for fabricating beehives (Moura et al. 2017). The ash may be utilized as fertilizer in shifting cultivation. Parts of *B. boehmii* are used in traditional medicine to treat a variety of diseases (Storrs 1995; Augustino et al. 2011; Chitemerere and Mukanganyama 2011; Amri and Kisangau 2012; Irondi et al. 2015; Chirisa and Mukanganyama 2016; Sibanda et al. 2022).

Piliostigma thonningii (Schumach.) Milne-Redh. also known as "camel foot" is a member of the family Fabaceae (Jimoh and Oladiji 2005; Omoregie and Oluyemisi 2010; Ighodaro and Omole 2012; Ebele et al. 2021). It occurs throughout tropical Africa, apart from Somalia (Storrs 1995; Orwa et al. 2009). The species is also found in certain parts of Asia (Ayisire et al. 2009). It *grows in open woodlands, grasslands, and wooded grasslands in low to medium elevations* (Storrs 1995; Jimoh and Oladiji 2005; Palgrave 2002). *P. thonningii* is a small- to medium-sized rounded evergreen tree that can reach 5 m in height (Palgrave 2002) and 15 m under ideal conditions (Storrs 1995). *The bark is* rough and dark brownish or grey in colour. The leaves are conspicuous, simple, two-lobed, and leathery, resembling a camel foot. *P. thonningii* fruits are large, thick, green when unripe, turning reddish brown as they ripen and do not split.

Piliostigma thonningii fixes nitrogen and contributes significantly to the cycling of nutrients in soils (Ayisire et al. 2009). *The wood is used as firewood* (Storrs 1995). *It is an excellent fodder tree species. Farmers collect pods and grind them into a meal to feed cattle during the dry season* (Storrs 1995; Jimoh and Oladiji 2005; Venter and Venter 2012). *Ground pods are also mixed with molasses and mineral salts to feed livestock* (Storrs 1995). *Pods and leaves are eaten by cattle* (Venter and Venter 2012) *and wild animals, especially antelope and elephant* (Palgrave 2002; Jimoh and Oladiji 2005; Venter and Venter 2012). *The tough inner bark is used in rope making* (Venter and Venter 2012). *Pods and seeds produce a blue dye that imparts a blackish colour to leather* (Venter and Venter 2012). *Piliostigma thonningii is an excellent garden that provides shade* (Venter and Venter 2012; Ayisire et al. 2009). *Almost parts of the tree (*bark, root, leaves, twigs, and fruits) are used in traditional medicine to treat various diseases (Akinniyi and Sultanbawa 1983; Akinpelu and Obuotor 2000; Egharevba and Kunle 2010; Jimoh and Oladiji 2005; Ayisire et al. 2009; Kwaji et al. 2010; Omoregie and Oluyemisi 2010; Ighodaro and Omole 2012; Asuzu and Nwaehujor 2013; Afolayan et al. 2018; Yakubu et al. 2021).

Both *Brachystegia boehmii* and *Piliostigma thonningii* are important native tree species that offer an array of goods and environmental services. Despite their significance, the two species have not been included in afforestation and reforestation initiatives in Botswana because there is insufficient or lack of knowledge on their propagation needs. It is necessary to identify native woody species with readily available seed and propagation techniques that can succeed in Botswana. Therefore, the goal of this study was to ascertain how different seed pre-sowing treatment techniques will affect germination of *B. boehmii* and *P. thonningii* seeds.

15.2 Materials and Methods

15.2.1 Study Site

The seed germination experiment was carried out at Botswana University of Agriculture and Natural Resources (BUAN) in Sebele, Gaborone, Botswana, in 2019. The research site is located at 23° 34′ S, 25° 57′ E, 994 m above sea level. BUAN is located 10 km away from the centre of Gaborone City, the capital city of Botswana, along the A1 North-South highway. With a hot, rainy season from November to March and a dry season from April to October, Gaborone has a semi-arid, subtropical climate. The average temperature of the coldest month (July) is 13.5 °C and that of the warmest month January is 26 °C. Gaborone receives an annual rainfall of 485 mm.

15.2.2 Collection of Seed Material

Seeds used in the study were collected from Kazuma Forest Reserve (18. 4259° S, 25.4970° E, altitude 997 m asl, Fig. 15.1) in the Chobe District in 2018. Kazuma Forest Reserve is one of the six forest reserves located in the Chobe District, north-eastern Botswana (Fig. 15.1). For *Brachystegia boehmii*, fresh pods were picked from the forest floor under mother trees. For *P. thonningii, pods were collected* by hand from mother trees selected at random. The gathered pods were placed inside paper bags and delivered to BUAN. A secateur was used to crush pods to extract

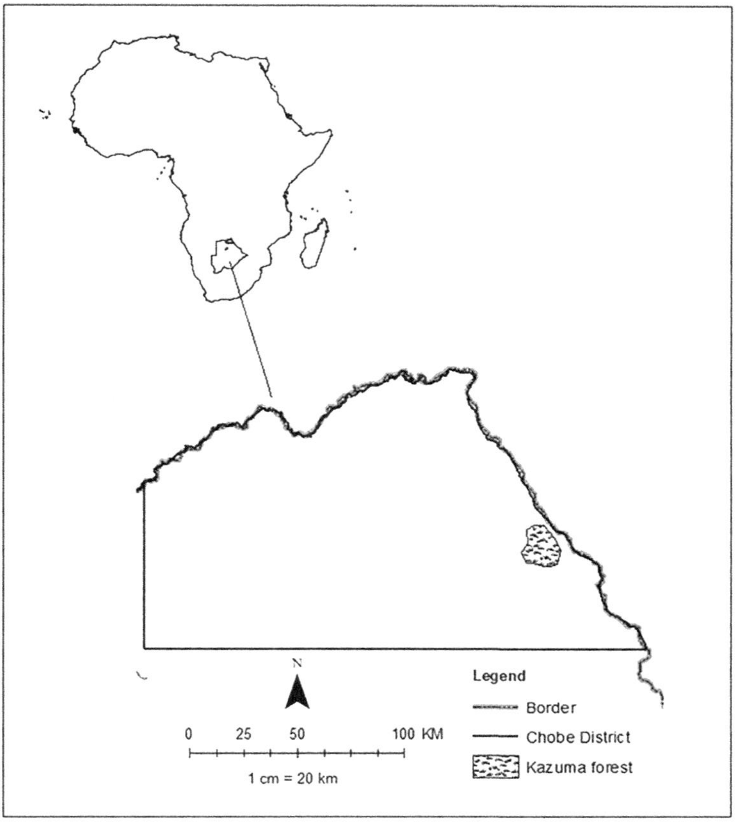

Fig. 15.1 A map showing Kazuma Forest Reserve. (Cartography: Keoikantse Sianga)

seeds in the laboratory. Thereafter, extracted seeds were stored in a refrigerator at 5 °C waiting for commencement of the experiment.

15.2.3 Experimental Design and Seed Pre-sowing Treatments

A completely randomized design (CRD) was used for this research. There were ten treatments in this study including control. The treatments were nicking/mechanical scarification, soaking seeds in boiling water for 1, 3, and 5 min, hot water (allowed to cool for 24 h) and concentrated sulphuric acid for 15, 30, 45, and 60 min. The germination experiment was conducted in the BUAN laboratory between January and February 2022. All the treatments had four replications of 25 seeds each. Seeds were germinated in 90 mm petri dishes lined with cotton wool which was continuously moistened with distilled water to avoid drying. Germination was recorded daily for a period of 1 month. Seeds were considered to have germinated if there was a visible radicle emergence (Tian et al. 2014). After counting and recording, germinated seeds were removed and thrown away. Using a floating method, the viability of the seeds was assessed prior to germinating seeds. This required soaking seeds in water for a short period of time. The seeds that floated were considered not viable and thrown away, while the ones that sank were considered viable. At the conclusion of the study, ungerminated seeds from each treatment were evaluated for viability using the cutting test. Viable seeds had a white firm embryo, while non-viable seeds had a brownish-yellow, squishy embryo (Fan et al. 2016).

Nicking

Effects of nicking on seed germination of the two study species were assessed using four replications of 25 seeds each. A nail cutter was used to nick seeds on the convex end opposite of the embryo. Care was taken during nicking to protect the embryo.

Concentrated Sulphuric Acid (H2SO4)

Using the method described by Teketay (1996a) seeds were soaked in sulphuric acid for 15, 30, 45, and 60 min, respectively. Four replications of 25 seeds for sulphuric acid treatments of each study species were put into four 100 mL, non-corrosive, heat-resistant, glass beakers containing sulphuric acid by making sure that all the seeds were covered by the acid. Seeds were stirred every 5 min during the specific exposure time to ensure uniform exposure to the acid. Following the designated soaking times, the acid was simultaneously drained off into a different beaker and the seeds were removed from the acid using an acid-resistant sieve. Seeds were then thoroughly washed and rinsed to remove acid using a running water tap first and, subsequently, using distilled water, successively.

Boiling Water

For the boiling water treatment, seeds of each study species were boiled for 1, 3, and 5 min, respectively. Four replications of 25 seeds of each boiling water treatment were placed into four separate coffee filter papers and immersed into a kettle with boiling water for the specified period. Thereafter, the seeds were removed from boiling water and placed in a bucket containing distilled water to cool them down for a few minutes.

Hot Water (Allowed to Cool for 24 h)

In this treatment, 250 mm beakers were filled with four replicates of 25 seeds of each of the two species; each replicate was placed in a separate coffee filter. After that, the beakers containing the seeds were filled with boiling water and allowed to cool for 24 h.

Control

Four replications of 25 untreated seeds of the two study species were used as control for all the experiments.

15.2.4 Data Analysis

The germination percentage (GP) was computed using data on seeds that germinated. The final germination was determined by applying the following formula:

$$\text{Germination percentage} = \frac{\text{Total mumber of seeds germinated}}{\text{Total number of seeds sown}} \times 100$$

Statistix software, Version 10 (Statistix 10, 1984–2003) was used to perform an analysis of variance (ANOVA) on the collected data. The germination data were arcsine transformed prior to the ANOVA to ensure that they met normalcy requirements (Zar 1996). Tukey's honestly significant difference (HSD) was used to test for significance in means at the significance level of $p < 0.05$.

15.3 Results

15.3.1 Seed Germination

Pre-sowing treatments had a highly significant ($p < 0.0001$) impact on *Brachystegia boehmii* seeds (Table 15.1). The percentage of germination of seeds that were soaked in sulphuric acid for 60 min and those that were boiled in water for 1, 3, and 5 min, respectively, were significantly lower than the percentage of untreated seeds (Table 15.1). The percentage of germination of untreated and nicked seeds as well as those that were soaked in sulphuric acid for 15, 30, and 45 min did not differ significantly. Both untreated and nicked B. *boehmii* seeds exhibited 100%

Table 15.1 Effect of seed pre-sowing treatment on percent germination of *B. boehmii* and *P. thonningii* (means ± standard error)

Treatment	Germination (%)	
	B. boehmii	*P. thonningii*
1. Control	100 ± 0^{a}	$6 \pm 2.58^{b,c}$
2. Nicking/mechanical scarification	100 ± 0^{a}	98 ± 1.15^{a}
3. Sulphuric acid (15 min)	$86.75 \pm 9.4^{a,b}$	99 ± 1^{a}
4. Sulphuric acid (30 min)	98.2 ± 1.75^{a}	100 ± 0^{a}
5. Sulphuric acid (45 min)	$94.75 \pm 1.75^{a,b}$	100 ± 0^{a}
6. Sulphuric acid (60 min)	75 ± 5.76^{b}	100 ± 0^{a}
7. Boiling water (1 min)	15 ± 5.76^{c}	23 ± 4.43^{b}
8. Boiling water (3 min)	0^{c}	1 ± 1^{c}
9. Boiling water (5 min)	0^{c}	3 ± 3^{c}
Hot water for 24 h	$85 \pm 9.6^{a,b}$	$9 \pm 5.26^{b,c}$

Means separated using Tukey's honestly significant difference (HSD) test at $p < 0.05$. Means with different letters as supercripts in each column are significantly different germination. The lowest germination (15%) was seen in seeds that were boiled in water for 1 min. There was no sign of germination in seeds boiled in water for 3 and 5 min.

Table 15.1 indicates a highly significant ($p < 0.0001$) impact of pre-sowing treatments on the percent germination of *Piliostigma thonningii.* In comparison with untreated seeds, percentage germination of nicked seeds and those that were soaked in sulphuric acid for 15, 30, 45, and 60 min showed significantly higher germination rates. The percentage of seeds that germinated after being boiled for 1, 3, and 5 min and those soaked in hot water (allowed to cool for 24 h) did not differ significantly from the control treatment. The boiling and hot water treatments resulted in low germination.

15.3.2 Rate of Seed Germination

According to the results of the present study untreated seeds of *B. boehmii* seeds and their counterparts subjected to sulphuric acid for 30 and 45 min as well as those nicked showed the fastest and most consistent germination (Fig. 15.2). The control and nicked seeds reached 100% cumulative germination in 8 days while those exposed to sulphuric acid for 30 and 45 min reached cumulative germination >90% within 7 and 8 days (Fig. 15.2), respectively. These were followed by those treated in hot water (allowed to cool for 24 h) which reached >80% cumulative germination

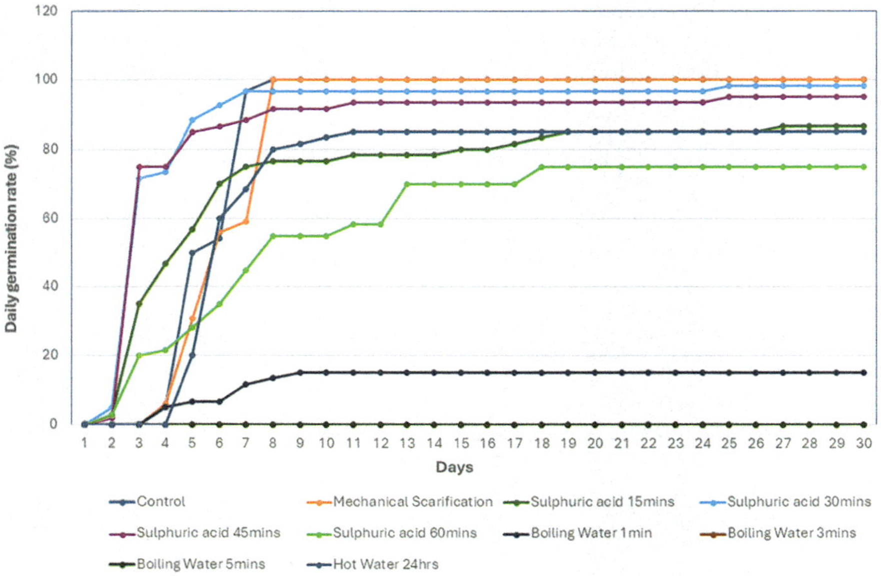

Fig. 15.2 Cumulative germination percentage of *B. boehmii* recorded for 30 days

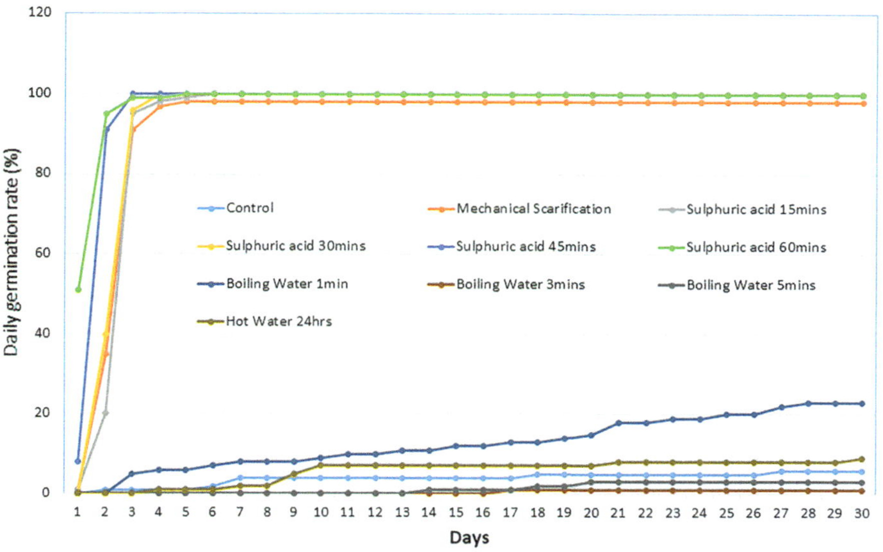

Fig. 15.3 Cumulative germination percentage of *P. thonningii* seeds recorded in 30 days

within 11 and 17 days, respectively. These were followed by those treated in sulphuric acid for 60 min which attained 75% cumulative germination within 18 days. On the other hand, untreated seeds and those treated with boiling water for 1, 3, and 5 min exhibited not only the lowest but the slowest germination.

The results showed that in the case of *P. thonningii*, nicked seeds and those soaked in sulphuric acid for 15, 30, 45, and 60 min exhibited the fastest and most uniform germination, reaching 100% cumulative germination within 5 days (Fig. 15.3). On the other hand, untreated seeds and those treated with hot water (allowed to cool for 24 h) and boiling water for 1, 3, and 5 min exhibited not only the lowest but also the slowest germination.

15.4 Discussion

Many leguminous tree and shrub species found in arid and semi-arid environments have seeds with impermeable hard coats that cause physical dormancy (Werker 1980). The hard seed coats have developed to withstand an array of hard field conditions, including intense solar heat, animals that disperse seeds, severe drought, and physical damage. Many leguminous plants impose seed dormancy, which is the suspension of germination in viable seeds because of the thick seed coats that are impermeable to uptake of oxygen and water. A hard seed coat allows the seed to store for a very long time but will require pre-sowing to allow water uptake and gaseous exchange or else germination rates will be poor. A common constraint that restricts the propagation of native woody plants in arid and semi-arid ecosystems for use in establishing forests (Smith et al. 2003) is poor seed germination which has been attributed to hard seed coats (Mojeremane et al. 2021).

The results of this study indicate that 100% of untreated *B. boehmii* seeds germinated. The germination was not significantly different from nicked seeds (100%) and those soaked in sulphuric acid for 5 (86.75%), 30 (98.2%), and 45 (94.75%) min as well as hot water allowed to cool for 24 h (85%). The results corroborate those reported by other authors for certain tree species belonging to the family Fabaceae, including *Colophospermum mopane* (J. Kirk ex Benth.) J. Kirk ex J. Léonard (Mojeremane and Kgathi 2005), *Baikiaea plurijuga* Harms (Botsheleng et al. 2014), *Brachystegia microphylla* Harms, *Brachystegia utilis* Hutch. & Burtt Davy, *Brachystegia bussei* Harms, *Julbernardia globiflora* (Benth.) Troupin (Nkengurutse et al. 2016), *Philenoptera violacea* (Klotzsch) Schrire (Setlhabetsi et al. 2019), *Guibourtia coleosperma* (Benth) J. Léonard (Rampart et al. 2021a), *Cassia abbreviata* Oliv., *Senegalia nigrescens* (Oliv.) P.J.H. Hurter (Rampart et al. 2021b), *Pericopsis angolensis* (Baker) Meeuwen (Nkengurutse et al. 2016; Mbailwa et al. 2023), and *Senegalia mellifera* (Benth.) Seigler & Ebinger (Madisa 2022). Seeds treated with sulphuric acid for 60 min attained germination of 75% which was significantly lower compared to untreated seeds. These results show that the seed coat of *B. boehmii* is impermeable to water and oxygen.

In this study, results also show that nicking and soaking *P. thonningii* seeds in sulphuric acid significantly enhanced germination compared to untreated seeds. The finding suggests that seeds of this species have hard impermeable seed coats. Impermeability prevents the embryo growth by restricting gas exchange and water entrance (Nasr et al. 2013). Nicking seeds prior to sowing probably created scars on the hard seed coat's surface and enhanced the embryo's ability to absorb oxygen and water. The findings corroborate those of previous studies (Koobonye et al. 2018; Mojeremane et al. 2021; Lujan Rocha et al. 2022) which revealed enhanced germination in tree seeds subjected to nicking.

When compared to untreated seeds, nicking and all sulphuric acid treatments enhanced the germination of *Piliostigma thonningii* (Table 15.1). Sulphuric acid softens the seed coat and speeds up the germination process. The softening of the seed coat increases the embryo's ability to absorb water by creating pores or cracks in the covering tissue. The enhanced germination rate suggests that *P. thonningii*'s hard seed coat, which is typical of many legume plants, is primarily to blame for its seed dormancy. This hard seed coat prevents the seed from absorbing the oxygen and water needed for germination (Baskin and Baskin 1998). The current findings are in line with those of previous studies conducted on other tree species, which reported that sulphuric acid enhanced germination (Teketay 1998; Gebre and Karam 2004; Zincirkiran et al. 2010; Taghizadeh and Sajadi 2023).

According to these results, hot water allowed to cool for 24 h achieved significantly lower germination when compared to untreated seeds. Similar results were reported in other woody species (Botumile et al. 2020; Mojeremane et al. 2017, 2020, 2021; Rampart et al. 2021a). It is possible that the observed poor germination may have been caused by the impermeable seed coat which was not softened by the hot water as it cooled down. The lowest germination in the two study species was recorded in seeds treated in boiling water. The reason for the observed low germination might be attributed to the seed's sensitivity to elevated temperatures. According to Kahaka et al. (2018), the boiling water's high temperature kills the seed embryo. Similar observations were noted by other researchers in different tree species (Mojeremane et al. 2021; Kahaka et al. 2018; Odirile et al. 2019; Botumile et al. 2020; Mmolutsi et al. 2020; Rampart et al. 2021a, b). For example, Botumile et al. (2020) did some work on *Senegalia robusta* (Burch.) Kyalangalilwa & Boatwright and found that germination decreased with increasing exposure in boiling water up to 5 min.

15.5 Conclusion

The results of this study demonstrated that *P. thonningii* seeds have an impermeable seed coat, which prevents entry of water and oxygen to initiate embryo growth. In contrast, *B. boehmii* seeds are not impermeable to water and oxygen, meaning they do not require pretreatment before sowing. Consequently, to achieve fast and uniform germination *P. thonningii* seeds require pre-sowing treatment prior to sowing. Germination of *P. thonningii* was enhanced by sulphuric acid and nicking. As a result, foresters and farmers aspiring to propagate this species are advised to use nicking which is not expensive and a less dangerous technique. Using sulphuric acid effectively increased germination and decreased germination mean time in *P. thonningii*. The high cost of the acid, its corrosive nature and possible danger to farmers and nursery personnel, and requirement for cautious handling procedures make sulphuric acid unsuitable for use in tree nurseries.

Acknowledgements The authors would like to acknowledge the financial support from the Research, Technology Development and Transfer Committee (RTDTC), Botswana University of Agriculture and Natural Resources. BUAN is also thanked for providing facilities for the experiments. They would also like to thank the Department of Forestry and Range Resources, Ministry of Environment and Tourism for their permission to collect seeds in Kazuma Forest Reserve under Research Permit [Ref. ENT 8/36/4 XXXX (29)].

References

Afolayan M, Srivedavyasasri R, Asekun OT, Familoni OB, Orishadipe A, Zulfiqar F, Ibrahim MA, Ross SA (2018) Phytochemical study of *Piliostigma thonningii*, a medicinal plant grown in Nigeria. Med Chem Res 27(10):2325–2330

Akinniyi JA, Sultanbawa MUS (1983) A glossary of names of plants with botanical names, distribution, and uses. Ann Borno 1:85–98

Akinpelu DA, Obuotor EM (2000) Antibacterial activity of *Piliostigma thonningii* stem bark. Fitoterapia 71:442–443

Akinyemi FO, Kgomo MO (2019) Vegetation dynamics in African drylands: an assessment based on the Vegetation Degradation Index in an agro-pastoral region of Botswana. Reg Environ Chang 19(7):2027–2039

Amoakoh OA, Nortey DDN, Sagoe F, Amoako PK, Jallah CK (2017) Effects of pre-sowing treatments on germination and early growth performance of *Pouteria campachiana*. For Sci Technol 13(2):83–86

Amri E (2010) The effects of pre-sowing seed treatments on germination of snake bean (*Swartzia madagascariensis*): a reported medicinal plant. Res J Agric Biol Sci 6(4):557–561

Amri E, Kisangau DP (2012) Ethnomedicinal study of plants used in villages around Kimboza forest reserve in Morogoro, Tanzania. J Ethnobiol Ethnomed 8:1. https://doi.org/10.1186/1746-4269-8-1

Asuzu IU, Nwaehujor FO (2013) The antidiabetic, hypolipidemic, and antioxidant activities of D-3-O-methulchiroinostol in alloxan-induced diabetic rats. Hygeia J Drugs Med 5:27–33

Augustino S, Hall JB, Makonda FBS, Ishengoma RC (2011) Medicinal resources of the Miombo woodlands of Urawa, Tanzania: plants and its uses. J Med Plants Res 5:6352–6372

Ayisire BE, Akinro LA, Amoo SO (2009) Seed germination and in vitro propagation of *Piliostigma thonningii*: an important medicinal plant. Afr J Biotechnol 8(3):401–404

Azad MS, Manik MR, Hasan MS, Matin MA (2011) Effect of different pre-sowing treatments on seed germination percentage and growth performance of *Acacia auriculiformis*. J For Res 22(2):183–188

Baskin CC, Baskin JM (1998) Seeds: ecology, biogeography, and evolution of dormancy and germination. Academic Press, San Diego

Bekele T (1994) Phytosociology and ecology of humid Afromontane Forest in the Central Plateaus of Ethiopia. J Veg Sci 5(1):87–98

Bingham MG, Willemen A, Wursten BT, Ballings P, Hyde MA (2023) Flora of Zambia: species information: *Brachystegia boehmii*. https://www.zambiaflora.com/speciesdata/species.php?species_id126610. Retrieved 20 February 2023

Botsheleng B, Mathowa T, Mojeremane W (2014) Effects of pre-sowing methods on the germination of pod mahogany (*Afzelia quanzensis*) and mukusi (*Baikiaea plurijuga*) seeds. Int J Innov Res Sci Eng Technol 3(1):8108–8113

Botumile A, Teketay D, Mojeremane W, Mathowa T (2020) Overcoming seed dormancy of *Senegalia galpinii* and *Vachellia* robusta through scarification pre-sowing treatments. Agric For 66(1):153–169

Bruschi P, Mancini M, Mattioli M, Michela M, Signorini MA (2014) Traditional uses of plants in a rural community of Mozambique and possible links with Miombo degradation and harvesting sustainability. J Ethnobiol Ethnomed 10:59. https://doi.org/10.1186/1746-4269-10-59

Carucci R (2000) Trees outside forests: an essential tool for desertification control in the Sahel. Unasylva 51(200):18–24

Cavendish W (2000) Empirical regularities in the poverty-environment relationship of rural households: evidence from Zimbabwe. World Dev 28:1979–2003

Central Statistics Office, CSO (2004) Forestry statistics. Central Statistics Office, Gaborone, 13 pp

Chirisa E, Mukanganyama S (2016) Evaluation of in vitro anti-inflammatory and antioxidant activity of selected Zimbabwean plant extracts. J Herbs Spices Med Plants 22(2):157–172

Chitemerere TA, Mukanganyama S (2011) In vitro antibacterial activity of selected medicinal plants from Zimbabwe. Afr J Plant Sci Biotechnol 5(1):1–7

Ebele OP, Ebere CC, Obinna O, Ginikachukwu U (2021) Pharmacognostic study of leaves of *Piliostigma thonningii* Schum (Caesalpiniaceae). J Adv Med Pharm Sci 11:1–9

Egharevba HO, Kunle FO (2010) Preliminary phytochemical and proximate analysis of the leaves of *Piliostigma thonningii* (Schumach.) Milne-Redhead. Ethnobot Leafl 14:570–577

Engel VL, Parrotta JA (2001) An evaluation of direct seeding for reforestation of degraded lands in central São Paulo State, Brazil. For Ecol Manag 152(1–3):169–181

Fan SG, Wang YR, Baskin CC, Baskin JM, Hu XW (2016) A rapid method to determine germinability and viability of *Agriophyllum squarrosum* (Amaranthaceae) seeds. Seed Sci Technol 44(2):410–415

Fox JT, Vandewalle ME, Alexander KA (2017) Land cover change in northern Botswana: the influence of climate, fire, and elephants on semi-arid savanna woodlands. Land 6:73. https://doi.org/10.3390/land6040073

Gebre GH, Karam SN (2004) Germination of C. Siliquastrum seeds in response to gibberellic acid and stratification. Seed Sci Technol 32(1):255–260

Holmes RJ, MacDonald JNAW, Juritz J (1987) Effects of clearing on seed bank of the alien invasive shrubs Acacia saligna and Acacia cyclops in South and South Western Cape, South Africa. J Appl Ecol 24:1045–1051

Hooper E, Condit R, Legendre P (2002) Responses of 20 native tree species to reforestation strategies for abandoned farmland in Panama. Ecol Appl 12(6):1626–1641

Ighodaro OM, Omole JO (2012) Effects of Nigerian *Piliostigma thonningii* species leaf extract on lipid profile in Wistar rats. Pharmacology 4. https://doi.org/10.5402/2012/387942

Iralu V, Upadhaya K (2018) Seed dormancy, germination, and seedling characteristics of *Elaeocarpus prunifolius* Wall. ex Müll. Berol.: a threatened tree species of north-eastern India. N Z J For Sci 48:16 10.1186/s40490-018-0121-y

Irondi EA, Oboh G, Akindahunsi AA (2015) Methanol extracts of *Brachystegia eurycoma* and *Detarium microcarpum* seeds flours inhibit some key enzymes linked to the pathology and complications of type 2 diabetes in vitro. Food Sci Human Wellness 4:162–168

Jimoh FO, Oladiji AT (2005) Preliminary studies on *Piliostigma thonningii* seeds: proximate analysis, mineral composition, and phytochemical screening. Afr J Biotechnol 4(12):1439–1442

Kahaka P, Mojeremane W, Teketay D, Mathowa T (2018) Effects of scarification treatments on seed germination of three leguminous species from Botswana. Silva Lusit 26(1–2):115–131

King SL, Keeland BD (1999) Evaluation of reforestation in the lower Mississippi River alluvial valley. Restor Ecol 7(4):348–359

Koobonye M, Maule BV, Mogotsi K (2018) Mechanical scarification and hot water treatments enhance germination of *Leucaena leucocephala* (Lam.) seeds. Livest Res Rural Dev 30(1). http://www.lrrd.org/lrrd30/1/kbmo30015.html

Kwaji A, Bassi PU, Aoill M, Nneji CM, Ademowo G (2010) Preliminary studies on *Piliostigma thonningii* Schum leaf extract: phytochemical screening and in vitro antimalarial activity. Afr J Microbiol Res 4(9):735–739

Latiwa AN, Makgobota K, Mojeremane W, Teketay D (2023) Seed characteristics and the influence scarification treatments on germination of *Pterocarpus angolensis* in Botswana. J Exp Biol Agric Sci 11(6):1021–1029

Lujan Rocha R, Khalil Y, Maity A, Beckie HJ, Ashworth MB (2022) Mechanical scarification technique breaks seed coat-mediated dormancy in wild oat (*Avena fatua*). Weed Technol 36:152–159

Luna T, Wilkinson KM, Kasten DR (2011) Chapter 9: Seed germination and sowing options. In: Wilkinson KM, Landis TD, Haase DL, Daley BF, Dumroese RK (eds) Tropical nursery manual: a guide to starting and operating a nursery for native and traditional plants. USDA Forest Service, Washington, DC, pp 163–183

Madisa F (2022) Contribution of seed germination of *Senegalia mellifera* to its invasion of natural habitats – evidence from laboratory experiments. Unpublished BSc Project. Department of Range and Forest Resources, Botswana University of Agriculture and Natural Resources, Gaborone

Malagnoux M, Sène EH, Atzmon N (2008) Forests, trees, and water in arid lands: a delicate balance. Unasylva-FAO 229:24

Marfoua M, Buisson E, Bissati-Bouafia S, Eddoud A, Benchettouh A (2023) Improving seed germination of native species for camel rangeland restoration: effects of cold stratification and temperature. Bot Lett. https://doi.org/10.1080/23818107.2023.2176356

Mbailwa YS, Chamshama SAO, Mwendwa BA, Mwambusi JN (2023) Effects of different pre-sowing on germination of *Pericopsis angolensis* seeds of Tabora Miombo woodlands, Tanzania. Tanzania J For Nat Conserv 92(1):202–213

Mmolutsi GG, Matsuane C, Mojeremane W, Mathowa T, Teketay D (2020) Application and use of pre-sowing treatment methods to improve germination of *Vachellia karroo* (Hayne) Banfi & Galasso. Agric For J 4(1):47–54

Mojeremane W, Kgathi T (2005) Seed treatments for enhancing germination of *Colophospermum mopane* seeds: a multipurpose tree in Botswana. J Biol Sci 5(3):309–311

Mojeremane W, Mathowa T, Teketay D, Stimela T, Kopong I, Rampart M (2017) Pre-sowing seed treatment method to overcome dormancy in seeds of *Vachellia rehmanniana* Schinz. Agric For 63(2):171–181

Mojeremane W, Mathowa T, Teketay D (2018) Effects of different pre-sowing treatments on germination of *Peltophorum africanum* Sond seeds from two provenances in Botswana. J Biodivers Environ Sci 12(2):230–236

Mojeremane W, Makgobota K, Teketay D, Rampart M, Mmolotsi R, Kopong I, Monekwe D, Makoi T, Lepetu J, Ne-Mahupeleng G, Kakambi O (2020) Germination studies on seeds of *Burkea africana* and *Erythrophleum africanum* from Kazuma Forest Reserve, Northern Botswana. Afr J Biotechnol 19(9):675–683

Mojeremane W, Makgobota K, Mathowa T, Obusitswe K, Rampart M, Teketay D, Neo-Mahupeleng G, Makoi T, Mmolotsi R (2021) Evaluation of dormancy breaking treatments on seed germination of two leguminous tree species from Chobe, District, Northern Botswana. For Ideas 27(2):380–395

Moura I, Maquia I, Rija AA, Ribeiro N, Ribeiro-Barros AI (2017) Biodiversity studies in key species in the African mopane miombo woodlands. In: Bitz L (ed) Genetic diversity. IntechOpen, pp 91–109

Müller FL, Raitt LM, Cupido CF, Chimphango SBM, Samuels MI, Boatwright JS (2017) Dormancy-breaking treatments in two potential forage legumes from semi-arid rangelands of South Africa. S Afr J Bot 113:133–136

Nasr SMH, Savadkooh SK, Ahmadi E (2013) Effect of different seed treatments on dormancy breaking and germination in three species in arid and semi-arid lands. For Sci Pract 15:130–136

Nkengurutse J, Khalid A, Mzabri I, Kakunze AC, Masharabu T, Berrichi A (2016) Germination optimization study of five indigenous Fabaceae tree species from Burundi miombo woodlands. J Mater Environ Sci 7(12):4391–4402

O'Brien L, Morris J (2014) Well-being for all? The social distribution of benefits from woodlands and forests in Britain. Int J Justice Sustain 19(4):356–383

Odirile O, Mojeremane W, Teketay D, Kashe K, Mathowa T (2019) Responses of seeds of *Vachellia erioloba* (E. Mey.) P.J.H. Hunter in Botswana to difference presowing treatments. Int J Biol Biotechnol 16(1):181–188

Omoregie EH, Oluyemisi KF (2010) Preliminary phytochemical and proximate analysis of the leaves of *Piliostigma thonningii* (Schumach.) Milne-Redhead. Ethnobot Leafl 14:570–577

Orwa C, Mutua A, Kindt R, Jamnadass R, Simons A (2009) Agroforestree Database: a tree reference and selection guide. Version 4

Owens MK, Wallace RB, Archer S (1995) Seed dormancy and persistence of *Acacia berlandieri* and *Leucaena pulverulenta* in a semi-arid environment. J Arid Environ 29:15–23

Palgrave CK (2002) Trees of southern Africa, 3rd edn, revised and updated by Palgrave, M.C. Struik publishers, Cape Town, 1212 pp

Palmer E, Pitman N (1972) Trees of Southern Africa, vol 2. A.A. Balkema, Cape Town

Pan C, Yao L, Yu L, Qiao Z, Tang M, Wei F, Huang X, Zhou Y (2023) Transcriptome and proteome analyses reveal the potential mechanism of seed dormancy release in *Amomum tsaoko* during warm stratification. BMC Genomics 24:99. https://doi.org/10.1186/s12864-023-09202-x

Portillo-Quintero C, Sanchez-Azofeifa A, Calvo-Alvarado J, Quesada M, Marcos do Espirito Santo, M. (2015) The role of tropical dry forests for biodiversity, carbon, and water conservation in the neotropics: lessons learned and opportunities for its sustainable management. Reg Environ Chang 15:1039–1049

Pouliot M, Treue T, Obiri BD, Ouedraogo B (2012) Deforestation and the limited contribution of forests to rural livelihoods in West Africa: evidence from Burkina Faso and Ghana. Ambio 41:738–750

Rampart M, Teketay D, Mojeremane W, Mmolotsi R, Neo-Mahupeleng G, Makgobota K, Obusitswe K, Makoi T, Kopong I, Monekwe D (2021a) Pre-sowing treatments for enhancing germination of *Guibortia coleosperma* and *Amblygonocarpus andongensis* seeds from Kazuma Forest Reserve, Botswana. J Hortic For 13(3):74–80

Rampart M, Teketay D, Makgobota K, Mojeremane W, Mmolotsi R, Makoi T, Kopong I (2021b) Enhancing germination of seeds of *Cassia abbreviata* and *Senegalia nigrescens* using pre-sowing seed treatments in Botswana. Curr Agric Res J 9(3):179–187

Rudel TK (2013) The national determinants of deforestation in sub-Saharan Africa. Philos Trans R Soc B368:20120405. https://doi.org/10.1098/rstb.2012.0405

Sampaio AB, Holl KD, Scariot A (2007) Does restoration enhance the regeneration of seasonal deciduous forests in pastures in central Brazil? Restor Ecol 15(3):462–471

Sebego R, Atlhopheng J, Chanda R, Mulale K, Mphinyane W (2019) Land use intensification and implications on land degradation in the Boteti area: Botswana. Afr Geogr Rev 38:32–47

Sedjo RA (2001) Forest carbon sequestration: some issues for forest investments (No. 1318-2016-103352)

Setlhabetsi OT, Mojeremane W, Mathowa T, Teketay D (2019) Breaking seed dormancy in Philenoptera violacea (Klotzsch) Schrire using different pre-sowing treatment methods. J Agric Environ Sci 8(1):104–111

Shackleton CM, Shackleton SE, Buiten E, Bird N (2007) The importance of dry woodlands and forests in rural livelihoods and poverty alleviation in South Africa. For Policy Econ 9(5):558–577

Sibanda S, Shoko R, Chishaya K, Chimwanda P, Nyoni S, Ndlovu J (2022) Antimicrobial effect of *Brachystegia boehmii* extracts and their green synthesised silver zero-valent derivatives on burn wound infectious bacteria. All Life 15(1):1117–1125

Sinhababu A, Banerjee A (2013) Optimization of seed germination of some multipurpose tree legumes by seed treatments. J Plant Physiol 114:170–175

Smith MT, Wang BSP, Msanga HP (2003) Dormancy and germination. In: Vozzo JA (ed) Tropical tree-seed manual, vol 5, no 1. United States Department of Agriculture/Forest Service, pp 149–176

Smýkal P, Vernoud V, Blair MW, Soukup A, Thompson RD (2014) The role of the testa during development and in the establishment of dormancy of the legume seed. Front Plant Sci 5:1–19

Storrs AEG (1995) Know your trees: some of the common trees found in Zambia. Regional Soil Conservation Unit, Nairobi

Taghizadeh M, Sajadi FS (2023) Effects of dormancy breaking methods on germination of *Cercis siliquastrum* and *Spartium junceum* and seedling growth. Ornam Hortic. https://doi.org/10.1590/2447-536X.v29i1.2528

Teketay D (1996a) The effect of different pre-sowing seed treatments, temperature, and light on the germination of five Senna species from Ethiopia. New For 11:155–171

Teketay D (1996b) Germination ecology of twelve indigenous and eight exotic multipurpose leguminous species from Ethiopia. For Ecol Manag 80(1–3):209–223

Teketay D (1998) Germination of *Acacia origena*, *Acacia pilespina,* and *Pterolobium stellatum* in response to different pre-sowing seed treatments, temperature, and light. J Arid Environ 38(4):551–560

Teketay D, Kashe K, Madome J, Kabelo M, Neelo J, Mmusi M, Masamba W (2018) Enhancement of diversity, stand structure and regeneration of woody species through area enclosure: the case of a mopane woodland in northern Botswana. Ecol Process 7:5. https://doi.org/10.1186/s13717-018-0116-X

Tian Y, Guan B, Zhou D, Yu L, Li G, Lou Y (2014) Responses of seed germination, seedling growth, and yield traits to seed pretreatments in maize (Zea mays L.). Sci World J. https://doi.org/10.1155/2014/834630

Venter F, Venter J-A (2012) Making the most of indigenous trees, 3rd Revised edn. Briza Publications, Pretoria, 320 pp

Vodouhè R, Dansi A, Avohou HT, Kpèki B, Azihou F (2011) Plant domestication and its contributions to in situ conservation of genetic resources in Benin. Int J Biodivers Conserv 3(2):40–56

Werker E (1980) Seed dormancy as explained by the anatomy of embryo envelopes. Isr J Bot 29(1–4):22–44

Wu S, Hou Y, Yuan G (2010) Valuation of forest ecosystem goods and services and forest natural capital of the Beijing municipality, China. Unasylva 61(1):28–36

Yakubu J, Sodipo OA, Abdulrahman FI, Balami VM (2021) Toxicity study and anticonvulsant effect of ethanol leaf extract of *Piliostigma thonningii* Milne-Redhead (Fabaceae). Niger J Pharm Res 17(1):63–70

Zar JH (1996) Biostatistical analysis, 3rd edn. Prentice Hall, Upper Saddle River, 662 pp

Zhaoa T, Yang B, Zheng H (2009) Assessment of the erosion control function of forest ecosystems based on GIS: a case study in Zhangjiajie National Forest Park, China. Int J Sust Dev World 16:356–361

Zincirkiran M, Tumsavas Z, Unal H (2010) The effects of different acid treatment stratification duration on germination of *C. siliquastrum* L. seeds. Bot Hortic Agron 38(1):59–163

Witness Mojeremane is Professor in Forest Sciences at the Botswana University of Agriculture and Natural Resources (BUAN). He holds a Diploma in Forestry, Diploma in Human Resources Management, Bachelor of Science (Honours) in Forestry, Master of Science in Tropical Environmental Forestry and PhD in Ecology and Resource Management. He joined BUAN as an instructor in 1994. His research interests include forest sciences and has been engaged in several consultancy and research Projects. Professor Mojeremane has published 52 articles in refereed journals, 16 book chapters, and 3 articles in refereed conference proceedings.

Ronnie Matlanaga Mmolotsi is a Professor of Agroforestry at the Botswana University of Agriculture and Natural Resources (BUAN). He holds a PhD in Agroforestry (Bangor University), MSc in Forest Science (Edinburgh University), BS in Forest Management (Colorado State University), and a Diploma in Agriculture (Botswana College of Agriculture). He has taught various courses at undergraduate level. His research interests include agroforestry, forestry ecology, and apiculture. He has been engaged in many consultancy and research projects. He has published several papers on agroforestry and forest ecology in high-impact peer-reviewed journals.

Melusi Rampart is a Professor in Forest Sciences in the Faculty of Natural Resources, BUAN. He holds BSc in Forest Management (Oregon State University), MSc in Environmental Forestry, University of Wales, Bangor, UK, and PhD in Forestry (Ecophysiology), University of Wales, Bangor (UK). His main areas of research are apiculture and forest sciences. Professor Melusi Rampart has been involved in teaching several forest and range science courses at both undergraduate and graduate levels. Professor Rampart has published 18 articles in peer reviewed journals and a book chapter. Prof. Rampart has been engaged in several consultancy and research projects.

Professor Demel Teketay holds a PhD in Forest Vegetation Ecology (Swedish University of Agricultural Sciences), an MSc in Plant Taxonomy (Reading University, UK), and a BSc in Agriculture (Ethiopia). He is a Professor in Forest Sciences at the Botswana University of Agriculture and Natural Resources (BUAN). He has served at various capacities in Africa and Europe. He is the founding Dean of the Faculty of Natural Resources, BUAN. He has been engaged in several consultancy and research projects. He has published 186 articles in peer reviewed journals and also has published 7 books, 2 edited books, 23 articles in conference proceedings, 24 abstracts, and 14 technical reports.

Mr. Obusitswe Kakambi works for the Department of Forestry and Range Resources, Ministry of Environment at Kasane, Chobe District. He holds Trade C and Trade B Certificate in Forestry from Serowe and Kgatleng Brigade, Mochudi, Botswana. Mr Kakambi is excellent in the identification of woody and herbaceous plant species. He is a member of Botswana National Inventory Team. He has forged a good relationship with the Faculty of Natural Resources, Botswana University of Agriculture and Natural Resources. Mr Kakambi has published three articles in refereed journals.

Topoyame Isaac Makoi holds a Master of Science in Environmental Science specializing in Geographical Information Science and Environmental Management and a Bachelor of Science in Land Management. He works for the Department of Range and Forest Resources, Botswana University of Agriculture and Natural Resources as a Senior Instructor. His research interests include the areas of range ecology, forest and woodland ecology, climatology, environmental management, vegetation mapping, geographical information systems (GIS), and remote sensing. He has been engaged in several research projects and has published five articles in refereed journals and one book chapter.

Mr. Ishmael Kopong works as Senior Technician in the Department of Range and Forest Resources, University of Agriculture and Natural Resources (BUAN). He holds a Certificate in Forest and Range Ecology (Botswana College of Agriculture), Higher Diploma in Forestry and Range Ecology (University of Botswana), and a Bachelor of Science in Range Science (BUAN). He has been assisting in teaching several undergraduate courses since he joined the University. He has been engaged in several research projects. Mr Kopong has published two book chapters and three articles in referred journals.

Mr. Kamogelo Makgobota is a Technician in the Department of Range and Forest Resources, Faculty of Natural Resources, Botswana University of Agriculture and Natural Resources. He is currently a student pursuing Master of Science in Crop Science. His research focuses on the effectiveness of beneficial rhizobia on indigenous wild legume establishments with poor natural regeneration. He has been involved in several consultancy and research projects. He has co-authored six articles in referred journals.

Thembinkosi Mathowa is a Chief Technician in the Department of Crop and Soil Sciences, Faculty of Agriculture, Botswana University of Agriculture and Natural Resources. He holds a Diploma in Agriculture, BSc in Agricultural Education, and Master of Science in Agriculture. He has been assisting in teaching several courses in the Department of Crop and Soil Sciences. He has participated in various research projects. His main research interests include environmental, plant, and soil sciences. He has published 36 articles in refereed journals, 2 book chapters, 2 refereed conference papers, 4 abstracts, and 6 conference posters.

Gosiame Neo-Mahupeleng is a lecturer in wildlife sciences at the Botswana University of Agriculture and Natural Resources (BUAN). He holds a Bachelor of Science in Biology and Environmental Science and Master of Science in Natural Resources Management and Sustainable Agriculture. He is currently a Doctor of Philosophy candidate at the Okavango Research Institute of the University of Botswana. Mr Neo-Mahupeleng research interests are large carnivore conservation, research techniques in wildlife management, environmental management and comanagement of biodiversity.

Dr. Keoikantse Sianga is a Senior Lecturer at the Botswana University of Agriculture and Natural Resource. He holds a BSc in Conservation Biology and Wildlife Management and an MPhil and a PhD in Natural Resource Management. His research interests include functional and vegetation heterogeneity and plant-wildlife interactions, fire ecology and management, human-wildlife interactions, and spatial ecology. Dr Sianga has published four articles in peer reviewed journals and seven book chapters.

Chapter 16
Impact of Invasive Alien Species on Biodiversity and Ecosystem Services

Shimane W. Makhabu, Koketso Tshireletso, Mpho R. Setlalekgomo, Zibanani Seletlo, Boipuso Legwatagwata, and Lawrence Akanyang

16.1 Introduction

Alien species, also known as exotic, nonnative, nonindigenous, or introduced species, are species found in areas outside their natural habitat (Minchin 2009; Shiferaw et al. 2018). Alien species are considered invasive when they produce reproductive offspring, often in very large numbers, at considerable distances from parent plants and thus have the potential to spread over a considerable area (Richardson et al. 2000; Ngweno et al. 2009; Shiferaw et al. 2018). Invasive alien species (IAS) are found in a variety of ecosystem types worldwide, including agricultural and range lands, parks and reserves, deltas, rivers, pans, dams, and roadsides. They differ in type and number from one habitat to another and threaten the biodiversity of many ecosystems (Shiferaw et al. 2018). IAS threaten ecosystem services to people, livestock, and wildlife (Sebua et al. 2012; Shiferaw et al. 2018). They have dire negative consequences on the health, food security, sustainability, financial services, and productivity of natural ecosystems (Mayfield III et al. 2021). Other costs include losses in agricultural production from arable land and livestock and forest products as well

S. W. Makhabu (✉) · M. R. Setlalekgomo · B. Legwatagwata
Department of Biological Sciences, Botswana University of Agriculture and Natural Resources, Gaborone, Botswana
e-mail: smakhabu@buan.ac.bw; msetlale@buan.ac.bw; blegwata@buan.ac.bw

K. Tshireletso · L. Akanyang
Department of Range and Forest Resources, Botswana University of Agriculture and Natural Resources, Gaborone, Botswana
e-mail: ktshirel@buan.ac.bw; lakanyang@buan.ac.bw

Z. Seletlo
Department of Animal Science and Production, Botswana University of Agriculture and Natural Resources, Gaborone, Botswana
e-mail: zseletlo@buan.ac.bw

W. L. Hambira et al. (eds.), *Environmental Change and Biodiversity Conservation in sub-Saharan Africa*, Advances in Global Change Research 75,
https://doi.org/10.1007/978-3-031-73136-5_16

as compromised health of humans and other animals (Crowl et al. 2008; Aukema et al. 2011; Mayfield III et al. 2021). The negative impact of IAS on biodiversity outweighs its positive impacts (de Wit et al. 2001; Crooks 2002; Dubale 2008; Wise et al. 2012; Shiferaw et al. 2018). Native species may become endangered and perhaps extinct due to alien species invasion (Mayfield III et al. 2021). The problem of IAS is more severe in developing countries than in developed nations due to the lack of resources to manage them (Tana and Milberg 2010).

In Botswana, documentation of IAS, their distribution and abundance requires due attention. Some recognised IAS in Botswana include *Cenchrus biflorus* Roxb., *Verbesina encelioides* (Cav.) Benth. & Hook. f. ex A. Gray, *Prosopis juliflora* (Swartz) DC, and *Salvinia molesta* D. Mitch. These four species are used in this chapter to illustrate how IAS affect biodiversity and ecosystem services, notably in Botswana. The four species used here were chosen by the authors since they are of significance to the biodiversity and ecosystem services in Botswana. This chapter discusses the distribution, impact on native species, and effects on ecosystem values and services (e.g. economic, aesthetic, and social benefits) that they provide to humans, livestock, and wild animals. This chapter is based on primary work and a review of the literature.

16.1.1 Reproduction and Establishment Capacity of Invasive Alien Species

Invasive alien species can invade, outcompete natives, and take over the new environment (Sebua et al. 2012; Shiferaw et al. 2018). They can be introduced to a new habitat intentionally or accidentally by human beings and/or by other forms (Shiferaw et al. 2018). Human activities that can introduce alien species into a new area intentionally include agro-forestry, horticulture, forestry, and animal husbandry. They can be brought in and cultivated for utility products or services or as ornamentals (O'Connor and van Wilgen 2020). Alien species provide valuable services, including forage and livestock shelter, soil protection, fuel and building materials, barrier plants, medicines, and pollination (O'Connor and van Wilgen 2020). IAS can also be introduced in an area through natural forces such as blown by winds or carried by water or animals (Moyo and Fatunbi 2010; Sebua et al. 2012; Shiferaw et al. 2018). Seeds can accidentally be brought into a place hidden in crop seeds, animal feeds, or fertilisers. Seeds of some IAS can reach a new area attached to the skins of domestic and wild animals and birds' feathers. Migratory birds can also bring in alien species seeds in their droppings and some attached to their feathers. Once they move into a new area, IAS have traits such as enhanced fertility, allelopathic nature, clonality, adaptability, rapid growth, increased dispersal, and excellent defence against herbivory (Kunwar and Acharya 2013) that allow them to establish themselves and become invasive (Fig. 16.1).

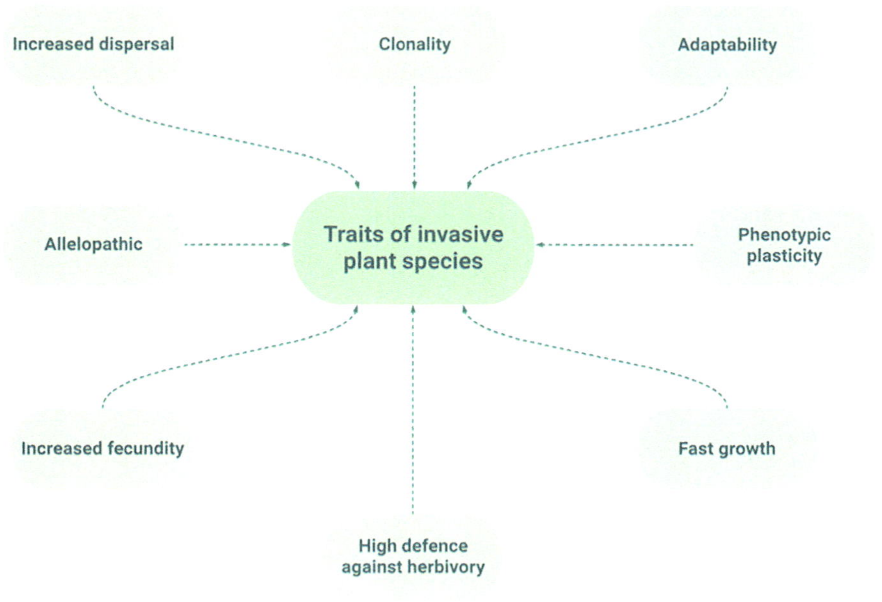

Fig. 16.1 A conceptual diagram depicting the possible traits of invasive alien plant species that make them thrive in natural ecosystems. (Adapted from Kunwar and Acharya 2013)

16.1.2 Some Invasive Alien Species of Concern in Botswana

Each country has a list of invasive alien species (Sebua et al. 2012). Some invasive alien species in Botswana are presented in Table 16.1 (Marumo 2018). According to Marumo's (2018) report, 16 invasive alien plant species were recorded in Gaborone alone from 2013 to 2017. These include *Agave angustifolia*, *Flaveria bidentis*, *Verbesina encelioides*, *Tecoma stans*, *Cylindropuntia imbricata*, *Ipomoea carnea*, *Ricinus communis*, *Parkinsonia aculeata*, *Malva parviflora*, *Melia azedarach*, *Eucalyptus camaldulensis*, *Argemone ochroleuca*, *Richardia brasiliensis*, *Nicotiana glauca*, *Lantana camara*, and *Senna occidentalis*. They were recorded from riverbeds, along roads, disturbed areas, and open spaces around residential and business areas in Gaborone.

According to Marumo (2018), invasive alien plant species have become a problem in Botswana. They have detrimental effects, such as choking rivers, reducing crop yields, destroying the natural beauty of landscapes, and consuming considerable amounts of underground water. The aquatic alien invasive plant species *Salvinia molesta* and *Pistia stratiotes* L. and the water hyacinth, *Eichhornia crassipes* (Mart) Solms-Laub reportedly threaten Botswana's aquatic habitats in Chobe and Limpopo Rivers (Kurugundla et al. 2016). Water hyacinth is said to be one of the world's most noxious aquatic weeds (Mujere 2015).

Table 16.1 Selected invasive alien species of concern in Botswana

No.	Scientific names	Family	Common names	Tswana names	Affected areas
1.	*Agave angustifolia*	Agavaceae	Kleingaringboom		
2.	*Argemone ochroleuca*	Papaveraceae	Mexican poppy	Lepero	
3.	*Cenchrus biflorus*	Poaceae	Indian sandbur	Tolela, Bojang jwa pitse, Bojang jwa Makunda	Ghanzi, Kweneng, and Central districts
4.	*Cylindropuntia imbricata*	Cactaceae	Imbricate prickly pear		
5.	*Eucalyptus camaldulensis*	Myrtaceae	Red river gum		
6.	*Flaveria bidentis*	Asteraceae	Smelter's bush		
7.	*Ipomoea carnea*	Convolvulaceae	Morning glory bush, potato bush		
8.	*Lantana camara*	Verbenaceae	Lantana		
9.	*Malva parviflora*	Malvaceae	Small mallow		
10.	*Melia azedarach*	Meliaceae	Syringa, Persian lilac	Morulwana, mosalaosi	
11.	*Nicotiana glauca*	Solanaceae	Wild tobacco		
12.	*Prosopis juliflora*	Fabaceae	Mesquite, *Prosopis*, Woyanezaf (Amh)	Sexanana	Kgalagadi and Ghanzi districts
13.	*Parkinsonia aculeata*	Fabaceae	Jerusalem thorn		
14.	*Richardia brasiliensis*	Rubiaceae	Tropical richardia		
15.	*Ricinus communis*	Euphorbiaceae	Castor oil plant	Mokhore	
16.	*Salvinia molesta*	Salviniaceae	Giant salvinia, kariba weed	Mochimbamo	Okavango Delta
17.	*Senna occidentalis*	Fabaceae	Stinking weed, wild coffee		
18.	*Tecoma stans*	Bignoniaceae	Yellow bells		
19.	*Verbesina encelioides*	Asteraceae	Wild sunflower, golden crownbeard		Kweneng, Central, Kgatleng, South–east
20.	*Eichhornia crassipes*	Pontederiaceae	Water hyacinth		Limpopo River
21.	*Pistia stratiotes*	Araceae	Water lettuce		Kwando and Chobe Rivers

Data sources: Marumo (2018) and Kurugundla et al. (2016)

16.1.2.1 *Salvinia molesta* Mitch

Salvinia molesta Mitch belongs to the family Salviniaceae. It is native to southeastern Brazil (Jacono et al. 2011). It is a floating fern that flourishes in wetlands or river systems. *Salvinia* is characterised by short, stalked leaves arranged in three whorls (Kurugundla et al. 2016). Two opposite leaves help the plant float, while the third leaves serve as roots to absorb water and nutrients (Kurugundla et al. 2016). In Botswana, *S. molesta* was first noticed in 1948 in the Chobe River at Kazungula and then in 1971 in the Kwando–Linyati–Chobe Rivers and reached the Okavango Delta after 1986 (Kurugundla et al. 2016). The species is now widely distributed in tropical and subtropical regions, dominating water bodies with slow-moving fresh water (Jacono et al. 2011).

16.1.2.2 *Prosopis* Species

Prosopis species belong to the family Fabaceae, subfamily Mimosoideae. Many *Prosopis* species are recognised, and the genus occurs in both arid and semiarid areas of Africa and elsewhere. It is native to Mexico, South America, and the Caribbean (Mehari 2015). It was introduced in South Africa in the 1800s from America (Shackleton et al. 2017). In Botswana, it was introduced to control desertification and the spread of sand dunes (Muzila et al. 2011). *Prosopis* have long tough thorns that are characteristic of some species in the Fabaceae family (Fig. 16.2). *Prosopis* species are reportedly deep-rooted (20–25 m deep), allowing them to flourish even in times of drought by tapping water from the deep soil layers (Mosweu

Fig. 16.2 *Prosopis juliflora* (*Prosopis* or Sexanana) trees in Tsabong Village in Kgalagadi District of Botswana. (Photo credits: S. W. Makhabu)

et al. 2013). A study by Mosweu et al. (2013) noted the presence of *Prosopis chilensis*, *Prosopis velutina*, *Prosopis juliflora* (widely distributed across the globe), *Prosopis glandulosa*, *P. pallida*, and hybrids of *P. juliflora* and *P. glandulosa* in the southern Kgalagadi district in Botswana. In Botswana's Kgalagadi district, *Prosopis* species are present in the villages of Tsabong, Bokspits, Rapelspan, Vaalhoek, and Struizendam (Muzila et al. 2011; Mosweu et al. 2013). *Prosopis* species are also found in the Ghanzi district of Botswana, particularly in Ghanzi township and Hanahai Valley (Chilume 2017).

16.1.2.3 *Cenchrus biflorus*

Cenchrus biflorus (Fig. 16.3) is an annual grass species in the family Poaceae. Common names include Indian sandbur, Bhurat in India, Kram-kram in Botswana, and Haskaneet in Sudan. In Botswana, it is called by various vernacular terms, such as "tolela", "bojang jwa pitse", and "bojang jwa Makunda". In Botswana, it is called "tolela" because of its bur (seeds/inflorescences that have hooks) characteristic that enables it to cling to clothes and furs/hairs of animals (Makhabu and Marotsi 2012). It is native to the Indian subcontinent and tropical Africa (Peerzada et al. 2016). In Botswana, it was first noticed around Nata village in the 1940s (Makhabu and Marotsi 2012). The species does well in disturbed areas. Like many invasive species, *C. biflorus* produces many seeds. In Botswana, the species is mainly found in the western and central parts of the country, such as the Ghanzi, Kweneng and Central districts (Makhabu and Marotsi 2012; Makhabu et al. 2021). This species largely invades the villages of Makunda, Charleshill, etc., in the Ghanzi district. Low abundance has been noticed in some Ghanzi and Western Sandveld farms (Makhabu and Marotsi 2012). It has invaded and entirely dominated croplands in some parts of Botswana. It is regarded as a harmful grass species because of its

Fig. 16.3 *Cenchrus biflorus* Roxb. spotted in a farm in the Western Sandveld Region, Botswana. (Photo credits: S. W. Makhabu)

objectionable burs that adhere to the coats of animals, people's clothing, and machinery, causing irritation when handled by people. Burs cause wounds to animals and people when they encounter them. The burs make weeding in arable fields difficult because it irritates and pierces people (Makhabu and Marotsi 2012). Botswana's "Fifth National Report on the Convention on Biological Diversity of 2015" mentions this species as one of the alien species of concern (DEA 2015). The grass is grazed by herbivores while still immature but becomes undesirable when mature (Makhabu and Marotsi 2012). The species does well in disturbed areas such as overgrazed rangelands, abandoned arable fields, sand dunes, old kraals, and along roads (Makhabu and Marotsi 2012; Peerzada et al. 2016). The species appears to be a poor competitor with indigenous species in undisturbed areas, as observed at Old Xade in Botswana, where its abundance decreased when the area was restored after people and their livestock relocated to a new village (Makhabu and Marotsi 2012). In May 2023, few individuals of this species were present in Old Xade, an indication that without disturbance, it is outcompeted by native species (Makhabu, personal observation) in disturbance areas.

16.1.2.4 *Verbesina encelioides*

Verbesina encelioides is an annual and flower-bearing plant in the family Asteraceae (Fig. 16.4), and its common name is "golden crownbeard" and wild sunflower. It originates in North America, particularly the southern and western USA and northern and central Mexico. It is an erect annual native to North and South America (Makhabu et al. 2021). Makhabu et al. (2021) reported having observed it along roads, arable fields, and overgrazed areas within villages in Botswana. It was seen in Gaborone, Kodibeleng, Ngware, Serowe, Tonota, Paje, Dimajwe, Lephephe, and

Fig. 16.4 *Verbesina encelioides* in Gaborone North, Botswana. (Photo credits: S. W. Makhabu)

Fig. 16.5 *Verbesina encelioides* growing as a weed in a field of maize near Gaborone, Botswana. (Photo credits: Zibanani Seletlo)

Molepolole. Other people have reported seeing it in other areas. In northern India, *V. encelioides* is a common weed in maize, barley, rice, peanut, and millet fields. In Botswana, it has been seen as a weed in some arable farms on the outskirts of Gaborone (Fig. 16.5). The ideal habitats for *V. encelioides* are open areas and disturbed habitats (PIER 2013; Makhabu et al. 2021).

16.2 Impacts of Invasive Alien Species on Biodiversity

Biodiversity, the variety of life forms, is a core concept in ecosystems. It typically defines ecosystems' health and ability to function correctly (Kernan 2015). Invasive species may affect plant and soil communities (Wang et al. 2022). Because of their aggressive and highly prolific nature, they can outcompete native species, thus reducing species richness and evenness. Lean (2021) noted that some researchers claim that invasive species create more valuable ecosystems (Haines-Young and Potschin 2010; Loreau et al. 2001; Mace et al. 2012). However, insights from research globally note the negative impacts of invasive species with concern. In contrast, other scholars highlight the interaction of multiple stressors on the effects of nonnative species on ecosystems (Kernan 2015).

Invasive alien species can affect the distribution, abundance, or function of native species through predation or competition for space and resources (Mayfield III et al. 2021). The effects can be through density or trait-mediated mechanisms (Mayfield III et al. 2021). The density-mediated indirect mechanism refers to the alteration of the abundance of a native species, which then alters the abundance of other native species (Ortega et al. 2006; Mayfield III et al. 2021). The trait-mediated mechanism is whereby the native species affects the interaction between two different species in ways that affect how strongly they interact. In the trait-mediated mechanism, invasive alien species change other species' traits (phenology, morphology, or behaviour) rather than affecting the abundance of those species, which alters how they interact (Mayfield III et al. 2021).

16.2.1 Salvinia molesta *Mitch*

Mechanisms by which invasive plants affect aquatic systems include reducing sunlight penetration, altering water temperature, pH, and nutrient concentrations, and reducing oxygen content (Mayfield III et al. 2021). Likewise, in aquatic environments, invasive plants of various growth forms (floating, submersed, or emergent) can considerably alter environmental conditions, ecosystem processes, plant and animal communities, and biological diversity (Mayfield III et al. 2021). In Botswana, it has been observed that *Salvinia* outcompeted water primrose *Ludwigia stolonifera* (Guill & Perr) and grass *Leersia hexandra* Swartz, *Utricularia gibba* L, *Utricularia stellaris* and *Utricularia foliosa* in the Chobe River and Okavango Delta (Kurugundla et al. 2016). However, upon control of *Salvinia*, these species were re-established (Kurugundla et al. 2016).

16.2.2 Prosopis *Species*

A study by Mosweu et al. (2013) in southern Kgalagadi of Botswana indicated diminished plant diversity due to the invasion of *Prosopis* species in the area. In the Ghanzi district, the *Prosopis* invasion has been associated with declining dominant native species heights (Chilume 2017). *Prosopis* species have an allelopathic effect that suppresses the growth of other plant species, reducing plant diversity in an area. Cattle and browsers consume and effectively disperse its pods, contributing to its spread (O'Connor and van Wilgen 2020).

16.2.3 Cenchrus biflorus

Cenchrus biflorus is regarded by many scientists as an invasive weed with a negative impact on agriculture and biodiversity in arid and semiarid environments (Peerzada et al. 2016). Its presence affects both microbes and animal and plant diversity. Livestock and other grazing animals graze *C. biflorus* in its juvenile stage and after the shedding of spiky burs but will not graze the prickly spikelets (Makhabu and Marotsi 2012; Peerzada et al. 2016). Thus, the presence of *C. biflorus* is important to the survival of some grazers, and it is used for shelter for some invertebrates and small vertebrates, thereby affecting the area's biodiversity.

16.2.4 Verbesina encelioides

The other IAS in Botswana, *V. encelioides,* has allelopathic effects that inhibit native plant growth. It primarily affects croplands by competing with crops for growth requirements. It has dominated areas and outcompeted native species that animals graze and even affects many agricultural lands (Makhabu et al. 2021).

16.3 Impacts of Invasive Alien Species on Ecosystem Services

Ecosystem services are the ecosystem's benefits to humanity (MEA 2005; Steinman et al. 2017; Pathak et al. 2021). Ecosystem services can be grouped into provisioning (e.g. food, fresh water), cultural (e.g. aesthetic, spiritual), regulating (e.g. climate, flood disease regulations), and supporting (e.g. nutrient cycling, primary production) (Costanza et al. 1997; Gallardo et al. 2019; MEA 2005; Pathak et al. 2021). Figure 16.6 shows more examples of these groups of ecosystem services.

16.3.1 Salvinia molesta *Mitch*

Similar to most invasive aquatic plants, *Salvinia* can significantly impede the use of water bodies by restricting navigation, impending water movement important for flood control, irrigation, or hydropower, disrupting recreational activities and decreasing aesthetics, property values, and tourist income (Mayfield III et al. 2021). Because of the massive coverage of *Salvinia* along the Okavango Delta, tourism activities were affected in 1986, 1992, 1999, and 2010 (Kurugundla et al. 2016). *Salvinia* can form mats that block river channels, thus affecting water flows. The formation of mats by *Salvinia* makes it difficult for animals to judge the depth of the water in a river or delta. The drowning of a large number of African buffaloes,

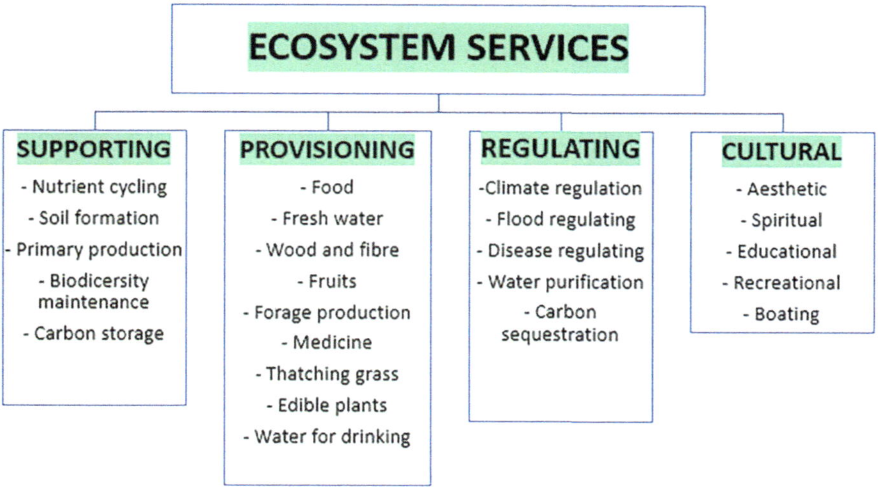

Fig. 16.6 Ecosystem services categories and examples. (Data sources: Costanza et al. 1997, Gallardo et al. 2019, Pathak et al. 2021)

Syncerus caffer, was blamed on invasive aquatic species that form mats, making it difficult for animals to see the depth of the water.

16.3.2 Prosopis *Species*

Various sources have indicated that the negative socioeconomic impacts of invasive species outweigh their benefits (Mosweu et al. 2013; Shackleton et al. 2015; O'Connor and van Wilgen 2020). For example, *Prosopis* species do not offer palatable foliage. However, one of the reasons for its introduction was for its edible pods by livestock (O'Connor and van Wilgen 2020) and for stabilising dunes. *Prosopis* species are sparingly used as fuel wood due to their timber's inferior quality compared with native species (Shackleton et al. 2015).

A review by Mosweu et al. (2013) noted many benefits of *Prosopis*. These benefits include improved soil fertility, use as feed for livestock, construction timber, honey production, medicinal purposes, and fuel energy, among others. However, because the species' impacts on rural livelihoods outweigh the benefits, it is considered an environmental nuisance in southern Kgalagadi (Mosweu et al. 2013). The allelopathic properties of the species suggest negative interactions with other species growing in proximity.

16.3.3 Cenchrus biflorus

Its invasion in an area makes it difficult to use as an arable field or grazing area and for the collection of fruits. It is a major weed in arid and semiarid cropping systems (Setshogo 2002), thus lowering food production. Its presence negatively affects the enjoyment of ecosystem services in the provisioning group (Makhabu and Marotsi 2012). In some cases, *C. biflorus* has been grown to combat desertification, and in other situations, it has been used for soil reclamation because it can persist until the end of the dry season, protecting the soil (Boudet 1977; Peerzada et al. 2016).

16.3.4 Verbesina encelioides

It has allelopathic effects inhibiting native plant growth (Makhabu et al. 2021). It can harbour pests that destroy other plants or transmit diseases to other organisms. It can dominate in an area outcompeting native species that animals graze. The plant species contains a toxic component called galegine, which poisons livestock, mainly sheep (Makhabu et al. 2021). It reduces crop yield when it grows as a weed in an arable field (Fig. 17.5).

16.4 Invasive Alien Species Control Measures

The impacts of invasive alien species can be reduced or eliminated by eradication, control, containment, exclusion, or impact mitigation (IUCN 2018). The methods available to control the spread of invasive alien species include biological, chemical, mechanical, physical or manual, and cultural methods (IUCN 2018; Finch et al. 2021; Mayfield III et al. 2021; USDA 2023). Examples of activities that can be applied to each method are presented in Fig. 16.7.

16.4.1 Biological Control Measures

Biological control is humans' intentional manipulation of natural enemies to control pests such as using prey targeting invasive species to reduce the population of the invasive species (USDA 2023). Natural enemies used in classical biological control of weeds include different organisms, such as insects, mites, nematodes, and pathogens (USDA 2023). Biological control agents are plant-feeding insects, of which beetles, flies, and moths are most commonly used. This option involves much research and testing to ensure that the prey targets only the intended invasive species. Biological control methods have been used with some level of success

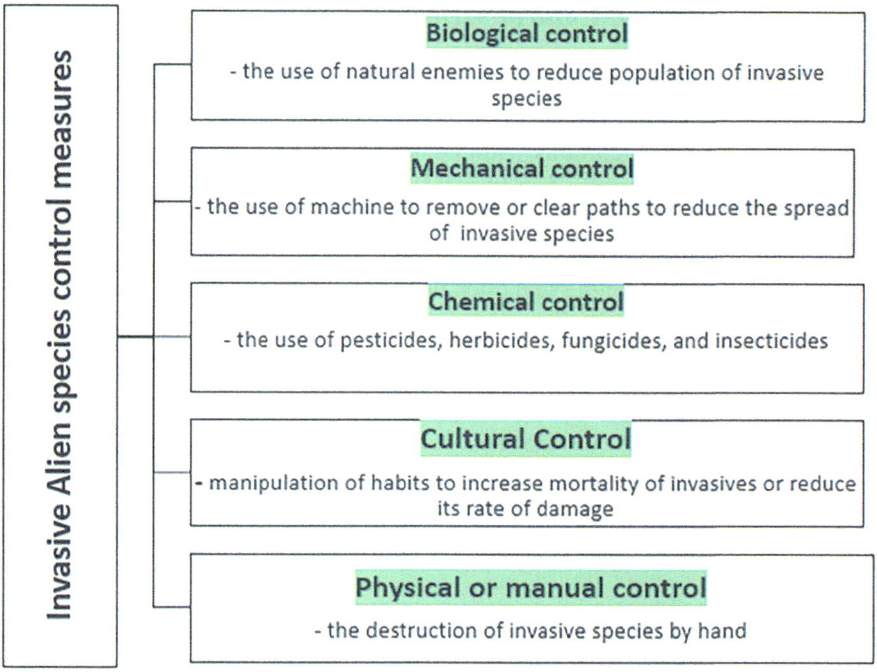

Fig. 16.7 Conceptual representation of invasive plant species control methods. (Data source: USDA 2023)

worldwide. Some insect pests have been used with noticeable results. Other studies reported success using goats (Holechek et al. 2001).

- To control *Prosopis* expansion, the use of *Algarobius prosopis* (LeConte), *Neltumius arizonensis* (Schaeffer), *Coelocephalapion gandolfoi* Kissinger, and *Asphondylia prosopidis* Cockerell are said to give promising results (Zachariades et al. 2011; Mosweu et al. 2013).
- *Salvinia*: Several biological agents have been used to control *Salvinia* with varying levels of success. In Botswana, the *Salvinia* species is contained and managed by the weevil *Cyrtobagous salviniae* Calder and Sands. The weevil population multiplied and fed on *S. molesta* Mitch. The problem with using weevils to control *S. molesta* is that some die during winter, thus reducing their effectiveness. Other species, such as *Cyrtobagous singularis* Hustache, the water lettuce moth *Samea multiplicalis* Guenée, and the *Salvinia* grasshopper *Paulinia acuminata* DeGeer, were tested elsewhere with little to no success (Kurugundla et al. 2016).

16.4.2 Mechanical Control Measures

Mechanical treatments complement herbicide (chemical) control and sometimes increase efficiency. Mechanical control involves the use of bulldozers, ploughs, cabling, mowing, hoeing, tilling, girdling, chopping, chaining, and disking, among others (Holechek et al. 2001; USDA 2023). Some species benefit from soil disturbance, resulting in the facilitation of species spreading; thus, measures should be based on species characteristics.

- For *Prosopis*, many of the suggested interventions promote the utilisation of the species. For example, seed pod harvesting, cutting, and uprooting were recommended in southern Kgalagadi (Mosweu et al. 2013). The cut material could be used as livestock feed, human food, fuel energy, and construction timber.
- *Salvinia* is rarely controlled mechanically due to the inaccessibility of machines in rivers and deltas. It can also entangle some machines, making the work laborious when removing entangled pieces.
- *Cenchrus biflorus* and *V. encelioides* can be controlled using mechanical measures such as hoeing and grading and then burning them to destroy the seeds.

16.4.3 Chemical Control Measures

Chemical control includes the use of pesticides, herbicides, fungicides, and insecticides. Although chemical use can be very effective, it can be dangerous to other species or the ecosystem (USDA 2023). Understanding chemicals is vital in vegetation management using herbicides. Similar to biological and mechanical control measures, chemical treatments are available. Herbicides can act by translocation, contact, selective, nonselective, and soil sterilant. The most commonly used herbicides include paraquat, tebuthiuron, and picloram (Holechek et al. 2001).

- *Prosopis*: Chemical control with Grazon DS herbicide, Garlon® 4 herbicide, and Access herbicides was recommended by Mosweu et al. (2013).
- *Salvinia*: Using chemicals in aquatic ecosystems is a challenge because of the possibility of affecting non-targeted species. Control has mainly been limited to biological and mechanical or manual methods. Although limited, attempts to spray with paraquat and glyphosate were made in the past in Chobe, Botswana.
- Herbicides can be used to control *C. biflorus* and *V. encelioides.*

16.4.4 Cultural Control

Cultural measures are aimed at changing human behaviour to address the issue of spreading invasives using opportunities to educate people about practices to increase awareness to prevent the spread of invasives (signage, public awareness campaigns) (USDA 2023). Cultural control includes manipulating habits to increase invasives' mortality or reduce damage rates (selecting pest-resistant crops, winter cover crops, changing planting dates). Cultural practices include mulching, soil solarisation with plastic film, thermal weed control (flaming, hot water, and steam), prescribed burning, water manipulation, and prescribed grazing with domesticated herbivores (cattle, sheep, goats, and horses). There needs to be more documentation of how the four species discussed in this chapter can be controlled using this method (USDA 2023).

16.4.5 Physical or Manual Control

Physical or manual control involves physical activities (harvesting) such as hand-pulling, digging, flooding, mulching, manual destruction or removal of nests, egg masses, or other life stages; generally, it includes destroying invasive species by hand (USDA 2023). Seedlings of *Prosopis* can be hand removed. The other three species can also be hand-pulled before the seed matures and burnt to reduce the spread of the species. This method can be used for the four species.

- *Salvinia*: Manual removal, booms and vegetation blockages, habitat alteration, fencing, and drying swamps were techniques used to control *Salvinia* in the Chobe and Okavango Delta, Botswana.
- The habitat alteration technique involved drawing down water to expose the *Salvinia* so it could dry. The dry Salvinia is then burnt to avoid regrowing. Fences and booms/vegetation blockages were meant to block the floating *Salvinia* from flowing downstream.

16.5 Prevention of Invasive Alien Species Introduction

The prevention of alien species introduction is the best control measure that needs to be followed. The screening of species in an area is vital to avoid introducing invasive alien species. Some countries require a permit before a species is brought into the country. During the screening process, different options need to be considered (Fig. 16.8). Such an option presented by Wittenberg and Cock (2001) can minimise the introduction of invasive alien species in a new environment. Another way to prevent the introduction of invasive species is monitoring and continuous assessment and follow-up and taking effective measures.

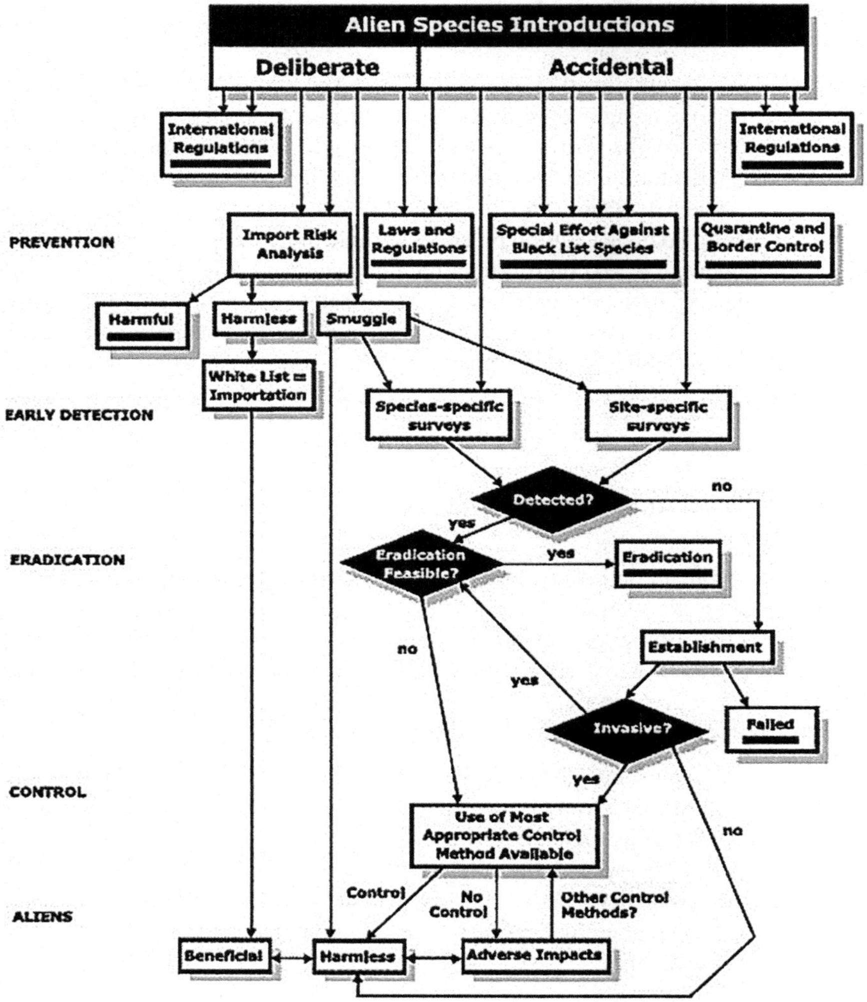

Fig. 16.8 Summary of options to consider when addressing alien species. Black bars mark the potential final stages of introduced alien species. Diamonds symbolise important bifurcations and decision points. (Adapted from Wittenberg and Cock 2001)

16.6 Conclusion and Recommendations

The four invasive species discussed in this chapter, *Cenchrus biflorus*, *Verbesina encelioides*, *Prosopis juliflora*, and *Salvinia molesta*, have benefits and drawbacks to biodiversity and ecosystem services. *Cenchrus biflorus* can be grazed by livestock when immature, as a soil cover and as a carbon sink. Animals eat *Prosopis* pods, and trees cover the soil, provide shade to animals and humans, act as a carbon sink, and are used as a source of fuelwood. *Verbesina encelioides* cover the soil to reduce erosion and act as a habitat for other organisms. *Salvinia molesta* acts as a carbon sink and food for some insects. However, all these species have some drawbacks to ecosystem services. Mature *C. biflorus* has seeds that irritate animals and people because its burs are harmful. *Verbesina encelioides* can be toxic to some livestock and can harbour pests. The allelopathic effects of some of these species impede the growth of native species. *Prosopis*' deep roots make it outcompete some native species for moisture. Some residents of the Kgalagadi and Ghanzi districts have blamed *Prosopis* species for drying their boreholes. *Salvinia molesta* has a drawback to the biodiversity and ecosystem services of some aquatic habitats, such as the Okavango Delta in Botswana. The disadvantages of invasive species outweigh the benefits; hence, there is a need to control alien species entering the country and the spread of those already in the country.

It is vital to strictly enforce the regulations of what to permit to enter the country. Permits should be sought before any alien plant is introduced into a country. Control measures such as biological, chemical, mechanical, cultural, and manual or physical measures are available to control the spread of invasive alien species. Research and eradication activities must be in place to control invasive alien species. Some species, such as *C. biflorus* and *V. encelioides*, perform well in disturbed areas, such as those created by overgrazing. Disturbances such as overgrazing and wildfires must be avoided to prevent the spread of *C. biflorus* and *V. encelioides*. This can be achieved by keeping livestock within the allowable carrying capacities of grazing areas. There is a need to identify invasive alien species and sensitise communities. Control of invasive aquatic weeds in our water bodies, water quality monitoring of rivers and wetlands, registration of boats, issuance of import permits, and boat spraying need to be enforced.

Acknowledgements We acknowledge the support our employer, Botswana University of Agriculture and Natural Resources (BUAN), gave us. The work was done using the resources and time of BUAN. Prof. Feyera Senbeta Wakjira and two anonymous reviewers provided helpful comments that were used in improving earlier drafts of this chapter.

References

Aukema JE, Leung B, Kovacs K, Chivers C, Britton KO, Englin J, Frankel SJ, Haight RG, Holmes TP, Liebhold AM, McCullough DG, Von Holle B (2011) Economic impacts of nonnative forest insects in the continental United States. PLoS One 6(9):e24587. https://doi.org/10.1371/journal.pone.0024587

Boudet G (1977) Cahiers ORSTOM Série biology: ecology and plant biology, vol 12, pp 293–300

Chilume T (2017) Quantifying the ecological impacts of the Woody invasive, *Prosopis juliflora* (mesquite) in Gantsi District, Botswana. MSc thesis, University of Auckland

Costanza R, d'Arge R, de Groot R, Farber S, Grasso M, Hannon B, Limburg K, Naeem S, RV ON, Paruelo J, Raskin RG, Sutton P (1997) The value of the world's ecosystem services and natural capital. Nature 387:253–260

Crooks JA (2002) Characterising ecosystem-level consequences of biological invasions: the role of ecosystem engineers. OKIOS 97(2):153–166. https://doi.org/10.1034/j.1600-0706.2002.970201.x

Crowl TA, Crist TO, Parmenter RR, Belovsky G, Lugo AE (2008) The spread of invasive species and infectious disease as drivers of ecosystem change. Front Ecol Environ 6(5):238–246. https://doi.org/10.1890/070151

de Wit M, Crookes D, van Wilgen BW (2001) Conflicts of interest in environmental management: estimating the costs and benefits of a tree invasion. Biol Invasions 3:167–178. https://doi.org/10.1023/A:1014563702261

DEA "Department of Environmental Affairs" (2015) Fifth National Report to the convention on biological diversity. Department of Environmental Affairs, Botswana

Dubale A (2008) Invasive plants and food security: the case of *Prosopis juliflora* in the Afar region of Ethiopia. IUCN, pp 1–12

Finch DM, Butler JL, Runyon JB, Fettig CJ, Kilkenny FF, Jose S, Frankel SJ, Cushman SA, Cobb RC, Dukes JS, Hicke JA, Amelon SK (2021) Effects of climate change on invasive species. In: Poland TM, Patel-Weynand T, Finch DM, Miniat CF, Hayes DC, Lopez VM (eds) Invasive species in forest and rangelands of the United States. Springer, pp 57–84. https://doi.org/10.1007/978-3-030-45367-1

Gallardo B, Bacher S, Bradley B, Comìn FA, Gallien L, Jeschke JM, Sorte CJB, Vilà M (2019) InvasiBES: understanding and managing the impacts of invasive alien species on biodiversity and ecosystem services. NeoBiota 50:109–122. https://doi.org/10.3897/neobiota.50.35466

Haines-Young R, Potschin M (2010) The links between biodiversity, ecosystem services and human well-being. In: Raffaelli D, Frid C (eds) Ecosystem ecology: a new synthesis. Cambridge University, Cambridge, pp 110–139. https://doi.org/10.1017/CBO9780511750458.007

Holechek JL, Pieper RD, Herbel CH (2001) Range management: principles and practices. Prentice Hall, Upper Saddle River

IUCN (2018) Guidelines for invasive species planning and management on islands. IUCN. viii + 40pp, Cambridge, Gland. https://doi.org/10.2305/IUCN.CH.2018.15.en

Jacono CC, Davern TR, Center TD (2011) The adventive status of *Salvinia minima* and *S. molesta* in the Southern United States and related distribution of weevil cyrtobagous salviniae. Castanea 66(3):214–226

Kernan M (2015) Climate change and the impact of invasive species on aquatic ecosystems. Aquat Ecosyst Health Manag 18(3):321–333. https://doi.org/10.1080/14634988.2015.1027636

Kunwar RM, Acharya RP (2013) Impact assessment of invasive plant species in selected ecosystems of Bhadaure Tamagi VDC, Kaski, Nepal. IUCN Nepal, Kathmandu, pp 1–84

Kurugundla CK, Mathangwane B, Sakuringwa S, Katorah G (2016) Alien invasive aquatic plant species in Botswana: historical perspective and management. Plant Sci J 9(1):1–40. https://doi.org/10.2174/1874294701609010001

Lean CH (2021) Invasive species increase Biodiversity and, therefore, services: an argument of equivocations. Conserv Sci Pract 3(e553):1–10. https://doi.org/10.1111/csp2.553

Loreau M, Naeem S, Inchausti P, Bengtsson J, Grime JP, Hector A, Tilman D (2001) Biodiversity and ecosystem functioning: current knowledge and future challenges. Science 294(5543):804–808. https://doi.org/10.1126/science.1064088

Mace GM, Norris K, Fitter AH (2012) Biodiversity and ecosystem services: a multilayered relationship. Trends Ecol Evol 27(1):19–26. https://doi.org/10.1016/j.tree.2011.08.006

Makhabu SW, Marotsi B (2012) Changes in herbaceous species composition in the absence of disturbance in a *Cenchrus biflorus* Roxb. Invaded area in Central Kalahari Game Reserve. Botsw Int J Ecol 2012:1–6. https://doi.org/10.1155/2012/174813

Makhabu SW, Tshireletso K, Setlalekgomo MR, Kgosiesele E, Akanyang L, Legwatagwata B, Nsinamwa M, Seletlo Z, Majaga S (2021) Herbaceous communities' heterogeneity in the Kalahari sandy soils environment: a case study of Kweneng, Kgatleng districts and Mahalapye sub-district of Botswana. J Biodivers Environ Sci 18(6):87–101

Marumo MK (2018) Invasive alien species plants of Gaborone. South African National Biodiversity Institute, Pretoria

Mayfield AE III, Seybold SJ, Haag WR, Johnson MT, Kerns BK, Kilgo JC, Larkin DJ, Lucardi RD, Moltzan BD, Pearson DE, Rothlisberger JD, Schardt JD, Schwartz MK, Young MK (2021) Impacts of invasive species in terrestrial and aquatic systems in the United States. In: Poland TM, Patel-Weynand T, Finch DM, Miniat CF, Hayes DC, Lopez VM (eds) Invasive species in forest and rangelands of United States. Springer, pp 5–39. https://doi.org/10.1007/978-3-030-45367-1

MEA, Millennium Ecosystem Assessment (2005) Ecosystem and human wellbeing synthesis. Island, Washington

Mehari ZH (2015) The invasion of Prosopis juliflora and Afa pastoral livelihoods in the middle wash area of Eithiopia. Ecol Process 4:13

Minchin D (2009) Introduction of alien species. In: Encyclopaedia of ocean sciences, 2nd edn. Academic Press, pp 332–344

Mosweu S, Munyati C, Kabanda T, Setshogo M, Muzila M (2013) *Prosopis* L. invasion in the south-western region of Botswana: the perceptions of rural communities and management options. Nat Resour 4:496–505. https://doi.org/10.4236/nr.2013.48061

Moyo HPM, Fatunbi AO (2010) Utilitarian perspective of the invasion of some South African biomes by Acacia mearnsii. Glob J Environ Res 4(1):6–17

Mujere N (2015) Water hyacinth: characteristics, problems, control options, and beneficial uses. In: Practice, progress, and proficiency in sustainability. https://doi.org/10.4018/978-1-4666-9559-7.ch015

Muzila M, Setshogo MP, Moseki B, Morapedi R (2011) An assessment of Prosopis L. in the Bokspits area, South-Western Botswana, based on morphology. African J Plant Sci Biotech 5(1):75–80

Ngweno CC, Mwasi MS, Kairu KJ (2009) Distribution, density and impact of invasive plants in Lake Nakuru National Park, Kenya. Afr J Ecol 48(4):905–913. https://doi.org/10.1111/j.1365-2028.2009.01191.x

O'Connor TG, van Wilgen BW (2020) The impact of invasive alien plants on rangelands in South Africa. In: van Wigen BW (ed) Biological invasions in South Africa, invading nature – Springer series in invasion ecology, p 14. https://doi.org/10.1007/978-3-030-32394-3_16

Ortega YK, McKelvey KS, Six DL (2006) Invasion of an exotic forb impacts reproductive success and site fidelity of a migratory songbird. Oecologia 149:340–351. https://doi.org/10.1007/s00442-006-0438-8

Pathak HN, Bhuju DR, Shrestha BB, Ranjitkar S (2021) Impacts of invasive alien plants on ecosystem services of Ramsar lake cluster in middle mountain Nepal. Glob Ecol Conserv 27:1–12. https://doi.org/10.1016/j.gecco.2021.e01597

Peerzada AM, Naeem M, Ali HH, Tanveer A, Javaid MM, Chauhan BS (2016) Cenchrus biflorus Roxb. (Indian sandbur), a blessing or curse in arid ecosystems: a review. Grass Forage Sci 72:179–192. https://doi.org/10.1111/gfs.12215

PIER (2013) Pacific islands ecosystems at risk. HEAR, University of Hawaii, Honolulu

Richardson, D.M., Bond, W.J., Dean, W.R.J., Higgens, S.I., Midgley, G.F., Milton, S.J., Powrie, L.W., Rutherford, M.C., Samways, M.J. & Schulze, R.E. 2000. Invasive alien species and global change: a South African perspective. In: Mooney, H.A. & Hobbs, R.J. (eds), Invasive species in a changing world, pp. 303–349. Island Press, Washington DC.

Sebua SS, Tshisikhawe MP, Potgieter MT (2012) Invasive alien plant species: a case study of their use in the Thulamela Local Municipality, Limpopo Province. S Afr Sci Res Essay 7(27):2363–2369. https://doi.org/10.5897/SRE11.2075

Setshogo MP (2002) Cenchrus biflorus: an alien invasive grass in the Botswana flora. SABONET News 7:181–182

Shackleton RT, Le Maitre DC, van Wilgen, Richardson DM (2015) The impact of invasive alien Prosopis species (mesquite) on native plants in different environments in South Africa. S Afr J Bot 97:25–31. https://doi.org/10.1016/j.sajb.2014.12.008

Shackleton RT, Le Maitre DC, van Wilgen BW, Richardson DM (2017) Towards a national strategy to optimise the management of a widespread invasive tree (*Prosopis* species; mesquite) in South Africa. Ecosyst Serv 27:242–252. https://doi.org/10.1016/j.ecoser.2016.11.022

Shiferaw W, Demissew S, Bekele T (2018) Invasive alien plant species in Ethiopia: ecological impacts on biodiversity a review paper. Int J Mol Biol 4(4):171–178. https://doi.org/10.15406/ijmboa.2018.03.00072

Steinman AD, Cardinale BJ, Munns WR Jr, Ogdahl ME, Allan JD, Angadi T, Bartlett S, Brauman K, Byappanahalli M, Doss M, DuPont D, Johns A, Kashian D, Lupi F, McIntyre P, Miller T, Moore M, Muenich RL, Poudel R, Price J, Provencher B, Rea A, Read J, Renzetti S, Sohngen B, Washburn E (2017) Ecosystem services in the Great Lakes. J Gt Lakes Res 43(3):161–168

Tana T, Milberg P (2010) Weed flora in arable fields of eastern Ethiopia with emphasis on the occurrence of *Parthenium hysterophorus* L. Weed Res 40(6):507–521. https://doi.org/10.1046/j.1365-3180.2000.00208.x

USDA (2023) U.S. Department of Agriculture. National Invasive Species Information Center. https://www.invasivespeciesinfo.gov/subject/control-mechanisms. Downloaded on 26 Mar 2023

Wang X, Wang X, Wang W, Wang J, Yu F (2022) Effects of invasive plant diversity on soil microbial communities. Diversity 14:1–18. https://doi.org/10.21203/rs.3.rs-2021161/v1

Wise RM, van Wilgen BW, Le Maitre DC (2012) Costs, benefits and management options for an invasive alien tree species: the case of mesquite in the Northern Cape. J Arid Environ 84:80–90. https://doi.org/10.1016/j.jaridenv.2012.03.001

Wittenberg R, Cock MJW (eds) (2001) Invasive alien species: a toolkit of best prevention and management practices. CAB International, Wallingford, p 228. https://doi.org/10.1079/9780851995694.0000

Zachariades C, Hoffmann JH, Roberts AP (2011) Biological control of mesquite (Prosopis species) (Fabaceae) in South Africa. Afr Entomol 19(2):402–415. https://doi.org/10.4001/003.019.0230

Prof Shimane W. Makhabu holds BSc, MSc, PhD degrees in biology fields. His area of specialization is Vegetation and Wildlife Ecology. Prof Makhabu is currently an Associate Professor of Ecology in the Department of Biological Sciences of the Botswana University of Agriculture and Natural Resources. He has held positions of Wildlife Biologist and Range Ecologist in the Government of Botswana. His research areas are vegetation inventory and mapping, wildlife movements, and restoration of degraded rangelands.

Koketso Tshireletso is a Senior Lecturer at the Botswana University of Agriculture and Natural Resources, Botswana, teaching both undergraduate and graduate students in the field of range ecology. His research is mainly focused on plant ecology and plant-animal interactions. His research interests include rangeland policy dynamics and its effects on human livelihoods.

Mpho R. Setlalekgomo is a Senior Lecturer at the Botswana University of Agriculture and Natural Resources. She holds BSc in Biology and Environmental Sciences, BSc Hons, MSc and PhD in Zoology. She is a member of Zoological Society of Southern Africa ZSSA and Herpetological Association of Africa HAA.

Zibanani Seletlo is a PhD (Natural Resource Management) student at the Okavango Research Institute—University of Botswana. Zibanani's research is focused on the impact of grazing management systems on soil-plant stoichiometry, herbaceous plant species diversity, composition and quality based on her research interest in animal-soil-plant interactions.

Boipuso Legwatagwata is a Laboratory Assistant in the Biological Sciences Department of the Botswana University of Agriculture and Natural Resources. His areas of specialization are plant identification on, herbarium management, analytical skills on plants, soils and water samples.

Lawrence Akanyang is a Lecturer in the Department of Range and Forest Resources (RFR), Faculty of Natural Resources (NR), Botswana University of Agriculture and Natural Resources, Botswana, teaching both undergraduate and graduate students in the field of range ecology. His research is mainly focused on plant ecology, wildlife and livestock interactions.

Chapter 17
Spatio-temporal Patterns of AVHRR GIMMS Normalized Vegetation Index (NDVI) Trends in Botswana 1982–2015

Linganani Kombani and Ednah Kgosiesele

17.1 Introduction

In recent decades, the world has been suffering from a global environmental change crisis from land cover change (Kertész 2009), land degradation (Conacher 2009) and soil erosion, leading to a decline in terrestrial productivity (Wronski et al. 2017). One of the results of this environmental change is changes in vegetation structure and condition. Vegetation plays an important role by mitigating soil erosion, reducing water erosion and reducing wind speeds (Billings 1952; Duveiller et al. 2018). Understanding the spatial and temporal trends of vegetation conditions and changes over time is necessary to inform decision-making on sustainable land management and support efforts towards vegetation restoration, particularly in semi-arid environments that are characterized by highly variable rainfall. In such environments, rainfall is the limiting factor for vegetation growth (Nicholson and Farrar 1994; Coe et al. 1976). Semi-arid regions are therefore characterized by highly variable vegetation cover, and the magnitude of change will also be variable.

L. Kombani (✉)
Faculty of Engineering and Technology, Department of Urban and Regional Planning, University of Botswana, Gaborone, Botswana
e-mail: kombanil@ub.ac.bw

E. Kgosiesele
Department of Wildlife and Aquatic Resources, Faculty of Natural Resources, Botswana University of Agriculture and Natural Resources, Gaborone, Botswana
e-mail: ekgosiesele@buan.ac.bw

W. L. Hambira et al. (eds.), *Environmental Change and Biodiversity Conservation in sub-Saharan Africa*, Advances in Global Change Research 75,
https://doi.org/10.1007/978-3-031-73136-5_17

Vegetation phenology and status is one of the priority focus areas in Botswana, since vegetation supplies important ecosystem services such as pasture for grazing, flood control and mitigation, thatching grass and providing various medicinal and cultural services to the local population. Despite its importance, vegetation in Botswana and elsewhere in the world is under pressure due to multiple stressors, such as increasing livestock numbers (Moleele and Perkins 1998; Dahlberg 2000), unsustainable land practices and conversions (Kalabamu 2019) and recent climate variability and change (Mberego 2017). Botswana, like any other semi-arid country, experiences erratic and deficient rainfall (Bhalotra 1987; Batisani and Yarnal 2010), pasture degradation and recurrent droughts. With all these climate-related factors coupled with anthropogenic processes, the future of vegetation status remains uncertain and requires research priority. Vegetation change in Botswana has been the subject of a number of more localized studies (Ringrose et al. 1999, 2003; Moleele et al. 2002; Akinyemi 2021), with few studies attempting to analyse the whole country. The most notable studies covering the whole country include Mberego (2017), Woods and Sekhwela (2003), Sebego et al. (2002), and Weare and Yalala (1971). However, of all the above-mentioned studies, only Mberego (2017) adopted a time series analysis, while the other studies adopted a description and classification of savanna vegetation and mopane woodland vegetation mapping.

Contrary to in situ methods, the use of remote sensing technology and earth observation data has improved efforts towards vegetation studies by providing high temporal and spatial resolution data to enable trend analysis. One of the vegetation indices (VIs) derived from spaceborne sensors is the normalized difference vegetation index (NDVI) (Rouse et al. 1974). The NDVI comprises the red and infrared spectral reflectance, is strongly correlated with photosynthetically active biomass and chlorophyll abundance and can therefore be used as a proxy for vegetation condition (Tucker et al. 2005). A time series analysis of several NDVIs combined allows for an objective analysis of vegetation dynamics (Gibbs and Salmon 2015). Furthermore, long-time series of NDVI data provides useful information for researching the history of vegetation, monitoring current conditions and expressing concerns about the future (Van Leeuwen et al. 2006; Yang et al. 2022). Since the first application of NDVI by Rouse et al. (1974), numerous vegetation indices and platforms have been developed to provide quality spatial and temporal time series datasets. In particular, the Global Inventory Monitoring System (GIMMS) NDVI3g dataset provides the longest time series and broadest coverage (Tucker et al. 2005; Pinzon et al. 2005; Yang et al. 2022). These data have been widely (Huntley 2019; Pausata et al. 2020; Hickler et al. 2005; Li et al. 2023) used to detect vegetation trends and study the relationship between vegetation and climate factors. This study therefore aimed to gain a better understanding of the changing spatial and temporal patterns of NDVI across the differing ecological zones of Botswana over the period of 1982–2015. Specifically, this study was carried out to identify long-term spatio-temporal GIMMS NDVI3g trends and patterns in Botswana to better understand the existing patterns and variations in vegetation cover under a changing climate for better ecological and environmental management practices.

17.2 Material and Methods

17.2.1 Study Area

The location of Botswana in southern Africa is shown in Fig. 17.1. Botswana (18° and 27° S of the equator and longitudes 20° and 29° E of Greenwich), with elevations ranging between 850 and 1300 m (Tsheko 2003) and an area of 582,000 km^2 is a landlocked country bounded by Namibia, Zimbabwe, South Africa and Zambia (Fig. 17.1). Approximately, 70% of the study area is covered by the Kalahari Desert (Batisani and Yarnal 2010; Wingqvist and Dahlberg 2008), and hence, its climate is classified as semi-arid (Bhalotra 1987; Nicholson and Farrar 1994). Botswana is characterized by highly variable, erratic, unreliable rainfall, which ranges from 250 mm in the south-western part to 650 mm in the north (Bhalotra 1987; Mberego 2017; Mogome-Ntsatsi and Adeola 1995). This rainfall distribution tended to influence the vegetation distribution in the area.

Vegetation distribution is therefore described as sparse on rolling dunes in the extreme south–west (low rainfall) and changing to forest in the north (high rainfall) portion of the study area (Weare and Yalala 1971; Woods and Sekhwela 2003; De Wit and Bekker 1990; Burgess 2006). A number of studies (Nicholson and Farrar 1994; Pickup 1995; Mberego 2017) have emphasized the importance of soil moisture in determining the distribution of vegetation in semi-arid regions. Vegetation in

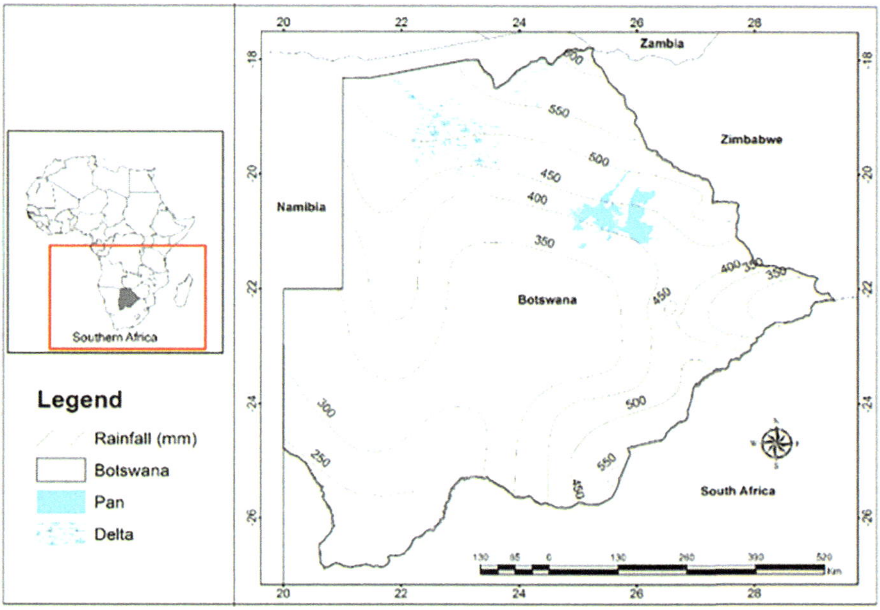

Fig. 17.1 Map showing the location of the study area within Africa and rainfall map of the study area

Botswana therefore varies both spatially and temporally, hence the need to understand the overall patterns of vegetation in the country to identify areas needing immediate attention.

17.2.2 Data

The study utilized the Global Inventory Monitoring and Modelling System (GIMMS), third generation (NDVI3g.v1) dataset generated by the Advanced Very High-Resolution Radiometer (AVHRR) sensor on board the National Oceanic and Atmospheric Administration (NOAA) satellites at a spatial resolution of 8 × 8 km and a biweekly temporal resolution (Tucker et al. 1991; Pinzon et al. 2005; Tucker 2015; Pinzon and Tucker 2014). The dataset was retrieved from http://poles.tpdc.ac.cn/en/data, which is openly available online (The National Center for Atmospheric Research 2018). These data were selected for three broad reasons: (i) they are the longest data record available online with a temporal resolution spanning from July 1981 to December 2015 (Tucker et al. 2005; Guan et al. 2021), (ii) they are the most accurate data applicable to semi-arid and arid regions and (iii) most importantly, the GIMMS dataset has been pre-processed by the GIMMS team using adaptive empirical mode decomposition (EMD) (Pinzon et al. 2005). This method addresses issues relating to sensor calibration, sensor degradation, view geometry, orbital drift, atmospheric contamination and sampling related to the AVHRR sensor (Pinzon et al. 2005) and eliminates any artefacts found in the NDVI time series (Ibrahim et al. 2015).

17.2.3 Methods

GIMMS NDVI3g.v1 data were downloaded in the scientific format Network Common Data Form, version 4 (NC4) and imported into the R statistical program following the pre-processing steps described by Detsch (2016). The analysis focused only on years with complete NDVI imagery; hence, the year 1981 was excluded because it only has half year (July to December) files. A total of sixty-eight (68) NC4 files were downloaded. The NC4 files were rasterized to tiff files and stacked together to create a time series from 1982 to 2015. The process of rasterizing uncompressed NC4 files to individual TIFF files. In total, eight hundred and sixteen (816) biweekly files (24 files per year for a period of 34 years) were analysed. The biweekly data were aggregated to monthly data using the maximum value composite (MVC) technique to further minimize the effects of clouds, sun angle, water vapour, aerosols and directional surface reflectance on the NDVI imageries (Holben 1986). The monthly data were then clipped to the study boundary (Botswana) and later aggregated to annual mean NDVI time series. The Savitzky–Golay filtering technique was then applied to smooth the NDVI time series (Savitzky and Golay

1964; Chen et al. 2004). Specific analysis codes were then developed to perform further statistical analysis on raster NDVI time series data in R. Pixelwise regression, Raster Kendall and Sen's slope were performed to visualize the spatial trends.

17.2.4 Statistical Analysis

The linear regression method was employed to analyse the monthly and annual GIMMS NDVI3.g trends across the country. The focus was on the slope of the regression, which is given by Eq. (17.1) below:

$$b = \frac{\Sigma(x-\overline{x})(y-\overline{y})}{\Sigma(x-\overline{x})^2} \tag{17.1}$$

where *b* is the slope and *x* and *y* are the sample means. The slope estimated by the linear regression method was used to determine the mean temporal change in NDVI. A positive slope indicates increasing trends, while negative slopes indicate decreasing trends.

Mann–Kendall (MK)

The Mann–Kendall test was used to test for the significance of the trend for AVHRR NDVI images. This method was proposed by Mann (1945) as a method that detects monotonic trends in time series data.

$$S = \sum_{i-1}^{n=1} \sum_{j=i+1}^{n} sgn(x_i - x_j)$$

where x_j and x_i are annual values in years *j* and *i*, *j* > *I*, respectively, *n* is the number of data points and sgn($x_j - x_i$) is calculated using Eq. (17.2) below:

$$sgn(x_j - x_i) = \begin{cases} 1 \text{if } x_j - x_i > 0 \\ 0 \text{if } x_j - x_i = 0 \\ -1 \text{if } x_j - x_i < 0 \end{cases} \tag{17.2}$$

A positive value of *S* indicates an upwards (increasing) trend, while a negative *S* indicates a downwards (decreasing) trend.

Sen's slope

Sen's slope has been found to be a powerful statistical method to develop linear relationships (Sen 1968). Sen's slope is a nonparametric statistical method that is used to evaluate the magnitude of a time series, and it assumes a linear trend in the data. Over the years, it has gained popularity in the study of vegetation dynamics (Tian et al. 2015; Fernandes and Leblanc 2005; Mishra and Mainali 2017; Li et al. 2019). It is given as:

$$\beta = \text{means}\left(\frac{x_j - x_i}{j - i}\right)$$

where β is Sen's slope estimate, x_j and x_i are the sequences of NDVI, and i and j are the years of the NDVI. $\beta > 0$ indicates an upwards trend in a time series; otherwise, the data series presents a downwards trend during the study period (Li et al. 2019).

17.3 Results

17.3.1 Spatial Patterns of NDVI

Mean monthly NDVI maps for Botswana are presented in Fig. 17.2. The general observable spatial patterns across all the months indicate the NDVI values were relatively low in the south-western while higher values occurred in the northern part of Botswana. However, negative values were recorded in at least 7 months, (January, July, August, September, October, November and December) around the pans, which is an indication of the presence of water in the area. In general, NDVI-based

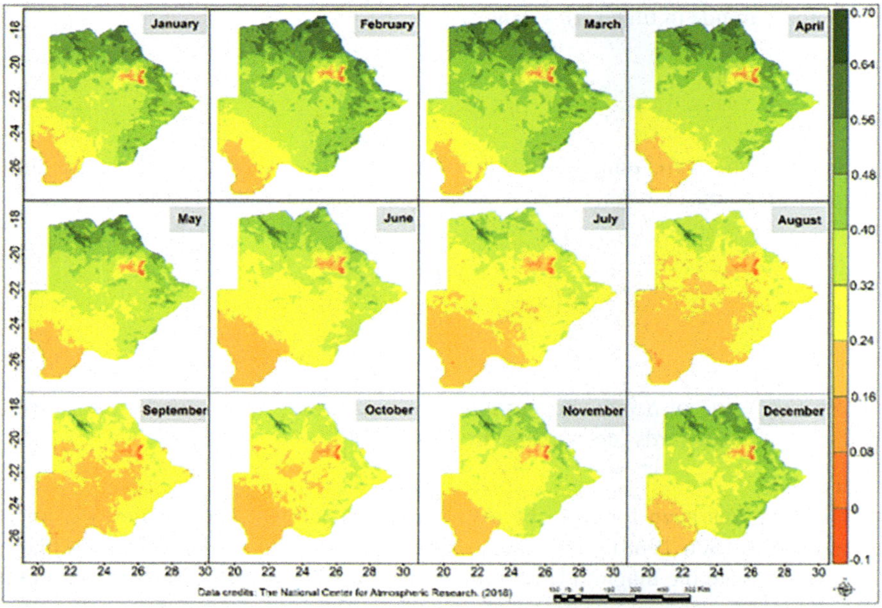

Fig. 17.2 Monthly mean NDVI maps for the study period 1982–2015

vegetation cover ranges from 0.02 to 0.70 with the minimum recorded in June and the maximum in April.

Much of the greening is experienced during the summer months especially January–April which corresponds with the rainfall season in Botswana. The winter months, on the other hand, experience very low NDVI values across the country with the exception of the area around the Okavango delta. In fact, the spatial distribution of vegetation in Botswana is affected by rainfall amounts which influence soil moisture and vegetation growth. Figure 17.3 shows the standard deviation of the monthly average NDVI from 1982 to 2015 to reflect the spatial variation in the study area. The results showed that spatial variability was high during the summer months especially February, and it decreases as we move towards the winter months. This observed spatial variation pattern can be attributed to the growth of vegetation during the summer months and vegetation loss or drying during the winter months.

Similar to the monthly spatial trends, the long-term annual patterns depict a south western to north increase as shown in Fig. 17.4. Low NDVI values were recorded in the south-western part of the study while the northern and the north-eastern parts recorded the highest. Negative values recorded around the Makgadikgadi pan indicate the presence of exposed water. The presence of these pans interferes with the observed spatial distribution of south-to-north increase creating a patch towards the north-eastern part of the country.

The pixel distribution indicates that most NDVI values are distributed around the mean pixel (0.33) and most of the pixels that are below the mean are found in the western portion of the study area.

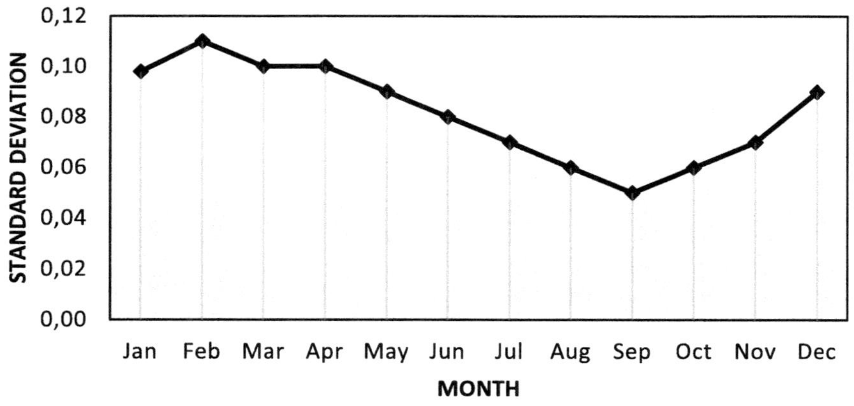

Fig. 17.3 Standard deviation for monthly NDVI 1982–2015

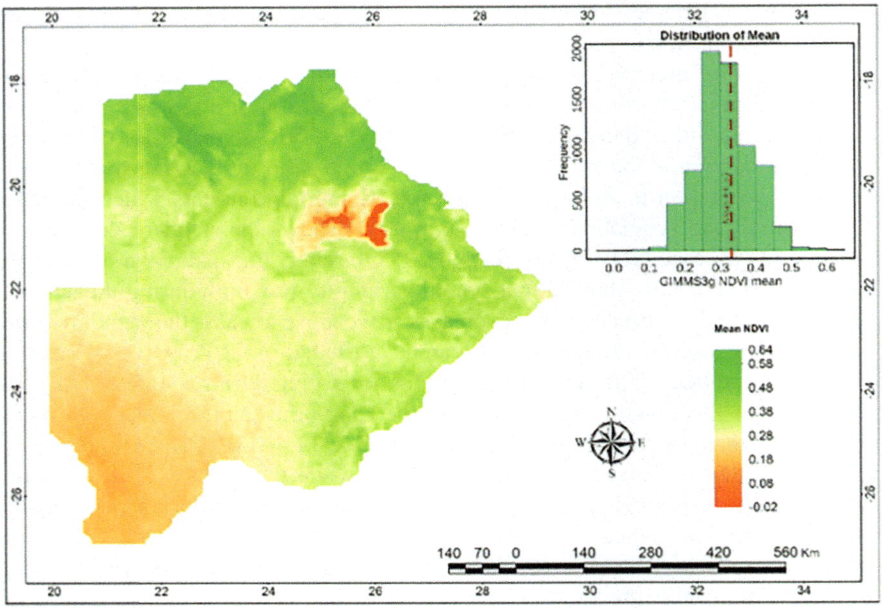

Fig. 17.4 Mean NDVI map for Botswana for the study period 1982–2015. (Data credit: The National Centre for Atmospheric Research (2018))

17.3.2 Trend Analysis

To understand the spatial patterns of AVHRR NDVI, three statistical (regression, Mann–Kendall and Thei–Sen's slope) methods were employed at the pixel level. The results of the analysis are presented in Fig. 17.5a–c. The slope of the annual NDVI was between −0.13–0.19, −0.49–0.43 and −0.0003–0.0004 for regression, Mann–Kendall and Sen's slope, respectively. In particular, the results demonstrate that the extreme northern part of the study area experienced decreasing trends between 1982 and 2015 (Fig. 17.5a). Compared to Mann–Kendal and Sen's slope results, the regression results (Fig. 17.5a) show that the highest increase happened around the central portion of the study area. On the other hand, the north-western part recorded the lowest values with most of the decrease taking place towards the border of the study area especially for regression slope and Sen's slope. Statistical distribution of NDVI trend is presented in Fig. 17.5i–iii and Table 17.1.

The study area experienced mostly weak increase trends across the different statistical analyses (Table 17.1), with Sen's slope detecting the highest (75.6) percentage. About 28% of the trends were classified as strong increases for the regression/linear slope. The percentage of pixels classified as strong decrease was less than 1% in all the statistical analysis.

Spatial trends were further calculated at $\alpha = 0.05$ to determine areas of significant spatial changes over time. The results indicate that most of the significant negative

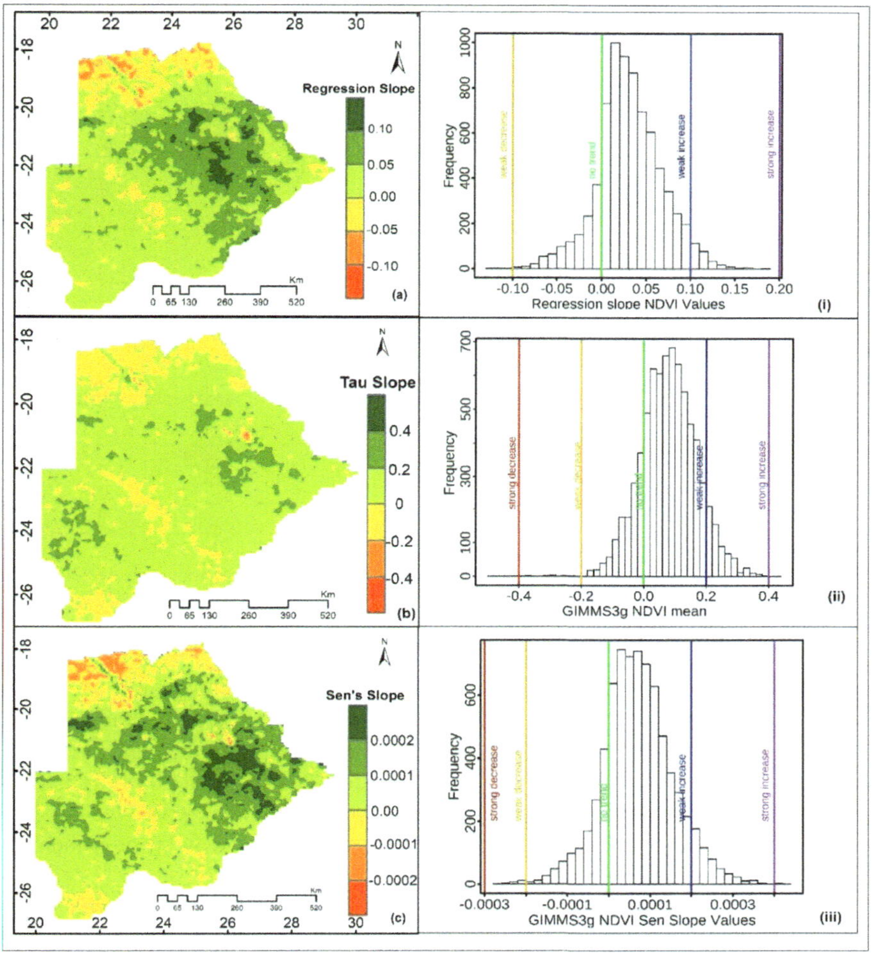

Fig. 17.5 Spatial GIMMS NDVI trends 1982–2015: (**a**) pixelwise regression, (**b**) Mann–Kendall trend analysis, and (**c**) Sen's slope estimator. (**i**) Pixelwise regression analysis trend distribution, (**ii**) Mann–Kendall analysis trend distribution, (**iii**) Sen's slope trend distribution

Table 17.1 Trend statistics for pixelwise regression, Mann–Kendall and Sen's Slope

	% Pixel count		
Trend	Regression slope	Tau slope	Sen's slope
Strong decrease	0.01	0.027	0.01
Weak decrease	1.7	0.18	0.3
No trend	12.7	17.2	17.1
Weak increase	58.0	73.9	75.6
Strong increase	27.6	8.7	7.0

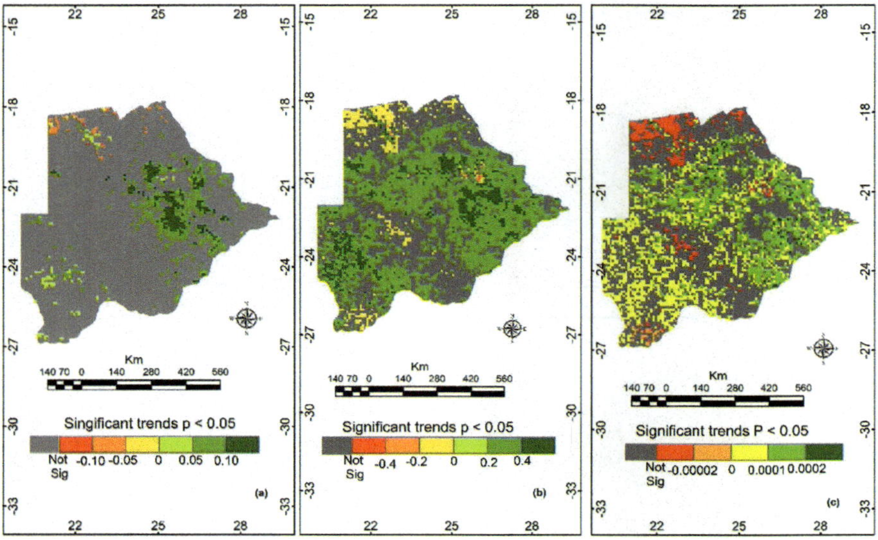

Fig. 17.6 Statistical test results showing significant trends $\alpha = 0.0$. (**a**) Linear trends, (**b**) Mann–Kendal's slope, and (**c**) Sen's slope within the study area between 1982 and 2015. Only trends significant at $\alpha = 0.05$ are shown in the figure

spatial changes occurred in the north-western part of the study area for Mann–Kendall and Sen's slope (Fig. 17.6a–c). However, compared to Mann–Kendall (3607 pixels) and Sen's slope (3607 pixels), the linear trend detected less (1157 pixels). In fact, Mann–Kendall and Sen's slope detected an equal number (3607) of pixels that were significant at $\alpha = 0.05$. Furthermore, negative trends were experienced in the south-western tip of the study area, and these were prominent for Mann–Kendall and Sen's slope.

The central part of the study is classified by high positive trends across all the statistics, with the Mann–Kendall analysis detecting most of the positive trends. On the other hand, the linear regression trends were very scattered and only clustered towards the centre of the study area. Quantitatively, most (93%, 89%, 89% for linear trends, Mann–Kendall and Sen's slope, respectively) significant trends were positive. Overall, most parts of the study area experienced greening as opposed to browning between 1982 and 2015.

17.3.3 Temporal NDVI Patterns and Trends

The Savitzky–Golay filtering technique was applied to the NDVI satellite images before analysis was carried out. The outcome of this process was then compared to the raw NDVI satellite images. As demonstrated by Fig. 17.7, the outputs between

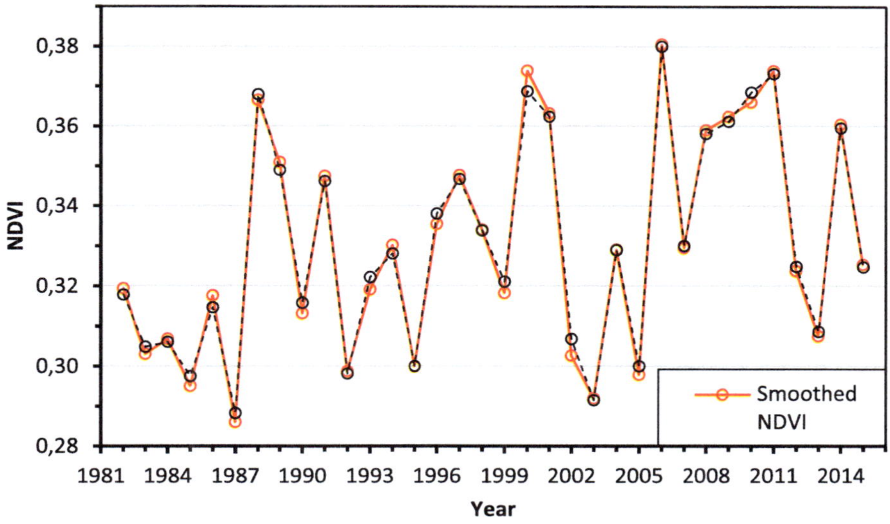

Fig. 17.7 Temporal trends of raw NDVI data and smoothed NDVI for the period 1982–2015

the two are almost a perfect fit. The filtered NDVI satellite images were then used for analysis, as these data are believed to have less noise and artefacts.

Temporal patterns and trends were carried out at both the monthly and long-term annual levels. The monthly results are presented in Fig. 17.8 and Table 17.2. The results indicate that NDVI trends were positive for all months in the estimation of both the linear and Sen's slopes. Based on the trend line fitted on the scatter plots, the upward trends varied across the months, with the first four (January, February, March and April) months registering relatively high increases, as indicated by the high slopes of 0.0012, 0.0016, 0.002 and 0.002, respectively. The lowest (0.0003) increase per year was recorded in September and October. A closer evaluation of the scatter plots indicates that NDVI values between 1982 and 1987 fluctuated below the monthly mean for the months from January to September. The remaining monthly (October to November) fluctuations below the mean were recorded between 1983 and 1992 consecutively. The month of December uniquely fluctuated below the mean from 1983 to 1986, which makes this period common to all the months across all the years.

The Mann–Kendall test further illustrates that there is indeed an upwards trend across all months between 1982 and 2015. However, this upwards trend is not significant in at least nine months, while only three (3) (March, April and May) months experienced significant (p values, 0.02, 0.02, 0.04, respectively) upwards trends with means and standard deviations of 0.41 ± 0.05 ± 0.04, 0.38 ± 0.04 and 0.35 ± 0.04, respectively. Table 17.2 shows the monthly temporal trends through three different statistical methods: p values, statistical interpretation means and standard deviations.

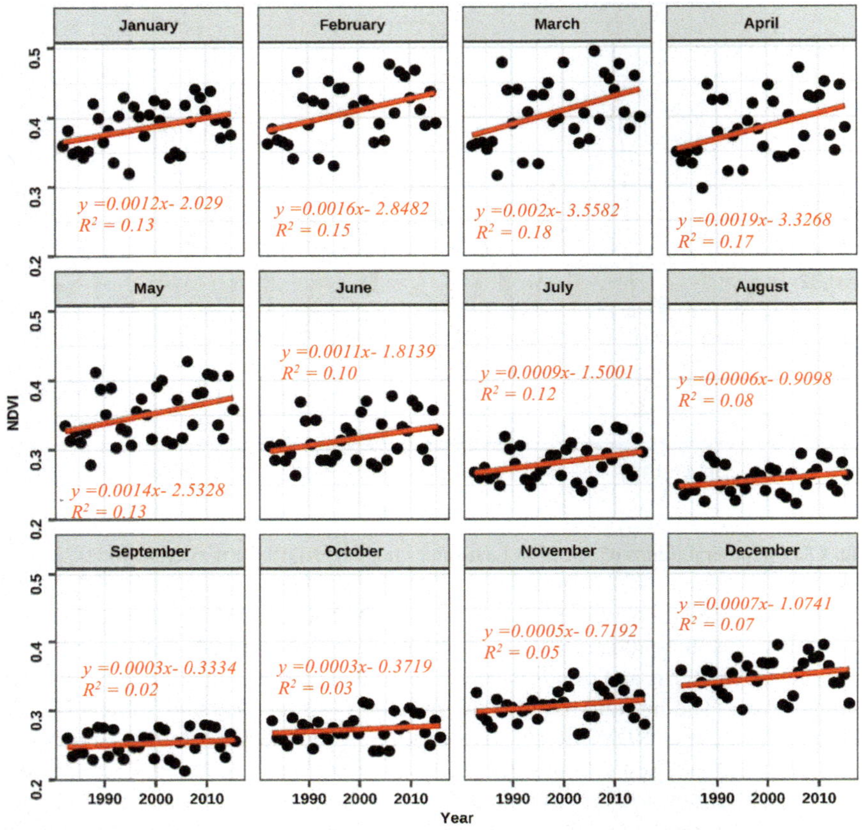

Fig. 17.8 Monthly temporal scatter plots

Table 17.2 Temporal trend statistics for linear slope, Sen's slope and Mann–Kendall

Month	Mean ± std	Linear slope	Sen's slope	Mann–Kendall Tau	*p* value
January	0.39 ± 0.03	0.0012	0.0012	0.212	0.080
February	0.41 ± 0.04	0.0016	0.0016	0.226	0.062
March	0.41 ± 0.05	0.0020	0.0019	0.280	**0.021**
April	0.38 ± 0.04	0.0019	0.0000	0.294	**0.015**
May	0.35 ± 0.04	0.0014	0.0015	0.244	**0.043**
June	0.32 ± 0.03	0.0011	0.0009	0.191	0.116
July	0.28 ± 0.03	0.0009	0.0009	0.223	0.066
August	0.26 ± 0.02	0.0006	0.0006	0.173	0.155
September	0.25 ± 0.02	0.0003	0.0003	0.112	0.343
October	0.27 ± 0.02	0.0003	0.0003	0.102	0.406
November	0. 31 ± 0.03	0.0005	0.0008	0.191	0.116
December	0.35 ± 0.038	0.0007	0.0008	0.184	0.131

Notes: *p* values in bold indicate significant trends

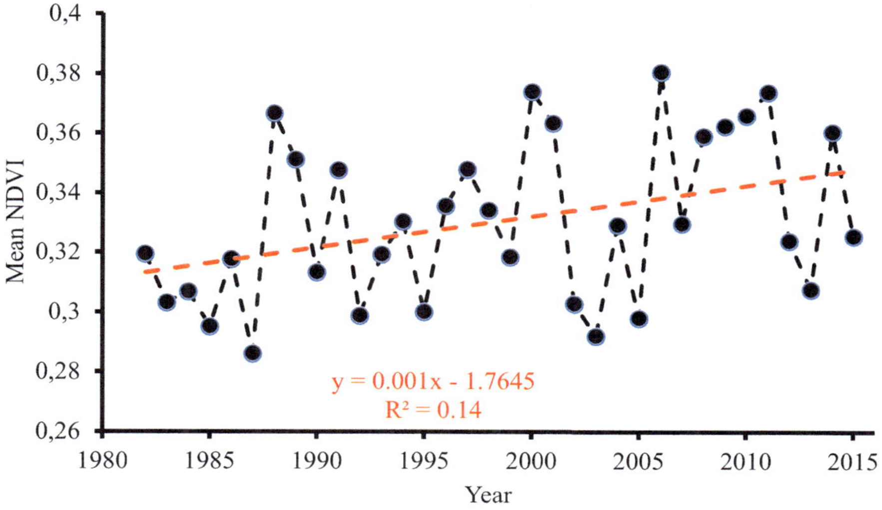

Fig. 17.9 Annual GIMMS NDVI variations scatter plot

Similar to the monthly trends, the annual trends demonstrated a statistically significant (*p* value 0.01) upwards trend with a linear slope of 0.001 and a Sen's slope of 0.00007. As indicated by Fig. 17.9, NDVI highly fluctuated with an increasing trend at a mean and standard deviation of 0.33 ± 0.02. The highest (0.38) NDVI was recorded in 2006, while the lowest (0.29) was recorded in 1987 and 2003. Of particular, the period between 2008 and 2011 is the only period that registered an increase in at least three consecutive years, with 2000–2003 being the only period that registered a decrease for the same number of years. NDVI variations are presented in Fig. 17.9.

NDVI trends from 1982 to 2015 demonstrate a common start and end of trend in that both experience a decrease. However, the start of the trend recorded a slightly lower (0.32) NDVI value than the end of the trend (0.33). A bulk (56%) of the years recorded NDVI values that were below the mean compared to 44% of the years that recorded NDVI values that were above the mean. NDVI patterns can therefore be said to have been mostly fluctuating below the mean more than half the time of the study period. The distribution of NDVI between 1982 and 2015 is given in Fig. 17.10.

The serial autocorrelation function (ACF) was performed on annual NDVI data to show AVHRR NDVI trends between 1982 and 2015. Fifteen (15 lags) were examined to determine if there is any autocorrelation in annual NDVI data. The results (Fig. 17.11) showed that there is insignificant autocorrelation present in the data as most of the spikes from the ACF analysis lie between the horizontal dotted signifying the absence of autocorrelation in the annual NDVI data. As a rule of thumb, if there are no autocorrelations present in the data, Mann–Kendall analysis can therefore be carried out to the data without any further corrections.

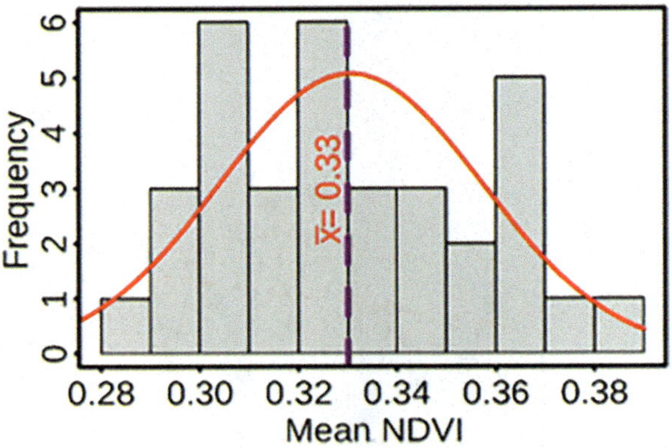

Fig. 17.10 GIMMS mean NDVI distribution 1982–2015

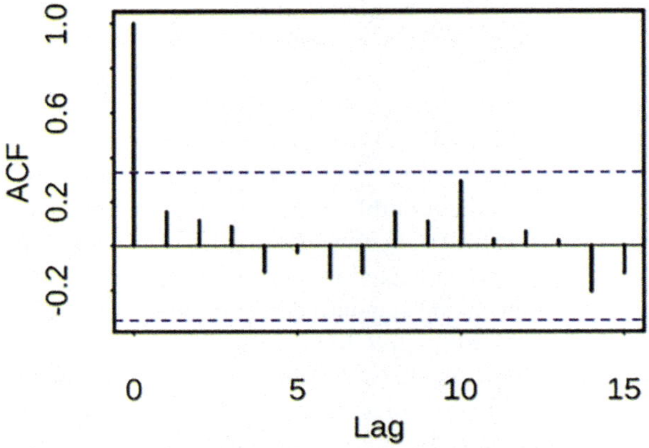

Fig. 17.11 GIMMS NDVI serial autocorrelation function (ACF) plot 1982–2015

17.3.4 Trend Decomposition

This is done in order to separate the data into three components which are trend, seasonal and remainder. These components constitute the data as indicated by Fig. 17.12. A cross-examination of the data shows that NDVI was highly fluctuating with no particular observable trend. Similarly, the season component shows a uniform fluctuation with no trend established. Figure 17.12 shows that the AVHRR NDVI trends were fluctuating with an increasing trend, this is demonstrated by the trend component.

The remainder demonstrates highly variable patterns with values ranging from −0.04 to 0.04. This component represents the residuals and shows that certain years such as 1988, 1992 and 1995 registered the highest negative residuals which implies that the NDVI values were fluctuating below the means. There is therefore a need to understand the causes of these high negative values in these years. The causes of the negative values or NDVI fluctuating below the means may be varied ranging from anthropogenic to climate-related issues.

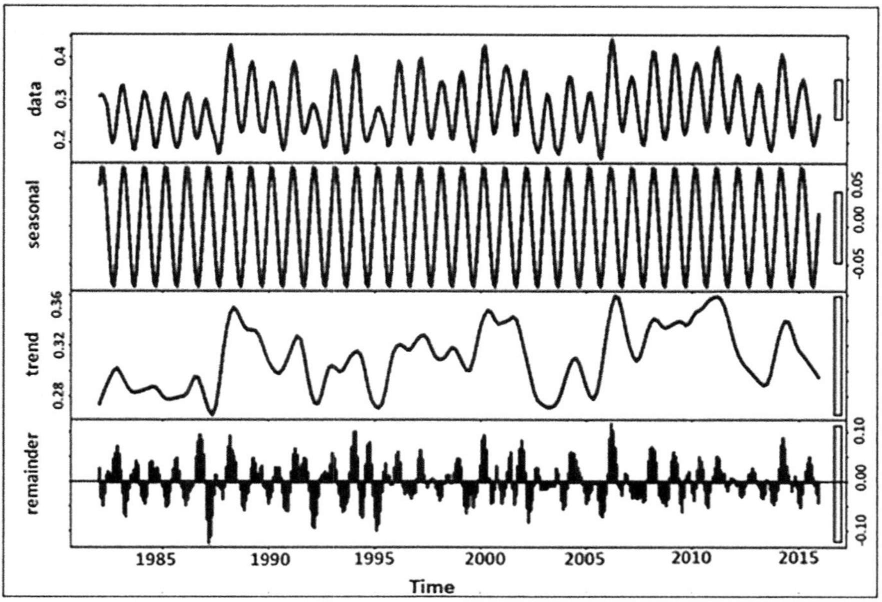

Fig. 17.12 GIMMS NDVI trend decomposition

17.4 Discussions

17.4.1 Spatial NDVI Trends

Spatial NDVI trends indicated a south-to-north gradual increase in vegetation cover across all the months and annual analysis. This observed spatial trend is consistent with the results from previous studies carried out in Botswana (Mberego 2017; Sebego et al. 2002) and other parts of sub-Saharan countries (Huntley 2019; Pausata et al. 2020). The south-western corner of Botswana has consistently recorded the lowest NDVI values throughout the study period. The decreasing vegetation trends imply that the south-western parts of the country experienced an overall decline in vegetative cover and biomass as influenced by rainfall and moisture availability. The observed spatial patterns are a true reflection of the rainfall and soil moisture being the limiting factor in vegetation growth. Coe et al. (1976) emphasized the direct linear relationship between rainfall and biomass hence the observed spatial trend is justified as rainfall patterns are a determining factor for vegetation growth and dynamics in Botswana. The areas around Sua Pan and Makgadikgadi pans recorded a negative NDVI trend, signifying areas of low to no vegetative cover, dominated by surface water bodies. Furthermore, the winter months (June, July, August and September) recorded very low NDVI values across the study area and very low spatial variation. This is so because photosynthesis activity is very low to non-existent during this period, and the vegetation changes from green to brownish colour.

Statistical pixelwise trend analysis indicates that the study area generally experienced more increasing than decreasing trends. More than 80% (86%, 82.6%, 82.6% for pixelwise regression, Mann–Kendall and Sen's slope, respectively) were classified as positive trends from 1982 to 2015. Different studies (Pausata et al. 2020; Hickler et al. 2005; Dardel et al. 2014) have used GIMMS AVHRR NDVI3g to assess trends in semi-arid and arid regions especially the Sahelian region and have determined that contrary to widely held notion of vegetation loss the region is actually greening. A study by Li et al. (2023) indicates that the proportions of increasing vegetation trends in Southern Africa are higher than the proportion of vegetation loss and that of no change. Hence, the results from the study about more increase as opposed to decreasing trends are in line with the general findings from the region and elsewhere in the African continent. Surprisingly, the northern part of Botswana (the areas area receiving the highest rainfall) is dominated by pixels classified as no trends across the three different statistical methods. Contrary to the other two methods, the pixelwise regression statistical method clearly shows decreasing trends in the northern part of the country, in particular, the area around the Okavango Delta.

17.4.2 Temporal NDVI Trends

Temporal trends show an upwards yet highly variable pattern across the different months and years. However, both the monthly and annual temporal trends show increasing trends across the different months and years. To better understand the temporal trends, particular attention was given to periods of unique trends within the overall study period. The following periods were examined: 1982–1987, 1983–1992 and 1983–1983. These periods had negative residuals when the monthly mean was subtracted from the NDVI value (Fig. 17.13); hence, they were fluctuating below the mean. The period corresponds to the classification by Mberego (2017) in which he classified the periods 1981/1982, 1984/1985, 1985/1986 as moderately dry and extremely dry 1991/1992 and 1994/1995. To further explain the observed temporal trend for the above time scales, Tsuaneng-Manthe (2014) explicitly states that the years between 1981 and 1987, 1990 and 1995 were drought periods in Botswana; hence, the negative anomalies experienced in the monthly time series are justified as biomass/vegetation in semi-arid areas rely heavily on rainfall amounts and variability.

Botswana shows a high temporal variability in vegetation productivity during the summer months which corresponds to the growing season and rainfall season (Akinyemi 2021). NDVI-based vegetation was therefore highly variable during the months of November–April with March being the highest with a mean and standard deviation of 0.41 ± 0.05. The variability gradually decreases during the winter with the lowest mean and standard deviation reaching lowest (0.25 ± 0.02) in September. One observation that can be deduced from the results is that the monthly standard deviations decrease until they rest at 0.02 from August to October then start picking from November. Similarly, the annual temporal variation shows a significant (p

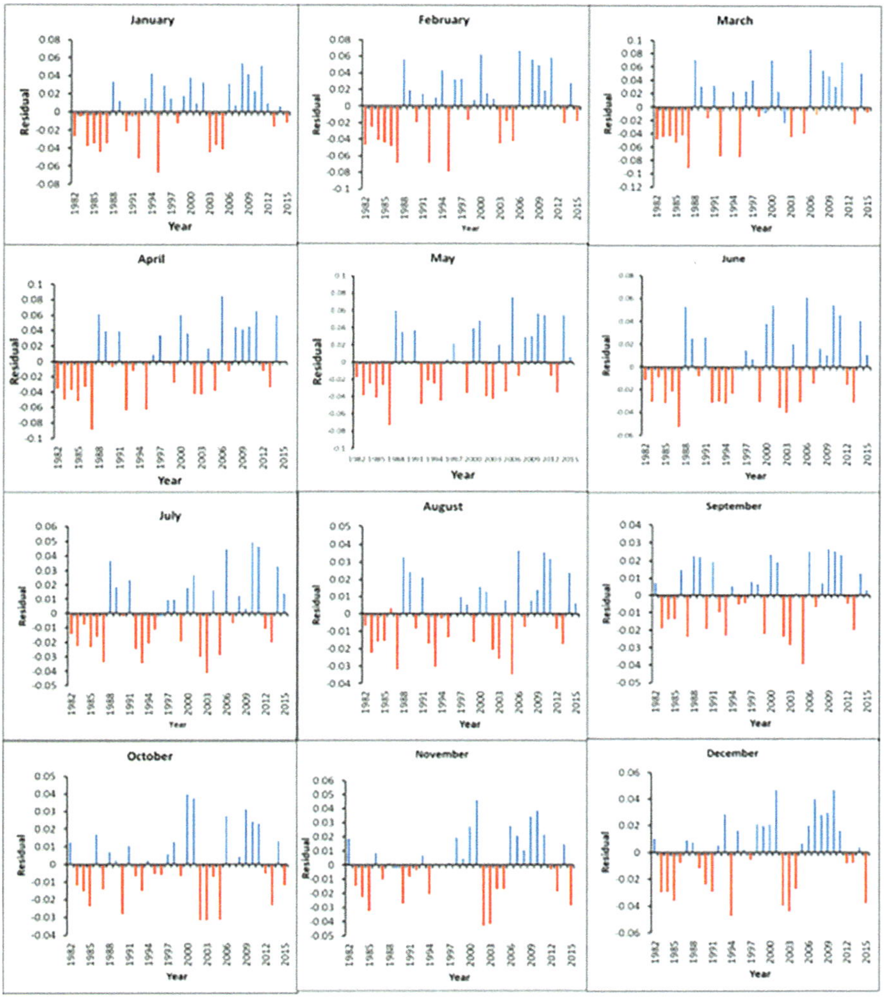

Fig. 17.13 Monthly GIMMS NDVI residual plots 1982–2015

value = 0.01, tau = 0.08214096, Sen's slope = 0.00007) upwards trend but yet highly variable. This observed trend is in line with findings from other studies from different parts of Africa such as the Sahara region (Pausata et al. 2020) and sub-Saharan Africa (Brandt et al. 2017). Previous research (Ruijsch et al. 2023; Brandt et al. 2017) argues that Africa has been greening over the past two decades, and this has been attributed to an increase in summer rainfall (Zeng et al. 2023) and an increase in woody vegetation (Brandt et al. 2017). A study by Ruijsch et al. (2023) found out that 2.1% of Africa has experienced greening over the past two decades and this area covers Botswana. Overall, recent studies tend to emphasize an opposing trend to the

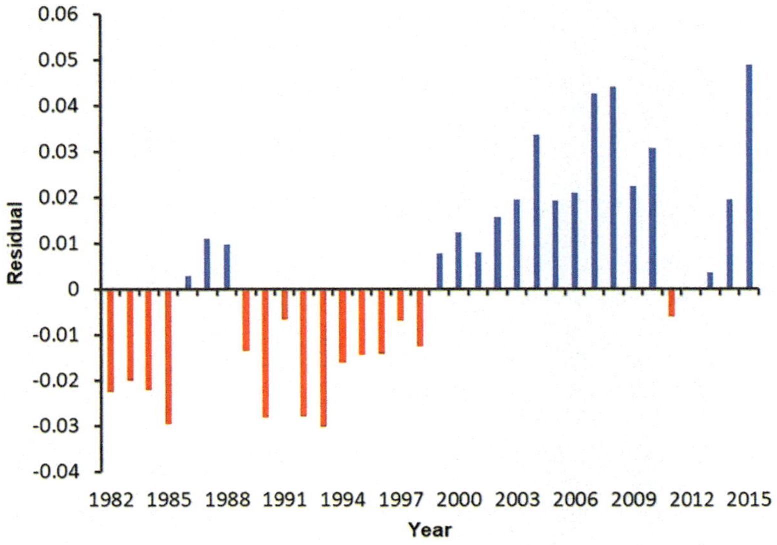

Fig. 17.14 Annual GIMMS NDVI residual plot 1982–2015

downward trend reported by earlier studies. However, the main reason for the opposing trend has been indicated as the increase in woody vegetation. Further investigations are therefore needed on the drivers for vegetation change and structure of vegetation in order to better understand the observed trend in the study area.

The findings of this study further indicate that about sixteen (16) years had their values fluctuating below the means with the period between 1989 and 1998 registering negative residual values consistently then the area between 1999 and 2009 experienced very high increase as the residuals are all positives (Fig. 17.14). Understanding the causes of the negative residuals between 1999 and 2009 requires bringing in the factors or drivers for vegetation change such as climate variability. This period (1999–2009) was preceded by the worst drought period (1981–1987) and also falls within yet another drought period (1990–1995) (Tsuaneng-Manthe 2014). The consecutive negative residuals are therefore justified as the period was predominantly characterized by droughts.

17.5 Conclusion

Vegetation cover in semi-arid areas is assumed to be particularly limited by rainfall and soil moisture. As a result, semi-arid areas are characterized by highly variable vegetation cover in response to interannual and seasonal rainfall. This study investigated the spatial patterns and trends of AVHRR NDVI in Botswana from 1982 to 2015. The results demonstrated that vegetation cover in the area varies both spatially and temporally. Three statistical methods (regression analysis, Mann–Kendall and Sen's Slope) were employed to investigate NDVI trends in the study area. The vegetation cover of Botswana like many other semi-arid environments gradually increases from the south-western portion to the northern part of the country following a moisture gradient. Very low NDVI values were recorded in some years corresponding with drought periods in the country, which clearly validates the argument presented by many authors both locally and internationally that rainfall is the determinant factor of vegetation cover in any given area. This study has provided insights into long-term spatial and temporal variations of vegetation in Botswana using freely available time series satellite data, with a potential to inform suitable future land management. While the association between NDVI trends and climatic variables has been well documented, it is difficult to attribute these patterns to specific changes because of the mixed effects of human activities and climate change and variability. Annual temporal variations showed an upwards trend which was highly variable. The results of this study therefore suggest that more research is needed to explore the variable correlation between climate change and NDVI trends, and accurately quantify and separate the impacts of natural factors and human influence on vegetation changes. The study further recommends the use of finer resolution scales such as Landsat and MODIS data to better capture vegetation variability in the study area.

References

Akinyemi FO (2021) Vegetation trends, drought severity and land use-land cover change during the growing season in semi-arid contexts. Remote Sens 13:836

Batisani N, Yarnal B (2010) Rainfall variability and trends in semi-arid Botswana: implications for climate change adaptation policy. Appl Geogr 30:483–489

Bhalotra YPR (1987) Climate of Botswana, Part II: elements of climate. Republic of Botswana, Department of Meteorological Services, Gaborone

Billings WD (1952) The environmental complex in relation to plant growth and distribution. Q Rev Biol 27(3):251–265

Brandt M, Rasmussen K, Peñuelas J, Tian F, Schurgers G, Verger A, Mertz O, Palmer JRB, Fensholt R (2017) Human population growth offsets climate-driven increase in woody vegetation in sub-Saharan Africa. Nat Ecol Evol 1:0081

Burgess J (2006) Country pasture/forage resource profiles. Food and Agriculture Organization of the United Nations (FAO), Rome

Chen J, Jönsson P, Tamura M, Gu Z, Matsushita B, Eklundh L (2004) A simple method for reconstructing a high-quality NDVI time-series data set based on the Savitzky–Golay filter. Remote Sens Environ 91:332–344

Coe MJ, Cumming DH, Phillipson J (1976) Biomass and production of large African herbivores in relation to rainfall and primary production. Oecologia 22:341–354

Conacher A (2009) Land degradation: A global perspective. New Zealand Geographer 65(2):91–94

Dahlberg AC (2000) Vegetation diversity and change in relation to land use, soil and rainfall—a case study from North-East District, Botswana. J Arid Environ 44:19–40

Dardel C, Kergoat L, Hiernaux P, Mougin E, Grippa M, Tucker CJ (2014) Re-greening Sahel: 30 years of remote sensing data and field observations (Mali, Niger). Remote Sens Environ 140:350–364

De Wit PV, Bekker RP (1990) Explanatory note on the land systems map of Botswana. Food and Agriculture Organization of the United Nations

Detsch F (2016) Gimms: download and process GIMMS NDVI3g data. R package version 1.0.0

Duveiller G, Forzieri G, Robertson E, Li W, Georgievski G, Lawrence P, ... & Cescatti A (2018) Biophysics and vegetation cover change: a process-based evaluation framework for confronting land surface models with satellite observations. Earth Syst Sci Data 10(3):1265–1279

Fernandes R, Leblanc G (2005) Parametric (modified least squares) and nonparametric (Theil–Sen) linear regressions for predicting biophysical parameters in the presence of measurement errors. Remote Sens Environ 95:303–316

Gibbs HK, Salmon JM (2015) Mapping the world's degraded lands. Appl Geogr 57:12–21

Guan X, Shen H, Wang Y, Chu D, Li X, Yue L, Liu X, Zhang L (2021) Fusing MODIS and AVHRR products to generate a global 1-km continuous NDVI time series covering four decades. Earth Syst Sci Data Discuss:1–32

Hickler T, Eklundh L, Seaquist JW, Smith B, Ardö J, Olsson L, Sykes MT, Sjöström M (2005) Precipitation controls Sahel greening trend. Geophys Res Lett 32

Holben BN (1986) Characteristics of maximum-value composite images from temporal AVHRR data. Int J Remote Sens 7:417–1434

Huntley BJ (2019) Angola in outline: physiography, climate and patterns of biodiversity. In: Biodiversity of Angola: science & conservation: a modern synthesis. Springer Nature, pp 15–42

Ibrahim YZ, Balzter H, Kaduk J, Tucker CJ (2015) Land degradation assessment using residual trend analysis of GIMMS NDVI3g, soil moisture and rainfall in Sub-Saharan West Africa from 1982 to 2012. Remote Sens 7:5472–5494

Kalabamu FT (2019) Land tenure reforms and persistence of land conflicts in sub-Saharan Africa–the case of Botswana. Land Use Policy 81:337–345

Kertész Á (2009) The global problem of land degradation and desertification. Hung Geogr Bull 58:19–31

Li Y, Xie Z, Qin Y, Zheng Z (2019) Estimating relations of vegetation, climate change, and human activity: A case study in the 400 mm annual precipitation fluctuation zone, China. Remote Sens 11:1159

Li Z, Li C, Gao D, Wang S (2023) Discrimination among climate, human activities, and ecosystem functional-induced land degradation in Southern Africa. Remote Sens 15:403

Mann HB (1945) Nonparametric tests against trend. Econometrica J Econometric Soc 13:245–259

Mberego S (2017) Temporal patterns of precipitation and vegetation variability over Botswana during extreme dry and wet rainfall seasons. Int J Climatol 37:2947–2960

Mishra NB, Mainali KP (2017) Greening and browning of the Himalaya: spatial patterns and the role of climatic change and human drivers. Sci Total Environ 587:326–339

Mogome-Ntsatsi K, Adeola OA (1995) Promoting environmental awareness in Botswana: the role of community education. Environmentalist 15:281–292

Moleele NM, Perkins JS (1998) Encroaching woody plant species and boreholes: is cattle density the main driving factor in the Olifants Drift communal grazing lands, south-eastern Botswana? J Arid Environ 40:245–253

Moleele NM, Ringrose S, Matheson W, Vanderpost C (2002) More woody plants? The status of bush encroachment in Botswana's grazing areas. J Environ Manag 64:3–11

Nicholson SE, Farrar TJ (1994) The influence of soil type on the relationships between NDVI, rainfall, and soil moisture in semiarid Botswana. I. NDVI response to rainfall. Remote Sens Environ 50:107–120

Pausata FSR, Gaetani M, Messori G, Berg A, De Souza DM, Sage RF, Demenocal PB (2020) The greening of the Sahara: past changes and future implications. One Earth 2:235–250

Pickup G (1995) A simple model for predicting herbage production from rainfall in rangelands and its calibration using remotely-sensed data. J Arid Environ 30:227–245

Pinzon JE, Tucker CJ (2014) A non-stationary 1981–2012 AVHRR NDVI3g time series. Remote Sens 6:6929–6960

Pinzon JE, Brown ME, Tucker CJ (2005) EMD correction of orbital drift artifacts in satellite data stream. World Scientific, Singapore, pp 167–186

Ringrose S, Lesolle D, Boshoma T, Gopolang B, Vanderpost C, Matheson W (1999) An analysis of vegetation cover components in relation to climatic trends along the Botswana Kalahari Transect. Botswana Notes Rec 31:33–52
Ringrose S, Matheson W, Wolski P, Huntsman-Mapila P (2003) Vegetation cover trends along the Botswana Kalahari transect. J Arid Environ 54(2):297–317
Rouse JW, Haas RH, Schell JA, Deering DW (1974) Monitoring the vernal advancement and retrogradation (greenwave effect) of natural vegetation. Progress Report, September 1972–March 1973. Type II report for the period September 1972–March 1973. Texas A & M University
Ruijsch J, Teuling AJ, Verbesselt J, Hutjes RWA (2023) Landscape restoration and greening in Africa. Environ Res Lett 18:064020
Savitzky A, Golay MJ (1964) Smoothing and differentiating of data by simplified least square procedures. Anal Chem 36:1627–1639
Sebego RJG, Arnberg W, Ringrose S (2002) Relation between cold cloud data, NDVI and MOPANE in eastern Botswana. Int Arch Photogramm Remote Sens Spat Inf Sci 34:W6
Sen PK (1968) Estimates of the regression coefficient based on Kendall's tau. J Am Stat Assoc 63:1379–1389
The National Center for Atmospheric Research (2018) Global GIMMS NDVI3g v1 dataset [Online]. Available: http://poles.tpdc.ac.cn/en/data. Accessed 16 Jan 2023
Tian H, Cao C, Chen W, Bao S, Yang B, Myneni RB (2015) Response of vegetation activity dynamic to climatic change and ecological restoration programs in Inner Mongolia from 2000 to 2012. Ecol Eng 82:276–289
Tsheko R (2003) Rainfall reliability, drought and flood vulnerability in Botswana. Water SA 29:389–392
Tsuaneng-Manthe M (2014) Drought conditions and management strategies in Botswana. Ministry of Environment, W. A. T., Gaborone
Tucker CJ (2015) NOAA's advanced very high resolution radiometer (AVHRR), NDVI3g.v1
Tucker CJ, Dregne HE, Newcomb WW (1991) Expansion and contraction of the Sahara Desert from 1980 to 1990. Science 253:299–300
Tucker CJ, Pinzon JE, Brown ME, Slayback DA, Pak EW, Mahoney R, Vermote EF, Saleous E (2005) An extended AVHRR 8-km NDVI dataset compatible with MODIS and SPOT vegetation NDVI data. Int J Remote Sens 26:4485–4498
Van Leeuwen WD, Orr BJ, Marsh SE, Herrmann SM (2006) Multi-sensor NDVI data continuity: uncertainties and implications for vegetation monitoring applications. Remote Sens Environ 100:67–81
Weare PR, Yalala A (1971) Provisional vegetation map of Botswana. Botswana Notes Rec 3:131–147
Wingqvist GO, Dahlberg E (2008) Botswana environmental and climate change analysis. University of Gothenburg, Stockholm
Woods J, Sekhwela MBM (2003) The vegetation resources of Botswana's savannas: an overview. S Afr Geogr J 85:69–79
Wronski T, Bariyanga JD, Sun P, Plath M, Apio A (2017) Pastoralism versus agriculturalism—how do altered land-use forms affect the spread of invasive plants in the degraded Mutara rangelands of north-eastern Rwanda? Plan Theory 6:19
Yang X, Yang Q, Yang M (2022) Spatio-temporal patterns and driving factors of vegetation change in the pan-third pole region. Remote Sens 14:4402
Zeng Y, Jia L, Menenti M, Jiang M, Barnieh BA, Bennour A, Lv Y (2023) Changes in vegetation greenness related to climatic and non-climatic factors in the Sudano-Sahelian region. Reg Environ Chang 23:92

Linganani Kombani is a Human Geographer with expertise in GIS and Remote Sensing. His area of expertise includes vegetation cover analysis, spatial analysis and modelling, application of geospatial sciences in addressing various environmental issues. Research interests include climate variability/change, vegetation analysis, land use modelling and predictions. He is a spatial analyst who focuses on the application of geospatial techniques across different fields, hydrology, ecology, land use/cover, climate, and EIA. He is an adept in the use of both commercial and open access packages such as ArcGIS, R, Google colab, Earth engine and Qgis.

Ms. Ednah Kgosiesele is a Lecturer at the Botswana University of Agriculture and Natural Resources (BUAN). She holds an undergraduate degree in Environmental Science from the University of Botswana and MSc in Geo-Information Science and Earth Observation from the University of Twente. Prior to joining the institution, she worked in an environmental consultancy company as a GIS/Remote Sensing specialist. Her core interests include the application of earth observation technology for agriculture and natural resources management with a focus on resource degradation and climate change, land use/land cover change, and spatial modelling.

Chapter 18
Effects of Bush Encroachment Control on Herbaceous Plant Species Composition, Ecological Status, and Grazing Value in a Semi-arid Savannah Ecosystem

Gaolathe Tsheboeng, D. Maseko, Meshack Nkosinathi Dludlu, and N. Nkonyane

18.1 Introduction

Globally, semi-arid savannah ecosystems are important for the ecosystem services and goods they provide. One such key role of savannah ecosystems is their provision of pasture to both livestock and game (Reynolds et al. 2007). This translates to about one-third of the global population deriving livelihoods from the savannahs (Reynolds et al. 2007). Despite this, there is no guarantee that the savannah ecosystem will continue to serve its critical role of supporting livelihoods by provisioning pasture to game and livestock. This is because these ecosystems are now threatened by the escalating proliferation of bush encroachment (D'Odorico et al. 2012; O'Connor et al. 2014; Ayelew and Mulualem 2018). Bush encroachment is the invasion of woody plant species into savannah and grassland ecosystems resulting in the increase of bushes and decline of the abundance of grass species that previously occupied that ecosystem (Ward 2005; Eldridge et al. 2011; O'Connor et al. 2014). An increase in woody abundance in savannahs reduces grass biomass and species richness which negatively affects their ecosystem functioning (Angassa and Baars 2000; Archer 2010). The carrying capacity of rangelands reduces as a result of bush encroachment and leads to a reduction in economic profitability (Mugasi et al. 2000; Espach 2006; Börner et al. 2007).

G. Tsheboeng (✉)
University of Botswana, Department of Biological Sciences, Gaborone, Botswana
e-mail: tsheboengg@ub.ac.bw

D. Maseko · M. N. Dludlu · N. Nkonyane
University of Eswatini, Department of Biological Sciences, Kwaluseni, Eswatini
e-mail: ndmaseko@uneswa.ac.sz; meshack@uneswa.ac.sz

W. L. Hambira et al. (eds.), *Environmental Change and Biodiversity Conservation in sub-Saharan Africa*, Advances in Global Change Research 75,
https://doi.org/10.1007/978-3-031-73136-5_18

Woody plant cover in northeastern Eswatini has increased over the last 70 years from 2% to >40% in recent years (Roques et al. 2001). Mbuluzi Game Reserve has experienced an increase in bush encroachment from 2% to >40% in the last 70 years with *D. cinerea* being the most encroaching species (Roques et al. 2001; Sirami and Monadjem 2012). The main woody plant species accounting for bush encroachment in Eswatini is *Dichrostachys cinerea* (Roques et al. 2001). Several factors including overgrazing (Bothma 2002; Sankaran et al. 2004) and fire have been associated with this increase of bush encroachment (Oba et al. 2000; Van Auken 2009). Overgrazing removes the balance between the grasses and trees, thus introducing competition (Briske et al. 2003; Sankaran et al. 2004). Cattle are responsible for opening up the grass sward and enabling woody plant species to have access to soil moisture that could otherwise have been utilised by the grasses (Tefera et al. 2008). Bush encroachment replaces palatable grasses with woody plants and shrubs which are often indigestible to herbivores (Ward 2005) leading to a decline in grazing capacity for both wildlife and domestic herbivores (Oba et al. 2000; Sheuyange et al. 2005; Angassa et al. 2012; Lukomska et al. 2014).

Fire is one of the most important management tools that can be used to control woody species in savannah ecosystems because it decreases the biomass of woody plants suppressing their recruitment and abundance (Higgins et al. 2000; Angassa et al. 2012; Grady and Hoffmann 2012). Fire typically reduces the height of savannah trees and in some species, it causes death (Hoffmann and Solbrig 2003). For instance, *Dichrostachys cinerea* and *Vachellia karroo* which are also part of the common encroachers seen in Eswatini savannahs are resistant to annual burns, but their heights are reduced by more frequent fires (Roques et al. 2001), therefore minimising the negative effect they have on the grass species. Bush encroachment can be controlled by increasing the fire frequency and intensity to increase the number of woody plants being killed (Midgley et al. 2010). While fire can help in controlling bush encroachment (Mphinyane et al. 2011), its poor use can lead to unintended outcomes of increased abundance of woody plants in rangelands. Regular burning suppresses woody plant growth by destroying the shrubs and juvenile trees and thus prevents their development into mature woody plants. Fire disturbs the seed development of woody species preventing them from maturing into adult plants therefore suppression of these fires may speed up the rate of bush encroachment (Bond and Van Wilgen 2012). Conversely, the reduction in fire frequency allows the invasion of woody evergreen species which are not easily destroyed by annual burns (Hoffmann 1998; Roques et al. 2001). In Eswatini for instance, rural communities often use fire to clear land for farming. This has implications for promoting bush encroachment. This is because the fire used is not fierce enough to kill woody saplings but can kill the grasses creating gaps for the establishment of woody species (Hessburg and Agee 2003).

Introducing selective browsers that will feed on woody plants can help reduce woody plant densities and help maintain grassland and savannah ecosystems (Staver et al. 2009). Browsing limits the establishment of woody plants by inhibiting the survival and recruitment of seedlings (Higgins et al. 2000). Woody plants have to survive and grow to maturity to produce seeds for sexual reproduction therefore selective browsing of woody species may decrease their ability to produce seeds (Goheen et al. 2007). The decrease in seed production consequently reduces the proportion of seedlings that may establish themselves (Goheen et al. 2007). Browsing may also reduce the ability of plants to compete with adjacent grass species because they lose nutrients (Gough and Grace 1998; Danell et al. 2006; Smith and Goodman 1986).

Bush encroachment can also be controlled through the selective cutting of woody species (Angassa et al. 2012; Mangani et al. 2022). In this approach, plants are physically cut and chemicals are applied to the stumps (Kellner et al. 2021; Mangani et al. 2022; Marquart et al. 2023). The main aim of bush control, whatever form it takes, is to shift the dominance of woody plants to herbaceous ones to improve grazing conditions (Angassa and Oba 2008; Mndela et al. 2022; Mangani et al. 2022). The change from woody dominance is expected to lead to an increase in abundance of plants that are intolerant of bush cover while those that prefer shade conditions associated with bush cover will decline (Clarke et al. 2005). These changes have the potential to lead to a reduction in plant biomass due to changes in plant structure, composition, and function (Angassa 2014; Kabeta et al. 2020).

Mbuluzi Game Reserve found in the Lowveld of Eswatini is one of the savannah ecosystems grappling with the proliferation of bush encroachment by *D. cinerea*. In 2015, the management of the reserve introduced a bush encroachment control program. This involved mechanically cutting down stands of *D. cinerea* and other encroaching woody plants in selected sites. To date, to the best of our knowledge, there is no study that sought to assess how bush encroachment in the Mbuluzi Game Reserve influenced herbaceous plant species composition, ecological status, and grazing value. This is despite this information being vital to inform the bush control strategy in this game reserve. Therefore, our study sought to address this knowledge gap by assessing the influence of bush encroachment control on herbaceous plant species composition, ecological status, and grazing value in the Mbuluzi Game Reserve. It was hypothesised that there would be a significant difference in plant species diversity between bush-encroached and bush-cleared sites. The information from this study will be critical in informing bush encroachment control strategies in the reserve and other areas in Eswatini.

18.2 Materials and Methods

18.2.1 Study Area

The study was conducted at Mbuluzi Game Reserve which is found within the Lubombo Conservancy (Fig. 18.1). Mbuluzi Game Reserve (26.1564° S, 31.9824° E) is privately owned and managed as a habitat for numerous wildlife species in Eswatini (Monadjem and Garcelon 2005).

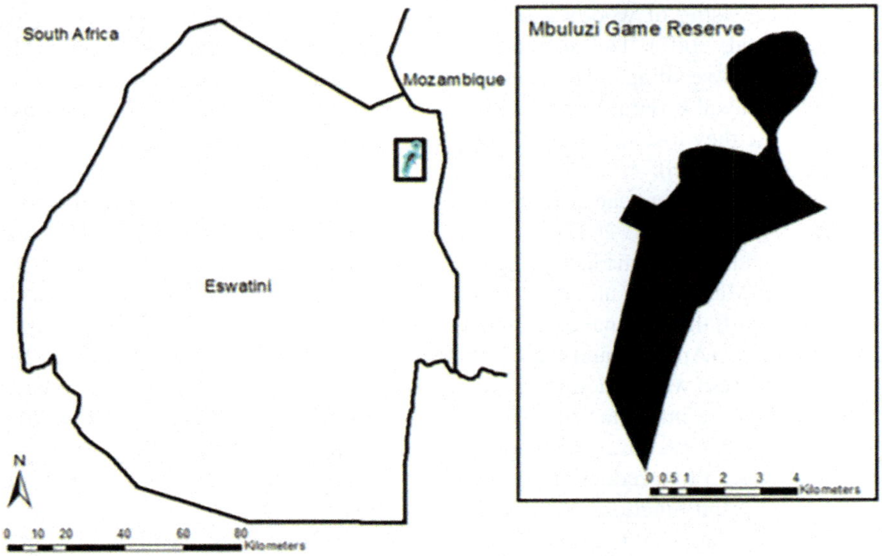

Fig. 18.1 The location of the Mbuluzi Game Reserve within Eswatini (Loggins et al. 2019)

Mbuluzi Game Reserve is located in the Lowveld region of Eswatini covering an area of 30 km^2 adjacent to the Lubombo Mountains, on basaltic soils (Loggins et al. 2019). It is characterised by natural heterogeneous savannah vegetation and a variety of game such as giraffe, kudu, nyala, zebra, and over 300 recorded bird species (Monadjem and Garcelon 2005). Common plant species in the Mbuluzi Game Reserve are *Senegalia nigrescens*, *Sclerocarya birrea*, *Themeda triandra*, *Megathyrsus maximus*, and *Dichrostachys cinerea* (Sirami and Monadjem 2012). The reserve is located in a semi-arid landscape that experiences seasonal changes between a wet summer starting from October to March with 75% precipitation and the dry winter from April to September with 25% precipitation (Matondo et al. 2005). The mean annual rainfall ranges between 550 and 725 mm (Matondo et al. 2005).

18.2.2 Vegetation Surveys

Vegetation surveys were conducted in March/April 2020 5 years after bush clearing. Bush clearing was done mechanically with encroaching species cut down and a chemical applied thereafter to inhibit coppicing. Herbaceous plants were identified along six transects in both controlled and encroached sites. We sampled from the sites that were mainly dominated by *Dichrostachys cinerea*. Other encroaching woody species had negligible cover with *D. cinerea* almost forming a monoculture. For each site, three transects measuring 50 m were established. From each transect, eleven 1 × 1 m quadrats were placed at 5 m intervals. In total, 66 quadrats were sampled. From each quadrat, plant species were identified and their percentage cover was estimated using the Braun-Blanquet cover/abundance scale (Ellenberg and Mueller-Dombois 1974). Grass species in each transect were identified using Van Oudtshoorn (2014), while herbs were identified using Manning (2013).

18.2.3 Statistical Data Analyses

The means of density, cover, species richness, evenness, and species diversity were compared for statistical significance with a Student *t*-test. Before a statistical analysis, the data were tested for normality and homogeneity of variance using Kolmogorov–Smirnov and Levene's tests respectively. The data met the assumptions of normality and homogeneity of variance.

18.2.3.1 Indicator Species and Ecological Status

Indicator species were determined by calculating indicator values for each species in each site. The indicator value was calculated as the sum of the relative abundance and relative frequency of each species in each treatment. This was done following the procedure by Dufrêne and Legendre (1997). The significance of indicator values was evaluated using the Monte Carlo test (Dufrêne and Legendre 1997). In terms of ecological status, plants were categorised as per the classes suggested by Trollope (1990) and Van Oudtshoorn (2014). The ecological status classes used were:

(i) Decreasers: Grasses that are abundant in good veld, but decrease in number when the soil veld is overgrazed or undergrazed.
(ii) Increaser I: Grasses that are abundant in underutilised veld. These are grasses that show that the veld is under poor conditions and cannot support animal production and is at risk of degradation during dry conditions.
(iii) Increaser II: Grasses that are abundant in overgrazed veld. They increase due to the disturbing effect of overgrazing and mostly include pioneer and subclimax species.
(iv) Increaser III: Grasses that are commonly found in overgrazed veld. They are usually unpalatable climax grasses and are strong competitors that increase in value when palatable grasses are weakened by overgrazing.

Van Oudtshoorn (2014) classification of grass species into palatability /grazing value groups and based on ecological status was followed. Herbs were not classified into palatability groups, as the literature does not have a definite classification for them.

18.3 Results

18.3.1 *Grass Cover, Density, Species Richness, Evenness, and Species Diversity*

Mean cover (p = 0.0001), density (p = 0.0006), and evenness (p = 0.0009) were significantly ($p < 0.05$) higher in bush-cleared treatment than in encroached one (Fig. 18.2). On the contrary, mean species richness (p = 0.003) was significantly higher in bush-encroached treatment than in cleared one (Fig. 18.2).

Mean species diversity was significantly (p = 0.006) higher in bush-cleared treatment than in the bush-encroached one (Fig. 18.3).

18.3.2 *Species Composition, Indicator Species, Ecological Status, and Ecological Status Response to Bush Clearing*

A total of 46 species were found in both bush-encroached and bush-cleared treatments (Table 18.1). These consisted of 29 herbs and 17 grasses. Of these, five species had significant ($p < 0.05$) indicator values in bush-encroached treatment as compared to three in cleared sites. Species with significant indicator values in bush-encroached treatment were *Melhania prostrata*, *Justicia flava*, *Ipomoea wightii*,

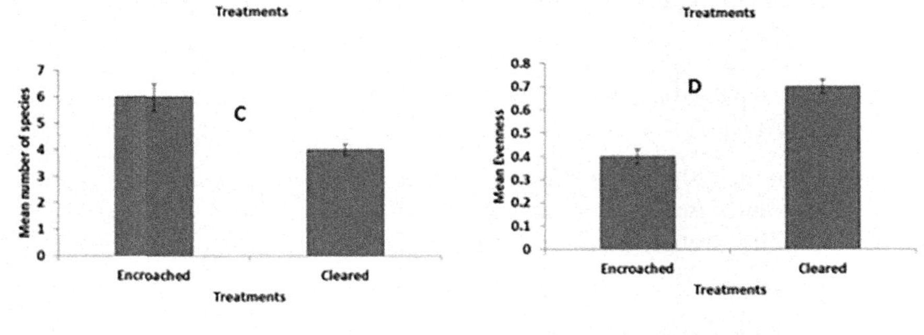

Fig. 18.2 Comparison of mean cover, density, species richness, and evenness between bush-encroached and bush-cleared treatments in Mbuluzi Game Reserve

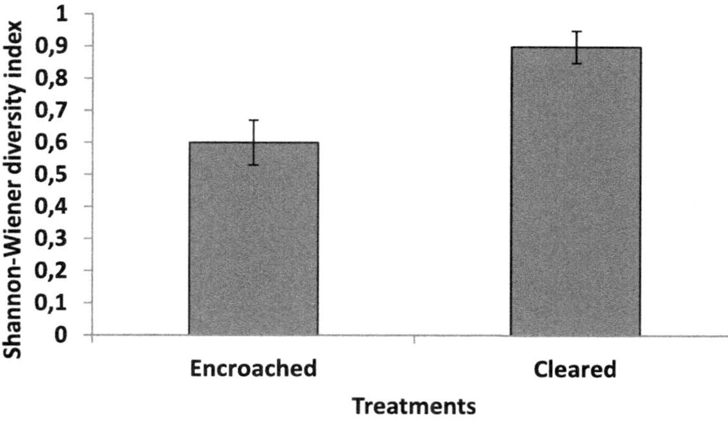

Fig. 18.3 Comparison of mean species diversity between bush-encroached and bush-cleared treatments in Mbuluzi Game Reserve

Oxygonum dregeanum, and *Megathyrsus maximus*. For bush-cleared treatment, species with significant indicator values were *Eragrostis rigidior*, *Megathyrsus maximus*, and *Eragrostis patentipilosa*. *Megathyrsus maximus* had the highest indicator value in the bush-cleared treatment than the bush-encroached one (Table 18.1). Two grass species, *Melinis repens* (Increaser II) and *Panicum natalense* (Decreaser), were found exclusively in bush-encroached treatment. Eight grass species occurred only in bush-cleared treatment. These were *Hemarthria altissima*, *Eriochloa meyeriana*, *Eragrostis patentipilosa*, *Digitaria diagonalis*, *Urochloa mosambicensis*, *Anthephora pubescens*, *Eragrostis heteromera*, and *A. argentea*. Of these, one species was classified as Increaser I (*Digitaria diagonalis*), three as Increaser II (*Urochloa mosambicensis, Eragrostis heteromera,* and *Eragrostis patentipilosa*), and four as Decreasers (*Hemarthria altissima*, *Eriochloa meyeriana* and *Echinochloa pyramidalis*, and *Anthephora pubescens*). Seven grass species were found in both treatments (Table 18.1). These were *Cynodon dactylon* (Increaser II), *Aristida bipartita* (Increaser II), *Heteropogon contortus* (Increaser II), *Eragrostis superba* (Increaser II), *Eragrostis rigidior* (Increaser II), *Themeda triandra* (Decreaser), and *Megathyrsus maximus* (Decreaser). In terms of grazing value, four species had low value, five had moderate value, and seven had high in all the treatments. In the bush-encroached treatment, there were four species with low grazing value, another four with moderate grazing value, and three with high grazing value. For the bush-cleared treatment, three species had low grazing values, while five and seven had moderate and high grazing values, respectively (Table 18.1).

Table 18.1 Comparison of indicator value, grazing value, and ecological status of plant species found in bush-encroached and bush-cleared treatments in Mbuluzi Game Reserve

Species	Life form	Bush encroached	Bush cleared	Grazing value/ palatability	Ecological status
Cynodon dactylon	Grass	50.0	12.2	High	Increaser II
Asparagus falcatus	Herb	50.0	–	–	Increaser III
Hoslundia opposita	Herb	49.6	–	–	Increaser III
Tephrosia purpurea	Herb	47.5	18.2	–	Increaser III
Indigofera confusa	Herb	37.9	5.3	–	Increaser III
Melinis repens	Grass	32.9	–	Low	Increaser II
Panicum natalense	Grass	17.8	–	Low	Decreaser
Indigofera zeyheri	Herb	16.7	5.3	–	Increaser III
Aristida bipartita	Grass	16.7	5.0	Low	Increaser II
Rhynchosia caribaea	Herb	15.8	–	–	Increaser III
Rhynchosia adenode	Herb	14.9	–	–	Increaser III
Heteropogon contortus	Grass	9.0	10.5	Low	Increaser II
Drimia elata	Herb	8.3	33.3	–	Increaser III
Tephrosia elongata	Herb	8.3	18.2	–	Increaser III
Asclepias gibba	Herb	8.3	–	–	Increaser III
Commelina africana	Herb	8.3	–	–	Increaser III
Commelina ecklonia	Herb	8.3	–	–	Increaser III
Hibiscus trionum	Herb	8.3	–	–	Increaser III
Eclipta prostrata	Herb	8.3	–	–	Increaser III
Ipomea alba	Herb	7.7	5.3	–	Increaser III
Rhynchosia totta	Herb	7.2	–	–	Increaser III
Eragrostis superba	Grass	5.3	10.5	Moderate	Increaser II
Protea simplex	Herb	5.3	5.3	–	Increaser III
Asparagus aethiopicus	Herb	5.3	–	–	Increaser III
Eragrostis rigidior	Grass	5.3	60.2[a]	Moderate	Increaser II
Polygala amatymbica	Herb	5.1	5.5	–	Increaser III
Themeda triandra	Grass	4.9	35.8	High	Decreaser
Bidens biternata	Herb	3.7	–	–	Increaser III
Vernonia natalensis	Herb	8.3	–	–	Increaser III
Hemarthria altissima	Grass	–	42.3	High	Decreaser
Eriochloa meyeriana	Grass	–	33.4	High	Decreaser
Digitaria diagonalis	Grass	–	33.3	Moderate	Increaser I

(continued)

Table 18.1 (continued)

Species	Life form	Bush encroached	Bush cleared	Grazing value/ palatability	Ecological status
Echinochloa pyramidalis	Grass	–	22.3	High	Decreaser
Urochloa mosambicensis	Grass	–	18.2	Moderate	Increaser II
Corchorus asplenifolius	Herb	–	17.9	–	Increaser III
Anthephora pubescens	Grass	–	10.5	High	Decreaser
Hemizygia teucriifolia	Herb	–	9.1	–	Increaser III
Melhania prostrata	Herb	46.5[a]	7.9	–	Increaser III
Eragrostis heteromera	Grass	–	7.2	Moderate	Increaser II
Justicia flava	Herb	97.4[a]	5.3	–	Increaser III
Acalypha grandifolia	Herb	–	5.3		Increaser III
Anthephora argentea	Grass	–	5.3	–	Increaser III
Vernonia centaureoides	Herb	–	5.3	–	Increaser III
Ipomoea wightii	Herb	49.1[a]	–	–	Increaser III
Oxygonum dregeanum	Herb	98.1[a]	–		Increaser III
Megathyrsus maximus	Grass	59.0[a]	75.5[a]	High	Decreaser
Eragrostis patentipilosa	Grass	–	60.2[a]	Low	Increaser II

[a]Significant indicator value ($p < 0.05$)

18.4 Discussion

This study showed that there was significant variation in plant density, cover, species richness, evenness, and species diversity between bush-cleared and bush-encroached treatments. Of all these, only species richness was significantly higher in bush-encroached treatment. This study indicates that bush encroachment control through mechanical clearing can improve rangeland conditions. In terms of grazing value, more species with moderate (*Eragrostis superba*, *E. rigidior*, *Digitaria diagonalis*, and *E. heteromera*) and high palatability (*Themeda triandra*, *Hemarthria altissima*, *Eriochloa meyeriana*, *Echinochloa pyramidalis*, and *Megathyrsus maximus*) were found in bush-cleared treatment. Highly palatable grasses are those that are preferred by ungulates while the less palatable are not preferred, and they only feed on them when grasses with high grazing value are depleted (Van Oudtshoorn 2014). Animals feeding on grass with low grazing value tend to lose body mass

(Mangani et al. 2022). While the bush-cleared treatment was also characterised by the presence of Increaser I and Increaser II indicating poor rangeland conditions (Van Oudtshoorn 2014), the majority of them except for *E. ridigior* and *E. patentipilosa* did not have significant indicator values. This calls for the management of the Mbuluzi Game Reserve to integrate bush clearing with other rangeland management strategies which can help to reduce the abundance of Increaser I and Increaser II species which are undesirable.

One of the interventions that can be used is fire. Prescribed burning helps to remove selected and undesirable species to improve the rangeland conditions (Mapiye et al. 2008). It helps in improving palatability by reducing the abundance of unpalatable species (Mapiye et al. 2008). Freshly germinated grasses after fire are palatable and rich in crude protein (Bebawi and Campbell 2002). However, it is worth noting that decreaser grasses had higher indicator values in bush-cleared treatment than in the encroached one. This further shows that bush clearing has a positive impact on improving rangeland conditions. The disparity in species composition between the two treatments can also be an indicator of the role played by soil seed banks (Solomon et al. 2006). This could suggest that before bush clearing the suppressed species which were intolerant of shade were excluded and only existed in the soil seed bank as seeds (Angassa et al. 2012). Bush clearing promoted their establishment as they were no longer suppressed by the shade from forest cover.

While it is acknowledged that bush clearing improved the rangeland conditions in the Mbuluzi Game Reserve, it should also be noted that bush-encroached treatment also had highly palatable grass species such as *Megathyrsus maximus* (Van Oudtshoorn 2014) which almost formed a monoculture in the bush-encroached treatment. *Megathyrsus maximus* is a shade-tolerant species (Van Oudtshoorn 2014); hence, it grew under the *D. cinerea* canopy. Furthermore, this species could be benefiting from *D. cinerea* through soil fertility amelioration and protection from grazers (Pierce et al. 2019), hence its significant indicator value in the bush-encroached treatment, indicating high abundance. Therefore, our results suggest that while the aim is to improve rangeland conditions through bush clearing this should be done with caution so that heterogeneity of plant cover with bush patches is maintained. Furthermore, the management of the Mbuluzi Game Reserve should explore ways in which *M. maximus* under *D. cinerea* thickets can be accessed by grazers. One of the possibilities is to harvest it, but this could be limited by the accessibility of the bush-encroached area in addition to the exercise being costly as it would require additional labour. Therefore, it can be suggested that there should be bush thinning in some sites enough to allow for the movement of the animals in the understory. This might also call for research to establish the extent of bush thinning to be carried out without upsetting the shade conditions required by *M. maximus*.

Mean plant cover and density were significantly higher in bush-cleared treatment than in encroached one. This could be a result of changing microclimatic conditions such as increased resource availability such as nutrients and water (Modungwa 2017). The indicator values of shade-intolerant plants were higher in bush-cleared sites indicating that they had higher plant cover and frequency of occurrence. In the bush-encroached site shade-intolerant species were excluded, resulting in low plant cover. These results are consistent with findings by Hare et al. (2021) in Ethiopian rangelands, Smit (2003), and Mndela et al. (2022) in South Africa who also found higher plant cover in terms of biomass in bush-cleared areas than bush-encroached ones.

Our study showed that plant species diversity was higher in the bush-cleared treatment than the bush-encroached one. This is in accordance with previous studies which showed that bush clearing increased species diversity (Ansley and Castellano 2006; Olson and Whitson 2002; Marquart et al. 2023). The higher species diversity in bush-cleared treatment is an indicator of increased functional diversity (Gladstone-Gallagher et al. 2019). It is also an indicator that bush encroachment suppresses plant diversity and their functional roles in an ecosystem.

Many semi-arid ecosystems have shown that there is a proportional relationship between plant species diversity and ecosystem functioning (Maestre et al. 2009). This is important as it provides more rangeland resources for both game and livestock, increasing the carrying capacity and promoting co-existence of different species feeding guilds. It also helps in averting degradation in rangeland areas that are at high risk of degradation (Safriel 2007). In our case, the study area is faced with the risk of degradation from the encroaching *D. cinerea*, and controlling it will help to restore the rangeland to a favourable state which can support diverse wildlife in the reserve. The dynamics of plant species diversity could be influenced by site conditions prevailing in both sites (Angassa et al. 2012). Bush cover provides microclimatic conditions which favour shade-tolerant species at the expense of intolerant ones (Angassa et al. 2012). This effect has the potential to result in an increase or reduction of species diversity under bush cover. In our study bush cover led to a reduction in plant species diversity. Bush-encroached sites were dominated by *P. maximum* with few co-existing herbaceous species. This is consistent with other studies elsewhere in which bush cover led to a reduction in plant species diversity (e.g. Kellner et al. 2021; Mangani et al. 2022; Marquart et al. 2023).

18.5 Conclusion

Our study showed that woody plant removal can improve the grazing condition of degraded rangelands. Therefore, it is recommended that rangeland managers should incorporate bush clearing as one of their strategies as it can help to improve grazing conditions. It was found that plant cover, density, evenness, and diversity were higher in bush-cleared sites. Only species richness was significantly higher in bush-encroached treatment than in the bush-cleared one. This could suggest that a mere species count is not a good indicator of rangeland conditions; instead, functional diversity should be incorporated (Utaile et al. 2021). Indeed our study did that when we compared grazing value and ecological status between bush-encroached and bush-cleared treatments. Using these parameters, it was still found that bush clearing improved rangeland conditions as indicated by two Increaser II species with significant indicator values in a bush-cleared treatment. Furthermore, it was found that more species with significant indicator values in bush-cleared treatment were of high palatability. We call for monitoring of rangeland conditions in the Mbuluzi Game Reserve and an increase in bush-cleared treatment. It is also suggested that the heterogeneous landscape with bush-cleared areas and bush-encroached areas should be maintained to cater to shade-tolerant species such as *Megathyrsus maximus*. Furthermore, it is suggested that fire be incorporated into rangeland management of the Mbuluzi Game Reserve to assist with controlling undesirable species and improve the rangeland condition.

References

Angassa A (2014) Effects of grazing intensity and bush encroachment on herbaceous species and rangeland condition in Southern Ethiopia. Land Degrad Dev 25(5):438–451

Angassa A, Baars RM (2000) Ecological condition of encroached and non-encroached rangelands in Borana, Ethiopia. Afr J Ecol 38(4):321–328

Angassa A, Oba G (2008) Herder perceptions on impacts of range enclosures, crop farming, fire ban, and bush encroachment on the rangelands of Borana, Southern Ethiopia. Hum Ecol 36:201–215

Angassa A, Oba G, Tolera A (2012) Bush encroachment control demonstrations and management implications on herbaceous species in savannas of Southern Ethiopia. Trop Subtrop Agroecosystems 15(1):173–185

Ansley RJ, Castellano MJ (2006) Strategies for savannah restoration in the southern Great Plains: effects of fire and herbicides. Restor Ecol 14(3):420–428

Archer SR (2010) Rangeland conservation and shrub encroachment: new perspectives on an old problem. In: Wild rangelands: conserving wildlife while maintaining livestock in semi-arid ecosystems. Wiley-Blackwell, pp 53–97

Ayelew S, Mulualem G (2018) A review on bush encroachment effect on cattle rearing in rangelands. J Rangel Sci 8(4):403–415

Bebawi FF, Campbell SD (2002) Impact of fire on bellyache bush (*Jatropha gossypiifolia*) plant mortality and seedling recruitment. Trop Grassl 36(3):129–137

Bond WJ, Van Wilgen BW (2012) Fire and plants, vol 14. Springer Science & Business Media

Börner J, Higgins SI, Kantelhardt J, Scheiter S (2007) Rainfall or price variability: what determines rangeland management decisions? A simulation optimization approach to South African savannahs. Agric Econ 37(2–3):189–200

Bothma JP (2002) Game ranch management, 4th edn. Van Schaik Publishers, Pretoria

Briske DD, Fuhlendorf SD, Smeins FE (2003) Vegetation dynamics on rangelands: a critique of the current paradigms. J Appl Ecol 40:601–614

Clarke PJ, Knox KJ, Wills KE, Campbell M (2005) Landscape patterns of woody plant response to crown fire: disturbance and productivity influence sprouting ability. J Ecol 93(3):544–555

D'Odorico P, Okin GS, Bestelmeyer BT (2012) A synthetic review of feedbacks feedback and drivers of shrub encroachment in arid grasslands. Ecohydrology 5(5):520–530

Danell K, Bergström R, Duncan P, Pastor J (eds) (2006) Large herbivore ecology, ecosystem dynamics, and conservation, vol 11. Cambridge University Press

Dufrêne M, Legendre P (1997) Species assemblages and indicator species: the need for a flexible asymmetrical approach. Ecol Monogr 67(3):345–366

Eldridge DJ, Bowker MA, Maestre FT, Roger E, Reynolds JF, Whitford WG (2011) Impacts of shrub encroachment on ecosystem structure and functioning: towards a global synthesis. Ecol Lett 14(7):709–722

Ellenberg D, Mueller-Dombois D (1974) Aims and methods of vegetation ecology. Wiley, New York, p 547

Espach C (2006) Rangeland productivity modelling: developing and customizing methodologies for land cover mapping in Namibia. Agricola 2006:20–27

Gladstone-Gallagher RV, Pilditch CA, Stephenson F, Thrush SF (2019) Linking traits across ecological scales determines functional resilience. Trends Ecol Evol 34(12):1080–1091

Goheen JR, Young TP, Keesing F, Palmer TM (2007) Consequences of herbivory by native ungulates for the reproduction of a savannah tree. J Ecol 95(1):129–138

Gough L, Grace JB (1998) Herbivore effects on plant species density at varying productivity levels. Ecology 79(5):1586–1594

Grady JM, Hoffmann WA (2012) Caught in a fire trap: recurring fire creates stable size equilibria in woody resprouters. Ecology 93(9):2052–2060

Hare ML, Xu XW, Wang YD, Yuan Y, Gedda AE (2021) Do woody tree thinning and season have an effect on grass species' composition and biomass in a semi-arid savannah? The case of a semi-arid savannah, in Southern Ethiopia. Front Environ Sci 9:692239

Hessburg PF, Agee JK (2003) An environmental narrative of inland Northwest United States forests, 1800–2000. For Ecol Manag 178(1–2):23–59

Higgins SI, Bond WJ, Trollope WS (2000) Fire, resprouting and variability: a recipe for grass–tree coexistence in savannah. J Ecol 88(2):213–229

Hoffmann WA (1998) Post-burn reproduction of woody plants in a neotropical savannah: the relative importance of sexual and vegetative reproduction. J Appl Ecol 35(3):422–433

Hoffmann WA, Solbrig OT (2003) The role of topkill top kill in the differential response of savannah woody species to fire. For Ecol Manag 180(1–3):273–286

Kabeta L, Dalle G, Tolera M, Kelboro G (2020) Bush encroachment and impacts on grass biomass in Senkelle Swayne's Hartebeest Sanctuary, Ethiopia. Biodiversity 21(4):217–226

Kellner K, Mangani RT, Sebitloane TJ, Chirima JG, Meyer N, Coetzee HC, Malan PW, Koch J (2021) Restoration after bush control in selected range-land areas of semi-arid savannahs in South Africa. Bothalia 51(1):1–13

Loggins AA, Shrader AM, Monadjem A, McCleery RA (2019) Shrub cover homogenizes small mammals' activity and perceived predation risk. Sci Rep 9(1):16857

Lukomska N, Quaas MF, Baumgärtner S (2014) Bush encroachment control and risk management in semi-arid rangelands. J Environ Manage 145:24–34

Maestre FT, Bowker MA, Puche MD, Belén Hinojosa M, Martínez I, García-Palacios P, Castillo AP, Soliveres S, Luzuriaga AL, Sánchez AM, Carreira JA (2009) Shrub encroachment can reverse desertification in semi-arid Mediterranean grasslands. Ecol Lett 12(9):930–941

Mangani T, Marquart A, Chirima G, Kellner K (2022) Grass species diversity response to brush packing in semi-arid rangelands of South Africa. J Arid Environ 207:104832

Manning J (2013) Field guide to wild flowers of South Africa. Penguin Random House South Africa

Mapiye C, Mwale M, Chikumba N, Chimonyo M (2008) Fire as a rangeland management tool in the savannahs of southern Africa: a review. Trop Subtrop Agroecosystems 8(2):115–124

Marquart A, Van Coller H, Van Staden N, Kellner K (2023) Impacts of selective bush control on herbaceous diversity in wildlife and cattle land use areas in a semi-arid Kalahari savannah. J Arid Environ 208:104881

Matondo JI, Peter G, Msibi KM (2005) Managing water under climate change for peace and prosperity in Swaziland. Phys Chem Earth (Pt A B,C) 30(11–16):943–949

Midgley JJ, Lawes MJ, Chamaillé-Jammes S (2010) Savannah woody plant dynamics: the role of fire and herbivory, separately and synergistically. Aust J Bot 58(1):1–11

Mndela M, Madakadze IC, Nherera-Chokuda FV, Dube S, Ramoelo A, Mangwane M, Tjelele JT (2022) Short-term responses of herbaceous vegetation to bush clearing in semi-arid rangelands of South Africa. Pastoralism 12(1):17

Modungwa RT (2017) The effect of brush packing on soil surface conditions that influence grass emergence and establishment on denuded soils. MSc dissertation. University of Pretoria, Pretoria

Monadjem A, Garcelon DK (2005) Nesting distribution of vultures in relation to land use in Swaziland. Biodivers Conserv 14:2079–2093

Mphinyane WN, Moleele NM, Sebego SJ (2011) Effect of fire burning interval on species composition, herbage yield, and bush control in the savannah of eastern Botswana

Mugasi SK, Sabiiti EN, Tayebwa BM (2000) The economic implications of bush encroachment on livestock farming in rangelands of Uganda. Afr J Range Forage Sci 17(1–3):64–69

O'connor TG, Puttick JR, Hoffman MT (2014) Bush encroachment in southern Africa: changes and causes. Afr J Range Forage Sci 31(2):67–88

Oba G, Post E, Syvertsen PO, Stenseth NC (2000) Bush cover and range condition assessments in relation to landscape and grazing in Southern Ethiopia. Landsc Ecol 15(6):535–546

Olson RA, Whitson TD (2002) Restoring structure in late-successional sagebrush communities by thinning with tebuthiuron. Restor Ecol 10(1):146–155

Pierce NA, Archer SR, Bestelmeyer BT, James DK (2019) Grass-shrub competition in arid lands: an overlooked driver in grassland–shrubland state transition? Ecosystems 22:619–628

Reynolds JF, Smith DMS, Lambin EF, Turner BL, Mortimore M, Batterbury SP, Downing TE, Dowlatabadi H, Fernández RJ, Herrick JE, Huber-Sannwald E (2007) Global desertification: building a science for dryland development. Science 316(5826):847–851

Roques KG, O'connor TG, Watkinson AR (2001) Dynamics of shrub encroachment in an African savannah: relative influences of fire, herbivory, rainfall, and density dependence. J Appl Ecol 38(2):268–280

Safriel UN (2007) The assessment of global trends in land degradation. In: Climate and land degradation. Springer, Berlin, pp 1–38

Sankaran M, Ratnam J, Hanan NP (2004) Tree–grass coexistence in savannahs revisited–insights from an examination of assumptions and mechanisms invoked in existing models. Ecol Lett 7(6):480–490

Sheuyange A, Oba G, Weladji BR (2005) Effects of anthropogenic fire history anthropogenic fire history on savannah vegetation in northeastern *Namibia.* J Environ Manag 75:189–198

Sirami C, Monadjem A (2012) Changes in bird communities in Swaziland savannahs between 1998 and 2008 owing to shrub encroachment. Divers Distrib 18(4):390–400

Smit GN (2003) The importance of *Salvadora australis* in relation to tree thinning in preserving herbaceous plants in a semi-arid *Colophospermum mopane* savannah. J Arid Environ 55(3):483–501

Smith TM, Goodman PS (1986) The effect of competition on the structure and dynamics of acacia savannahs in southern Africa. J Ecol 74:1031–1044

Solomon TB, Snyman HA, Smit GN (2006) Soil seed bank characteristics in relation to land use systems and distance from water in a semi-arid rangeland of Southern Ethiopia. S Afr J Bot 72(2):263–271

Staver AC, Bond WJ, Stock WD, Van Rensburg SJ, Waldram MS (2009) Browsing and fire interact to suppress tree density in an African savannah. Ecol Appl 19(7):1909–1919

Tefera S, Dlamini BJ, Dlamini AM, Mlambo V (2008) Current range condition in relation to land management systems in semi-arid savannahs of Swaziland. Afr J Ecol 46(2):158–167

Trollope WSW (1990) Veld management with specific reference to game ranching in the grassland and savannah areas of South Africa. Koedoe 33(2):77–86

Utaile YU, Honnay O, Muys B, Cheche SS, Helsen K (2021) Effect of *Dichrostachys cinerea* encroachment on plant species diversity, functional traits, and litter decomposition in an East-African savannah ecosystem. J Veg Sci 32(1):e12949

Van Auken OW (2009) Causes and consequences of woody plant encroachment into western North American grasslands. J Environ Manage 90(10):2931–2942

Van Oudtshoorn FV (2014) Guide to grasses of southern Africa. Briza publications

Ward D (2005) Do we understand the causes of bush encroachment in African savannahs? Afr J Range Forage Sci 22(2):101–105

Dr Gaolathe Tsheboeng is a Senior Lecturer/Researcher in Plant Ecology at the University of Botswana in the Department of Biological Sciences. His research interests include phytosociology, plant population structure, and soil seedbanks dynamics.

D. Maseko is a Technologist at the University of Eswatini in the Department of In-Service Education. Her research interests include plant ecology and conservation.

Dr Meshack Nkosinathi Dludlu is a Botanist, with a special interest in Systematics, Ecology, and the Bio-geography of higher plants. He is currently a Senior Lecturer at the University of Eswatini in the Department of Biological Sciences.

Ms N. Nkonyane 's research interests include plant ecology and conservation.

Chapter 19
Contribution of Non-timber Forest Products to the Livelihood of Marginalised Communities Around the Okavango Delta, Northern Botswana

Joyce Lepetu, Hesekia Garekae, Gofaone Rammotokara, and Topoyame Isaac Makoi

19.1 Introduction

Globally, non-timber forest products (NTFPs) have remained crucial in sustaining livelihoods for most people (Rai et al. 2019; Beyene et al. 2020). Although dependence on NTFPs was largely associated with rural areas, a growing body of knowledge has demonstrated the prevalence of reliance on NTFPs in urban areas as well (Mollee et al. 2017; Garekae and Shackleton 2020). Therefore, the significance of NTFPs to livelihoods is gaining wide recognition. NTFPs have been widely defined in the literature, with a consensus definition not yet reached (Rai et al. 2019). Hence, NTFPs can mean different things to different people. The most widely cited definition is offered by De Beer and McDermott (1989: 24), who defined NTFPs as 'all biological materials other than timber which are extracted from forests for human use'. Shackleton et al. (2011) developed a compendium of NTFPs that expounded on the evolution of the concept, conceptual and definitional ambiguities, and concluded by proposing a holistic definition that incorporates the different facets of NTFPs. Various synonyms have been used to denote forest products other than timber, such as 'other', 'minor', 'secondary', 'non-wood', or 'non-timber' forest products (FAO 1992; Delgado et al. 2016), but for the context of this study, NTFP is adopted since it is the most widely used phrase.

The global prevalence of NTFPs is provided, although some estimations were beset with deficiency, and errors and generally surpassed by time (Shackleton and

J. Lepetu · G. Rammotokara · T. I. Makoi
Department of Range and Forest Resources, Botswana University of Agriculture and Natural Resources, Gaborone, Botswana
e-mail: jlepetu@buan.ac.bw; grammotokara@buan.ac.bw; tmakoi@buan.ac.bw

H. Garekae (✉)
Department of Environmental Science, University of Botswana, Gaborone, Botswana

W. L. Hambira et al. (eds.), *Environmental Change and Biodiversity Conservation in sub-Saharan Africa*, Advances in Global Change Research 75,
https://doi.org/10.1007/978-3-031-73136-5_19

de Vos 2022). However, a more recent and comprehensive projection is offered by Shackleton and de Vos (2022), who estimated dependence on NTFPs to be within the ranges of 3.5–5.8 billion people worldwide. NTFPs provide key provisioning ecosystem services, such as food, medicine, energy, construction materials, and arts and crafts materials (Lepcha et al. 2019; Beyene et al. 2020). These products are integral to fulfilling household needs, such as income, energy, nutrition, and health (Olesen et al. 2022; Pelane et al. 2023). In regard to energy, NTFPs provide fuelwood, which is an indispensable source of domestic energy in rural areas. Over three-quarters of households in sub-Saharan Africa (SSA) rely on fuelwood as a primary source of energy (Eba'a Atyi et al. 2016; Hoffmann et al. 2016). However, this figure is poised to increase given the socio-economic hardships befallen the continent, which constrains access to alternative energy sources. Forest resources play a vital role in diversifying diets and nutrition (Chamberlain et al. 2020). Empirical evidence has established a positive association between forest ecosystems and improved nutrition and food security (Luna-González and Sørensen 2018; Rasmussen et al. 2020), albeit at the micro-level. Forest-adjacent communities have been found to be dietary diverse compared to those living far away. Moreover, food sourced from the forests greatly reduces households' expenditure on food (Chamberlain et al. 2020). Therefore, households can invest income saved in food expenditures in other productive assets.

In addition, NTFPs also contribute to household income. The global value of NTFPs is approximately 90 billion dollars (Suleiman et al. 2017). A sustained body of knowledge has quantified the proportion of forest income over total household income (Malleson et al. 2014; Beyene et al. 2020). Forest income can contribute as little as 10% to as high as 70% of the total household income (Cavendish 2000; Kamanga et al. 2009; Le and Nguyen 2020). Over and above, NTFPs have been found to be critical for fulfilling the livelihood needs of the poor (Malleson et al. 2014). Empirical evidence has demonstrated that NTFPs play an important role in cushioning households against shocks and stressors, e.g. poverty, food insecurity, income loss, illness, and poor agricultural output (Shackleton and Shackleton 2004; Rai et al. 2019; Shackleton et al. 2024). Regarding poverty, NTFPs have been found to keep livelihoods afloat and reduce the vulnerability of deepening into abject poverty (Le and Nguyen 2020; Miller et al. 2022). Since some NTFPs can withstand unfavourable climatic conditions, they can be relied upon during times of food shortages and deficits (Mukul et al. 2016).

Despite the significance of NTFPs, their contribution to livelihoods tends to differ across space and over time. This variation is a function of several factors, but key among them are socio-economic, and demographic parameters (Suleiman et al. 2017; Garekae et al. 2017). Furthermore, other contextual factors may come into play in the forest–livelihood nexus. For example, some communities living adjacent to protected areas face marginalisation and depredation (Carson et al. 2018; Koné 2019). They have been alienated and dispossessed from the landscapes they inhabited for so long (Tropp 2003; Torri 2011). Moreover, these communities are generally excluded from the mainstream economy, an unbearable situation that presents limited opportunities for sustenance. Hence, forest resources such as NTFPs may be the only dependable source of livelihood (FAO and UNHCR 2018). Therefore, the prevalence and patterns of reliance on NTFPs are bound to differ based on these contextual aspects. This dynamic underscores the importance of conducting context-specific studies to unravel the contribution of NTFPs to livelihoods, particularly in marginalised communities. This paper invokes the definition of marginalised communities offered by IPBES (2021), who defined it as '... those experiencing discrimination and exclusion because of unequal power relationships across economic, political, social and cultural dimensions'. In the context of the study area, the Basarwa of Shaikarawe are deemed marginalised because they have been excluded from social, political, and economic processes. Moreover, their settlement is ungazetted—with limited access to basic services and amenities for their overall well-being. Against this backdrop, this study examines the role of forest products in the livelihoods of marginalised communities in the Okavango Delta, northern Botswana. Specifically, the study assessed the level of dependency on NTFPs and the collection and use patterns of NTFPs and determined the importance of NTFPs to rural livelihoods. Chambers and Conway (1992: 25) defined a livelihood as a '... means of living, and the capabilities, assets and activities required for it'.

19.2 Methods

19.2.1 Study Area

This study was conducted in the then Okavango sub-district, Ngamiland District, Botswana. However, the sub-district has now been upgraded into a full administrative authority, the Okavango District Council. Okavango district is multi-ethnic, with Bayei and Bambukushu being the most dominant ethnic groups found in the region (Arntzen 2005). The region boasts a diversity of wildlife and aquatic resources. The renowned Okavango Delta is situated in the region, a 1000th UNESCO World Heritage Site. The Okavango Delta provides a suite of wetland ecosystem services that are integral to human and environmental sustainability. The adjacent communities rely on the delta to sustain their livelihoods. The delta is endowed with riverine resources that provide, among others, food, medicine, water, and raw materials (King et al. 2016).

The study was conducted in the Shaikarawe settlement (Fig. 19.1). Shaikarawe is entirely composed of the Basarwa community. Approximately 400 people live in this settlement. Traditionally, Basarwa are hunters and gatherers. They relied on the natural resources endowed in their proximate environment to sustain their livelihoods. Recently, the Basarwa have somewhat diversified their livelihoods through government welfare programmes aimed at improving their socio-economic conditions. Although ungazetted, the Shaikarawe settlement has few infrastructures offering essential services, such as Mobile Stop Clinic, Primary School, *Kgotla*, and Village Development Committee office. Moreover, the community is in pursuit of

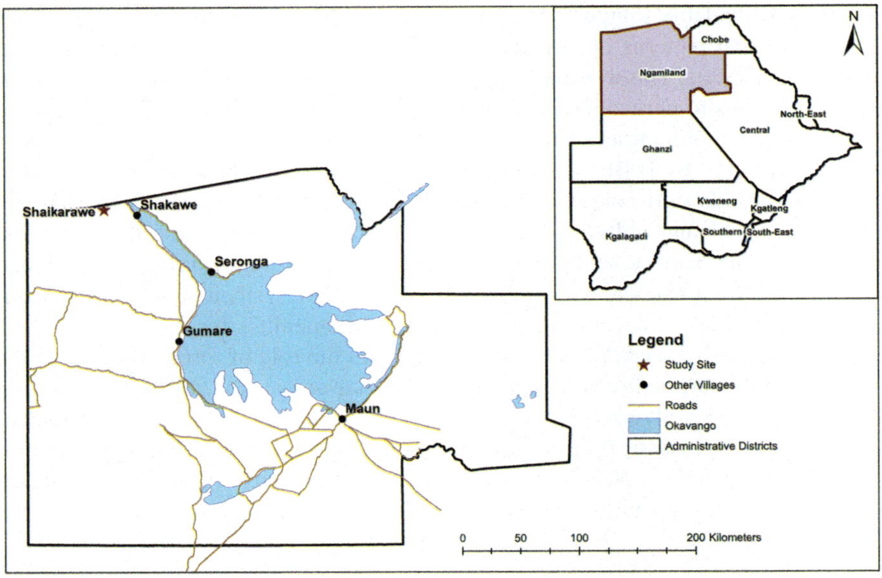

Fig. 19.1 Location of the study settlement

proclaiming a remnant forested area into a community forest reserve. This signifies the communities' positive attitudes towards the conservation of forest resources found in their area.

19.2.2 Data Collection and Analysis

Since Shaikarawe is fairly a small settlement, a census of all households was conducted to understand the contribution of forest-related products to rural livelihoods. Approximately 50 households were sampled in this study. In the context of this study, a household denotes people who live, eat, and pool resources together in the same unit (Beall and Kanji 1999). Primary data were solicited through the administration of a structured questionnaire comprising both open and close-ended questions. The questionnaires were administered through face-to-face interviews. The questionnaire was structured into three sections, which broadly elucidated on the following: respondents' socio-economic and demographic factors, livelihood activities, and forest resource use. The questionnaire was administered to the household head; in their absence, any member who was 18 years and above was considered. The data were collected in November 2021.

Descriptive statistics were used for analysing the data. Frequencies, means, measures of central tendency, and dispersion were used for summarising the data. Pie charts and bar graphs were used for visualising data graphically.

19.3 Results

19.3.1 Respondents' Profile

About half of the respondents were females (52%) (Table 19.1). The majority (44%) were the youth, with their ages ranging between 19 and 39 years. Approximately one-third (32%) of the respondents were middle aged (40–57 years), while 24% were elderly (≥58 years). The village of Shaikarawe is predominantly inhabited by the Basarwa ethnic group. Almost all the respondents identified themselves as Basarwa. Only two respondents belonged to different ethnic groups, Bambukushu and Bayei. The majority of the respondents (60%) acknowledged that they had not attained any formal education, while less than one-fifth had achieved primary and junior secondary education (Table 19.1). However, most of the respondents expressed that they dropped out of school, hence the high number of people without formal education.

Unemployment is rife in the study village. An overwhelming majority (86%) of the respondents were unemployed. Household monthly income averaged BWP 604.21 ± 637.40. The income was mostly earned from ad hoc employment opportunities such as the Poverty Eradication Programme (Ipelegeng) and, to a lesser extent, old age pension. The mean household size was 5.38 ± 2.84, ranging from 1 to 14 persons. Most of the respondents grew up in Shaikarawe, as evidenced by the mean length of residency of 39.06 ± 20.87. Similarly, over three-quarters of the

Table 19.1 Respondents' profile

Variable	Category	Frequency (%)	*M* (SD)
Gender	Male	48	
	Female	52	
Age	Youth	44	
	Middle	32	
	Elderly	24	
Ethnicity	Basarwa	96	
	Bahambukushu	2	
	Bayeyi	2	
Education	None	60	
	Primary	18	
	Secondary	20	
	Tertiary	2	
Employment	Full-time	8	
	Part-time	4	
	Self-employed	2	
	Unemployed	40	
	Other	46	
Household size			5.38 (2.842)
Monthly household income			604.21 (637.404)

respondents opined that they had always lived in Shaikarawe since birth. The few who migrated to Shaikarawe cited, among others, family matters and livelihood activities as the primary push factors.

19.3.2 Livelihood Activities

About one-third (34%) of the participants were somewhat engaged in agriculture. Crop production and livestock keeping were the most common agricultural activities. Approximately 12% indicated that they cultivate crops, while 28% keep livestock. Crops were cultivated either within homesteads or small gardens averaging 1 ha or less in size. Four respondents cultivated crops in small gardens, while only two cultivated in their homesteads. The typical crops cultivated in small gardens were maize, beans, sweet reeds, watermelon, and pumpkins, while vegetables were favoured in homesteads. Livestock mostly comprised small stock, such as goats and chicken, which were mentioned by two and seven respondents, respectively. Only one and two respondents had cattle and donkeys, respectively. However, the contribution of the aforementioned activities to livelihoods was generally insignificant, given the very low output and stocking rate reported by the respondents. Moreover, the majority of the respondents do not have a dependable source of income (Fig. 19.2). Most of them were subsisting on income earned from intermittent employment opportunities, such as the poverty eradication programme—commonly

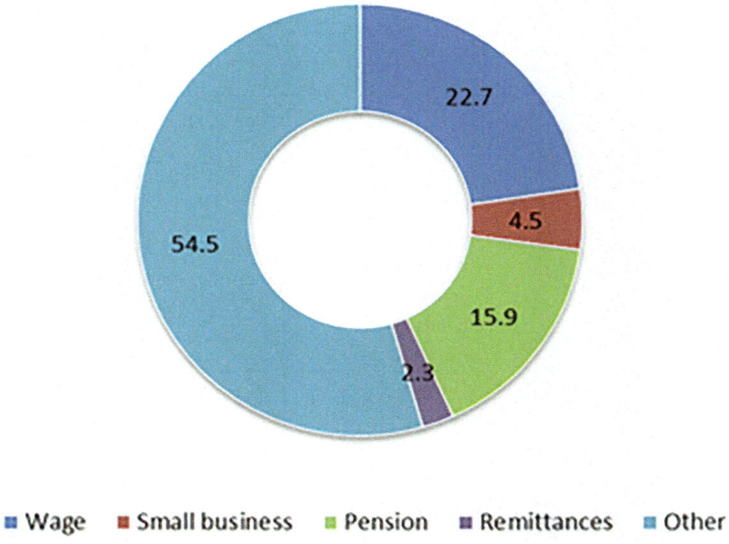

Fig. 19.2 Sources of household income

known as *Ipelegeng*. Given the limited livelihood activities of the Shaikarawe community, forest resources play an important role in the livelihoods of most dwellers.

19.3.3 Level of Dependence on Forest Resources

All but two respondents affirmed having visited the adjacent forest to Shaikarawe settlement. The primary reason cited for visiting the forests was the collection of forest products. In addition, few respondents mentioned culture and religion as some of the driving factors. Approximately, two-thirds (66%) of the respondents concurred that they visited the forest on a regular basis, at least 4–7 times per week. A total of 18% acknowledged visiting the forest occasionally, while 12% rarely did so. However, 4% opined that they never visited the forest.

Based on the foregoing, an overwhelming majority (94%) of the respondents were extremely dependent on forest resources for their overall livelihoods and well-being. A small proportion of the respondents were moderately (4%) and least (2%) dependent on forest resources, respectively. This finding implies that forest resources were an integral part of livelihood for the respondents and the community at large. Disaggregating dependence against resource type provided deeper insights. Generally, dependency was slightly higher for fuelwood and wild foods, than for other resources. However, the level of forest dependency did not differ with respondent age or gender. This demonstrates that reliance on forest resources transcends socio-demographic backgrounds. In this study, forest products were primarily collected for subsistence purposes. Although there is potential for commercialisation of some forest products in the study area, the opportunity remains untapped. Only one respondent acknowledged selling forest products as an alternative income-generating activity.

19.3.4 Collection and Use Patterns of NTFPs

A total of 65 plant species were utilised by the community for various purposes. The most frequently used species were *Garcinia livingstonei*, *Terminalia sericea*, *Baikiaea plurijuga*, *Dialium engleranum*, *Strychnos cocculoides*, *Strychnos pungens*, *Eragrostis pallens*, *Burkea africana*, *Amaranthus thunbergii*, and *Ricinodendron rautenii*, in order of frequency. These plant species provide multiple uses, such as firewood, food, medicine, construction materials, and crafts. The findings revealed that wild fruits, firewood, wild vegetables, thatching grass, and poles were the most common NTFPs collected and used by the households, in order of frequency (Fig. 19.3). However, wild fruits were the only product that was collected by all respondents, while firewood, wild vegetables, thatching grass, and poles were collected by over three-quarters of the respondents (Fig. 19.3). In addition, other forest products collected from the forest included medicine, handicraft materials,

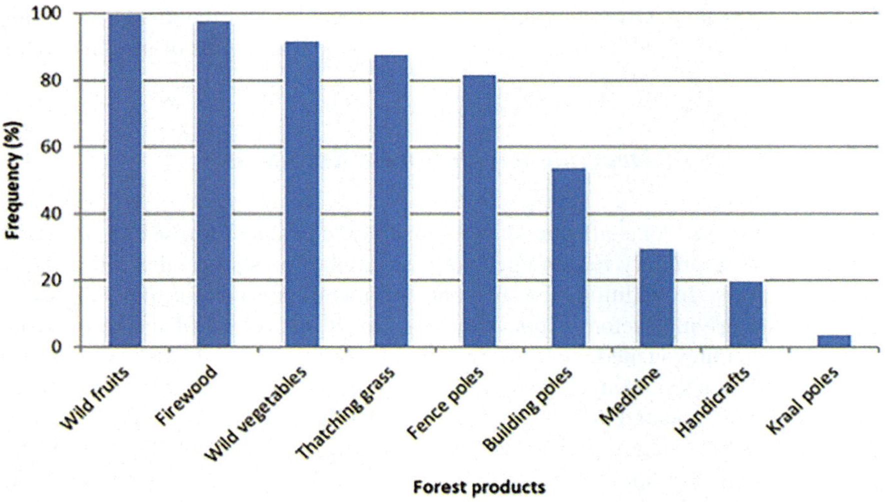

Fig. 19.3 Frequency of collection of forest products

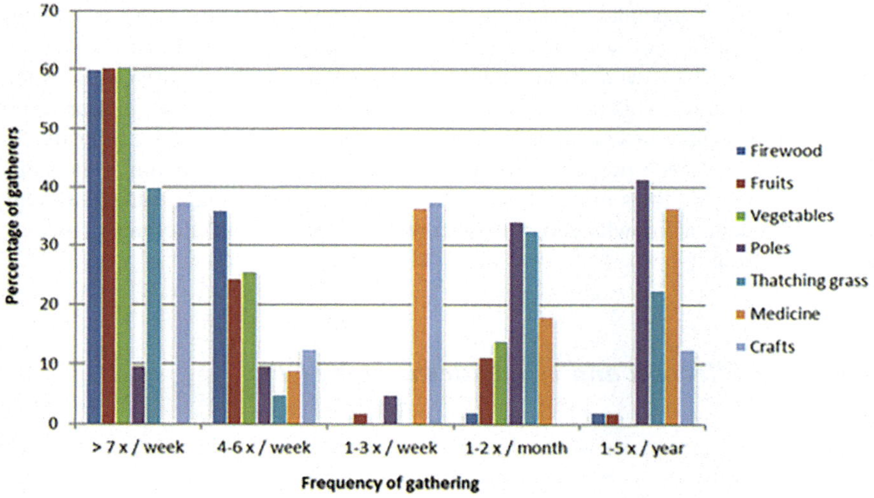

Fig. 19.4 Frequency of collecting forest resources

and poles for erecting livestock kraals. Most of these products were collected at irregular times and often when needs arose.

The findings indicate that forest resources were collected on a frequent basis, albeit with differences across the resource types (Fig. 19.4). Firewood, wild fruits, and wild vegetables emerged as the top three and most collected resources on a daily basis, while poles, thatching grass, medicine, and crafts were collected on an irregular basis, at least a few times per week or month. This is not surprising since the latter resources are mostly gathered as and when needs arise, e.g. medicine. Other resources, such as poles and thatching grass, are only needed either when constructing a new homestead or when renewing existing structures, which might still be done after a prolonged period of time. Overall, the findings suggest that forest resources were the main source of subsistence for the Shaikarawe community, particularly firewood and wild foods.

19.3.5 Seasonal Variation in the Collection and Consumption of NTFPs

Although most of the resources were collected throughout the year, the intensity of collection of some resources varied with the season (Fig. 19.5). Firewood was collected every month, but high peaks were recorded during the winter season between the months of May and July. This is an expected observation since more firewood is needed during the winter season for heating and keeping warm. The collection and consumption of wild foods (fruits and vegetables) were relatively high during the rainy season compared to the dry season (Fig. 19.5). The incidence of wild food collection peaked from November to March. Thatching grass was mostly collected during the dry seasons, between May and August. The collection of poles was seemingly uniform across the year but with slight peaks during the winter season. The collection of medicine and craft materials was not uniform throughout the year.

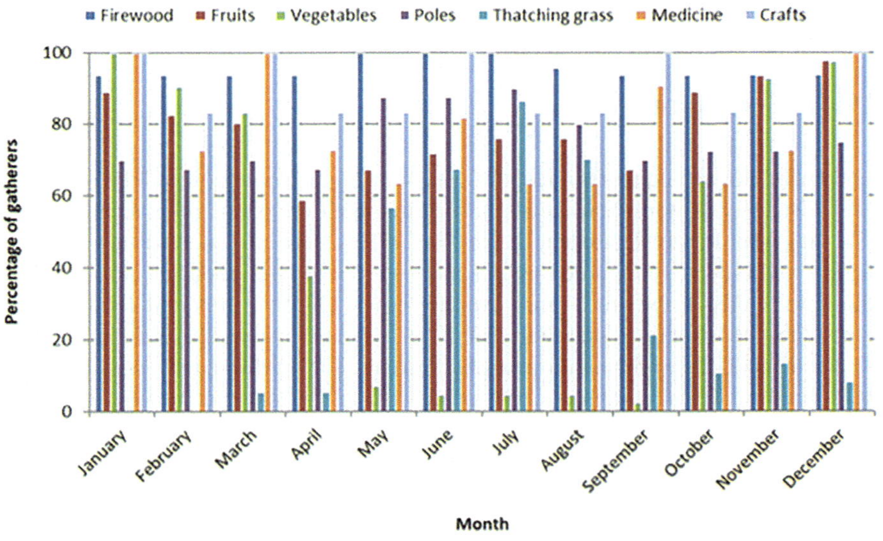

Fig. 19.5 Seasonal variation in the collection of forest resources

Table 19.2 Frequency distributions (%) of the level of importance of forest products to household livelihood

Forest products		Scale 0	1	2	3	4	5	*n*
Fuelwood					2	2	96	50
Poles	Kraal	96		4				50
	Fencing	18		2	6	20	54	50
	Building	46		4	4	4	40	50
Handicraft materials		80	4	2	6	6	2	50
Thatching grass		12	2	4	6	22	54	50
Wild fruits					4	22	74	50
Wild vegetables		10	2	2	2	20	64	50
Medicinal plants		68	12		6	2	12	50

Likert-type scale options: 0, not applicable; 1, extremely unimportant; 2, unimportant; 3, neutral; 4, important; and 5, extremely important, respectively

19.3.6 The Importance of NTFPs to Rural Livelihoods

The majority of respondents concurred that NTFPs are critical to livelihoods. NTFPs are an important livelihood strategy, and they can either play a role as primary or complementary sources of livelihood. When asked to evaluate the level of importance of individual NTFPs, fuelwood, wild fruits, and wild vegetables were considered extremely important to livelihoods, in order of frequency (Table 19.2). This finding mirrors those on the level of dependency on NTFPs, as the same resources were highly reliant upon. Therefore, the aforementioned resources are integral to livelihoods in the Shaikarawe settlement.

19.4 Discussion

The findings revealed that the respondents were highly reliant on NTFPs for fulfilling a diversity of household needs. Over three-quarters of the respondents opined that they were highly dependent on NTFPs. NTFPs were crucial in sustaining livelihoods, and they were regarded as a primary source of livelihood for most if not all the households in the study area. Among others, NTFPs provided food, energy, medicine, construction materials, and to a lesser extent, agricultural inputs. Therefore, NTFPs were an indispensable resource for fulfilling a diversity of household needs. These findings are consistent with local studies conducted in central (Pelane et al. 2023) and northern Botswana (Lepetu et al. 2009; Garekae et al. 2020). In these studies, NTFPs were found to be an important component of rural livelihoods. In Pelane et al.'s (2023) study, NTFPs were an important source of income and accounted for almost half of the total household income. Moreover, the findings corroborate previous studies from elsewhere (Dalu et al. 2021; Duong et al. 2021).

The level of dependence varied with resource type, being markedly high for fuelwood and wild foods compared to other resources. This is the case because the resources were collected and/or consumed almost on a daily basis.

Fuelwood was the primary source of energy for all households, which was used for cooking, heating, and keeping warm. Shaikarawe is a remote settlement with limited or lack of infrastructure and services. Therefore, the respondents have limited access to diversified energy sources. Moreover, the poor socio-economic conditions befallen most households constrain their ability to afford alternative energy sources such as gas (liquefied petroleum gas). Fuelwood is the dominant source of household energy in Botswana, particularly for cooking in rural areas. Statistics Botswana (2017) estimates that about 90% of rural households are highly dependent on fuelwood to meet their household energy needs. Although nearly one-third of rural households have access to electricity, they are still reliant on fuelwood as their primary energy source (IRENA 2021). Electricity is mostly used for lighting and powering electric appliances, while for cooking it is fuelwood. Furthermore, fuelwood usage is widespread across the region. Approximately, 70% of households in sub-Saharan Africa (SSA) depend on fuelwood to meet their domestic energy needs (Hoffmann et al. 2016; Masekela and Semenya 2021).

The dominance of fuelwood as a household energy source in the study area and elsewhere could be attributed to an array of factors, but key among them being affordability and accessibility. Firewood is normally harvested from the woodlands/forested areas fringing the rural communities, and it is mostly collected for free, save for those using vehicles as the mode of transport. Moreover, firewood can be readily available upon need. Against this backdrop, fuelwood remains the most viable energy source for most households in rural areas. Therefore, policies and regulations geared towards forest conservation should take into account the integral role of fuelwood in household energy needs. Moreover, the high level of dependence on firewood calls for the adoption of alternative domestic energy sources to relieve the pressure on the forest, e.g. biogas. In addition to being an energy source, fuelwood is an alternative income-generating activity for many households in rural and recently urban areas. Mendako et al. (2022) quantified the value of firewood in Bas-Uélé Province in the Democratic Republic of Congo (DRC), and it accounted for 22% of the total forest income, which contributed to 32% of the total household income. Income earned from firewood can be used to fulfil other household's basic needs such as food.

Wild foods were an important source of food in the study area. Wild fruits and vegetables are essential in providing dietary and nutritional diversity. Wild vegetables were a key source of relish in Shaikarawe, and they were commonly served with 'stiff porridge', a staple dish in the area and the rest of the country. Empirical evidence has demonstrated that wild foods are rich in nutrients and micronutrients essential for human health and development (Yang and Keding 2009; Mavengahama et al. 2013). Likewise, wild foods could be instrumental in fulfilling the minimum daily intake/requirements of vegetables and fruits recommended by the World

Health Organisation (Smith and Eyzaguirre 2007). Although not established in this study, wild foods have been positively associated with improved dietary diversity in eastern and western Africa (Fungo et al. 2016; Maseko et al. 2017). Wild foods can also be a source of income for many households.

A comprehensive study extending over 24 countries spanning three continents found that 77% of the sampled households were dependent on wild foods, one-fifth of which were for cash income (Hickey et al. 2016). This demonstrates the potential of wild foods in diversifying income sources for some households. Beyond dietary and nutrition, wild foods promote cultural identity and instil a sense of belonging to the community (Ahmed et al. 2022). Furthermore, wild foods were found to be resilient to unprecedented climatic changes (Legwaila et al. 2011; Mukul et al. 2016; Rai et al. 2019). Therefore, wild foods can also serve as a coping and adaptation measure during times of uncertainty, such as climate change and variability. Other NTFPs that were collected included construction materials (thatching grass, poles), medicine, and handicraft materials. Despite their importance, these products were collected at irregular times and often when needs arose.

Although NTFPs were collected and used throughout the year, the intensity varied with the season. The seasonality of collection was brought about by the increased demand for the particular product during the respective seasons. Moreover, seasonality could be linked with cyclical events such as flowering and fruiting, especially for wild fruits. However, this study falls short of elucidating alternative resources/products that households resort to during seasonal variation, which can be erratic.

Drawing upon the above, NTFPs contribute to rural livelihoods and well-being in the Shaikarawe settlement. As well known, Basarwa are traditionally hunters and gatherers who have subsisted on wild foods since time immemorial (Madzwamuse 2005). With subsistence hunting curtailed and/or restricted in the Okavango Delta and elsewhere in Botswana, wild foods remain the only primary source of subsistence for most Basarwa. Therefore, wild foods continue to be an integral part of Basarwa's diet and overall food security. In contrast, Basarwa have experienced marginalisation from the mainstream economy (Nthomang 1999; Lucas 2008). Despite the government's efforts of integration, Basarwa are still secluded and situated in areas with limited infrastructure, services, and economic opportunities.

Given these prevailing socio-economic conditions in the area, dependency on NTFPs is anticipated to remain constant or perhaps increase in the foreseeable future. Therefore, this calls for sustainable management of forest resources to ensure sustainability. Over and above, dependence on NTFPs serves as a source of empowerment for the Basarwa community in the Shaikarawe settlement. Through NTFPs, the Basarwa can engage in self-food production to fulfil their dietary needs. Moreover, there is potential for value addition of certain forest resources, such as *Ricinodendron rautenii*, which may provide various economic opportunities across the value chain of production. This has the potential to generate revenue for the community. Having realised the importance of NTFPs to livelihoods and well-being, the community has formed a community trust, Shaikarawe Community Trust, which is entrusted with managing the forest resources found in their proximate environment. Likewise, the community is in the process of establishing a community forest reserve to manage and conserve the remnant forested area in the vicinity of their settlement. This move will promote the sustainable utilisation of forest resources. Moreover, sustainable forest management activities such as agroforestry, identification and protection of valuable ecological species, domestication of valuable species for commercialisation, and value addition of NTFPs are recommended.

19.5 Conclusion

This study examined the role of forest products in the livelihoods of marginalised communities in the Okavango Delta, northern Botswana. The findings revealed that NTFPs were an integral part of rural livelihoods. On this note, NTFPs emerged as the only source of dependable livelihood strategies for most, if not all, households in the study area. Among others, NTFPs provided firewood, food, medicine, construction materials, and crafts. Firewood and wild foods were the most common NTFPs, typically collected frequently, at least seven times per week. This finding demonstrates the essential role played by NTFPs in livelihoods. Shaikarawe is located in a remote area, with inadequate infrastructure and limited economic opportunities for subsistence. Therefore, NTFPs remain the primary source of livelihood for most households. Against this backdrop, it is important for regulations to acknowledge and recognise the social use value of NTFPs to rural livelihoods. Moreover, this study calls for sustainable utilisation of forest resources for posterity.

In regard to household domestic energy, alternative approaches that harness the forest for producing energy are recommended. Fuelwood energy-saving stoves and biogas are recommended to address the challenges of unsustainable wood harvesting in forest-dependent communities. These alternative energy forms are contextual to the environmental settings of most rural dwellers, which are beset with inadequate infrastructure such as the Shaikarawe settlement. Over and above, the domestication of valuable species can greatly assist with decreasing pressure on forest resources.

References

Ahmed S, Warne T, Stewart A, Byker Shanks C, Dupuis V (2022) Role of wild food environments for cultural identity, food security, and dietary quality in a rural American State. Front Sustain Food Syst 6. https://doi.org/10.3389/fsufs.2022.774701

Arntzen J (2005) Livelihoods and biodiversity in the Okavango Delta, Botswana. Centre for Applied Research, Gaborone

Beall J, Kanji N (1999) Households, livelihoods and urban poverty. Retrieved from https://www.ucl.ac.uk/dpu-projects/drivers_urb_change/urb_society/pdf_liveli_vulnera/DFID_ESCOR_Beall_Households_Livelihoods.pdf. 23 Aug 2023

Beyene AD, Mekonnen A, Hirons M, Robinson EJZ, Gonfa T, Gole TW, Demissie S (2020) Contribution of non-timber forest products to the livelihood of farmers in coffee growing areas: evidence from Yayu Coffee Forest Biosphere Reserve. J Environ Plan Manag 63(9):1633–1654. https://doi.org/10.1080/09640568.2019.1679615

Carson SL, Kentatchime F, Nana ED, Njabo KY, Cole BL, Godwin HA (2018) Indigenous Peoples' concerns about loss of forest knowledge: implications for forest management. Conserv Soc 16(4):431–440

Cavendish W (2000) Empirical regularities in the poverty-environment relationship of rural households: evidence from Zimbabwe. World Dev 28(11):1979–2003

Chamberlain JL, Darr D, Meinhold K (2020) Rediscovering the contributions of forests and trees to transition global food systems. Forests 11. https://doi.org/10.3390/f11101098

Chambers R, Conway C (1992) Sustainable rural livelihoods: practical concepts for the 21st century, IDS discussion paper 296. Institute of Development Studies, Brighton

Dalu MTB, Gunter AW, Makatu M, Dowo GM, Dondofema F, Dalu T (2021) 'Contribution of natural forest products to rural livelihoods at Mavunde and Sambandou Villages, Vhembe Biosphere Reserve, South Africa. Sustainability 13(8). https://doi.org/10.3390/su13084252

De Beer JH, McDermott MJ (1989) The economic value of non-timber forest products in South East Asia. The Netherlands Committee for IUCN, Amsterdam

Delgado TS, McCall MK, López-Binqüist C (2016) Recognized but not supported: assessing the incorporation of non-timber forest products into Mexican forest policy. For Policy Econ 71:36–42. https://doi.org/10.1016/j.forpol.2016.07.002

Duong TMP, Lobry de Bruyn L, Kristiansen P, Marshall GR, Wilkes J (2021) Nature and level of NTFP reliance: a case study in the buffer zone of Cat Tien National Park, Vietnam. For Trees Livelihoods 30(2):116–132. https://doi.org/10.1080/14728028.2021.1891976

Eba'a Atyi R, Poufoun JN, Awono JPM, Manjeli AN, Sufo-Kankeu R (2016) Economic and social importance of fuelwood in Cameroon. Int For Rev 18(1):52–65

FAO (1992) Forest, trees and food. Food and Agricultural Organization of the United Nations FAO, Rome

FAO & UNHCR (2018) Managing forests in displacement settings: guidance on the use of planted and natural forests to supply forest products and build resilience in displaced and host communities, Rome

Fungo R, Muyonga J, Kabahenda M, Kaaya A, Okia CA, Donn P, Mathurin T, Tchingsabe O, Tiegehungo JC, Loo J, Snook L (2016) Contribution of forest foods to dietary intake and their association with household food insecurity: a cross-sectional study in women from rural Cameroon. Public Health Nutr 19(17):3185–3196

Garekae H, Shackleton CM (2020) Urban foraging of wild plants in two medium-sized South African towns: people, perceptions and practices. Urban For Urban Green 49. https://doi.org/10.1016/j.ufug.2020.126581

Garekae H, Thakadu OT, Lepetu J (2017) Socio-economic factors influencing household forest dependency in Chobe enclave, Botswana. Ecol Process 6:40. https://doi.org/10.1186/s13717-017-0107-3

Garekae H, Lepetu J, Thakadu OT (2020) Forest resource utilisation and rural livelihoods: insights from Chobe enclave, Botswana. S Afr Geogr J 102(1):22–40. https://doi.org/10.1080/03736245.2019.1606730

Hickey GM, Pouliot M, Smith-Hall C, Wunder S, Nielsen MR (2016) Quantifying the economic contribution of wild food harvests to rural livelihoods: a global-comparative analysis. Food Policy 62:122–132. https://doi.org/10.1016/j.foodpol.2016.06.001

Hoffmann H, Brüntrup M, Dewes C (2016) Wood energy in Sub-Saharan Africa: how to make a shadow business sustainable. German Development Institute/Deutsches Institut für Entwicklungspolitik (DIE), Bonn

IPBES (2021) Marginalised community. Retrieved from https://www.ipbes.net/glossary-tag/marginalized-community#:~:text=Marginalized%20communities%2C%20peoples%20or%20populations,Centre%20for%20Determinants%20of%20Health. 23 Aug 2023

IRENA (2021) Renewables readiness assessment: Botswana. International Renewable Energy Agency, Abu Dhabi

Kamanga P, Vedeld P, Sjaastad E (2009) Forest incomes and rural livelihoods in Chiradzulu District, Malawi. Ecol Econ 68(3):613–624. https://doi.org/10.1016/j.ecolecon.2008.08.018

King B, Shinn JE, Crews KA, Young KR (2016) Fluid waters and rigid livelihoods in the Okavango Delta of Botswana. Land 5(2). https://doi.org/10.3390/land5020016

Koné L (2019) Challenges and opportunities in the adoption of community forestry by local communities and indigenous peoples in the Republic of Congo. Forest Peoples Programme, Organisation for Development and Human Rights in Congo

Le HD, Nguyen TTK (2020) The contribution of non-timber forest products to the livelihoods of forest-dependent people: a case study in Hoa Binh province, Vietnam. For Trees Livelihoods 29(3):143–157. https://doi.org/10.1080/14728028.2020.1770131

Legwaila GM, Mojeremane W, Madisa ME, Mmolotsi RM, Rampart M (2011) Potential of traditional food plants in rural household food security in Botswana. J Hortic For 3(6):171–177

Lepcha LD, Shukla G, Pala NA, Pal VPK, Chakravarty S (2019) Contribution of NTFPs on livelihood of forest-fringe communities in Jaldapara National Park, India. J Sustain For 38(3):213–229. https://doi.org/10.1080/10549811.2018.1528158

Lepetu J, Alavalapati J, Nair PK (2009) Forest dependency and its implication for protected areas management: A case study from Kasane forest reserve, Botswana. Int J Environ Res 3(4):525–536

Lucas TB (2008) Basarwa: genuine targets of empowerment or pawns in a political power game? Botswana Notes Rec 40(1):124–134

Luna-González DV, Sørensen M (2018) Higher agrobiodiversity is associated with improved dietary diversity, but not child anthropometric status, of Mayan Achí people of Guatemala. Public Health Nutr 21(11):2128–2141

Madzwamuse MS (2005) Adaptive livelihood strategies of the Basarwa: a case of Khwai and Xaxaba, Ngamiland District, Botswana. MSc dissertation, Department of Environmental Science, Rhodes University, Makhanda, South Africa

Malleson R, Asaha S, Egot M, Kshatriya M, Marshall E, Obeng-Okrah K, Sunderland T (2014) Non-timber forest products income from forest landscapes of Cameroon, Ghana and Nigeria – an incidental or integral contribution to sustaining rural livelihoods? Int For Rev 16(3):261–277. https://doi.org/10.1505/146554814812572449

Masekela ME, Semenya K (2021) Factors influencing the use of firewood post-electrification in rural South Africa: the case of Ga-Malahlela village. J Energy S Afr 32(3):24–40

Maseko H, Shackleton CM, Nagoli J, Pullanikkatil D (2017) Children and wild foods in the context of deforestation in rural Malawi. Hum Ecol 45(6):795–807

Mavengahama S, McLachlan M, de Clercq W (2013) The role of wild vegetable species in household food security in maize based subsistence cropping systems. Food Secur 5(2):227–233

Mendako RK, Tian G, Ullah S, Sagali HL, Kipute DD (2022) Assessing the economic contribution of forest use to rural livelihoods in the Rubi-Tele hunting Domain, DR Congo. Forests 13(1). https://doi.org/10.3390/f13010130

Miller DC, Cheek JZ, Mansourian S, Wildburger C (2022) Forests, trees and the eradication of poverty. For Policy Econ 140:102753. https://doi.org/10.1016/j.forpol.2022.102753

Mollee E, Pouliot M, McDonald MA (2017) Into the urban wild: collection of wild urban plants for food and medicine in Kampala, Uganda. Land Use Policy 63:67–77

Mukul SA, Rashid AZMM, Uddin MB, Khan NA (2016) Role of non-timber forest products in sustaining forest-based livelihoods and rural households' resilience capacity in and around protected area: a Bangladesh study. J Environ Plan Manag 59(4):628–642

Nthomang K (1999) The Basarwa/San/Bushman question in Botswana. Community Economic Development (CED): creating opportunities for empowerment. J Soc Dev Afr 14(1):53–67

Olesen RS, Hall CM, Rasmussen LV (2022) Forests support people's food and nutrition security through multiple pathways in low- and middle-income countries. One Earth 5(12):1342–1353. https://doi.org/10.1016/j.oneear.2022.11.005

Pelane GL, Mulalu MI, Mmopelwa G, Stone MT (2023) Rural community dependency on woodland resources: the case of Foley and Makomoto settlements in Central-East Botswana. S Afr Geogr J 1–20. https://doi.org/10.1080/03736245.2023.2167855
Rai RK, Neupane BK, Sapkota K (2019) Non-timber forest product and its impacts on livelihood in the Middle Hill: a case of Lamjung district, Nepal. J Geogr Geol 11(4):29–37
Rasmussen LV, Fagan ME, Ickowitz A, Wood SLR, Kennedy G, Powell B, Baudron F, Gergel S, Jung S, Smithwick EAH, Sunderland T, Wood S, Rhemtulla JM (2020) Forest pattern, not just amount, influences dietary quality in five African countries. Glob Food Secur 25:100331. https://doi.org/10.1016/j.gfs.2019.100331
Shackleton CM, de Vos A (2022) How many people globally actually use non-timber forest products? For Policy Econ 135:102659. https://doi.org/10.1016/j.forpol.2021.102659
Shackleton CM, Shackleton SE (2004) The importance of non-timber forest products in rural livelihood security and as a safety net: a review of evidence from South Africa. S Afr J Sci 100:658–664
Shackleton CM, Delang CO, Shackleton SE, Shanley P (2011) Non-timber forest products: concept and definitions. In: Shackleton SE, Shackleton CM, Shanley P (eds) Non-timber forest products in the global context. Springer, Berlin/Heidelberg, pp 3–21. https://doi.org/10.1007/978-3-642-17983-9_1
Shackleton CM, Garekae H, Sardeshpande M, Sinasson-Sanni G, Twine WC (2024) Non-timber forest products as poverty traps: fact or fiction? For Policy Econ 158. https://doi.org/10.1016/j.forpol.2023.103114
Smith FI, Eyzaguirre P (2007) African leafy vegetables: their role in the World Health Organization's global fruit and vegetables initiative. Afr J Food Agric Nutr Dev 7(3) https://www.ajfand.net/Volume7/No3/Smith-IPGRI1-1.pdf
Statistics Botswana (2017) Botswana environment statistics 2016. Statistics Botswana, Gaborone
Suleiman MS, Wasonga VO, Mbau JS, Suleiman A, Elhadi YA (2017) Non-timber forest products and their contribution to households income around Falgore Game Reserve in Kano, Nigeria. Ecol Process 6(23):1–14
Torri MC (2011) Conservation, relocation and the social consequences of conservation policies in protected areas: case study of the Sariska Tiger Reserve, India. Conserv Soc 9(1):54–64
Tropp J (2003) Displaced people, replaced narratives: forest conflicts and historical perspectives in the Tsolo District, Transkei. J S Afr Stud 29(1):207–233. https://doi.org/10.1080/0305707032000060458
Yang RY, Keding GB (2009) Nutritional contributions of important African indigenous vegetables. In: Shackleton CM, Pasquini MW, Drescher AW (eds) African indigenous vegetables in urban agriculture. Earthscan, pp 105–143

Joyce Lepetu is an Associate Professor in the Department of Range and Forest Resources and Dean of Faculty of Natural Resources at Botswana University of Agriculture and Natural Resources. She has a PhD in Forest Science, University of Florida, USA. Her research interests include forest conservation and livelihoods, protected area management, agroforestry, and climate change adaptation.

Hesekia Garekae is a Senior Lecturer of Human Environment at the Department of Environmental Science, University of Botswana. He has a PhD in Environmental Science, Rhodes University, South Africa. His research interests intersect forests, livelihoods, and conservation.

Gofaone Rammotokara is a Technician in the Department of Range and Forest Resources at the Botswana University of Agriculture and Natural Resources. She has a Bachelor's degree in Range Sciences, Botswana University of Agriculture and Natural Resources. Her research interests include range ecology, ecological restoration and sustainability, economics of environmental issues, as well as the application of geographical information systems and remote sensing in forest ecology and management.

Topoyame Isaac Makoi holds a Master of Science in Environmental Science specializing in Geographical Information Science and Environmental Management and a Bachelor of Science in Land Management. He works for the Department of Range and Forest Resources, Botswana University of Agriculture and Natural Resources as a Senior Instructor. His research interests include areas of range ecology, forest and woodland ecology, climatology, environmental management, vegetation mapping, geographical information systems (GIS) and remote sensing. He has been engaged in several research projects and has published five articles in refereed journals and one book chapter.

Chapter 20
Economic Valuations, Levels of Involvement, and Impacts of Participatory Forest Management Practices on Livelihoods: The Cases of Sheka and Kafa Forests, South-Western Ethiopia

Endale Difabachew, Abdi Khalil Edriss, Jema Haji Mohamed, Belaineh Legesse, and Mengistu Ketema

20.1 Introduction

The world's forest cover is approximately 4.06 billion hectares, with natural forests covering 93% of that area, or 3.7 billion hectares (FAO 2020; UN 2021). Unfortunately, the world's forest cover area has been declining alarmingly over the last quarter of a century (FAO 2010a, b). Africa has the second-highest deforestation rate (FAO 2010a, 2010b). Similarly, Ethiopia has seen significant forest loss. According to estimates, the nation's forests have been disappearing at a pace of 1.25% since 2010 (FAO 2015). Realizing this, the Ethiopian government has

E. Difabachew
Department of Agricultural Economics, Mizan–Tepi University, Mizan-Aman, Ethiopia

A. K. Edriss (✉)
Department of Agricultural and Applied Economics and Africa Centre of Excellence (Agricultural Policy Analysis), Lilongwe University of Agriculture and Natural Resources (LUANAR), Lilongwe, Malawi

J. H. Mohamed
School of Agricultural Economics, Haramaya University, Dire Dawa, Ethiopia

B. Legesse
Frontieri Consult PLC, Addis Ababa, Ethiopia
e-mail: b.legesse@frontieri.com

M. Ketema
School of Agricultural Economics, Haramaya University, Dire Dawa, Ethiopia

W. L. Hambira et al. (eds.), *Environmental Change and Biodiversity Conservation in sub-Saharan Africa*, Advances in Global Change Research 75,
https://doi.org/10.1007/978-3-031-73136-5_20

created various forest-protected areas and forest conservation management techniques to protect forests.

Ethiopia has established 21 national parks, two sanctuaries, three wildlife reserve areas, 20 controlled hunting areas, six open hunting areas, six community conservation areas, and 58 national forest priority areas, 37 of which are protected forests (Yemiru 2011). Sheka and Kafa forests were selected as national forest priority areas from the Ethiopian protected forest areas, and they are a part of community conservation forests. The two zones of forests, Moist Montane forests and Afromontane forests in Ethiopia, occur nearby under similar climatic conditions. These forests provide vital products to local communities, including forest and non-forest products, such as medicinal plants, coffee, honey, and wild fruits. Participatory forest management (PFM) was introduced in the mid-1990s by international non-governmental organizations (NGOs) to encourage sustainable management and conservation of forests and improve the livelihoods of people living in or around these resources in Ethiopia (Siraj et al. 2018).

PFM is a mechanism to protect forests and enhance the livelihoods of communities using and benefiting from them. PFM was meant to avert deforestation's persistent problems and deliver better social and economic outcomes than the former centralized command-and-control resource management approaches (Alemayehu et al. 2015). In addition, the program designed two approaches to sustainable forest management to achieve sustainable forest management objectives while generating livelihood benefits.

Many countries have developed policies, strategies, and regulations for forest resource management in general and participatory forest management (PFM) in particular. PFM is widely promoted throughout the world tropics as a policy that leads to participation and poverty alleviation (Adam and Eltayeb 2016). This is linked to the argument that forest ecosystems contribute significantly to the socio-economic development of forest-dwelling communities (Mislimshoeva et al. 2016). One of the challenges for successful management to achieve sustainable forest management in the PFM is the need for members to participate in forest conservation and livelihood activities. However, there are still controversies among rural households regarding the values of community-based conservation practices, participation programs, and the impacts of participatory forest management practices. We examined these issues using the PFM members' value placement lenses, the framework that captures the economic, ecological, and institutional features. We analysed the PFM contributions based on estimating PFM livelihood outcome indicators and their involvement in the PFMP of the Sheka and Kafa conservation in Ethiopia.

PFM was meant to avert deforestation's persistent problems and deliver better social and economic outcomes (Alemayehu et al. 2015). The broad objective is to improve participatory forest sustainability and achieve development and environmental goals. An issue for policymakers, particularly in developing countries with limited resources, is that deforestation and forest degradation are increasing in many regions, and primary forests are under particular threat. Forest communities that have been using the forest for centuries are often faced with the wrong choice between conservation and development. There are many ways to generate income from forests through sustainable activities that recognize and capture the benefits of forest ecosystem services beyond timber. Effectively engaging individual or community responsibility in conservation management and planning is complex and takes a long time to achieve. It requires a holistic approach, bringing together conservation and social science aspects to find common ground. Participatory forest management (PFM) is a mechanism to protect forests and enhance the livelihoods of communities using and benefiting from them. Many studies have demonstrated that PFM contributes to forest protection (e.g. Blomley et al. 2008; Takahashi and Todo 2012; Thoms 2008).

However, welfare implications related to household involvement in PFM remain poorly understood despite their importance for sustainability. Regarding harvesting limits, PFM frequently imposes more rules and regulations on forest-related livelihood options that could result in a decline in forest-based incomes (Larson and Pulhin 2012; Schreckenberg and Luttrell 2009). The restricted access was balanced in some PFM programs that introduced income-generating activities through adding value to forests (e.g. ecotourism), developing additional earnings opportunities including forest products (such as woodlots), or compensating for the losses (Gobeze et al. 2009). However, these added benefits or compensation welfare implications remain poorly understood.

Therefore, this study analysed the economic valuation of forest conservation, level of involvement, and membership duration impact on the consumption expenditure status of participating rural households in the study areas. Furthermore, the study estimated households' willingness to pay (WTP) for forest conservation practices, identified factors affecting participation in forest management practices, level of involvement, and factors affecting forest users' management, and evaluated the impact of PFMP on forest users' livelihoods.

20.2 Methodology

20.2.1 Description of the Study Areas

This study was conducted in the Sheka and Kafa zones of the Southern Nations, Nationalities, and People's Regional State (SNNPRS), where participatory forest reserves are located, and the PFM project was implemented (Fig. 20.1). The forests in south-west Ethiopia are an essential natural resource, rendering households high-value and low-value non-timber forest products (NTFPs). High-value products include commercial items such as coffee, spices, and medicinal plants. Agricultural practices are the sole livelihood source for most inhabitants. *Enset* (banana products) and maize are major subsistence crops. Honey, spices (turmeric and black pepper), and coffee are primary cash income sources (Wikipedia 2021).

According to CSA (2007), the Sheka zone has a total population of 192,970 and an area of approximately 2175 km^2. The administrative centre of the Sheka zone is

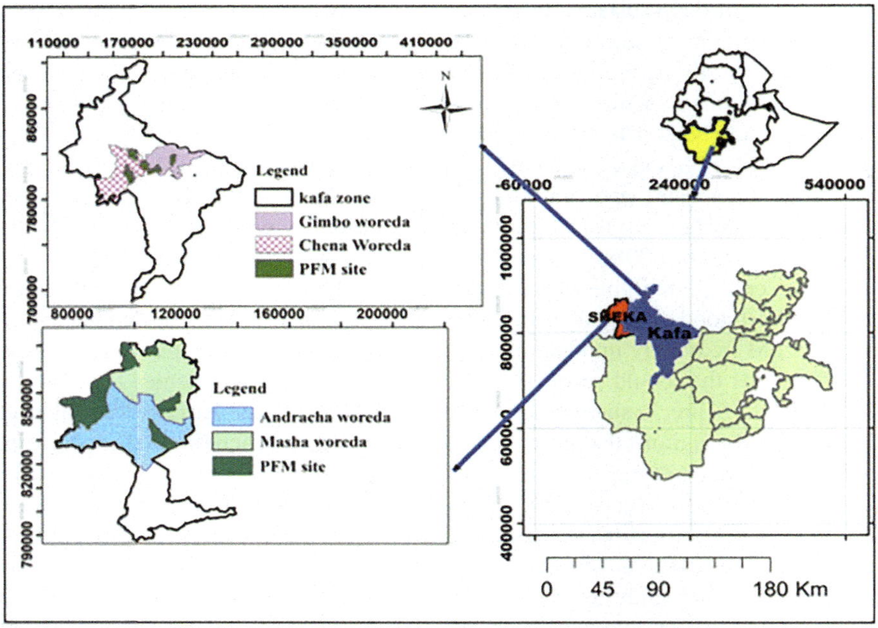

Fig. 20.1 Map of the study area

located 676 km south-west of Addis Ababa. Geographically, the area lies between 7°24′–7°52′ N latitude and 35°13′–35°35′ E longitude and consists of three districts, namely, Masha, Andracha, and Yeki. The Oromia Regional State borders the zone to the North, the Gambella Regional State to the West, the Kafa zone to the East, and the Bench Maji zone to the South. In total, there are 45 rural and 2 *urban kebeles* (*Kebele* is the minor administrative hierarchy in Ethiopia) in the two districts, Masha and Yeki (town name, Teppi).

The Kafa zone is in the SNNPRS, Ethiopia's most ethnically and linguistically diverse region. Bonga is the administrative town of the area located 450 km away from Addis Ababa.

The zone is mainly covered with evergreen Montane forest and is part of the Eastern Afromontane Biodiversity Hotspot. The site has a total land area of 10,602.7 square kilometres, of which 23.1% is cultivable, 31.54% forestland, 6.03% grazing land, 24.9% fallow land, and the remaining 14.43% is uncultivable (swampy and wetlands). According to the 2007 census, the total population of the zone is 858,600, with a population density of 90 persons per square kilometre. Its altitude ranges from 500 to 3500 m above sea level with a mean annual rainfall of 1001–2200 mm and temperature of 10.1–27.5 °C. The agro-ecological conditions of the Kafa zone are very suitable for growing coffee, tea, spices, and other crops (Wikipedia 2021).

20.2.2 Data Collection and Sampling Technique

Primary data were collected through a face-to-face interview encompassing demographic, socioeconomic, institutional services, social capital, networking, and forest management issues. Information was collected from respondents using a structured questionnaire, and key informants were also interviewed using checklists. Information was also collected through unstructured interviews—key informant interviews with government officials and development agents in each sample *Kebele* of the Sheka and Kafa zones. Secondary data were collected from government and non-governmental organizations (NGOs) to supplement the primary data. The primary sources of secondary data include the Bureau of Forest and Environmental Protection, regional *Woreda* experts of Sheka and Kafa Zones, and the Bureau of Agriculture and Rural Development, including their words, Tepi National Spics Research Centre, and Farm Africa and SOS Sahel NTFP at Masha.

Sampling Techniques

The study applied multistage sampling techniques to select a sample household for an interview. In the first stage of the sampling procedure, four *Woredas (or districts)* from two zones were selected based on the PFMP-targeted forest areas actively working *Woredas,* namely, Masha and Andracha from the Sheka zone and Gimbo and Chena from the Kafa zone. In the second stage, the selection of *Kebeles* from the respective sample *Woredas* with each *Kebele is* given an equal probability of being selected as a sample. Three *Kebeles* from Masha, two from Andracha, three from Gimbo, and two from Chena *Woreda* were chosen randomly. In the third stage, 384 sample households were randomly and proportionally selected from participatory forest *kebeles* in the Sheka and Kafa zones (Table 20.1).

Table 20.1 Distribution of the sample zones, districts (Woredas), kebeles, and households

Zone	District	Sample Kebeles	PFM Households	Sample Size
Kafa	Gimbo	Bayemo	278	46
		Michiti	264	44
		Tula	196	33
	Chena	Beko	305	50
		Dehera	260	43
Sheka	Andracha	Yokichichi	204	34
		Chegecha	164	28
	Masha	Gatimo	253	42
		Uwa	209	35
		Keja	174	29
Total			**2307**	**384**

20.2.3 Choice Experiment (CE) Model

The choice experiment is used in estimating willingness to pay (WTP) functions to investigate the economic valuation for improved forest conservation in terms of different attributes of forest conservation management alternatives. Thus, the random theory of the econometric model was used to analyse the results of the discrete choice experiment. A participant selects the option from the selection set with the highest utility value. As shown in Eq. (20.1), the indirect utility function (*U*), which represents the satisfaction that household (*n*) receives from the attributes of forest conservation benefits by alternative (*j*), is:

$$U_{nj} = V_{nj} + \varepsilon_{nj} \tag{20.1}$$

Here, the indirect function is composed of two parts: V_{nj} represents the deterministic element of individuals' utility, specified as a linear index of the attributes (*X*) of the *j* alternatives in the choice set; and ε_{nj} the stochastic disturbance term represents all the unobserved influences on an individual's choice (Adamowicz and Boxall 2002).

Notably, the primary assumption should be that everyone perceives a level of utility (*U*) or "attractiveness" from each alternative *j* and then selects the alternative that delivers the highest utility level from those available on that survey card. In this way, the probability (*Pnj*) that households *n* choose option (*i*) in preference to any other alternative (*j*) can be expressed as the probability that the utility associated with choice (*i*) is greater than that associated with all different options. This is shown in Eq. (20.2):

$$\begin{aligned} P_{ni} &= \text{Prob}\left(V_{ni} + \varepsilon_{ni} > V_{nj} + \varepsilon_{nj}\right), \forall j \epsilon C, i \neq j \\ P_{ni} &= \text{Prob}\left(\varepsilon_{nj} - \varepsilon_{ni} < V_{ni} - V_{nj}\right), \forall j \epsilon C, i \neq j \end{aligned} \tag{20.2}$$

where *C* is the set of all possible alternatives. The multinomial logit (MNL) model is derived by assuming that the individual error terms are independently and identically distributed and follow a Gumbel distribution (McFadden and Train 2000). This gives rise to the following choice probability, which is a general form of the MNL model:

$$P_{ni} = e^{\beta' X_{ni}} / \sum_j e^{\beta' X_{nj}} \tag{20.3}$$

β' is a scale parameter that is inversely proportional to the standard deviation of the error distribution and is typically assumed to be 1 (Ben-Akiva and Lerman 1985). The analyst can capture the observed taste heterogeneity in this model by interacting the preference for the choice attributes or an alternative-specific constant with observable data such as farmers' socioeconomic characteristics (including age, gender, and annual income) and the particular farm factors. Assuming linear parameters in the indirect utility function, V_{it}, we can characterize the following as in Louviere et al. (2010):

$$V_{it} = \text{ASC}_i + \sum_k \beta_{ik} X_{ik} \tag{20.4}$$

where ASC_i is an alternative-specific constant (ASC), X_{ik} is the k attribute value of alternative i, and the coefficient associated with the Kth attribute of option i. X variables reflect the effect of characteristics in the choice set. At the same time, the *ASC* captures any systematic variations in choice observations that are not explained by the attribute variation.

WTP estimation "implicit price" or part-worth of the attribute is calculated as follows:

Defining non-monetary (non-price attribute) as *nma* and monetary attribute as *ma,* the (part-worth) WTP is obtained as parameter estimates from the estimated choice experiment (CE) model that can be used to calculate WTP for each attribute in the utility function. Marginal WTP is estimated by Hanley et al. (2002):

$$\text{WTP} = \left(\frac{\beta_{\text{nma}}}{\beta_{\text{ma}}} \right) \tag{20.5}$$

where β_{nma} is the coefficient of the non-price attribute, and β_{ma} is the coefficient of the price attribute. Confidence intervals associated with marginal willingness to pay estimates can be obtained using methods such as Monte Carlo simulation, the delta method, or the Krinsky and Robb method (Hole 2007). Where estimates of WTP (part-worth) are made on a "Ceteris Paribus", they are estimates of willingness to pay respondents for an increase in the attribute of concern, given that everything else is held constant.

Random parameter logit (also called multinomial mixed logit, MXL) models were used to determine respondents' WTP for alternative forest conservation management plans (with interactions), from the most straightforward but stringent model to the most complex and flexible model. The RPL model treats a good or service's attributes (or characteristics) as a random variable. Each parameter includes a systematic random component; thus, the model estimates each distribution's mean and standard deviation. Most model applications utilize the typical distribution assumption for the model parameters (Bennett and Blamey 2001). The random utility function for the random parameter logit model takes the following form:

$$U_{it} = V_{it} + \varepsilon_{it} = X_i \left(\beta + \eta_t \right) + \varepsilon_{it} \tag{20.6}$$

where U_{it} is the total utility for respondent t from choosing alternative I in the choice set. It is assumed that the utility function consists of a systematic component (V_{it})

and a stochastic component (ε_{it}). The indirect utility is a function of the attribute vector X through β (including socioeconomic and environmental attitudinal variables). In contrast to the MNL model, a stochastic component of utility may be correlated among alternatives and across the sequence of choices via the expected influence $\eta_{it.}$ The probability that individual t picks an option from each choice set is then presented as:

$$P_{it} = \frac{e^{X_{it}(\beta+\eta_t)}}{e^{X_{jt}(\beta+\eta_t)}} \tag{20.7}$$

The estimated deterministic (indirect) utility function generally had the following form:

$$V_{it} = \mathrm{ASC} + \Sigma \beta_k X_k + \Sigma \gamma_m S_m \tag{20.8}$$

ASC is an alternative-specific constant that captures the effect of an attribute (not included in the choice-specific characteristics) on utility, k is the number of features, and m is the number of socioeconomic variables. Note that socioeconomic factors are constant across choice occasions for any respondent and can only be entered as interaction terms with the attributes.

20.2.4 Data for Attribute Selection and Choice Set Construction

Choice experiments are grounded in the idea that a good can be valued based on its attributes. Four stages are necessary for CE implementation (Hoyos 2010; Blasch et al. 2012): (1) establishment and definition of CE attributes and levels; (2) selection of choice sets; (3) setting up the structure and assembly of the questionnaire; and (4) data collection. The CE method involves surveying individuals to determine their preferences regarding their value to environmental aspects (Delibes-Mateos et al. 2010; Guimarães et al. 2014). Thus, a set of alternatives is presented to respondents. Different levels of providing good or service attributes are described, and interviewees are asked to choose one. In this way, choosing one alternative over the others shows that the interviewee considers it a higher utility than the rest (McFadden 1974).

However, it is essential to believe that one of these alternatives is the current situation or status quo (SQ), including the price for implementing each alternative as one more attribute is fundamental. The CE method provides the willingness to pay (WTP) for each attribute considered, a proxy for its implicit value. The transcendence of the CE method allows an analysis of the provision of different attributes. In this study, CE design decided that the included attributes are biodiversity, cultural ecosystem services, the facility of ecotourism employment, NTFP harvesting, and research and education.

20.2.5 Attributes and Levels Used in the CE

The choice of attributes and their levels of provision were made by reviewing the literature on the area under study and direct interviews with households involved in the forest conservation management of the study areas. These sources of information allowed the establishment of the existing situation of forest conservation management and potential management alternatives. In the main section that elicited respondents' preferences for the different forest conservation management characteristics, we introduced two alternative policy options and a business-as-usual option. The options were described regarding payment attributes and five forest conservation management characteristics. The level of the attributes captured the range of current and potential future trends in the Sheka and Kafa forests. The characterization of each attribute and its levels, represented by specific management measures, are shown in Table 20.2.

20.2.6 Choice Set Construction

Combining attributes and levels generated 180 alternatives, which must be reduced to construct choice sets cards. To this end, the orthogonal design was applied to achieve an optimal stage design through D efficiency (Hensher et al. 2005). Thus, a total of 36 choice sets were obtained. They were grouped into three blocks; each respondent considered twelve choice sets. Each set comprised three alternatives:

Table 20.2 Attributes and levels used to define choice alternatives

Attributes	Descriptions	Levels
Biodiversity	To promote habitat and species conservation	Current protection (SQ) Medium (10%), High, (20%)
Cultural Ecosystem services	Promoting the preservation of cultural ecosystem services through forest conservation is considered	Current levels (SQ), Moderate (10%), Aggressive preservation (20%)
Research and Education	Capacity building, students, and school children to learn about nature and research opportunities.	Low: current levels Medium (10%) High: Improve by (20%)
Ecotourism employment	The facility of ecotourism employment and business activities	The current number, 15, increased to 50, 70, and 150
NTFPs Harvesting	Promote NTFPs harvesting level	Low: current harvesting (SQ), Medium (3%) optimum:(6%) of harvesting level.
Payment	Contribution to the improved alternative FC management in a month	0 Birr, 25 Birr, 86 Birr, 246 Birr

Table 20.3 Sample choice card

Attributes	Alternative 1	Alternative 2	Status quo
Biodiversity improvement	High	Medium	Low
Research and Education	Low	High	Low
Ecotourism employment	75	150	15
Harvesting NTFPs	3%	0%	0
Preserving Cultural Ecosystem services	Less	Aggressively	Less use
Payment (Monthly contribution)	Birr 25	Birr50	Birr 0

two related to the forest, the improved conservation management, and a third referring to the current situation (SQ). The alternatives of each group were presented generically, such as "Alternative 1" and "Alternative 2" (Blamey et al. 2000), as shown in Table 20.3.

20.3 Factor Analysis and Seemingly Unrelated Regression

In assessing the level of involvement in the PFM program, the estimation strategy adopted a two-step strategy. First, we assessed and estimated the local peoples' level of participation in the different activities of a PFM program. In the second stage, we examined the factors influencing participation. In assessing the level of involvement, the study measured the levels of involvement in PFM using a program activities index computed using 12 participation indicators. The analytical approach was used following previous studies (Adhikari et al. 2014; Hedge et al. 2014; Dolisca et al. 2006). Participation scores were constructed for the dependent variables using the exploratory factor analysis method. In the first stage, determining the number of factors to retain for a study is critical in EFA and over-factoring (removing too many factors), resulting in non-interpretable or unreliable elements (Gebru et al. 2018). The following structural matrix equation describes the relationship between the observed and latent variables (underlying variables):

$$Z = \lambda F + \varepsilon \tag{20.9}$$

Z is a $px1$ vector of variables, a $p \times m$ matrix of factor loadings, F is a $mx1$ vector of factors, and ε is a $px1$ vector of error or residual elements (Sharma 1996).

20.3.1 Determinants of the Level of Involvement in the PFM Program

In the second stage, a seemingly unrelated regression (SUR) model was used to assess determinants of the level of involvement in the PFM program. The elements (explanatory or independent variables) affect involvement. This study's dependent variables are governance and decision-making, forest conservation, and perceived incentives. Hence, the primary purpose of applying this model was to identify the determinant factors that affect the level of involvement. The study categorized participation scores given their fractional nature and the fact that they are typically bounded within the unity interval (0, 1) (Wooldridge 2008).

$$Y = \begin{cases} 1 & \text{if the predicted score of factor analysis is} > 0.5 \\ 0 & \text{if the predicted score of factor analysis is} < 0.5 \end{cases}$$

Given the predicted score values, the analysis was used in modelling the SUR model. Since members' levels of involvement are measured multidimensionally, their responses toward the level of participation in different activities of the PFMP are a combination of characteristics of individual households (Dolisca et al. 2006). Given the expected heterogeneous level of the involvement of respondents' age, sex, education level, family size, farm size, distance from the forest, frequency of extension contacts, access to the social network, training, member to cooperative, and training facility were included to estimate their determinant factors on the level of involvement.

The classic SUR model assumes that for each observation i, there are m dependent variables $y_{i1}, y_{i2}, \ldots., y_{im}$ available, each with its linear regression model and with the usual stacking of observations over;

$$Y_j = X_j \beta_j + E_j \tag{20.10}$$

For $j = 1, \ldots\ldots, M$ where y_i and E_j are N-vectors and X_j an $N * K_j$ matrix where $K_j = \dim(\beta_j)$ the number of regressors for the jth regressions. Then, the three-factor equations estimated using the SUR model are specified below:

$$\begin{aligned} \text{Govt}_i &= X_1 \beta_1 + \varepsilon_{1i} \\ \text{Conserv}_i &= X_2 \beta_2 + \varepsilon_{2i} \\ \text{Resilin}_i &= X_3 \beta_3 + \varepsilon_{3i} \end{aligned} \tag{20.11}$$

where Govt_i, Conserv_i, and Resilin_i are members' participation in household i, forest governance, forest conservation, and perceived resilience, respectively. X_1 s are the respective vectors of the covariate determining the dependent variables, and the β_i s are their vectors of parameter coefficient.

20.3.2 Generalized Propensity Score Matching (GPS)

The econometric model, generalized propensity score matching (GPS), is the potential outcome approach Hirano et al. (2004) developed and is now widely used in different intervention evaluation literature. Assuming a random sample of units from a large population and indexed by $i = 1, \ldots, N$. For each unit under treatment level t, there exists a set of potential outcomes $\{Y_i(t)\}$ for $t \epsilon T$, referred to as the unit-level dose–response function (DRF). In the binary treatment case, $T = \{0, 1\}$. However, in continuous treatment cases such as participatory forest management (PFM) participation with different participation intensities (household's years of membership with PFM), T is in an interval $[t_0, t_1]$, with $t_0 > 0$ (Hirano et al. 2004).

The main objective in such cases is to estimate the average potential outcome or the average dose–response function (ADRF), $\mu(t) = E[Y_i(t)]$. It represents the function of the average potential livelihood outcome over all possible PFM participation levels, and $Y_i(t)$ is a livelihood indicator of annual household per capita consumption expenditure for a household that is a member of the PFM program.

The observable variables for each unit i are a vector of covariates, the level of the treatment unit i receives, and the potential outcome corresponding to the treatment received. As the GPS application focuses on ADRF and marginal treatment functions for households that are members of the PFM program, families that did not participate in the PFMP are not included in the model. Hirano et al.'s (2004) fundamental assumption generalizes unconfoundedness and the balancing property assumptions similar to the binary treatment of Rosenbaum and Rubin (1983) to the continuous impact study. It states that after controlling for observable characteristics X_i, any remaining difference in treatment intensity T_i across units is independent of the potential outcomes $Y_i(t)$, Eq. (20.12).

$$Y_i(t) \perp \mathrm{T}_i / \mathrm{X}_i \quad \text{for all } t \epsilon \mathrm{T} \tag{20.12}$$

Steps in the GPS Method

Following Hirano et al. (2004), the GPS method has been implemented in three steps.

In the first step, the parameters of the treatment duration function, that is, β_0, β_1. Moreover, δ^2 (conditional distribution of membership duration) are estimated using maximum likelihood or ordinary least squares regression according to Eq. (20.13).

$$\mathrm{T}_i / \chi_i \sim \mathrm{N}\left[\beta_0 + \beta_1 \mathcal{X}_i, \delta^2\right] \tag{20.13}$$

After GPS estimation, its balancing property is tested before proceeding to step two. Following the estimation of the parameters of the treatment function in Eq. (20.14), GPS can be estimated in Eq. (20.15).

$$\hat{R}_i = \frac{1}{\sqrt{2\pi\delta^2}} \exp\left[-\frac{1}{2\hat{\delta}^2}\left(T_i - \hat{\beta}_0 - \hat{\beta}_1' X_i\right)^2\right] \tag{20.14}$$

In the second step, we calculated the conditional expectation function of the outcome (Y_i), given modelled as a flexible function (polynomial approximation) of observed treatment (T_i) and estimated GPS (R_i), for the analytical approach, using the quadratic approximation followed in Eq. (20.16):

$$\mathrm{E}\left(\left[Y_i / \mathrm{T}_i, \widehat{\mathfrak{R}}_i\right]\right) = \alpha_0 + \alpha_1 \mathrm{T}_i + \alpha_2 \mathrm{T}_i^2 + \alpha_3 \widehat{\mathfrak{R}}_i + \alpha_4 \widehat{\mathfrak{R}}_i^2 + \alpha_5 \mathrm{T}_i \widehat{\mathfrak{R}}_i \qquad (20.15)$$

The third step estimates the average dose–response function at a particular value of the treatment t was calculated by averaging the (estimated) conditional expectation E over the GPS at that length of membership duration using Eq. (20.16):

$$\mu(t) = E\left[\widehat{Y}(t)\right] = \frac{1}{N}\sum_{i=1}^{N} g^{-1}\left[\begin{matrix}\widehat{\alpha}_0 + \widehat{\alpha}_1 .t + \widehat{\alpha}_2 .t^2 + \widehat{\alpha}_3 .\widehat{r}\left(t, X_i\right)\\ +\widehat{\alpha}_4 .\widehat{r}\left(t, X_i\right)^2 + \alpha_5 .t\widehat{r}\left(t_i, X_i\right.\end{matrix}\right)\Bigg] \qquad (20.16)$$

is the vector of parameters estimated in Step Two, and $\hat{r}(t, x_i)$ is the predicted value of $r(t, x_i)$ at level t of the treatment. The entire dose–response function can then be obtained by estimating the average potential outcome for each level of the treatment. As the results of GPS analysis are presented graphically, plots of the average dose–response functions and marginal treatment effect functions, defined as derivatives of the corresponding dose–response functions, should be shown graphically.

20.3.3 Data for Measuring the Impacts of Forest Management Practices on Livelihoods

Treatment Variable Duration of membership in PFMP in years is the treatment variable used in the GPS estimation and is a continuous variable.

Outcome variables We used four key livelihood outcome indicators as proxy variables in measuring a household's livelihood. These are the four key livelihood outcome indicators: household per capita consumption expenditure (HPCExp), annual household income from non-timber forest products (IncmNTFP), livestock production (IncmLivstk), and crop production (IncmCrop).

To reduce the mismatching and biased estimators of GPS in the program, we discarded observations with membership duration that stayed for 6 months up to 2 years (mainly recently entered the program), since such short durations arguably do not imply a severe effect on outcome variables. Durations above 12 years are also discarded since very few (5%) observations are available. Numerically, membership duration was assigned to measure the long-term continuous impacts of the outcome variables, and we used 70% of members in duration in programs for this analysis Eddy et al. 2017). During the duration, members learn specific skills needed for harvesting NTFPs, training, and workshops. They obtain different support benefits and incentives through participation in PFMP (e.g. agricultural-aided design for their production) or receive general qualifications for forest-related use.

20.4 Results and Discussion

20.4.1 Perceptions of Households Regarding Forest Resource Deforestation

The forest cover of south-western Ethiopia has been declining at an alarming rate. The survey results revealed that respondents stated that the areas' forest resources are depleting, where tree species, wildlife, NTFP harvesting, and forest coverage are affected. Even though the PFM owns the woods, it failed to manage the forest sustainably and the forest resources. The findings confirm the consensus on forest degradation and deforestation drivers in the Sheka and Kafa forests. As reported in Table 20.4, the respondents observed that the drivers of deforestation were an investment in coffee and tea (78.7%), settlement (81%), illegal cutting (76.7%), harvesting for fuel wood and charcoal (68.6%), and urban expansion (51%).

As reported by Minahan (2000), the growing population demands ever-larger forest supplies, and increasing numbers of people are settling and recreating the primary forest resources for large cities. Today, settlement is the most pressing land use issue affecting forest quality and quantity in the study area due to the increasing demand for the above elements. It is widely argued that forest areas managed by or with rural communities are likely to have lower levels of forest disturbance and improved forest conditions than areas under exclusive state management or open access regimes (Blomley and Iddi 2009).

This result was in agreement with different findings that more than 60% of the forestland covers were changed to investment land (plantation of coffee and tea), agriculture expansion, settlement, fuelwood extraction, illegal logging for timber, and farm tools in forests of south-western Ethiopia (Tadess and Teketay 2017). The ongoing urbanization processes in other regions of Ethiopia significantly push other

Table 20.4 Perceptions of households regarding deforestation

Causes of forest degradation	Perceptions of households (%)
Poor forest management/use	64.2
Agricultural land expansion	46.7
Settlement/population pressure	81.0
Illegal cutting	76.7
Harvesting for fuel wood and charcoal	68.6
Investment (Coffee and Tea)	78.7
Urban expansion in another region of Ethiopia	51
Forest grazing	14.4

causes of deforestation in the Sheka and Kafa forests. This result was supported by a projection made by the United Nations (2015), which indicated that the proportion of the world's population living in urban areas will increase from 54% in 2014 to 66% by 2050. This implies that continued urban development has adverse effects on the natural environment, such as ecosystem destruction and unsustainable wood and food production.

20.4.2 Preferences of Households on the Effects of Existing Forest Conservation

The impact of the current forest conservation efforts in the Sheka and Kafa zones was evaluated through questions centred on respondents' preferences. Approximately, 82% of respondents agreed that the current conservation efforts impacted their harvesting amount. Table 20.5 reports the perceptions of respondents about forest conservation in these zones.

Local perceptions about the effects of forest conservation on the PFM forest area and acceptance of existing management were low among the interviewed groups. The average value of the Likert scale is 4, indicating that respondents residing in the forest areas perceived an undesirable way of conserving the forest (Table 20.5). Insights from key informant interviews and group discussions indicate that most of the respondents from the study areas agreed to manage forests by contributing money to address the improved conservation forest management program.

They were ranking attributes assigned to the planned intervention of forest conservation. Interviewees were asked to rate the predefined features on a five-item Likert scale, ranging from 1 = least important to 5 = most important. The results show that the most critical attributes are improving the non-timber harvesting level (4.4), whereas research and education development (2.3) is the least important. What is apparent from the average Likert scale results is that the respondents preferred forest management option(s) to improve the economic aspects of their livelihoods; it is the backbone of firm incorporation with additional income obtained from harvesting NTFPs for home and livestock consumption. The sample respondents preferred to participate in a management option(s). They were ranked as improving harvesting in NTFPs (4.4), preserving cultural ecosystem services (4.1)

Table 20.5 Perceptions of sample households about forest conservation

Perception	Freq.	Percent	Likert Index
Strongly agree	127	33.2	1.66
Agree	186	48.5	1.936
Indifferent	17	4.4	0.132
Disagree	37	9.6	0.192
Strongly disagree	17	4.4	0.044
	384	100	3.964

and biodiversity protection (3.7), employment opportunity for ecotourism (3.3), and facility of research and education (2.3). They were ranked 1st, 2nd, 3rd, 4th, and 5th, respectively. Furthermore, respondents recognize the positive contribution of forest conservation to harvesting in NTFPs, preserving cultural ecosystem services.

20.4.3 Econometric Results on Values of Forest Conservation Practices

Based on the random parameters logit (RPL) model estimates, all attributes with positive signs (Table 20.6) imply an increase in the choice of these attributes with a probability of choosing improved scenarios. In other words, the respondents gave more weight to the economic aspects of the characteristics and quality of preserving

Table 20.6 Results from the random parameter logit model

Attributes	Coefficients	Std. Err	*p*-value
ASC education	1.080***	0.198	0.000
Payment	−0.079***	0.007	0.000
Medium biodiversity	1.768***	0.345	0.000
High biodiversity	2.512 ***	0.465	0.000
Medium access to reduction	0.499*	0.289	0.084
Higher access reduction	0.499 **	0.289	0.026
Medium ecotourism employment	0.906***	0.293	0.000
High ecotourism employment	2.781***	0.353	0.000
Medium harvesting level NTFPs	1.598***	0.254	0.000
Optimum harvesting level NTFPs	2.400***	0.277	0.000
Moderate Preserve CES	1.180**	0.262	0.003
Aggressive Preserve CES	1.607***	0.383	0.000
ASC*Age	0.274**		
ASC*Sex	0.623*		
ASC*Family size	0.487***		
ASC*EdU	0.773***		
ASC*Income	0.331		
Log-Likelihood	−1415.236		
Pseudo-R^2	0.7064		
Observation	13,824		

Note: ***, **, and * denote significant statistical coefficients at 1%, 5%, and 10%, respectively cultural ecosystem services (CES) of the forest. The positive sign of the NTFP harvesting attribute is consistent with the results of other valuation studies (e.g. Babulo et al. 2009). The CES feature has also been positively favourable and consistent with the results of previous studies (e.g. Nath and Makoto 2010). The estimated results of alternative-specific attributes affect the household's indirect utility choice decisions, whereby forest conservation management attributes are determined by biodiversity, CES, NTFP harvesting, tourism facility, research, and education.

As shown in Table 20.6, the positive signs of the coefficients indicate that improving these attributes can increase the respondents' utility. In other words, estimated coefficients with a positive sign imply that changing from the status quo option to the corresponding attribute increases the probability of choosing improvement plans over the status quo. Analysing the attribute according to the proposed categories of the conservation management attribute result ensures that the magnitude of the coefficient, the economic aspect, the optimum level of harvesting of NTFPs, and the higher level of ecotourism employment are enormously significant. The results also show that, on average, sample households prefer forest management options to enhance the harvesting of NTFPs and ecotourism employment.

Moreover, participating in ecological aspects also had a higher magnitude of positive signs. Cultural ecosystem preservation and improvement in biodiversity are the highest positive FC attributes delivered by forest households in the study areas. Stating simply, a management option that commonly enhances ecosystems and restorations of biodiversity in their localities is the most preferred. The significance of the standard deviation of the parameters for biodiversity conservation, preserving cultural ecosystem services (CES), research and ASC education, ecotourism employment, and NTFP harvesting suggests heterogeneity in preferences for those attributes.

The findings from the random parameter logit model inform the main factors determining respondents' willingness to pay for the forest conservation of the attributes of the Sheka and Kafa forests. Random taste variations among respondents were explained through interaction terms between alternative-specific features and socioeconomic characteristics of the respondents (Adamowicz et al. 1998) (see the lower part of Table 20.6). The results of the interaction variables show that respondents with higher incomes are more likely to choose improvement scenarios because of the positive coefficient of the ASC*Income variable ($p < 0.01$). Similar to Kelbessa and Destoop (2007), older respondents are more likely to select the status quo alternative than younger respondents since the interaction variable of ASC*Age has a significant negative effect ($p < 0.01$). It can be explained that young respondents are exposed to more knowledge about the current situation and environmental problems than older ones; as such, they have better awareness and more willingness to pay for the conservation project. The coefficient of the ASC* where education is significantly positive ($p < 0.01$) shows that respondents with higher education are more likely to support forest conservation, which is similar to the study by Diafas et al. (2017).

From the results, respondents strongly preferred that forest conservation affects their economic benefits and are more likely to contribute to forest management option(s) that would improve the harvesting of NTFPs and ecotourism employment facilities. In addition, in the study areas, the solid preference of respondents for ecotourism employment facilities seems to influence them to participate in forest management option(s), positively affecting their benefit. Additionally, the results suggest that awareness or communication on both existing and improved intervention effects is noteworthy in sustainable forest management. The preference for forest biodiversity and CES services positively affects households' decisions.

In summary, it is highlighted that every family in the study area does not always choose the unity of interest to promote forest conservation. Planning and implementing sustainable forest management options may increase the likelihood of commitment of local community members if preference heterogeneity is considered. Therefore, the results suggest that no single management option is accepted and implemented in the study areas. Instead, integrated and community-specific forest management options (Shackleton et al. 2016; Villegas-palacio 2016) specific to Ethiopia should be pursued. In general, the study areas in particular needed in line with the recently launched (in January 2017) national strategy on forest management, which calls for a coordinated and integrated management approach by different government and nongovernment institutions in the country.

20.4.4 Estimation of Marginal Willingness to Pay

The estimates of marginal WTP for PFM in the study areas are presented in Table 20.7. Using Eq. (20.5), marginal WTP was calculated as the ratio between the mean of the random parameter of the particular forest conservation attribute and the payment coefficient. The standard error of marginal WTP and confidence intervals associated with marginal willingness to pay estimates were obtained using the Krinsky and Robb method (Hole 2007). Using the coefficient of the results of the attribute RPL model, marginal willingness to pay (WTP), which is calculated as the ratio of the coefficients' detailed interest and that of the monetary feature, was estimated by using Eq. (20.5), as shown in Table 20.7. Note also that marginal willingness to pay is the indicative amount of money a customer is willing to pay for a particular feature of a product (i.e. how much a customer is ready to pay for an upgrade from feature A to feature B, in addition to the price the customer pays now).

As seen in Table 20.7, the payments for all attributes are positive and significant, implying that sample households have a positive marginal WTP for an increase in the results of each attribute. For the marginal WTP for each increase in harvesting level from the status quo (US$0) to medium and optimum levels, the respondents are willing to pay Birr 23.37 (US$0.84) and Birr 41.12 (US$1.47) per month, respectively. This indicates that the respondents are willing to pay these prices for them to be allowed to harvest more than they have been harvesting. Some improvement scenarios of forests would be from 0% to 10%, and the significant improvement scenario would be up to 20% at the harvesting level of the forest and preservation with culture.

Table 20.7 Marginal WTP for management alternative attributes per month

Attributes	Birr/month	Std. Err	Confidence interval
Medium biodiversity	21.90	4.045***	(29.83, 13.98)
High biodiversity	31.15	4.951***	(40.85, 21.45)
Medium access reduction	4.58	2.292**	(8.880, 0.104)
Higher access to reduction	10.98	3.930**	(12.10, 2.90)
Medium ecotourism employment	11.47	3.970***	(19.25, 3.69)
High ecotourism employment	35.21	5.491***	(45.97, 24.45)
Medium harvesting level NTFPs	23.37	5.089***	(33.34, 13.39)
Optimum harvesting level NTFPs	41.12	6.874***	(60.59, 33.65)
Moderate preserve CES	14.94	3.702***	(22.19, 7.68)
Aggressive preserve CES	20.34	4.747***	(29.65, 11.04)

Note: *** and ** denote significant statistical coefficients at 1% and 5%, respectively

The marginal WTP results indicate that respondents are willing to pay to harvest more under the new management plans, which also improves forest attributes. This is because they assume that expanding and enhancing the harvesting level of NTFPs would increase their current consumption and better awareness about the benefits of forest conservation attributes.

Based on the results of the best-fitting model (RPL with correlations), the study found that respondents prefer moving away from the status quo and implementing conservation actions in the Sheka and Kafa forests. The program is considered more valuable (respondents are more inclined to accept an alternative) if it offers the most significant improvements for the availability of NTFPs, preserving cultural ecosystem services (CES), ecotourism employment, biodiversity protection, and research and education. The mean monthly marginal WTPs for increasing the availability in harvesting NTFPs to medium and higher levels were estimated at Birr23.37 (US$0.84) and Birr41.12 (US$1.47) per person per month.

On average, households' choice to contribute to managing forests is determined by biodiversity, harvesting in NTFPs, CES services, research and education, and facility of ecotourism employment. Heterogeneity exists for all five attributes of forest conservation in the study areas. Furthermore, it was found that in the study areas, local preference for the positive effects of forests on harvesting NTFP availability significantly affects the likelihood of contributing to forest management option(s). The results also reveal that the program's outcome for each indicator approaches an optimum year of membership doses of 12 years with Birr4800.00 (US$171.43) for annual household per capita consumption spending, 11 years with approximately Birr 3300 (US$117.86) for income from NTFPs, 7 years with approximately Birr 2300 (US$82.14) for crop income, and 12 years with approximately Birr7800 (US$278.57) for livestock income. This finding agrees with an argument that humans play significant roles in different characteristics of forest conservation management, enjoying the benefits, bearing the costs, and making decisions in managing (Gobeze et al. 2009). Local stakeholders' preference for the impacts of interventions plays a fundamental role in sustainably managing forests (Villegas-palacio 2016; UNDP 2012). How local community members understand an intervention's benefits and adverse effects implicitly affects the success of their livelihoods (Jones-Walters and Mulder 2009; Jumbe and Angelsen 2006).

20.4.5 Estimation of Welfare Measures

As noted, economic welfare measurement involves an investigation of the difference between the individual's utilities that could be achieved under the status quo and changed scenario alternatives (Bennett and Blamey 2001). One of the benefits of the choice experiment method is that the estimated coefficients of the attributes can assess the economic welfare improvement due to a particular policy scenario (Ben-Akiva et al. 2002). The results of the CE exercise can further be used to estimate the compensating surplus for a change from the initial situation (status quo) to different improved forest conservation management scenarios. In this study, two alternative methods are proposed: forest management options for improving attributes to some level of progress and significant refinement. They consider the case that the PFM program forest conservation has improved at some improvement levels, that is, a 10% increase in biodiversity, a 10% increase in the preservation of cultural ecosystem services, an increase of 3% in the harvesting level of NTFPs, a 10% increase in access to research and education opportunities and ecotourism employment practices from 15 to 70. The total amount that respondents are willing to contribute to some improvement options is estimated to be Birr 76.26 (Birr 21.9 0 + 4.58 + 11.47 + 23.37 + 14.94) (=US$2.72).

For the significant level of improvements, a 20% increase in biodiversity, a 20% increase in the preservation of cultural ecosystem services, a 6% increase in the harvesting level of NTFPs, a 20% increase in access to research and education opportunities, and 150 ecotourism employment. The total amount that respondents are willing to contribute to the project is estimated to be Birr138.80 (= Birr 31.15 + 10.98 + 35.21 + 41.12 + 20.34) (=US$4.96). The reported marginal WTP for the scenario indicates the amount respondents are willing to pay to experience an improvement in their utility (which results from a movement from the status quo to the changed alternative scenarios). Thus, the respondents are willing to pay Birr76.26 (US$2.72) to significantly improve the attributes from the status quo and Birr138.80 (US$4.96) to improve substantially from the status quo (US$0). Note that the level of forest management in the significant improvement scenario is almost twice the moderate level. In contrast, the level of plantation in the aggressive scenario improvement is five times larger than the average.

20.4.6 Households' Level of Involvement in PFM Activities and Determinants

The analysis undertaken on 12 participatory activities shows that the first three factors have eigenvalues more significant than the first, leading to three elements. Factor analysis of the 12 variables was expected to explain the level of involvement within forestry management and determine the most critical factors affecting the participation of members in forestry management. The rotated factor loadings indicated how variables were weighted for each aspect and showed the correlations between the variables and the elements. All 12 variables were correlated with the selected three factors with correlation coefficient values of 0.5 and above, indicating that factors explained most of the variation in the original variables.

The factor analysis yielded three unique factors for this participation in questions. Rotated factor loadings on the three factors that emerged are presented in Table 20.8. Each aspect clustered a group of related variables that revealed particular types of activities. The first dimension, "Forest governance and decision-making", included characteristics associated with participation activities, program design, policy-making engagement, community service contributions, whether rules to allocate forest benefits are distributed fairly, and capacity-building activities. It accounted for approximately 49.23% of the relative contribution of each factor in terms of shared variance. The second dimension, "forest conservation", included local community involvement in making extraction rules and enforcement rules, preparing tree seeds for planting, and preserving unique parts of the forests and explained approximately 28.91% of the shared variance loading relative contribution of factors.

Based on the benefits and service flow and the various activities related to forest management experienced by the households in the PFMP, reconciling wildlife-human conflicts also had some loading on participation component 3, which accounted for approximately 24.71% of the involvement in the participatory forest management approach. Members living on the edge of forests responded negatively, facing various risks of crop damage, including the occurrence of wildlife-human conflict and extreme events in the interactions between them. Therefore, factor three labelled incentives and resilience as promoting capacities that enable members to be resilient to risks.

Table 20.8 Factor loading of 12 indicators of member involvement

Activities	Governance and decision	Forest conservation	Resilience incentive
Governance and decision-making indicators			
Forest benefits allocation	**0.885**	−0.010	−0.015
Policy-making engagement	**0.742**	0.051	0.111
Community service activities	**0.862**	−0.035	0.061
Elections and PFM meetings	**0.566**	0.060	−0.169
Program design-specific factors	**0.774**	−0.024	−0.0002
Forest conservation			
Making rules of extraction	0.003	**0.824**	−0.063 −0.0685 0.1425 −0.0821
Preparing tree seeds and planting	−0.043 −0.0189 0.0196 −0.0492 0.0686 0.0040	**0.50**	−0.068
Preservation of unique parts of the forest	−0.019 0.0196 −0.0492 0.0686 0.0040	**0.738**	0.143
Controlling illegal forest harvesting	0.020	**0.543**	−0.082
Economic incentives and resilience			
Reconciling wildlife-human conflict.	−0.049	0.121	**0.451**
Compensation for crop damage	0.069	−0.018	**−0.824**
Livelihood incentive scheme	0.004	0.012	**0.737**
Eigenvalue	3.01	1.77	1.52
% of variance	**49.07%**	**28.91%**	**24.71%**

Notes: Kaiser–Meyer–Olkin Measure of Sampling Adequacy = 0.73; Bartlett's Test of Sphericity Approximate Chi-square = 1231.34, which is significant at the 1% level

20.4.7 Determinants of Member's Level of Involvement

Seemingly unrelated regression models were estimated for three dependent variables, forest governance and decision-making, forest conservation, and incentives and resilience, to identify factors affecting the members' involvement. As reported in Table 20.9, regression estimates indicated that demographic, socioeconomic, cooperative membership, administrative contact, membership duration, and institutional factors influenced forest users' participation in the PFM program.

Table 20.9 The seemingly unrelated regression results of the level of participation in PFMP activities

Governance and Economic incentives	Decision-making		Forest Conservation		Resilience	
	Coeff.	Std. Err.	Coeff.	Std. Err.	Coeff.	Std. Err.
Age of the household head	0.005**	0.003	0.008**	0.002	0.011**	0.002
Sex the household head	0.053**	0.050	0.074**	0.041	0.041**	0.044
Education of the household head	0.006	0.006	0.004	0.005	0.003	0.005
Family size of the household	0.033**	0.138	0.334**	0.011	0.018**	0.012
Farm size	0.092	0.050	0.040	0.041	0.033	0.044
Forest distance	0.002	0.001	0.003	0.001	−0.003**	0.001
Access to social network	0.0001	0.016	−0.021	0.014	0.001	0.014
Training facility in management	0.006	0.049	0.409***	0.046	−0.025	0.043
Frequency of extension contact	0.204***	0.057	0.251***	0.046	0.371	0.049
Membership duration	0.027**	0.011	0.007**	0.008	0.001**	0.146
Constant	−0.414**	−0.167	−0.464***	0.137	−0.376	0.147
R^2	0.546		0.475		0.340	

chi2(1) = 82.21***, chi2(2) = 303.00***, chi2(3) = 167.05***; Number of observations = 384, Breusch–Pagan test of independence: chi2(3) = 61.292***
Note: ***, **, * represent levels of significance at 1%, 5%, and 10%, respectively. Source: Own computation result based on survey data (2018)

The results of the SUR model showed that different variables significantly influenced the level of involvement across the three components. The results show that the residuals from the three dependent variables were positively and significantly correlated ($p < 0.01$). The null hypothesis that they are independent was rejected, suggesting that some common underlying factors influenced forest governance and decision-making, conservation, economic incentives, and resilience. As a result, the significant variables in a model are presented and discussed as follows.

- *Age of the household head*: Respondent age is significantly and positively associated with all three factors of level of involvement, that is, forest governance and decision-making, forest conservation, and economic incentives and resilience. This suggests that older respondents tend to be better involved (governance, conservation, financial incentives, and stability) than younger respondents. This result is not surprising, as the elderly have the tradition and custom of conserving the forest for various reasons. For example, there is a tradition of managing cast-based violent conflicts among members inside the forest, which is mainly the responsibility of older people. With the rise in age, a person's status in society to engage in forestry programs is relatively increased, and their degree of governing management becomes more remarkable. The younger population's lower level of involvement in forest governance and decision-making may be because their relatively better physical strength for active labour suggests that young people are less likely to be involved in conservation and governance.
- *Sex of the household head*: The sex of the household head positively influenced people's involvement in forest governance and forest conservation under participatory forests. As shown in Table 20.9, sex is found to significantly impact the participation of PFM members in management, forest conservation, and economic incentives, indicating that males are expected to be more involved in forest governance, protection, and economic incentives than females. Male members were more involved than female members in forest governance, conservation, and economic incentive activities, such as policy-making engagement, enforcement of rules of extraction, controlling illegal forest harvesting, meeting attendance, and participation in meetings.
- *Family size*: Family size was found to be positively and significantly correlated with the level of involvement in the forest conservation activities of the program. This indicates that increased family size could increase forest members' participation in conservation governance and economic activities. According to Gobeze et al. (2009), households with many families are more involved in forest management than smaller ones. Larger families provided more workers. This finding is also consistent with that of Dolisca et al. (2006), who reported that households with fewer members participated less in social forestry activities in Haiti. One possible explanation for this finding is that households with large families have a greater demand for forest products such as firewood and cut grass. Households with large household sizes collect and use more forest products than smaller households because larger homes have more labour and more people to feed the family (Walle and Nayak 2022).

- *Forest distance*: Distance from respondents' homes to the PFM forest area has a significant negative association with economic incentives and resilience. This implies that respondents who live closer to the forest have better involvement than those who live far away. The reason might be that households living in or adjacent areas may frequently incur direct costs associated with wildlife-human conflict and crop damage. Such hazards damage their crops and livestock and occasionally cause loss of human life. A similar problem was reported in several Latin American countries, including Madagascar in Africa (International Union of Forest Research Organization, IUFRO 2010).
- *Extension contact*: A seemingly unrelated regression model estimated that extended exposure was positively and significantly associated with forest governance and protection in PFMP ($p < 0.01$). The positive effect of extension contact on involvement in forest conservation implies that respondents who have regular contact with extension personnel tend to engage in forest conservation more than those with less extension contact. Frequent contact facilitates the flow of information between the extension agent and the members, thereby giving room for forest governance and the conservation of forests. This is because effective extension services could enhance local people's knowledge of sustainable forest resource management, improving their perceptions and attitudes about it and, hence, more participation in community forest management practices. Consistent with this study result, Bakala et al. (2021) revealed that households having access to extension services are expected to participate more in forest management programs. Hence, to increase the local community's participation in the PFM program, information sharing and environmental education are required to enhance community awareness.
- *Membership duration*: It is positively and significantly correlated with the level of involvement in forest governance, forest conservation, and economic incentives. Membership duration indicates that an increase in years of membership duration increases the sense of ownership and could increase the probability of forest members' levels of participation in different activities in the PFM; in support of this study, various studies show that without PFM, deforestation rates may have been much higher (e.g. Kelbessa and Destoop 2007). Additionally, local communities have developed a strong sense of ownership. However, this has not yet been supported by a secure tenure regime and exhibits a more positive attitude toward conserving forests (Tesfaye et al. 2015). This finding is similar to Cottle and Howard (2012), who emphasized that viewing early local involvement leads to less conflict and greater project acceptance. Essential aspects of effective involvement include shared learning and satisfaction with the outcomes of the processes (Table 20.10).

- *Training*: This refers to the training given to members on improved forest conservation practices or controlling illegal forest harvesting. It is a dummy variable that takes a value of one if a member receives training on improved tree seeds, silvicultural techniques, or proper technical assistance; otherwise, it is zero. Participation in training significantly and positively affected the level of involve-

Table 20.10 Dose–response estimation for household per capita consumption expenditure

Consumption Expenditure	Coef.	Std. Err.	*t* value
MDurn	−271.03	179.78	−1.53
MDurn square	37.36***	12.01	3.11
GPS	−8656.27	6454.54	−1.34
GPS square	41,368.4*	22,359.51	1.85.
MDurn *GPS	222.93	441.75	0.54
Intercept	2695.14***	676.46	3.98
Adj R-squared	0.44		
The number of observations	240		

Note: *** and * denote significant statistical coefficients at 1% and 10%, respectively

ment in PFMP ($p < 0.05$). Most PFM members who planted trees on their farms had a high level of participation in PFM. The story of PFM involvement benefits trees growing on farms positively. Regression estimates showed that members' household engagement in forest conservation increases involvement factors. These situations in their environments encouraged the local people to participate in the reforestation project by anticipating immediate improvements. Russell et al. (2013) and Nath and Makoto (2010) identified the existence of synergies between local people's participation and their awareness of natural resource-related problems. Furthermore, Alemayehu et al. (2015) indicated that understanding the aim of management can affect the level of involvement, access to training, and experience-sharing programs and positively influence a person's decision on forest management activities.

The final step of impact analysis using the GPS method is estimating the average DRF, which shows causal inference. The moderate impact of membership duration on household per capita consumption expenditure at a particular year of membership dose was estimated using Eq. (20.6).

Following Hirano and Imbens (2004), the average potential outcome was assessed based on ten values of duration levels 3, 4, 5… 12. The results from GPS estimations indicate that the optimum average annual household per capita consumption expenditure was attained at approximately 12 years of membership duration. The corresponding optimum values of yearly average household per capita consumption expenditures at these optimum years of membership duration are at a dose–response of Birr4800.00 (UD$171.43). The concern here is whether the desired optimum membership duration should be maintained. Forest management reduces transaction costs for its members and improves the coherence of forestry actions at the product scale. Increasing the size of the management unit improves market positioning, permits more technical management, and improves the corresponding household per capita consumption expenditures.

Figure 20.2 shows the results of estimated average dose–response function curves and marginal effects curves characterized by changes in a causal treatment analysis. The shape of the dose–response curves shows increasing marginal effects.

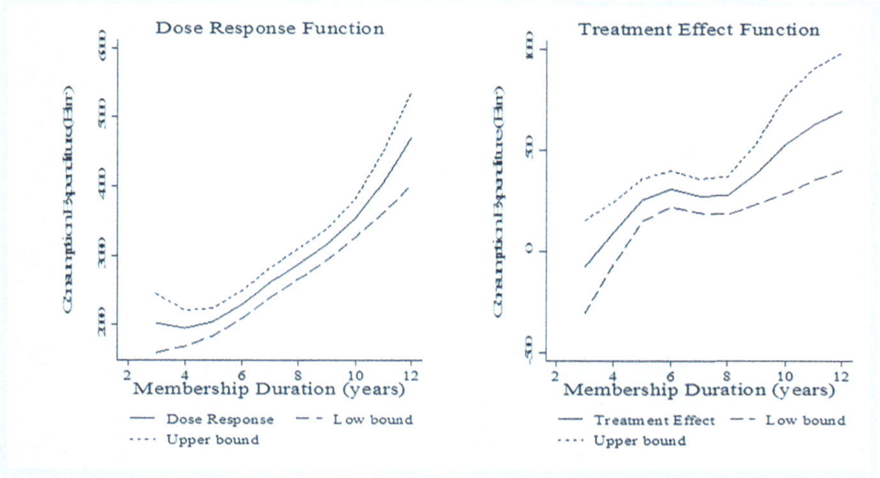

Fig. 20.2 DR consumption expenditure

The values of the results exhibit a marginal impact, with the length of membership in the treatment significantly affecting consumption expenditures. When the number of stay years of members in the program increases, households have a better marginal value of consumption expenditures. The slope of the dose–response curve is flatter at higher levels of treatment. The plot with the first 6 years of membership increases at an increasing rate and reaches the maximum, indicating an association.

The association of causal inference specification for respective outcome variables (Fig. 20.3) in the forest area uses NTFPs for household consumption and as a source of cash income. Individuals who lived in the area for long years of exposure to the program and had excellent experience using NTFPs benefited. Similar to this study, Rai et al. 2016 emphasize that household involvement may have made them more viable and produced more profits because older forest user groups provide more gains to households than more recently established ones. The income from selling NTFPs demonstrates that the forest is vital to household incomes. Most NTFPs (forest coffee, honey, and spices) were collected for sale and contributed 47% of the annual household income. From the right-hand side of Fig. 20.2, it is observed that the maximum marginal effect is attained at a membership duration of 11 years dose with an income level of around Birr 33,000.00 per year.

Annual household income from crop production has a positive causal relationship with membership beyond 4 years (Fig. 20.4). However, the figure shows a negative relationship below this duration dose (38% of the sample households fell). From the right-hand side of Fig. 20.3, it is observed that the maximum marginal effect is attained at a membership duration of 7 years dose with an income of around Birr 23,000.00 (US$822.02). Annual household income from livestock production has a positive causal relationship with membership beyond 4 years (Fig. 20.5). The effect of the entry rate concerning the continuous years of membership duration in

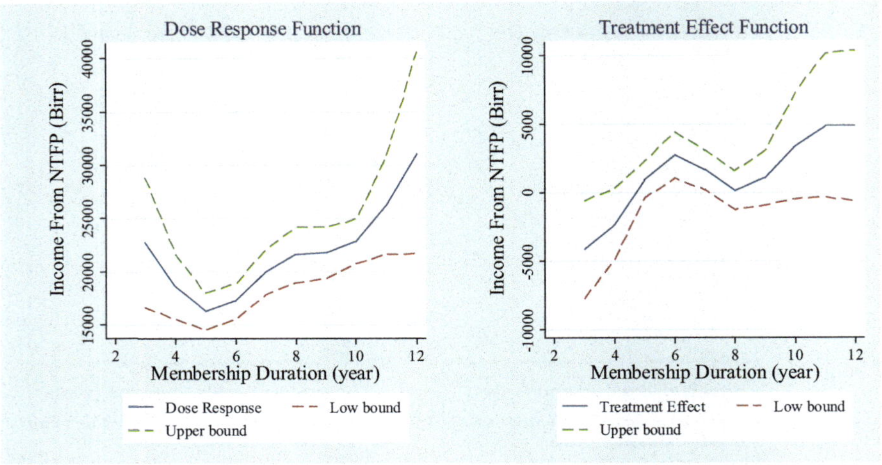

Fig. 20.3 DR income from NTFP

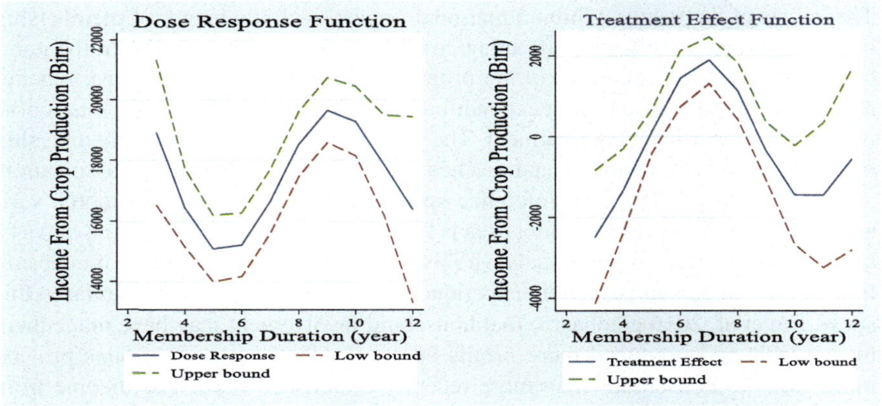

Fig. 20.4 Dose–response crop income

the program was firmly incorporated with additional income obtained from forests. A longer membership duration can survive based on their livestock; they diversify their livelihood by engaging in nonfarm income-generating activities (Getachew et al. 2006). From the right-hand side of Fig. 20.5, it is observed that the maximum marginal effect is attained at a membership duration of 12 years dose with an income level of around Birr7800.00 (US$278.77). In sum, in the study areas, all four livelihood indicators have relationships with membership duration in the PFMP dose. These results confirm and strengthen the impacts of membership duration on members' yearly consumption expenditure, income from NTFP, crop income, and

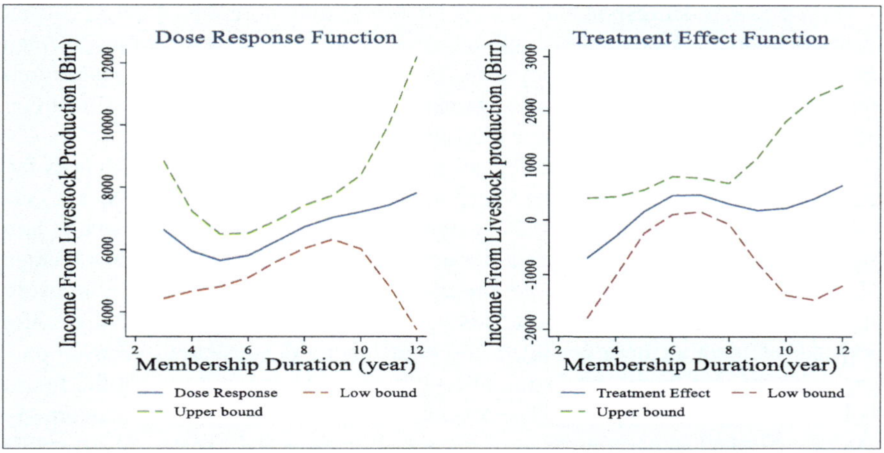

Fig. 20.5 Dose–response livestock income

livestock income. The shape of the dose–response curves shows increasing marginal effects. The values of the results exhibit a marginal impact, with the length of treatment membership significantly affecting livelihoods.

20.5 Conclusion and Recommendations

20.5.1 Conclusion

Policymakers in Ethiopia and elsewhere view the forest sector as essential to ensuring the sustainability of PFM practices and whether the envisioned benefits supplied by the PFM program are economically and ecologically valuable for local communities. The overall objective of the study was to generate information on the economic valuation of forest conservation, level of involvement, and livelihood impacts of participatory forest management practices in the participatory forest areas of the Sheka and Kafa zones of southern nations, nationalities, and people's regional (SNNPR) state of Ethiopia. First, it analysed preferences and their contribution to the valuation of economic, institutional, and ecological aspects in the Kafa and Sheka forests. The results revealed that households are willing to pay Birr76.26 ($2.72) to improve the attributes of the status quo and Birr 138.80 (US$4.96) to improve the status quo significantly. The conditions of conservation interventions show that respondents paid much attention to and supported the proposed forest conservation management practices to conserve and develop forests in southwestern Ethiopia. In addition, the results showed that higher education and income increase the likelihood of selecting the intervention.

In contrast, older respondents choose the status quo more than younger respondents. Second, the factor analysis model was used to identify factors of involvement in the PFM members in the study area. The results identify three basic local community involvements in forest management, that is, forest governance, forest conservation, and perceived resilience incentives.

Third, the seemingly unrelated regression (SUR) model was used to identify factors affecting the level of involvement in PFMP. The results show that age, sex, family size, distance from the forest, extension service, training, membership duration, and social network service are the significant factors affecting the involvement level. Membership durations in the program create a sense of ownership in involving members in PFMP. Fourth, the study attempted to estimate the average causal impacts of member duration in the program on rural households' consumption expenditures and found the optimum length of membership at which the households' benefits are maximized. The results confirm that the program's outcome for the indicator had an optimum year of membership doses of 12 years and a Birr4800 (US$171.43) for annual household per capita consumption spending.

Additionally, there is a strong negative and significant effect of forest distance on perceived resilience incentives in members' households. Members near the edge of forests are highly affected by wildlife conflict and crop damage. The impact of PFMP increases with the length of membership duration. This implies that members of longer membership durations have higher returns on their contribution than those of shorter-duration households. Therefore, social protection through diversified livelihood activities, enterprise implementation, and scaling up the linkages of NTFP marketing are further insightful research needed to absorb the unused potential we have in the participatory forests in the Sheka and Kafa forests.

20.5.2 Recommendations

Based on the results, the following policy recommendations are put forward. First, concerning objective one, the study suggests that designing a set of forest conservation attributes related to the social protection approach would sustainably keep the participating smallholders from conserving the surrounding forest. The economic and ecological values of forests are inefficient. Policy mechanisms are therefore required to make socially optimal forest management and biodiversity conservation. Policymakers should focus on which attributes effectively increase local communities' probability of effectively adopting forest conservation. Second, the study recommends that policymakers and concerned bodies develop strategies so that the program should improve young group and female participation in the forest management of natural resource conservation. The responsible authority should be able to increase households' awareness about the importance of creating incentives or support to compensate for damage. The study recommends creating and promoting a better commercialization environment and supporting the value addition of economically valuable forest products, which is an important option to increase forest product value to local people and thereby enhance the involvement of local people in PFM.

Optimizing the productivity in the participatory activities in study districts, when broadly viewed as enhancing the benefits gained from PFMP operations, is limited in volume due to farmers in the study area. Thus, the study recommends that strict follow-up work be done on the participating forest community members to encourage new incomes and integrate the socioeconomic network more closely with the forest ecosystem and biodiversity. Consider the new entrants to the programs.

Finally, management decisions should consider forest governance sectors of rural households with suitable interventions and built social protection for the local forest community, improve level harvesting and ecotourism employment to enhance agricultural production, particularly crop and livestock production, and alternative livelihood strategies promoting resource-dependent poor households. Nevertheless, suppose all the challenges in participatory forest management are set promptly. In that case, the study suggests promoting forest sectors, specifically in the study area, uplifting and redeeming society from the long-term structural problem of PFM in Ethiopia.

Acknowledgements We very much value the financial support from the Ministry of Education of the Federal Democratic Republic of Ethiopia for the research work, and Mizan-Tepi University for awarding a Ph.D. scholarship to Endale to pursue his studies at Haramaya University, Ethiopia.

References

Adam YO, Eltayeb AM (2016) Forestry decentralization and poverty alleviation: review. Forest Policy Econ. https://doi.org/10.1016/j.forpol.2016.05.009

Adamowicz PC, Boxall WL (2002) Understanding heterogeneous preferences in random utility models: a latent class approach. Environ Resour Econ 23:421–446

Adamowicz W, Boxall P, Williams M, Louviere J (1998) Stated preference approaches for measuring passive use values: choice experiment and contingent valuation. Am J Agric Econ 80:64–75

Adhikari S, Kingi T, Ganesh S (2014) Incentives for community participation in the governance and management of common property resources: the case of community forest management in Nepal. For Policy Econ 44:1–9

Alemayehu N, Ayana N, Vandenabeele, Arts B (2015) Performance of participatory forest management in Ethiopia: institutional arrangement versus local practices. Crit Policy Stud 10:247–253

Babulo B, Muys B, Nega F, Tollens E, Nyssen J, Deckers J, Mathijs E (2009) The economic contribution of forest resource use to rural livelihoods in Tigray, Northern Ethiopia. For Policy Econ 11:109–117

Bakala F, Asfaw M, Tadesse B (2021) Factors influencing household participation in a participatory forest management scheme: Bench-Sheko Zone, Southwest Ethiopia Forest Management Scheme: Bench-Sheko Zone, Southwest. J Sustain For 00(00):1–13

Ben-Akiva M, Lerman SR (1985) Discrete choice analysis: theory and application to travel demand. MIT Press, Cambridge

Ben-Akiva M, Walker J, Adriana T, Bernardino DA, Gopinath TM, Polydoropoulou A (2002) Integration of choice and latent variable models. In: Perpetual motion: travel behavior research opportunities and application challenges. Elsevier, Amsterdam, pp 431–470

Bennett J, Blamey R (2001) The choice modeling approach to environmental valuation. Elgar Publishers, New York. ISBN: 1840643048
Blamey, R.K., Bennett, J.W., Louviere, J. J., Morrison, M.D., and Rolfe, J. (2000). A test of policy labels in environmental choice modelling studies. Ecological Economics. 32(2): 269–286
Blasch E, Yang C, Kada I, Chen G, Bai L (2012) Overview of design of experiments performance assessment of multi-target algorithms. In: Proceedings Volume 8392, Signal processing, sensor fusion, and target recognition XXI, p 839207. https://doi.org/10.1117/12.919031
Blomley T, Iddi S (2009) Participatory forest management in Tanzania: 1993–2009. Lessons learned and experiences to date. Ministry of Natural Resources and Tourism, Forestry and Beekeeping Division, Dar es Salaam
Blomley T, Pfliegner K, Isango J, Zahabu E, Ahrends A, Burgess N (2008) Seeing the wood for the trees: towards an objective assessment of the impact of participatory for management on forest condition in Tanzania. Oryx 42(2):1–12
Cottle MA, Howard TE (2012) Forest policy and economics conflict management and community support for conservation in the northern forest: case studies from Maine. Forest Policy Econ 20:66–71
CSA (Central Statistics Agency of Ethiopia) (2007) Summary and statistical report of population and housing census of 2007. Federal Democratic Republic of Ethiopia Population Census Commission, Addis Ababa
Delibes-Mateos M, Aguilar MAF, Olivero J, Vargas JM (2010) Land-use changes as a critical factor for long-term wild Rabbit conservation in the Iberian Peninsula. Environ Conserv 37(2):169–176. https://doi.org/10.1017/S0376892910000214
Diafas I, Barkmann J, Mburu J (2017) Measurement of bequest value using a non-monetary payment in a choice experiment: the case of improving forest ecosystem services for the benefit of local communities in rural Kenya. Ecol Econ 140:157–165. https://doi.org/10.1016/j.ecolecon.2017.05.006
Dolisca F, Carter DR, Mc Daniel JM, Shannon DA, Jolly CM (2006) Factors influencing farmers' participation in forestry management programs: a case study from Haiti. For Ecol Manag 236(2):324–331
Eddy BN, Dieter O, Vehaest D, Baert S (2017) Does student work really affect educational outcomes? A review of the literature. Discussion paper. IZA Institute of Labour Economics
FAO (2015) Global forest resources assessment: desk reference. Food and Agriculture Organization of the United Nations (FAO), Rome
FAO (Food and Agricultural Organization) (2010a) Global forest resources assessment 2010 (FAO Forestry Paper 163). Rome
FAO (Food and Agricultural Organization) (2010b) Policy on Indigenous and tribal peoples; Working Paper-10. Rome. ISBN 978-92-5-106689-8
FAO (Food and Agriculture Organization) (2020) Global forest resources assessment 2020–key findings. http://www.fao.org/3/CA8753EN/CA8753EN
Gebru GW, Ichoku HE, Phil-Eze PO (2018) Determinants of livelihood diversification strategies in Eastern Tigray Region of Ethiopia. Agric Food Secur 7(1):1–9. https://doi.org/10.1186/s40066-018-0214-0
Getachew A, Amare G, Woldeyesus S (2006) Yield performance and land-use efficiency of barley and faba bean mixed cropping in Ethiopian highlands. Eur J Agron. https://doi.org/10.1016/j.eja.2006.05.002
Gobeze T, Bekele M, Lemenih M, Kassa H (2009) Participatory forest management impacts on livelihood and forest status: the case of Bonga forest in Ethiopia. Int For Rev 11:346–358
Guimarães M, Munguía-Steyer R, Doherty PF, Martins M, Sawaya RJ (2014) Population dynamics of the critically endangered golden lancehead pitviper, *Bothrops insularis*: stability or decline? PLoS One 9(4):e95203. https://doi.org/10.1371/journal.pone.0095203
Hanley N, Wright RE, Koop G (2002) Modelling recreational demand using choice experiment climbing in Scotland. Environ Resour Econ 22:449–466

Hedge R, Bull GQ, Wunder SVEN, Kozak R (2014) Household participation in a Payment for Environmental Services Programme: the Nhambita Forest Carbon Project (Mozambique). Environ Dev Econ 43:1–18
Hensher D, Shore N, Train K (2005) Households' willingness to pay for water service attributes. Environ Resour Econ:509–531. https://doi.org/10.1007/s10640-005-7686-7
Hirano. K., and Imbens, G. W. (2004). The Propensity Score with Continuous Treatments (Chapter 7). In: Applied Bayesian Modeling and Causal Inference from Incomplete-Data Perspectives: An Essential Journey with Donald Rubin's Statistical Family. Editors: Andrew Gelman, Xiao-Li Meng https://doi.org/10.1002/0470090456.ch7. Wiley Series in Probability and Statistics
Hole AR (2007) A comparison of approaches to estimating confidence intervals for willingness to pay measures. Health Econ 16:827–840
Hoyos D (2010) Using discrete choice experiments for environmental valuation. RePEc. https://www.researchgate.net/publication/46450082_Using_discrete_choice_experiments_for_environmental_valuation#fullTextFileContent
IUFRO (International Union of Forest Research Organizations) (2010) Climate change impacts on African forests and people. IUFRO Occasional Paper No. 24, Vienna
Jones-Walters L, Mulder I (2009) Valuing nature: the economics of biodiversity. J Nat Conserv 17:245–247
Jumbe CBL, Angelsen A (2006) Do the poor benefit from devolution policies? Evidence from Malawi's forest co-management program. Land Econ 82:562–581
Kelbessa E, Destoop C (2007) Participatory Forest Management (PFM): biodiversity and livelihoods in Africa. In: Proceedings in international conference, 19–21 Mar 2002, Addis Ababa
Larson AM, Pulhin JM (2012) Enhancing forest tenure reforms through more responsive regulations. Conserv Soc 10:103–113
Louviere JJ, Flynn TN, Carson RT (2010) Discrete choice experiments are not conjoint analysis. J Choice Modell 3(3):57–72. https://doi.org/10.1016/S1755-5345(13)70014-9
McFadden (1974) The measurement of urban travel demand. J Public Econ 3:3030–3328. North-Holland Publishing Company
McFadden D, Train K (2000) Mixed MNL models for discrete response. J Appl Econ 15(5):447–470
Minahan S (2000) Crafting organization. Working paper series, Faculty of Business and Economics, Monash University, Australia
Mislimshoeva B, Herbst P, Koellner T (2016) Current pathways towards good forest government governance for ecosystem services in the former Soviet Republic Tajikistan. Forest Policy Econ 63:67–74
Nath TK, Makoto I (2010) Impacts of participatory forestry on livelihoods of ethnic people experience from Bangladesh. Soc Nat Resour 23:1093–1107
Rai RK, Neupane P, Dhakal A (2016) Is the contribution of community forest users financially efficient? A household level benefit-cost analysis of community forest management in Nepal. Int J Commons 10(1):142–157. https://doi.org/10.18352/ijc.594
Rosenbaum PR, Rubin DB (1983) The Central role of the Propensity score in observational studies for causal effects. Biometrika 70(1):41–55. https://doi.org/10.1093/biomet/70.1.41
Russell R, Guerry AD, Balvanera P, Gould RK, Basurto X, Chan KMA, Klain S, Levine J, Tam J (2013) Humans and nature: how knowing and experiencing nature affect well-being. Annu Rev Environ Resour 38:473–502
Schreckenberg K, Luttrell C (2009) Participatory forest management: a route to poverty reduction? Int For Rev 11:221–238
Shackleton CM, Shackleton SE, Buiten E, Bird N (2016) The importance of dry woodlands and forests in rural livelihoods and poverty alleviation in South Africa. For Policy Econ 9:558–577
Sharma S (1996) Applied multivariate techniques. John Wiley and Sons Inc., New York
Siraj M, Zhang K, Xiao W, Bilal A, Gemechu S, Geda K, Yonas T, Xiaodan L (2018) Does participatory forest management save the remnant forest in Ethiopia? Proc Natl Acad Sci India Sect B Biol Sci 88(1):1–14

Tadess B, Teketay A (2017) Protection: evidence from an aid-funded project in Ethiopia. Environ Manag 10:396–404

Takahashi R, Todo Y (2012) Impact of community-based forest management on forest protection: evidence from an aid-funded project in Ethiopia. J Environ Manag 50(3):396–404

Tesfaye Y, Roos A, Campbell BM, Bohlin F (2015) Livelihood strategies and the role of forest income in participatory-managed forests of Dodola area in the bale highlands, Southern Ethiopia. Forest Policy Econ 13(4):258–265

Thoms CA (2008) Community control of resources and the challenge of improving local livelihoods: a critical examination of community forestry in Nepal. Geol Forum 39:1452–1465

UN (United Nations) (2021) The global forest goals report. United Nations Department of Economic and Social Affairs, United Nations Forum on Forests Secretariat. Available in: 2021/08/Global-Forest-Goal s-Report-2021.pdf

UNDP (2012) United Nations conference on sustainable development, Ethiopia. (Rio+20) National Report of Ethiopia Federal Democratic Republic of Ethiopia Environmental Protection Authority, Addis Ababa

United Nations, Department of Economic and Social Affairs, Population Division (2015) World urbanization prospects: the 2014 revision, (ST/ESA/SER.A/366)

Villegas-palacio C (2016) Lessons from the integrated valuation of ecosystem services in a developing country: three case studies on ecological, socio-cultural and economic valuation. Ecosyst Serv 2:1–12

Walle Y, Nayak D (2022) How do local communities evaluate forest conservation through participatory management; a case of Amhara region, Ethiopia. Int J Glob Environ Issues 20(1):80–99

Wikipidia (2021) Ethiopia, Sheka, and Kafa Zones. https://en.wikipedia.org/wiki/Sheka_Zone

Wooldridge JM (2008) Introductory econometrics: modern approach, 4th edn. South-Western College Pub, Mason

Yemiru T (2011) Participatory forest management for sustainable livelihoods in the Bale Mountains, Southern Ethiopia. Faculty of Forestry. Department of Forest Products. Uppsala Doctoral Thesis, Swedish University of Agricultural Sciences

Dr. Endale Difabachew holds a Diploma, BSc, MSc, and PhD in Agricultural Economics from Haramaya University, Ethiopia. He also graduated with a Diploma in Manufacturing Technology from Adama (former Nazareth) Technical College. He worked for Dire Dawa Food Complex Share Company and served as a mechanic and accountant for 5 years as the head of the market research and promotion division for 4 years, as the head of the plan and statistics services, and as ISO system facilitator for 1 year. He is currently a Lecturer at MizanTepi University, Ethiopia.

Professor Abdi Khalil Edriss holds a dual BA. BSc (Maths & Econ), MSc (Stats), MA, and PhD (Econ), Canada. He is currently a Professor of Applied Economics and Statistics and a Big Data Coordinator for the Africa Centre of Excellence for Agriculture Policy Analysis at the Lilongwe University of Agriculture and Natural Resources, Malawi. He is instrumental in capacity building, mentorship, and postgraduate supervision. He has supervised hundreds of MSc and PhD students across Africa and beyond. Professor Edriss has authored 17 books and published over 180 journal research articles. He pioneered regional postgraduate programs in the COMESA and SADC regions. With his vast international research and teaching experiences, he is the recipient of several awards including the *Quality of Life Award* from the Association of Commonwealth Universities, as well as, World Agricultural Research Champion Awards.

Professor Jema Haji Mohamed holds a BSc and MSc (Mathematics) from Addis Ababa University and MSc and PhD (Economics) from the Swedish University of Agricultural Sciences (SLU). He is currently a Professor of Agricultural Economics at Haramaya University, Ethiopia. He has supervised over 350 MSc and 45 PhD students, as well as internationally published over 120 journal research articles. He is involved in teaching postgraduate students in COMESA and

SADC regions. He was the Students Research Director, SMIP Capacity Builder, Associate Dean, Faculty of Education, and Head of the Agricultural Economics Department at Haramaya University, Ethiopia.

Dr. Belaineh Legesse holds a BSc and an MSc in Agricultural Economics from Haramaya University, as well as a PhD in Rural Development from the Swedish University of Agricultural Sciences. He was the Dean of the Faculty of Continuing and Distance Education, Vice President for Business and Development, and Vice President for Administration and Student Services at Haramaya University, Ethiopia. Dr. Legesse is an Associate Professor of Agricultural Economics and has published several research articles in reputable international journals. He supervised several postgraduate students, as well as served as a board member in several institutes across Africa. He is currently a Managing Partner at *Frontieri Consult PLC*, Ethiopia.

Professor Mengistu Ketema holds a BSc and MSc in Agricultural Economics from Haramaya University (Ethiopia), as well as a PhD in Agricultural Economics from Justus-Liebig University of Giessen, Germany. He worked as a Researcher and Head of the Socioeconomics Research Division at *Sinana* Agricultural Research Center, Ethiopia. He is currently a Professor of Agricultural and Resource Economics and Chief Executive Officer for the Ethiopian Economics Association. Professor Ketema has supervised over 70 MSc and 20 PhD students, as well as published 7 book chapters, 25 Journal articles, and over 20 proceedings. He is a member of the Econometrics Society, the Agricultural Economics Society of Ethiopia, and *Tropentag* (Society of Tropical Agriculture).

Chapter 21
Conclusion: Pathways for a Sustainable Future—*Synthesis on Environmental Change and Biodiversity Conservation Efforts in Sub-Saharan Africa*

Victor K. Muposhi, Wame L. Hambira, Ditiro B. Moalafhi, Ayana Angassa Abdeta, and Ketlhatlogile Mosepele

21.1 Overview

Sub-Saharan Africa's biodiversity is critical to global conservation efforts due to its unique and diverse ecosystems that host a significant proportion of the world's flora and fauna, many of which are endemic and threatened (Leisher et al. 2022; Okello et al. 2021). The region plays a critical role in sustaining global biodiversity, contributing to climate regulation, ecosystem services, and supporting local livelihoods and cultural heritage, thus necessitating urgent and concerted conservation actions (Saliu et al. 2023). This edited volume, 'Environmental Change and Biodiversity Conservation in Sub-Saharan Africa', offers a comprehensive overview of the region's conservation challenges and natural dynamics within the continent's diverse ecosystems, framed around the dual context of biodiversity conservation amidst the Anthropocene's challenges and the pathways towards sustainable management of forest and rangeland ecosystems. It highlights the necessity of integrating contemporary conservation science with traditional knowledge and developing adaptive strategies for tackling the complex effects. The synthesis underscores the critical need for collective action in biodiversity conservation, essential for ecosystem sustainability and the socio-economic resilience of communities across sub-Saharan Africa. This collective work emphasizes the need for a holistic approach that integrates the science-policy interface with local community engagement to navigate the complexities of climate change, human activity, and land-use dynamics in an era of unprecedented change.

V. K. Muposhi (✉) · W. L. Hambira · D. B. Moalafhi · A. A. Abdeta · K. Mosepele
Botswana University of Agriculture and Natural Resources, Gaborone, Botswana
e-mail: vmuposhi@buan.ac.bw; whambira@buan.ac.bw; dmoalafhi@buan.ac.bw; aabdeta@buan.ac.bw; kmosepele@buan.ac.bw

W. L. Hambira et al. (eds.), *Environmental Change and Biodiversity Conservation in sub-Saharan Africa*, Advances in Global Change Research 75,
https://doi.org/10.1007/978-3-031-73136-5_21

21.2 Biodiversity Conservation in the Anthropocene

The Anthropocene presents significant challenges for biodiversity conservation. The complex interplay between environmental changes, habitat loss, and species extinction necessitates innovative solutions to protect the Earth's biodiversity (Caro et al. 2014; Corlett 2015). The first part of this volume offers a comprehensive and nuanced exploration of the various factors impacting biodiversity within the current geological epoch, characterized by pronounced human influence on the planet's geology and ecosystems. This section meticulously unpacks the intricacies of these challenges, drawing attention for its methodical integration of diverse disciplinary perspectives. It is noteworthy for its integration of various disciplinary viewpoints and comprehensive approach, combining ecological, socio-economic, and technological perspectives to address the multifaceted nature of biodiversity conservation in today's world. Each chapter contributes profoundly to our comprehension by presenting a broad range of research findings and conservation strategies that underscore the urgency and feasibility of safeguarding biodiversity amidst escalating anthropogenic pressures (Upreti 2023).

Chapter 2 explores the complex and seemingly contradictory relationship between nature-based tourism (NBT) and biodiversity conservation efforts in sub-Saharan Africa (Jones 2022). The authors undertake a comprehensive meta-synthesis, demonstrating that NBT, although economically advantageous, unintentionally contributes to biodiversity depletion through mechanisms such as climate change, land-use modification, and biological invasion. This chapter is particularly significant due to the growing reliance on tourism for economic development in numerous African countries (Simo-Kengne 2022; Gidebo 2023). The contrast between the economic benefits of tourism and its ecological costs presents a persuasive argument for re-evaluating current tourism practices to better align with sustainability goals (Samal and Dash 2023). In Chap. 3, the authors introduce Nexus Thinking as an innovative framework for navigating the complex interactions among biodiversity conservation, human livelihoods, and sustainable development. Highlighting the importance of systemic thinking and localized initiatives, this approach is especially pertinent to sub-Saharan Africa, given its abundant ecological resources (Oteng-Abayie et al. 2022). The chapter's core strength is its advocacy for a holistic strategy that acknowledges the interconnectedness of ecological and socio-economic systems, underlining the imperative for innovative and inclusive sustainable solutions to address the biodiversity crisis (Solangi and Jianguo 2023; Tartaruga et al. 2024).

In Chap. 4, the authors provide critical insights into the migratory patterns of the Plains zebra in Botswana, emphasizing the impact of climate change on these movements. Migratory corridors and dispersal routes are fundamental components in the conservation of species within human-mediated ecosystems, acting as vital conduits for genetic exchange, seasonal migration, and species dispersal (Joly et al. 2021; Kauffman et al. 2021b). These pathways facilitate the movement of species across fragmented landscapes, mitigating the impacts of habitat loss, environmental change, and spatial isolation (Kauffman et al. 2021a). In the context of increasing anthropogenic pressures, the identification, protection, and restoration of these ecological networks are crucial for maintaining biodiversity, ecosystem resilience, and the provision of ecosystem services (Xu et al. 2021). The integration of wildlife movement data into spatial planning is pivotal for facilitating connectivity crucial to species' adaptation and resilience against environmental changes, thereby ensuring the preservation of biodiversity and ecosystem health (Allen and Singh 2016; Beger et al. 2022; Jeltsch et al. 2013). It is therefore critical to acknowledge the significance of incorporating animal movement patterns into conservation planning to maintain ecological networks essential for species survival and ecosystem functionality in a changing world.

Similarly, Chap. 5 examines the spatial ecology of roan antelope in human-mediated ecosystems using geospatial modelling techniques. The chapter demonstrates how advancements in the spatial ecology of wildlife species within human-mediated ecosystems have been markedly propelled by cutting-edge modelling techniques (Klappstein et al. 2022; Northrup et al. 2022; Serva et al. 2024). Satellite and GPS telemetry, remote sensing, and machine learning techniques offer advanced tools and approaches for monitoring wildlife responses to anthropogenic disturbances, habitat alterations, and climate change-induced habitat shifts (Adams-Hosking et al. 2011; Rumiano et al. 2020; Wilson et al. 2020; Wobler 2023). Overall, these approaches enhance our understanding of the complex interactions between animal movement, habitat selection, and the influence of human activities (Delaney and Larson 2024; Osterhout et al. 2024; Wessely et al. 2017). The chapter further illustrates how anthropogenic disturbances impedes natural habitat selection and utilization, forcing wildlife species towards suboptimal habitats (Keeley et al. 2017; Tablado and Jenni 2017). This shift not only disrupts their ecological roles but also risks creating ecological traps where preferred habitats become maladaptive (Joly et al. 2021; Merkle et al. 2022; Morrison et al. 2021; Zuñiga-Palacios et al. 2021). Therefore, the integration of geospatial modelling in studying the spatial ecology of wildlife species in human-mediated landscapes provides critical insights into the mechanisms driving species habitat selection and associated conservation. This underscores the need for adaptive management strategies that consider the dynamic nature of ecosystems and human impacts in the Anthropocene.

Area enclosures within urban ecosystems play a pivotal role in conserving plant species by creating microhabitats that shield flora from urban stressors such as pollution, trampling, and invasive species (Lindley et al. 2018; Marvin and Rutherford 2018; Rodenbiker 2020). In Chap. 6, the authors use a botanical garden as a case study to examine the effectiveness of area enclosures in preserving plant species within urban ecosystems. These designated spaces not only facilitate the survival and proliferation of native plant species but also contribute to biodiversity, ecological stability, and green infrastructure in metropolitan areas (Cardone et al. 2023; Collins et al. 2017). They serve as integral components of urban planning, aligning with climate action goals by enhancing urban biodiversity, sequestering carbon, and improving air quality (Butt et al. 2018; Girma et al. 2019). These enclosures provide strategic approaches to integrating nature with urban development through green spaces. Therefore, legislation should mandate the integration of green spaces in urban planning, as they are crucial for sustainable and resilient cities that value ecological health and human well-being.

Chapter 7 offers an elaborate examination of fish community stability within the seasonal floodplain of the Okavango Delta, leveraging advanced sampling and analysis approaches. The authors demonstrate that in the seasonal floodplains such as the Okavango Delta, fish community stability is intrinsically linked to the hydrological pulse, which dictates habitat availability, food resources, and breeding opportunities, thus driving species diversity and abundance (Bokhutlo et al. 2021; Mosepele et al. 2017). The stability of fish communities within seasonal floodplains serves as a crucial indicator for ecological health, reflecting the resilience of these systems to environmental variability and anthropogenic pressures (Borgwardt et al. 2019; McCluney et al. 2014; Rockström et al. 2014). By understanding these aquatic ecosystem dynamics, we can infer that similar floodplain ecosystems globally might exhibit parallel patterns of community stability, where the predictability of water cycles supports resilient and biodiverse aquatic communities (Abrial et al. 2019; Siziba et al. 2013). This principle underscores the importance of integrated, ecosystem-based approaches for sustaining fish biodiversity in floodplain environments as well as supporting local livelihoods dependent on these aquatic resources (Lynch et al. 2023; Mosepele et al. 2022). To safeguard the ecological integrity and socio-economic value of floodplain environments worldwide, it is imperative to adopt and prioritize integrated, ecosystem-based management strategies that maintain hydrological regimes, support biodiversity, and uphold community livelihoods, drawing on insights from the Okavango Delta's fish community dynamics.

With a different perspective, the authors in Chap. 9 explore the social media backlash and calls for tourism boycotts following Botswana's decision to lift its trophy hunting ban, highlighting significant concerns over the impact on the country's tourism and conservation efforts. The analysis reveals polarized sentiments, with critics arguing that the decision undermines Botswana's conservation image and proponents asserting the economic and conflict mitigation benefits. The study underscores the need for a balanced approach to conservation that integrates environmental, social, and economic sustainability, informed by empirical data and scientific evidence.

In Chap. 8, the authors evaluate community participation in Mozambique's natural resource management landscape, highlighting the complexity of community-led conservation initiatives. The chapter underscores the significance of adopting inclusive conservation models that effectively integrate all stakeholders, encompassing the private sector, to ensure the sustainability and success of conservation efforts (Berkes 2004; Kashwan et al. 2021; Raymond et al. 2022). Community-led conservation initiatives present a complex interplay of ecological, social, and economic factors that require nuanced understanding and adaptable frameworks (Armitage et al. 2012). These initiatives present opportunities for biodiversity conservation and community empowerment as has been observed and advocated for in other sub-Saharan countries (Gatiso et al. 2022; Heffernan 2022; Petriello et al. 2021).

However, aligning conservation and livelihood goals, securing sustainable funding, and managing governance complexities are crucial challenges for their long-term success. This complexity necessitates robust, multidisciplinary frameworks that are responsive to changing ecological and social landscapes, ensuring conservation efforts are both effective and equitable (Oldekop et al. 2016). We argue and conclude that the complexity of these efforts further underscores the importance of integrated approaches that leverage local knowledge, foster stakeholder collaboration, and prioritize adaptive management to navigate the multifaceted dimensions of conservation and sustainability. Therefore, in the Anthropocene, integrating sustainability issues into community-led conservation initiatives is imperative to address the interconnected challenges of biodiversity loss, climate change, and human well-being.

21.2.1 Emerging Themes on Biodiversity Conservation in the Anthropocene

The chapters in Part I are intricately linked, forming a cohesive narrative that underscores the complex interplay between human activities and biodiversity conservation in the Anthropocene (Fig. 21.1). Here, we present a synthesis of the broad themes which came out strongly through the chapters.

1. *Human Influence and Ecological Impact*

   The profound impact of human activities on biodiversity is evident through Chaps. 2, 3, and 5. Chapter 2 discusses the possible effects of nature-based tourism in sub-Saharan Africa, where economic benefits come at the cost of ecological disturbances. This relates to the Nexus Thinking, highlighted in Chap. 3 emphasizing the necessity of a holistic approach that harmonizes human socio-economic needs with ecological integrity and ecosystem health. Chapter 5 further explores how anthropogenic factors shape animal habitat selection and spatial behaviour. We argue that in socio-ecological systems, adaptive management is crucial for sustainability, especially within human-dominated landscapes. This approach enables the iterative learning from outcomes to inform future actions, ensuring resilience and the ability to sustain ecological functions and services amidst anthropogenic pressures.
2. *Climate Change and Migration Patterns*

   Climate change emerges as a critical factor influencing biodiversity as detailed in Chaps. 4 and 7. Chapter 4 correlates climate-induced habitat alterations with changes in the Plains zebra migratory patterns in Botswana. This underscores the need to safeguard dispersal routes and corridors for species resilience and survival. In contrast, Chap. 7 explores the impact of climate variability on fish community stability in the Okavango Delta, stressing the importance of grasping ecological dynamics amidst environmental fluctuations.

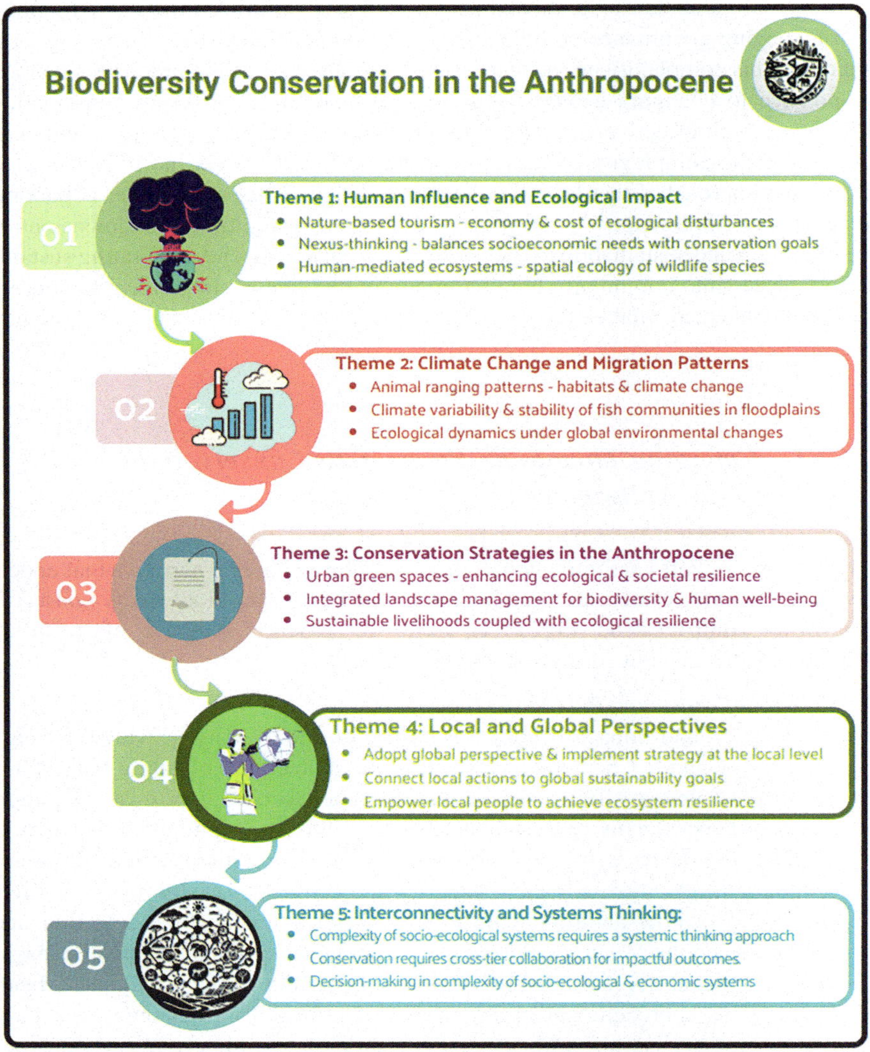

Fig. 21.1 Broad thematic areas on biodiversity conservation in the Anthropocene

3. *Conservation Strategies in the Anthropocene*

   Chapters 4, 5, 6, and 8 collectively underscore the significance of innovative conservation strategies in the Anthropocene. These include leveraging technology for comprehensive ecological monitoring to bolster species and habitat conservation efforts and emphasizing community—approaches as crucial for harnessing local expertise and fostering sustainable ecosystem-based socio-economic balances. Additionally, urban green spaces act as carbon sinks and potential biodiversity hotspots, highlighting the need for creatively integrating nature into urban planning to enhance conservation efforts and human well-being.

4. *Local and Global Perspectives*

   Effectively tackling conservation challenges requires adopting a global perspective while actively implementing strategies locally to align with overarching sustainability goals. This perspective of 'thinking globally, acting locally' has been highlighted through Chaps. 3, 4, 7, and 8. By connecting local actions to these broader objectives, we can enhance biodiversity conservation across various scales. Similarly, empowering local communities is essential for ecosystem resilience and achieving sustainability goals, demonstrating a coherent strategy for global conservation efforts.

5. *Interconnectivity and Systems Thinking*

   Part I of this volume consistently emphasizes the interdependence of ecological, social, and economic systems, advocating for a systemic approach to conservation. Nexus Thinking in Chap. 3 explicitly promotes this perspective, which echoes across discussions on tourism's biodiversity impact (Chap. 2), animal movement in human-modified landscapes (Chaps. 4 and 5), and community-led conservation (Chap. 8).

21.3 Sustainable Management of Forest and Rangeland Ecosystems

Part II of this volume underscores the dynamic interplay between human activities and forest ecosystems in sub-Saharan Africa, highlighting the need for innovative, sustainable management approaches. In Chap. 10, the authors explore the e-extension model and its utility in enhancing sustainable forest management. Information and Communication Technology (ICT) plays a pivotal role in sustainable forest management by facilitating real-time monitoring and data analysis of forest conditions, enabling proactive conservation measures (Gavilanes Montoya et al. 2023). ICT empowers rural communities through access to market information and sustainable practices, significantly improving their livelihoods and economic resilience (Zhao et al. 2024). Moreover, ICT tools aid in the dissemination of knowledge on sustainable agriculture and forestry, fostering community engagement and participation in conservation efforts, thereby ensuring the long-term health of forest ecosystems (Brown 2021). Strengthening rural digital infrastructure and promoting digital forestry models are crucial. There is a need to escalate efforts to nurture specialized digital skills in rural areas to boost farmers' digital literacy, and foster ecological values and sustainable practices among them, enhancing community livelihoods and well-being.

In Chap. 11, the authors provide insights into the anthropogenic impacts on the Okavango Delta's forests, using Normalized Difference Vegetation Index (NDVI and Land Surface Temperature (LST) metrics. Utilizing NDVI and LST metrics offers an insightful perspective on the complex interplay between vegetation cover and human activities, particularly in forest ecosystems (Shrama et al. 2022; Soubry et al. 2021). NDVI facilitates the monitoring and management of forest ecosystems by identifying changes in vegetation health and temperature anomalies that may indicate unsustainable land use or climate change impacts (Pandey et al. 2022). By applying these metrics, forest managers can implement targeted conservation strategies, such as reforestation or controlled burns, to maintain ecosystem balance and biodiversity. This approach offers a data-driven framework for sustainable ecosystem management and serves as a replicable model for other regions grappling with the impacts of human activities on natural ecosystems.

Chapter 12 illustrates how Ethiopia's coffee forests characterize the widespread challenges of biodiversity conservation and ecosystem service provision on the continent. Threatened by agricultural encroachment and climate change (Abera et al. 2021), these ecosystems underscore the necessity for participatory management strategies that reconcile human livelihoods with ecological integrity. Scaling up agroforestry practices and ecological restoration initiatives is essential at enhancing both carbon sequestration and biodiversity conservation. Therefore, there is a need to implement adaptive co-management practices, which integrate traditional knowledge with modern conservation techniques could offer robust solutions to mitigate impacts, enhance resilience, and sustain ecosystem services (Notaro et al. 2022). More so, establishing payments for ecosystem services (PES) schemes could incentivize conservation by directly linking the well-being of local communities with the health of these vital ecosystems (González-Chaves et al. 2023; Schultner et al. 2021). These efforts could be achieved by promoting landscape-level planning and fostering international collaboration to achieve sustainability across these vital coffee forest ecosystems.

Forest fires not only result in the loss of biodiversity and habitat destruction but also disrupt the vital ecological processes that sustain ecosystem services. In Chap. 13, the authors highlight the urgent need for integrating advanced technologies and comprehensive policies into management strategies to mitigate the adverse effects of forest fires. It is crucial to adopt proactive measures to align human management techniques with fire disturbance regimes to promote ecological resilience (Sample et al. 2022; Wollstein et al. 2022). Addressing this challenge requires a multifaceted strategy that includes community engagement in fire prevention (Manley et al. 2024), the development of early warning systems, and the application of sustainable land management practices that reduce fire risk (Shuman et al. 2022).

Chapter 14 highlights the critical challenge of woody plant encroachment in arid rangelands, underscoring its detrimental impact on biodiversity conservation and rangeland productivity. It has detrimental effects on ecosystem structure, function, and biodiversity. Woody plant encroachment has far-reaching consequences, poses significant challenges to rangeland health, resilience, and the livelihoods of communities dependent on these ecosystems (Geißler et al. 2024). To address woody plant encroachment in arid rangelands, policymakers should prioritize implementing integrated management approaches that combine mechanical, chemical, and biological control methods (Mudau et al. 2022; Musekiwa et al. 2022). Similarly,

Chap. 18 evaluates bush encroachment control measures and their impacts on herbaceous plant species composition in a semi-arid savannah ecosystem. It profiles practical insights into the management of these ecosystems for enhanced biodiversity and grazing value. We suggest that fostering collaboration among stakeholders, including landowners, researchers, and government agencies, is essential for developing and implementing effective strategies tailored to local ecological contexts. Policymakers can foster sustainable land management practices in arid rangeland regions by investing in research, monitoring, and capacity building. This support can help mitigate the adverse impacts of woody plant encroachment while simultaneously enhancing ecosystem resilience and socio-economic well-being. Chapter 15's investigation into seed germination treatments for native species in Botswana complements the discourse on woody plant encroachment by highlighting sustainable management practices. By emphasizing the significance of native plant regeneration processes, this study contributes to ecosystem restoration efforts, offering potential solutions to mitigate the impacts of woody encroachment and promote rangeland resilience.

The impacts of invasive alien species on rangelands were examined in Chap. 16. Invasive species have detrimental impacts ranging from displacing native vegetation to altering ecosystem dynamics and degrading habitat quality. Due to these impacts, they pose significant threats to biodiversity, ecosystem services, and the livelihoods of communities dependent on rangeland resources (Yapi et al. 2023). Addressing these challenges requires interdisciplinary collaboration, informed by robust research and monitoring efforts (Bennett et al. 2023). By implementing measures focused on prevention, early detection, and control of invasive species, stakeholders can safeguard rangeland integrity, bolster resilience, and sustain the ecological and socio-economic functions of these vital landscapes for future generations.

Chapter 17 provides perspective long-term vegetation cover changes amidst evolving climatic and anthropogenic pressures using NDVI trends in Botswana. Whereas Chaps. 19 and 20 explored the multifaceted role of non-timber forest products and participatory forest management in enhancing the livelihoods of marginalized communities. The land cover change dynamics underlines the complex interplay between climate change, human impacts, and the dynamics of natural ecosystems (Malhi et al. 2020; Ortiz et al. 2021). As we observe shifts in vegetation cover over time, it becomes increasingly evident that proactive measures are essential to mitigate the adverse effects of these changes. Policymakers and land managers must prioritize the implementation of adaptive management strategies that account for both climatic variability and anthropogenic pressures, ensuring the preservation of biodiversity and ecosystem services (Kariuki et al. 2021; Roy et al. 2022).

Moreover, the potential of non-timber forest products (NTFPs) to support sustainable livelihoods highlights the importance of integrating community-centred approaches into forest management practices (Gurung et al. 2021; Meinhold et al. 2022). By empowering marginalized communities to actively participate in decision-making processes and benefit from the sustainable use of forest resources (Piabuo et al. 2022; Tieminie et al. 2021), we can foster socio-economic resilience and environmental stewardship at the local level. It is imperative that stakeholders across sectors collaborate to develop inclusive policies and management frameworks that prioritize the conservation of natural resources while promoting socio-economic well-being. However, this requires a concerted effort to invest in capacity building, facilitate knowledge exchange, and support community-led initiatives aimed at achieving holistic and equitable outcomes.

21.3.1 Complex Interdependence of Ecological, Sociological, and Economic Systems

Part II of this volume highlights the urgent need for innovative, integrated, and participatory management and conservation strategies for forest and rangeland ecosystems in sub-Saharan Africa. The chapters provide key insights for policymakers, conservationists, and local communities striving to harmonize ecological integrity with socio-economic growth.

1. *Technological Innovation and Environmental Monitoring*: Chapters 11 and 12 highlight the critical role of technological advancements in sustainable forest management. The e-extension model proposed in Chap. 11 exemplifies how ICT can bridge the knowledge gap in forest resource utilization, thus fostering sustainable practices that enhance rural livelihoods. Similarly, Chap. 12 showcases the application of remote sensing and GIS technologies, particularly through NDVI and LST metrics, to monitor and understand anthropogenic impacts on forest ecosystems. These technological approaches are crucial for real-time, accurate assessment and management of natural resources, facilitating informed decision-making.
2. *Anthropogenic Impacts and Adaptive Management*: The chapters underscore the pervasive influence of human activities on forest ecosystems. Chapter 14 focuses on the impacts of forest fires in Botswana, while Chap. 13 discusses the threats to Ethiopia's coffee forests from agricultural expansion and climate change. These examples illustrate the need for adaptive management strategies that consider the ecological, economic, and social dimensions of environmental change. The chapters advocate for policies that balance conservation goals with human needs, highlighting the necessity of integrating scientific knowledge with local expertise.
3. *Participatory Management and Community Engagement*: Chapters 13, 19, 20, and 21 emphasize the importance of community involvement in forest management. These chapters advocate for participatory approaches, where local communities play an active role in decision-making processes. This is particularly evident in the context of the coffee forests of Ethiopia and the bush encroachment studies in savannah ecosystems. Engaging communities not only fosters a sense of ownership and responsibility but also leverages local knowledge and practices that are vital for the sustainable use and conservation of natural resources.

4. *Biodiversity, Ecosystem Services, and Invasive Species Management*: The chapters collectively highlight the intricate relationship between biodiversity, ecosystem services, and human well-being. Chapter 17's focus on invasive alien species in Botswana exemplifies the challenges posed to biodiversity and ecosystem services. The discussions in Chapters 15 and 19 on woody plant encroachment and its control further illustrate how changes in ecosystem composition can affect landscape structure, biodiversity, and the value of ecosystem services. These themes underscore the necessity of conserving biodiversity not only for its intrinsic value but also for the essential services it provides to human societies.
5. *Economic Aspects and Sustainable Livelihoods*: Finally, Chaps. 20 and 21 delve into the economic aspects of forest resources, particularly non-timber forest products (NTFPs) and participatory forest management practices. These chapters demonstrate how sustainable management of forest resources can contribute to poverty alleviation, food security, and economic resilience of marginalized communities. They also highlight the potential of economic valuation and market-based mechanisms in promoting conservation and sustainable use of forest resources.

The integration of technological innovation, understanding of anthropogenic impacts, participatory management, focus on biodiversity and ecosystem services, and emphasis on economic aspects provides a multifaceted approach to environmental conservation and sustainable livelihoods. This volume illustrates the multidisciplinary nature of the approaches requisite for addressing conservation challenges through the lens of complex systems theory, recognizing the intricate interdependence of ecological, social, and economic systems (Fig. 21.2). We outline four key aspects broadly emphasized across chapters:

1. *Complex Systems Understanding*: Emphasizes the importance of acknowledging the interconnected nature of ecosystems, human communities, and economic activities. Recognizing this interconnectedness is critical for developing long-term, effective, and sustainable conservation strategies.
2. *Enhanced Stakeholder Engagement*: Stresses the integration of social and economic contexts into conservation strategies to boost local community and stakeholder engagement, thereby improving prospects for successful and sustainable outcomes.
3. *Sustainability and Resilience*: Advocates for a systemic approach that safeguards natural resources and ensures their sustainable utilization, thus enhancing ecosystem resilience. This entails fostering strategies that promote biodiversity conservation while concurrently advancing human well-being, addressing the root causes of biodiversity loss and ecosystem degradation.
4. *Adaptive Management*: Highlights the need to understand the interconnectedness of systems to enable adaptive management practices that can respond to changing conditions, such as climate change and socio-economic shifts. This approach values flexibility and is essential for maintaining ecosystem health amidst global changes.

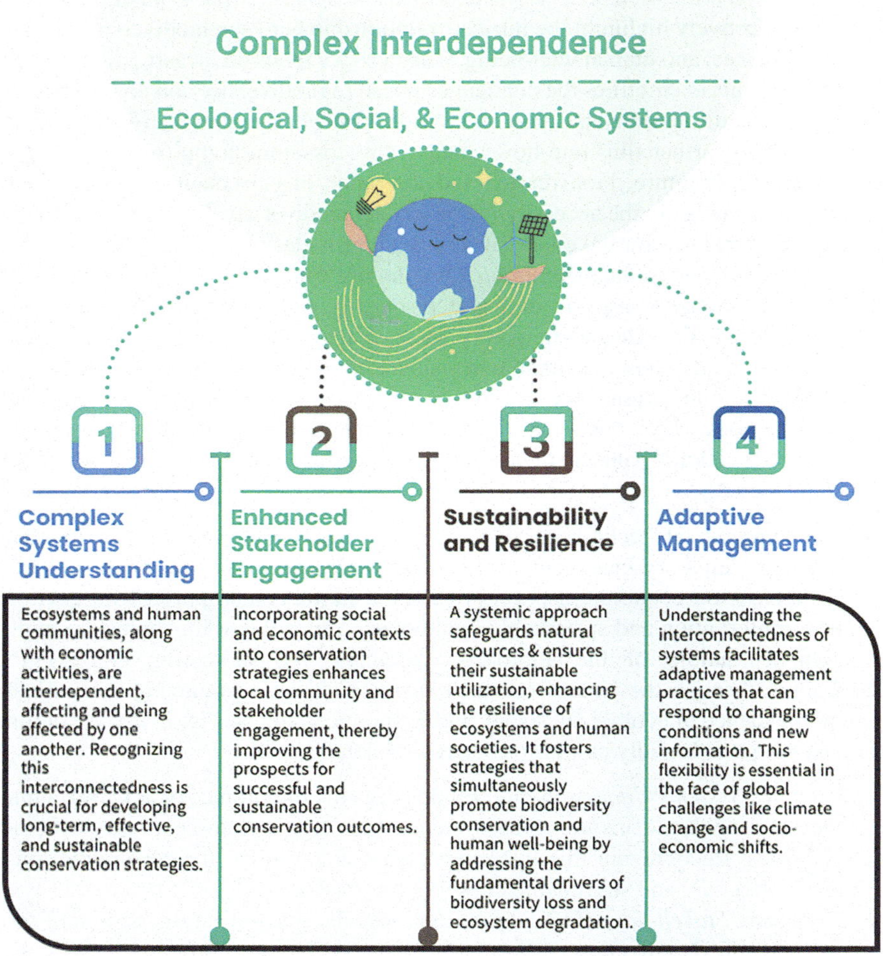

Fig. 21.2 Complex interdependence of ecological, sociological, and economic systems underpinning conservation efforts

21.4 Strategic Pathways to a Sustainable and Resilient Future in Sub-Saharan Africa

Effective environmental policy and conservation action rely on strong governance and institutions. To align with global sustainability efforts, sub-Saharan Africa needs to strengthen governance frameworks to address both direct and indirect drivers of biodiversity loss. This involves establishing transparent, accountable, and participatory institutions at all levels that can effectively implement, monitor, and enforce conservation action plans. The establishment of multi-level governance structures is critical, from local councils to regional conservation alliances, to ensure that biodiversity and climate policies are coherent and mutually reinforcing. Legal and institutional reforms should aim at mainstreaming environmental sustainability within all sectors of the economy, promoting sustainable land-use practices, and integrating biodiversity conservation into national development agendas.

We suggest a suit of strategic interventions aimed at fostering a sustainable development trajectory for sub-Saharan Africa, encompassing multidisciplinary approaches that integrate ecological resilience, socio-economic equity, and robust governance frameworks. These measures are designed to harmonize the region's rich biodiversity with the well-being of its local communities, ensuring a balance between conservation efforts and the pursuit of sustainable livelihoods. Suggested action measures for a sustainable path for sub-Saharan Africa include, but are not limited to, the following:

1. Increase regional cooperation through the creation of comprehensive, coordinated transboundary conservation plans that address ecological connectivity, ecosystem-wide ecological challenges, and biodiversity loss. By pooling resources, sharing best practices, and harmonizing policy frameworks, these strategies can facilitate coordinated efforts across national borders to effectively manage shared natural resources and promote biodiversity conservation.
2. Enhance and mobilize sufficient, sustainable funding for biodiversity conservation by leveraging both domestic resources and international financing mechanisms, aligned with global biodiversity and climate goals.
3. Prioritize and ensure inclusive conservation and environmental governance by involving local communities and indigenous peoples, aligning with their needs and knowledge systems for equitable benefit sharing and meaningful participation, which is vital for managing ecosystems and invasive species.
4. Invest in capacity building for the application of technological advances in ecological modelling and spatial ecology to inform decision-making.
5. Foster an interdisciplinary approach in conservation science, integrating insights from ecological, social, and economic disciplines to address complex environmental challenges.
6. Promote inter-sectoral collaboration to acknowledge the interconnection between biodiversity conservation and sectors such as agriculture, forestry, fisheries, and urban development, while fostering strong institutions capable of executing complex environmental mandates with efficacy, transparency, and accountability.
7. Encourage cross-sectoral collaboration to acknowledge the interdependence of biodiversity conservation with agriculture, forestry, fisheries, and urban development, while fostering strong institutions capable of executing complex environmental mandates with efficacy, transparency, and accountability.

In the pursuit of a sustainable future for sub-Saharan Africa and beyond, the region is poised to play a significant role in the global endeavours aimed at halting and reversing the trend of biodiversity loss. By leveraging its unique ecosystems and rich biodiversity, sub-Saharan Africa can provide invaluable insights and strategies that contribute significantly to global conservation efforts, thereby ensuring a resilient and biodiverse planet for future generations. This commitment would enhance the resilience of local communities and economies against environmental challenges.

References

Abera W, Tamene L, Kassawmar T, Mulatu K, Kassa H, Verchot L, Quintero M (2021) Impacts of land use and land cover dynamics on ecosystem services in the Yayo coffee forest biosphere reserve, southwestern Ethiopia. Ecosyst Serv 50:101338. https://doi.org/10.1016/j.ecoser.2021.101338

Abrial E, Espínola LA, Rabuffetti AP, Eurich MF, Paira AR, Blettler MCM, Amsler ML (2019) Variability of hydrological connectivity and fish dynamics in a wide subtropical–temperate floodplain. River Res Appl 35(9):1520–1529. https://doi.org/10.1002/rra.3517

Adams-Hosking C, Grantham HS, Rhodes JR, McAlpine C, Moss PT (2011) Modelling climate-change-induced shifts in the distribution of the koala. Wildl Res 38(2):122–130. https://doi.org/10.1071/WR10156

Allen AM, Singh NJ (2016) Linking movement ecology with wildlife management and conservation. Front Ecol Evol 3. https://doi.org/10.3389/fevo.2015.00155

Armitage D, De Loë R, Plummer R (2012) Environmental governance and its implications for conservation practice. Conserv Lett 5(4):245–255. https://doi.org/10.1111/j.1755-263X.2012.00238.x

Beger M, Metaxas A, Balbar AC, McGowan JA, Daigle R, Kuempel CD, Treml EA, Possingham HP (2022) Demystifying ecological connectivity for actionable spatial conservation planning. Trends Ecol Evol 37(12):1079–1091. https://doi.org/10.1016/j.tree.2022.09.002

Bennett J, Marandure T, Hawkins H-J, Mapiye C, Palmer A, Lemke S, Wu L, Moradzadeh M (2023) A conceptual framework for understanding ecosystem trade-offs and synergies, in communal rangeland systems. Ecosyst Serv 61:101533. https://doi.org/10.1016/j.ecoser.2023.101533

Berkes F (2004) Rethinking community-based conservation. Conserv Biol 18(3):621–630. https://doi.org/10.1111/j.1523-1739.2004.00077.x

Bokhutlo T, Keppeler FW, Winemiller KO (2021) Seasonal hydrology influences energy channels in food webs of rivers in the lower Okavango Delta. Environ Biol Fish 104(10):1303–1319. https://doi.org/10.1007/s10641-021-01156-x

Borgwardt F, Robinson L, Trauner D, Teixeira H, Nogueira AJA, Lillebø AI, Piet G, Kuemmerlen M, O'Higgins T, McDonald H, Arevalo-Torres J, Barbosa AL, Iglesias-Campos A, Hein T, Culhane F (2019) Exploring variability in environmental impact risk from human activities across aquatic ecosystems. Sci Total Environ 652:1396–1408. https://doi.org/10.1016/j.scitotenv.2018.10.339

Brown HCP (2021) Youth, migration and community forestry in the Global South. For Trees Livelihoods 30(3):213–225. https://doi.org/10.1080/14728028.2021.1958065

Butt N, Shanahan DF, Shumway N, Bekessy SA, Fuller RA, Watson JEM, Maggini R, Hole DG (2018) Opportunities for biodiversity conservation as cities adapt to climate change. Geo: Geogr Environ 5(1):e00052. https://doi.org/10.1002/geo2.52

Cardone B, D'Ambrosio V, Di Martino F, Miraglia V, Rigillo M (2023) Analysis of the ecological efficiency increase of urban green areas in densely populated cities. Land 12(3):Article 3. https://doi.org/10.3390/land12030523

Caro T, Darwin J, Forrester T, Ledoux-Bloom C, Wells C (2014) Conservation in the Anthropocene. In: Wuerthner G, Crist E, Butler T (eds) Keeping the wild. Island Press/Center for Resource Economics, pp 109–113. https://doi.org/10.5822/978-1-61091-559-5_9

Collins R, Schaafsma M, Hudson MD (2017) The value of green walls to urban biodiversity. Land Use Policy 64:114–123. https://doi.org/10.1016/j.landusepol.2017.02.025

Corlett RT (2015) The Anthropocene concept in ecology and conservation. Trends Ecol Evol 30(1):36–41. https://doi.org/10.1016/j.tree.2014.10.007
Delaney JT, Larson DM (2024) Using explainable machine learning methods to evaluate vulnerability and restoration potential of ecosystem state transitions. Conserv Biol 38:e14203. https://doi.org/10.1111/cobi.14203
Gatiso TT, Kulik L, Bachmann M, Bonn A, Bösch L, Freytag A, Heurich M, Wesche K, Winter M, Ordaz-Németh I, Sop T, Kühl HS (2022) Sustainable protected areas: synergies between biodiversity conservation and socioeconomic development. People Nat 4(4):893–903. https://doi.org/10.1002/pan3.10326
Gavilanes Montoya AV, Castillo Vizuete DD, Marcu MV (2023) Exploring the role of ICTs and communication flows in the forest sector. Sustain For 15(14):Article 14. https://doi.org/10.3390/su151410973
Geißler K, Blaum N, von Maltitz GP, Smith T, Bookhagen B, Wanke H, Hipondoka M, Hamunyelae E, Lohmann D, Lüdtke DU, Mbidzo M, Rauchecker M, Hering R, Irob K, Tietjen B, Marquart A, Skhosana FV, Herkenrath T, Uugulu S (2024) Biodiversity and ecosystem functions in southern African savanna rangelands: threats, impacts and solutions. In: von Maltitz GP, Midgley GF, Veitch J, Brümmer C, Rötter RP, Viehberg FA, Veste M (eds) Sustainability of southern African ecosystems under global change: science for management and policy interventions. Springer International Publishing, pp 407–438. https://doi.org/10.1007/978-3-031-10948-5_15
Gidebo HB (2023) Linking livelihood and biodiversity conservation in protected areas: community based tourism development perspective from developing country. Tour Hosp Res 23(3):361–375. https://doi.org/10.1177/14673584221102699
Girma Y, Terefe H, Pauleit S (2019) Urban green spaces use and management in rapidly urbanizing countries:-the case of emerging towns of Oromia special zone surrounding Finfinne, Ethiopia. Urban For Urban Green 43:126357. https://doi.org/10.1016/j.ufug.2019.05.019
González-Chaves AD, Carvalheiro LG, Piffer PR, d'Albertas F, Giannini TC, Viana BF, Metzger JP (2023) Evidence of time-lag in the provision of ecosystem services by tropical regenerating forests to coffee yields. Environ Res Lett 18(2):025002. https://doi.org/10.1088/1748-9326/acb161
Gurung LJ, Miller KK, Venn S, Bryan BA (2021) Contributions of non-timber forest products to people in mountain ecosystems and impacts of recent climate change. Ecosyst People 17(1):447–463. https://doi.org/10.1080/26395916.2021.1957021
Heffernan A (2022) Development, conservation, empowerment: the trilemma of community-based natural resource management in Namibia. Environ Manag 69(3):480–491. https://doi.org/10.1007/s00267-021-01589-1
Jeltsch F, Bonte D, Pe'er G, Reineking B, Leimgruber P, Balkenhol N, Schröder B, Buchmann CM, Mueller T, Blaum N, Zurell D, Böhning-Gaese K, Wiegand T, Eccard JA, Hofer H, Reeg J, Eggers U, Bauer S (2013) Integrating movement ecology with biodiversity research—exploring new avenues to address spatiotemporal biodiversity dynamics. Mov Ecol 1(1):6. https://doi.org/10.1186/2051-3933-1-6
Joly K, Gurarie E, Hansen DA, Cameron MD (2021) Seasonal patterns of spatial fidelity and temporal consistency in the distribution and movements of a migratory ungulate. Ecol Evol 11(12):8183–8200. https://doi.org/10.1002/ece3.7650
Jones P (2022) Tourism and biodiversity: a paradoxical relationship. Athens J Tour 9(3):Article 3. https://doi.org/10.30958/ajt.9-3-2
Kariuki RW, Munishi LK, Courtney-Mustaphi CJ, Capitani C, Shoemaker A, Lane PJ, Marchant R (2021) Integrating stakeholders' perspectives and spatial modelling to develop scenarios of future land use and land cover change in northern Tanzania. PLoS One 16(2):e0245516. https://doi.org/10.1371/journal.pone.0245516
Kashwan P, Duffy RV, Massé F, Asiyanbi AP, Marijnen E (2021) From racialized neocolonial global conservation to an inclusive and regenerative conservation. Environ Sci Policy Sustain Dev 63(4):4–19. https://doi.org/10.1080/00139157.2021.1924574
Kauffman MJ, Aikens EO, Esmaeili S, Kaczensky P, Middleton A, Monteith KL, Morrison TA, Mueller T, Sawyer H, Goheen JR (2021a) Causes, consequences, and conservation of

ungulate migration. Annu Rev Ecol Evol Syst 52(1):453–478. https://doi.org/10.1146/annurev-ecolsys-012021-011516

Kauffman MJ, Cagnacci F, Chamaillé-Jammes S, Hebblewhite M, Hopcraft JGC, Merkle JA, Mueller T, Mysterud A, Peters W, Roettger C, Steingisser A, Meacham JE, Abera K, Adamczewski J, Aikens EO, Bartlam-Brooks H, Bennitt E, Berger J, Boyd C et al (2021b) Mapping out a future for ungulate migrations. Science 372(6542):566–569. https://doi.org/10.1126/science.abf0998

Keeley ATH, Beier P, Keeley BW, Fagan ME (2017) Habitat suitability is a poor proxy for landscape connectivity during dispersal and mating movements. Landsc Urban Plan 161:90–102. https://doi.org/10.1016/j.landurbplan.2017.01.007

Klappstein NJ, Potts JR, Michelot T, Börger L, Pilfold NW, Lewis MA, Derocher AE (2022) Energy-based step selection analysis: modelling the energetic drivers of animal movement and habitat use. J Anim Ecol 91(5):946–957. https://doi.org/10.1111/1365-2656.13687

Leisher C, Robinson N, Brown M, Kujirakwinja D, Schmitz MC, Wieland M, Wilkie D (2022) Ranking the direct threats to biodiversity in sub-Saharan Africa. Biodivers Conserv 31(4):1329–1343. https://doi.org/10.1007/s10531-022-02394-w

Lindley S, Pauleit S, Yeshitela K, Cilliers S, Shackleton C (2018) Rethinking urban green infrastructure and ecosystem services from the perspective of sub-Saharan African cities. Landsc Urban Plan 180:328–338. https://doi.org/10.1016/j.landurbplan.2018.08.016

Lynch AJ, Cooke SJ, Arthington AH, Baigun C, Bossenbroek L, Dickens C, Harrison I, Kimirei I, Langhans SD, Murchie KJ, Olden JD, Ormerod SJ, Owuor M, Raghavan R, Samways MJ, Schinegger R, Sharma S, Tachamo-Shah R-D, Tickner D et al (2023) People need freshwater biodiversity. WIREs Water 10(3):e1633. https://doi.org/10.1002/wat2.1633

Malhi Y, Franklin J, Seddon N, Solan M, Turner MG, Field CB, Knowlton N (2020) Climate change and ecosystems: threats, opportunities and solutions. Philos Trans R Soc B Biol Sci 375(1794):20190104. https://doi.org/10.1098/rstb.2019.0104

Manley PN, Long JW, Scheller RM (2024) Keeping up with the landscapes: promoting resilience in dynamic social-ecological systems. Ecol Soc 29(1). https://doi.org/10.5751/ES-14563-290103

Marvin S, Rutherford J (2018) Controlled environments: an urban research agenda on microclimatic enclosure. Urban Stud 55(6):1143–1162. https://doi.org/10.1177/0042098018758909

McCluney KE, Poff NL, Palmer MA, Thorp JH, Poole GC, Williams BS, Williams MR, Baron JS (2014) Riverine macrosystems ecology: sensitivity, resistance, and resilience of whole river basins with human alterations. Front Ecol Environ 12(1):48–58. https://doi.org/10.1890/120367

Meinhold K, Dumenu WK, Darr D (2022) Connecting rural non-timber forest product collectors to global markets: the case of baobab (*Adansonia digitata* L.). Forest Policy Econ 134:102628. https://doi.org/10.1016/j.forpol.2021.102628

Merkle JA, Abrahms B, Armstrong JB, Sawyer H, Costa DP, Chalfoun AD (2022) Site fidelity as a maladaptive behavior in the Anthropocene. Front Ecol Environ 20(3):187–194. https://doi.org/10.1002/fee.2456

Morrison TA, Merkle JA, Hopcraft JGC, Aikens EO, Beck JL, Boone RB, Courtemanch AB, Dwinnell SP, Fairbanks WS, Griffith B, Middleton AD, Monteith KL, Oates B, Riotte-Lambert L, Sawyer H, Smith KT, Stabach JA, Taylor KL, Kauffman MJ (2021) Drivers of site fidelity in ungulates. J Anim Ecol 90(4):955–966. https://doi.org/10.1111/1365-2656.13425

Mosepele K, Kolding J, Bokhutlo T (2017) Fish community dynamics in an inland floodplain system of the Okavango Delta, Botswana. Ecohydrol Hydrobiol 17(2):89–102. https://doi.org/10.1016/j.ecohyd.2017.01.005

Mosepele K, Kolding J, Bokhutlo T, Mosepele BQ, Molefe M (2022) The Okavango Delta: fisheries in a fluctuating floodplain system. Front Environ Sci 10. https://doi.org/10.3389/fenvs.2022.854835

Mudau HS, Msiza NH, Sipango N, Ravhuhali KE, Mokoboki HK, Moyo B (2022) Veld restoration strategies in South African semi-arid rangelands. Are there any successes?—a review. Front Environ Sci 10. https://doi.org/10.3389/fenvs.2022.960345

Musekiwa NB, Angombe ST, Kambatuku J, Mudereri BT, Chitata T (2022) Can encroached rangelands enhance carbon sequestration in the African Savannah? Trees Forest People 7:100192. https://doi.org/10.1016/j.tfp.2022.100192

Northrup JM, Vander Wal E, Bonar M, Fieberg J, Laforge MP, Leclerc M, Prokopenko CM, Gerber BD (2022) Conceptual and methodological advances in habitat-selection modeling: guidelines for ecology and evolution. Ecol Appl 32(1):e02470. https://doi.org/10.1002/eap.2470

Notaro M, Gary C, Le Coq J-F, Metay A, Rapidel B (2022) How to increase the joint provision of ecosystem services by agricultural systems. Evidence from coffee-based agroforestry systems. Agric Syst 196:103332. https://doi.org/10.1016/j.agsy.2021.103332

Okello M, Lamo J, Ochwo-Ssemakula M, Onyilo F (2021) Challenges and innovations in achieving zero hunger and environmental sustainability through the lens of sub-Saharan Africa. Outlook Agric 50(2):141–147. https://doi.org/10.1177/0030727020975778

Oldekop JA, Holmes G, Harris WE, Evans KL (2016) A global assessment of the social and conservation outcomes of protected areas. Conserv Biol 30(1):133–141. https://doi.org/10.1111/cobi.12568

Ortiz AMD, Outhwaite CL, Dalin C, Newbold T (2021) A review of the interactions between biodiversity, agriculture, climate change, and international trade: research and policy priorities. One Earth 4(1):88–101. https://doi.org/10.1016/j.oneear.2020.12.008

Osterhout MJ, Stewart KM, Wakeling BF, Schroeder CA, Blum ME, Brockman JC, Shoemaker KT (2024) Effects of large-scale gold mining on habitat use and selection by American pronghorn. Sci Total Environ 921:170750. https://doi.org/10.1016/j.scitotenv.2024.170750

Oteng-Abayie EF, Duodu E, Mensah G, Frimpong PB (2022) Natural resource abundance, environmental sustainability, and policies and institutions for environmental sustainability in sub-Saharan Africa. Resour Policy 79:103097. https://doi.org/10.1016/j.resourpol.2022.103097

Pandey S, Kumari N, Dash SK, Al Nawajish S (2022) Challenges and monitoring methods of forest management through geospatial application. In: Advances in remote sensing for forest monitoring. John Wiley & Sons, Ltd, pp 289–328. https://doi.org/10.1002/9781119788157.ch13

Petriello MA, Redmore L, Sène-Harper A, Katju D (2021) Terms of empowerment: of conservation or communities? Oryx 55(2):255–261. https://doi.org/10.1017/S0030605319000036

Piabuo SM, Hoogstra-Klein M, Ingram V, Foundjem-Tita D (2022) Community forest enterprises (CFEs) as social enterprises: empirical evidence from Cameroon. Forest Policy Econ 135:102664. https://doi.org/10.1016/j.forpol.2021.102664

Raymond CM, Cebrián-Piqueras MA, Andersson E, Andrade R, Schnell AA, Battioni Romanelli B, Filyushkina A, Goodson DJ, Horcea-Milcu A, Johnson DN, Keller R, Kuiper JJ, Lo V, López-Rodríguez MD, March H, Metzger M, Oteros-Rozas E, Salcido E, Sellberg M et al (2022) Inclusive conservation and the post-2020 global biodiversity framework: tensions and prospects. One Earth 5(3):252–264. https://doi.org/10.1016/j.oneear.2022.02.008

Rockström J, Falkenmark M, Allan T, Folke C, Gordon L, Jägerskog A, Kummu M, Lannerstad M, Meybeck M, Molden D, Postel S, Savenije HHG, Svedin U, Turton A, Varis O (2014) The unfolding water drama in the Anthropocene: towards a resilience-based perspective on water for global sustainability. Ecohydrology 7(5):1249–1261. https://doi.org/10.1002/eco.1562

Rodenbiker J (2020) Urban ecological enclosures: conservation planning, peri-urban displacement, and local state formations in China. Int J Urban Reg Res 44(4):691–710. https://doi.org/10.1111/1468-2427.12913

Roy PS, Ramachandran RM, Paul O, Thakur PK, Ravan S, Behera MD, Sarangi C, Kanawade VP (2022) Anthropogenic land use and land cover changes—a review on its environmental consequences and climate change. J Indian Soc Remote Sens 50(8):1615–1640. https://doi.org/10.1007/s12524-022-01569-w

Rumiano F, Wielgus E, Miguel E, Chamaillé-Jammes S, Valls-Fox H, Cornélis D, Garine-Wichatitsky MD, Fritz H, Caron A, Tran A (2020) Remote sensing of environmental drivers influencing the movement ecology of sympatric wild and domestic ungulates in semi-arid savannas, a review. Remote Sens 12(19):Article 19. https://doi.org/10.3390/rs12193218

Saliu AO, Komolafe OO, Bamidele CO, Raimi MO (2023) The value of biodiversity to sustainable development in Africa. In: Izah SC, Ogwu MC (eds) Sustainable utilization and conservation of Africa's biological resources and environment. Springer Nature, pp 269–294. https://doi.org/10.1007/978-981-19-6974-4_10

Samal R, Dash M (2023) Ecotourism, biodiversity conservation and livelihoods: understanding the convergence and divergence. Int J Geoheritage Parks 11(1):1–20. https://doi.org/10.1016/j.ijgeop.2022.11.001

Sample M, Thode AE, Peterson C, Gallagher MR, Flatley W, Friggens M, Evans A, Loehman R, Hedwall S, Brandt L, Janowiak M, Swanston C (2022) Adaptation strategies and approaches for managing fire in a changing climate. Climate 10(4):Article 4. https://doi.org/10.3390/cli10040058

Schultner J, Dorresteijn I, Manlosa AO, von Wehrden H, Hylander K, Senbeta F, Fischer J (2021) Ecosystem services from forest and farmland: present and past access separates beneficiaries in rural Ethiopia. Ecosyst Serv 48:101263. https://doi.org/10.1016/j.ecoser.2021.101263

Serva D, Cittadino V, Bernabò I, Biondi M, Iannella M (2024) Habitat suitability and connectivity modelling predict a latitudinal-driven expansion in the Mediterranean basin for a historically introduced reptile. Eur J Wildl Res 70(2):27. https://doi.org/10.1007/s10344-024-01780-9

Shrama L, Gupta R, Verma RK (2022) Efficacy of advanced remote sensing (hyperspectral and LIDAR) in enhancing forest resources management. In: Research anthology on ecosystem conservation and preserving biodiversity. IGI Global, pp 1714–1733. https://doi.org/10.4018/978-1-6684-5678-1.ch083

Shuman JK, Balch JK, Barnes RT, Higuera PE, Roos CI, Schwilk DW, Stavros EN, Banerjee T, Bela MM, Bendix J, Bertolino S, Bililign S, Bladon KD, Brando P, Breidenthal RE, Buma B, Calhoun D, Carvalho LMV, Cattau ME et al (2022) Reimagine fire science for the anthropocene. PNAS Nexus 1(3):pgac115. https://doi.org/10.1093/pnasnexus/pgac115

Simo-Kengne BD (2022) Tourism growth and environmental sustainability: trade-off or convergence? Environ Dev Sustain 24(6):8115–8144. https://doi.org/10.1007/s10668-021-01775-5

Siziba N, Chimbari MJ, Masundire H, Mosepele K, Ramberg L (2013) Variation in assemblages of small fishes and microcrustaceans after inundation of rarely flooded wetlands of the lower Okavango Delta, Botswana. Environ Manag 52(6):1386–1399. https://doi.org/10.1007/s00267-013-0183-9

Solangi YA, Jianguo D (2023) Examining challenges and solutions for environmental and natural resource management with a focus on mineral resources. Resour Policy 86:104085. https://doi.org/10.1016/j.resourpol.2023.104085

Soubry I, Doan T, Chu T, Guo X (2021) A systematic review on the integration of remote sensing and GIS to forest and grassland ecosystem health attributes, indicators, and measures. Remote Sens 13(16):Article 16. https://doi.org/10.3390/rs13163262

Tablado Z, Jenni L (2017) Determinants of uncertainty in wildlife responses to human disturbance. Biol Rev 92(1):216–233. https://doi.org/10.1111/brv.12224

Tartaruga I, Sperotto F, Carvalho L (2024) Addressing inclusion, innovation, and sustainability challenges through the lens of economic geography: introducing the hierarchical regional innovation system, pp 1–12. https://www.sciencedirect.com/science/article/pii/S2666683923000639

Tieminie RN, Loh CE, Tieguhong JC, Nghobuoche MF, Mandiefe PS, Tieguhong MR (2021) Non-timber forest products and climate change adaptation among forest dependent communities in Bamboko forest reserve, southwest region of Cameroon. Environ Syst Res 10(1):20. https://doi.org/10.1186/s40068-020-00215-z

Upreti G (2023) Valuation of biodiversity, ecosystem services, and natural capital. In: Upreti G (ed) Ecosociocentrism: the earth first paradigm for sustainable living. Springer Nature Switzerland, pp 163–188. https://doi.org/10.1007/978-3-031-41754-2_8

Wessely J, Hülber K, Gattringer A, Kuttner M, Moser D, Rabitsch W, Schindler S, Dullinger S, Essl F (2017) Habitat-based conservation strategies cannot compensate for climate-change-induced range loss. Nat Clim Chang 7(11):823–827. https://doi.org/10.1038/nclimate3414

Wilson MW, Ridlon AD, Gaynor KM, Gaines SD, Stier AC, Halpern BS (2020) Ecological impacts of human-induced animal behaviour change. Ecol Lett 23(10):1522–1536. https://doi.org/10.1111/ele.13571
Wobler, E. (2023). Machine learning for classification of satellite geolocation data: addressing the path segmentation challenge in animal movement ecology. https://orbilu.uni.lu/handle/10993/58863
Wollstein K, Creutzburg MK, Dunn C, Johnson DD, O'Connor C, Boyd CS (2022) Toward integrated fire management to promote ecosystem resilience. Rangelands 44(3):227–234. https://doi.org/10.1016/j.rala.2022.01.001
Xu W, Barker K, Shawler A, Van Scoyoc A, Smith JA, Mueller T, Sawyer H, Andreozzi C, Bidder OR, Karandikar H, Mumme S, Templin E, Middleton AD (2021) The plasticity of ungulate migration in a changing world. Ecology 102(4):e03293. https://doi.org/10.1002/ecy.3293
Yapi TS, Shackleton CM, Le Maitre DC, Dziba LE (2023) Local peoples' knowledge and perceptions of Australian wattle (Acacia) species invasion, ecosystem services and disservices in grassland landscapes, South Africa. Ecosyst People 19(1):2177495. https://doi.org/10.1080/26395916.2023.2177495
Zhao X, Zhang K, Zhao R (2024) Sustainable forest development in the digital era: the impact of internet use on the happiness of forest farmers' families in ecologically fragile ethnic areas of China. Forests 15(3):Article 3. https://doi.org/10.3390/f15030564
Zuñiga-Palacios J, Zuria I, Castellanos I, Lara C, Sánchez-Rojas G (2021) What do we know (and need to know) about the role of urban habitats as ecological traps? Systematic review and meta-analysis. Sci Total Environ 780:146559. https://doi.org/10.1016/j.scitotenv.2021.146559

Victor K. Muposhi is a Conservation Scientist and Associate Professor of Wildlife Conservation in the Department of Wildlife and Aquatic Resources at the Botswana University of Agriculture and Natural Resources. His research interests include biodiversity, ecosystems dynamics, spatial ecology, socio-ecological systems, human-wildlife coexistence, and sustainability science.

Wame L. Hambira (PhD) is an Associate Professor of Environmental Management at the Botswana University of Agriculture and Natural Resources (BUAN). Prof Hambira has served the *Intergovernmental Science-Policy Platform for Biodiversity and Ecosystem Services (IPBES) as a scoping* expert (the *nexus assessment of the interlinkages among biodiversity, water, food and health in the context of climate change as well as* the global assessment of the second biodiversity and ecosystem assessment). She is also a Review Editor for the Solution pathways for transformation of environmental systems chapter of the seventh Global Environment Outlook Report, Environment Program (UNEP).

Ditiro B. Moalafhi is a hydrologist and water resources management professional with over 20 years of experience. He holds a PhD in Civil and Environmental Engineering (hydro-climatology) from the University of New South Wales in Sydney, Australia, MPhil in Environmental Science, Post Graduate Diploma in Education and Bachelor of Science all from University of Botswana. Professor Moalafhi is currently an Associate Professor of water resources management at the Botswana University of Agriculture and Natural Resources, where he is the Head of Wildlife and Aquatic Resources department. He is also the SASSCAL Research Chair in Water Resource Quantity and Availability.

Professor Ayana Angassa Abdeta is a Professor in the Department of Range and Forest Resources—Botswana University of Agriculture and Natural Resources. He received BSc and MSc from the Alemaya University of Agriculture, Ethiopia, PhD from the Norwegian University of Life Sciences, Department of International Environment and Development Studies, Postdoctoral training from both the University of Hohenheim (Food Security Centre) and University of Cologne (Botanical Institute) in Germany, Certificate (2015) in Disaster Risk Management from the

Humboldt University of Berlin, Certificate (2020) in Project Management Foundations from the University of Cape Town, and Certificate (2022) in Agroforestry Systems from the University of Freiburg, Germany.

Ketlhatlogile Mosepele (PhD) is the Vice Chancellor of the Botswana University of Agriculture and Natural Resources. He holds the degree of Doctor of Philosophy (PhD) in Fisheries Biology and Management from the University of Bergen, Norway. He also holds the degree of Master of Philosophy (MPhil) and a Postgraduate Diploma in Fisheries Biology and Management from the University of Bergen, Norway. Professor Mosepele obtained his degree of Bachelor of Science (BS) in Agriculture majoring in Fisheries Science from the New Mexico State University, USA. He previously served as a Professor of Fisheries Biology and Management and Deputy Director of the Okavango Research Institute, University of Botswana.

Index

W. L. Hambira et al. (eds.), *Environmental Change and Biodiversity Conservation in sub-Saharan Africa*, Advances in Global Change Research 75,
https://doi.org/10.1007/978-3-031-73136-5

The manufacturer's authorised representative in the EU is Springer Nature Customer Service Centre GmbH, Europaplatz 3, 69115 Heidelberg, Germany. If you have any concerns regarding our products, please contact ProductSafety@springernature.com

Printed and bound by CPI Group (UK) Ltd, Croydon, CR0 4YY
01/10/2026
02225871-0015